MW00844891

LECTURES ON LIGHT

Lectures on Light

Nonlinear and Quantum Optics using the Density Matrix

Second Edition

Stephen C. Rand

University of Michigan, USA

Great Clarendon Street, Oxford, OX2 6DP,
United Kingdom

Oxford University Press is a department of the University of Oxford.
It furthers the University's objective of excellence in research, scholarship,
and education by publishing worldwide. Oxford is a registered trade mark of
Oxford University Press in the UK and in certain other countries

© Stephen C. Rand 2016

The moral rights of the author have been asserted

First Edition published in 2010
Second Edition published in 2016
First published in paperback 2019

Impression: 1

All rights reserved. No part of this publication may be reproduced, stored in
a retrieval system, or transmitted, in any form or by any means, without the
prior permission in writing of Oxford University Press, or as expressly permitted
by law, by licence or under terms agreed with the appropriate reprographics
rights organization. Enquiries concerning reproduction outside the scope of the
above should be sent to the Rights Department, Oxford University Press, at the
address above

You must not circulate this work in any other form
and you must impose this same condition on any acquirer

Published in the United States of America by Oxford University Press
198 Madison Avenue, New York, NY 10016, United States of America

British Library Cataloguing in Publication Data
Data available

Library of Congress Cataloging in Publication Data
Data available

ISBN 978–0–19–875745–0 (Hbk.)
ISBN 978–0–19–883590–5 (Pbk.)

Printed and bound by
CPI Group (UK) Ltd, Croydon, CR0 4YY

Links to third party websites are provided by Oxford in good faith and
for information only. Oxford disclaims any responsibility for the materials
contained in any third party website referenced in this work.

Preface to the First Edition

This book, and the course from which it sprang, attempts to bridge the enormous gap between introductory quantum mechanics and the research front of modern optics and other fields that make use of light. This would be an impossibly daunting task, were it not for the fact that most of us love to hear things again and again that we already know. Taking that into account, this book uses a single approach repeatedly to tackle progressively more exciting topics in the science of light, moving systematically and swiftly from very basic concepts to sophisticated topics.

The reader should be aware from the outset that the approach taken here is unconventional. It is highly selective instead of encyclopedic, and it teaches the reader only how to use the density matrix on new problems. Nowadays scientific answers are being sought to an ever-expanding array of problems across numerous disciplines. The trend in textbooks on quantum optics is understandably to cover an increasing number of topics comprehensively, and to familiarize students with an ever-widening array of analytic tools, exhaustively. There are many fine texts that fulfill this function, but this book is not one of them. Instead, important topics and alternative methods of analysis have been omitted here to keep it as brief as possible, with a single-minded pedagogical purpose in mind. The main objective of the book is to provide students and researchers with one reliable tool, and the confidence that comes from practice, to analyze new optical phenomena in their chosen field successfully and rigorously.

Thus, the one and only analytic tool developed here for attacking research-level problems in optical science is the density matrix. A systematic procedure is applied to representative problems of "critical subjects"—usually only one example each—to show how virtually any problem can be analyzed with the density matrix. Each successive example adds one system property at a time, with the result that one qualitatively new feature appears in the dynamics each time. By using a systematic "building-up" principle to approach complicated interactions, students begin to recognize what terms in the analysis are associated with particular changes in the dynamics. Following two slow-paced introductory chapters on review material, the text shifts focus to the development of insights as to when atomic motion, or multi-level structure, or coherence effects dominate the behavior of complicated systems. It ends with fast-paced coverage of selected topics and applications.

The organization of the book follows the original sequence of lectures on light prepared for applied physics graduates with typical undergraduate physics, chemistry and engineering preparation, expected to handle interdisciplinary research topics during their careers. Present day graduate students who use light often face important problems that no longer fall into the neat categories of unique models from the early days of quantum mechanics, and require broad perspective and reliable mathematical tools to handle new cross-disciplinary topics quickly. This course therefore embraces not only students from traditional subject areas that make use of light (physics, chemistry, electrical engineering, materials science), but also the biophysicist who needs laser tweezers, the photochemist who wants coherent control, the biomedical engineer who needs to image through scattering media, the mechanical engineer interested in molecular design of new materials, and others. The greatest theoretical challenge faced by most students is to make appropriate connections between standard models and the bewildering landscape of new research questions on intersecting boundaries of "hyphenated" subjects like biophysics, biomedicine, photochemistry, etc. For them, the systematic progression of *Lectures on Light* offers an approach to quantum optical analysis that should help bridge the gaps.

Why choose the density matrix? After all, there are many mathematical tools available to treat nonlinear and quantum optics. The answer is that the density matrix has features that make it a natural choice. For example it permits one to ignore parts of a problem that appear to be irrelevant and to focus mathematically on the dynamics of interest to the researcher. Also, if desired, it can be reduced to a rate equation treatment—a familiar approach to analysis that all students of science encounter. In addition, it is particularly well suited for dealing with coherence in isolated or interactive systems.

This makes it an excellent point of departure for anyone who wishes to use light either to probe or control systems about which little is known. By focusing on this adaptable tool, readers can cover a lot of intellectual ground with the minimum investment in mathematical complexity. What emerges is a reliable analytic framework for use in any research where light probes or controls or alters a system, regardless of whether the problems are classical or quantum.

Another hurdle that is addressed explicitly in this book, and that is very important, is the persistent issue of whether simplified models provide reliable representations of complicated systems. That is why the last part of Chapter 3 examines in detail the question of whether sodium atoms can ever legitimately be viewed as two-level systems given the complexity of their energy level structure. It turns out the experimentalist is much more in control of the effective number of levels than one might guess. By the end of this course, graduate students in interdisciplinary science are able to exercise considerable judgment in the creation of useful models for their own frontier research problems and analyze them by drawing on solid examples and the explicit methodology of the text.

Some familiarity with introductory quantum mechanics is assumed. However, advanced preparation in optics is not essential to learn and use this material. Over the years, students from disciplines as diverse as the ones mentioned above—mechanical engineering, materials science, electrical, and biomedical engineering together with physics, applied physics, and chemistry—have found it to provide the essential insights and analysis they need for immediate application in their research. Some material that is ordinarily omitted from advanced quantum mechanics texts is included to set the stage for the broadest possible applicability. An example of this is third order perturbation theory. This topic provides an important bridge to understanding nonlinear behavior of even the simplest quantum mechanical systems. Since self-saturation effects, cross-saturation, four-wave mixing processes, and other third order phenomena are encountered much more commonly in pump–probe experiments than one might expect, it is important that readers become familiar with third order effects early in the course of their research. Another unusual feature of this book is that a few procedural errors are presented early on to illustrate what can go wrong when quantum mechanical calculations are formulated inconsistently. These humbling examples remind us all that in research, and most especially in cross-disciplinary work, there is no substitute for the use of common sense and no shortcut to pioneering science.

This book progresses as rapidly as possible from a simple and easy review to challenging modern applications. One layer of conceptual or computational complexity is added in each new section. The technical material begins with some uncommon examples of introductory quantum mechanics that force students from the start to revisit the basic physical principles of optics and quantum mechanics studied as undergraduates. Potential pitfalls are also pointed out that arise from the inclusion of relaxation processes in system dynamics. Once the density matrix approach to dynamics is motivated, formulated and understood, the course progresses at an accelerated rate through important applications. Hence the material is best studied in sequence.

This lecture material will be most useful to students interested in acquiring rigorous, broadly-applicable analysis quickly. However, the systematic application of one mathematical tool to many forefront topics in nonlinear and quantum optics will be of interest to seasoned researchers as well. The heavy reliance in late chapters on the insights and dynamic effects described in earlier chapters helps to keep the treatment short. Much of the course relies on the semi-classical description in which only the atoms are quantized—the light field is not. This is intended to encourage intuitive thinking to as late a stage as possible. However, in the last two chapters several topics are covered where both the atoms and the light field are quantized and intuitive notions are sometimes poor guides. Finally, the selected research topics of Chapter 7 not only illustrate the power of systematic density matrix analysis but give students confidence that, as they approach the exciting frontiers of their own research, the combination of density matrix analysis and common sense perspectives developed throughout this course will facilitate success.

I would like to acknowledge all the help I have received during various stages of preparation of this monograph. First and foremost, I am grateful for the comments and questions of students who took this course over a period of two decades. Along with my own graduate students, they helped to make the presentation compact by forcing me to provide concise answers about confusing notions. I am indebted to Philbrick Bridgess of Roxbury Latin School for imparting to me his respect for analytic geometry, which ultimately led to the discovery of transverse optical magnetism, covered in Section 7.4 of this book. On a few topics I have drawn liberally from existing texts, but most especially from *Elements of Quantum Optics* by Meystre and Sargent, *Quantum Electronics* by Yariv, *Quantum Optics* by Zhubary and Scully, *Laser Physics* by Sargent, Scully, and Lamb, *Optical Resonance and Two-Level Atoms* by Allen and Eberly, and *Foundations of Laser Spectroscopy* by Stenholm. I am thankful for their fine examples of concise pedagogy. Also I thank my colleagues at the University of Michigan for creating the intellectual environment that made this book possible. Support was provided by the Department of Electrical Engineering and Computer Science for typing of a rough draft by Ruby Sowards, Nick Taylor, and Susan Charnley. The graduate student course itself was offered through the Department of Physics. I owe a debt of gratitude to Kevin Rand for preparing many original illustrations and for the adaptation of published figures. The cover diagram was furnished by William Fisher. I am particularly grateful to Boris Stoicheff, Richard Brewer, Art Schawlow, Ted Hansch, and Juan Lam for their friendship and for sharing what they knew. Their examples were inspirational. Finally, I am deeply indebted to my family—especially my wife Paula who patiently endured the taxing process of finalizing the manuscript.

S.C. Rand
February 27, 2010

Preface to the Second Edition

In the years following the first edition of this book, some topics have been added in response to student interest in research advances. These have now been incorporated in the new edition, making it more comprehensive in its coverage of advanced research. However by continuing to emphasize compact descriptions it has been possible to keep the book close to its original length even though the number of problems at the ends of chapters has doubled. Many students and colleagues have provided suggestions and corrections to improve the presentation and I gratefully acknowledge their input. I would especially like to thank Hope Wilson, Alex Fisher, Hamed Razavi, and Austin Tai for help preparing new figures in Chapters 5, 6, and 7.

S.C. Rand

October 29, 2015

Contents

Appendices

1 Basic Classical Concepts

1.1 Introduction

In classical physics, light interacts with matter as described by a compact set of equations formulated by James Clerk Maxwell in 1865, namely Maxwell's equations. Solutions for the propagation of electromagnetic fields in arbitrary media can generally be built up from solutions to these equations for individual frequency components, and are particularly easy to find if the fields have slowly varying electric and magnetic field magnitudes. The medium through which light passes must also be uniform and the timescale of interest must greatly exceed the optical period. Fortunately, these conditions are not terribly restrictive. They encompass most (though not all) distinctive phenomena in classical as well as quantum optics. Perhaps more importantly, they form a useful framework for lectures on light without unnecessarily limiting our horizons.

Nowadays, the optical characteristics of media can often be engineered to enhance specific interactions deliberately. For example, quasi-phase-matching crystals can be prepared with periodically inverted domain structure to allow the build-up of new frequency components in the optical field. Similarly, in so-called metamaterials, deliberate variation in microstructure from point to point within a medium is intended to permit light fields to evolve in complicated but controllable ways. It has been shown in recent years that metamaterials can be designed to distort the way electromagnetic waves move through an occupied region of space, in such a way as to render objects located there effectively invisible. Yet even the passage of light through these non-uniform media is again entirely predictable using Maxwell's equations and modern computational tools. The relative maturity of the subject of electromagnetism and the power of modern computers have put us in a position to predict in great detail how light moves, and how it is attenuated, emitted, amplified, or scattered as it progresses through practically any kind of matter. So why is it that light is still such a vital topic today, and when do we have to treat problems quantum mechanically? How can it be that so many new marvels have emerged from the study of electromagnetism and optical science in the last decade or two? The pace of major discoveries continues unabated. How can light tantalize and surprise experts in the twenty-first century with an ever-expanding landscape of discovery and applications at a time when Maxwell's equations are 150 years old?

Lectures on Light. Second Edition. Stephen C. Rand.
© Stephen C. Rand 2016. Published in 2016 by Oxford University Press.

In addition to being a manual for applications of the density matrix, this book seeks to answer this question. One of its goals is to teach students how to explore and analyze the unknown without already knowing everything. For this journey, the density matrix is a perfect companion, since one finds that with it the wavefunction of complex systems is no longer needed to calculate most things of interest in optical interactions. The content of this course has been presented now for two decades as an advanced lecture course on light for graduate students at the University of Michigan—primarily those with backgrounds in physics, chemistry, and engineering. It seeks to give students familiarity with a single analytic tool, the density matrix, by applying it systematically to a great many forefront problems in modern optics. It presents a concise, broad (though admittedly incomplete) picture of active research fronts in optical science that students can absorb in a single semester. It does not pretend to be a comprehensive reference for research in any specialty area from which one or more examples may have been drawn. Instead, it shows students how to get started on virtually any research problem using a standard toolbox, and gives them the requisite perspective on what is physically possible and essential in the analysis. Students acquire a sense for when classical, semi-classical, or quantum optical approaches need to be applied, when coherence plays an important role and when it does not, when an exact solution is required and when it is not, through an approach that adds one concept at a time systematically with a single mathematical tool.

Among the more advanced topics that are included in this course are free induction decay, photon echoes, nutation, spectral and spatial hole-burning, light shifts, two- and three-level coherence, Zeeman coherence, coherent population transfer, electromagnetically induced transparency, slow light, high-order perturbation theory, laser cooling, optical magnetism, squeezed light, dressed atoms, quantum computation, and cavity quantum electrodynamics. Although some of these subjects are typically omitted from standard textbooks on light, experience has shown they can be handled by intermediate or beginning graduate students when the progression through these topics is presented in sufficiently small steps. Moreover, this process succeeds in providing a framework for understanding optical phenomena in a way that is quite different from that of books that emphasize a sequence of different mathematical techniques to handle quantum mechanics. For this reason, most of the book utilizes a "semi-classical" approach that treats the system under study as a quantum system but avoids the introduction of operators to describe the electromagnetic field itself. This has the merit of postponing operator mathematics required for the quantization of the electromagnetic field until the final chapters when they are really needed.

Analysis of many forefront topics in optics with a single approach gives readers confidence that they can proceed into virtually any developing field where light is used to probe or control dynamics, and confidently formulate an initial attack on *their* research problem. Naturally it also has the disadvantage of being just one approach. Many of us are familiar with the mathematical handicap that can result from a bad choice of coordinate system, or an ill-suited choice of variables. However, the density matrix is a remarkably complete and forgiving tool that accomplishes the essential things. First, it incorporates the all-important phases of fields and polarizations responsible for some of the surprising phenomena encountered in optical science. It can describe dephasing, coherent control, and relaxation processes in a manner consistent with the occupation of various states of the system. Second, it eliminates the need for detailed wavefunctions in the prediction of most of the important dynamics in new systems. In beginning courses in quantum mechanics we are taught that the wavefunction contains all the details of the quantum system and is essential for understanding or predicting its behavior. Unfortunately, solutions are available for wavefunctions of only the simplest systems like atomic hydrogen. Consequently it is fortunate, to say the least, that the density matrix provides a method for computing all the "interesting" dynamics of new systems given only limited information on energy levels and symmetries. One can even ignore parts of multi-component systems through the use of the reduced density matrix. In this way it becomes possible to analyze complicated systems like macromolecules, for which analytic wavefunctions are not likely to be available at any time in the foreseeable future.

1.2 Electric and Magnetic Interactions

1.2.1 Classical Electromagnetism

The interaction of electromagnetic waves with matter is more complex at optical frequencies than at radio or microwave frequencies. There are innumerable possibilities for resonance in the optical range, which do not exist at frequencies below 10 GHz, for example. Such resonances result in large changes to absorption, dispersion, and scattering when only small changes in frequency are made. Also, relatively large variations of constitutive parameters, namely permittivity ε and permeability μ, over the frequency ranges between resonances can be exploited to cause energy exchange between waves of different frequencies. This emphasizes the importance of finding a general approach to optical analysis that incorporates resonant, dispersive, and quantum mechanical character of atomic and molecular interactions with light. So we shall develop a formalism that combines Maxwell's equations and quantum properties of matter from the outset, and seek perspective by applying it systematically to a variety of problems.

1.2.2 Maxwell's Equations

The fundamental equations relating time-varying electric and magnetic fields are:

$$\bar{\nabla} \times \bar{H} = \bar{J} + \frac{\partial \bar{D}}{\partial t}, \tag{1.2.1}$$

$$\bar{\nabla} \times \bar{E} = -\frac{\partial \bar{B}}{\partial t}, \tag{1.2.2}$$

$$\bar{\nabla} \cdot \bar{D} = \rho_v, \tag{1.2.3}$$

$$\bar{\nabla} \cdot \bar{B} = 0. \tag{1.2.4}$$

Constitutive relations describe the response of charges to applied electric or magnetic fields in real materials. The displacement field $\bar{D}$ in Maxwell's equations is

$$\bar{D} = \varepsilon\bar{E} = \varepsilon_0\bar{E} + \bar{P}, \tag{1.2.5}$$

and the magnetic flux density in Maxwell's equations is

$$\bar{B} = \mu\bar{H} = \mu_0\left(\bar{H} + \bar{M}\right). \tag{1.2.6}$$

At optical frequencies or in non-magnetic systems, we typically assume $\bar{M} = 0$, because magnetic Lorentz forces are small compared to electric forces at high frequencies. In media with linear response the polarization is $\bar{P} = \varepsilon_0\chi_e^{(1)}\bar{E}$, $\chi_e^{(1)}$ being the linear electric susceptibility. Thus, the displacement field is

$$\bar{D} = \varepsilon_0(1 + \chi_e^{(1)})\bar{E}, \tag{1.2.7}$$

and the relative, linear dielectric constant $\varepsilon_r = \varepsilon/\varepsilon_0$ is given by $\varepsilon_r = 1 + \chi_e^{(1)}$.

In nonlinear or strongly excited media, there are higher order terms that introduce field-induced effects. That is, $P = P^{(1)} + P^{(2)} + \ldots = \varepsilon_0(\chi_e^{(1)}E + \chi_e^{(2)}E^2 + \ldots)$. We deal with nonlinear response when multi-photon transitions of isolated atoms are considered in later chapters. Nonlinear response can also arise from the finite response time of bound electrons or charge motion in the case of free carriers. Either mechanism can produce a "nonlocal" relationship between the polarization and the field. That is, $P(\bar{r},t) \approx \varepsilon_0\chi_e(t-t', \bar{r}-\bar{r}')E(\bar{r}',t')$. If we are interested only in slow dynamics in dielectrics we can often ignore such effects. However, finite response times cannot be ignored on ultrafast timescales in semiconductors, plasmas, or organic electronic materials. They also cannot be ignored in photorefractive media, where charges diffuse slowly from place to place, or in multi-level media where long-lived states lead to time-delayed response.

1.2.3 The Wave Equation

The key equation describing propagation of classical light is the wave equation, obtained by applying the curl operator to combine Eqs. (1.2.1) and (1.2.2). Consistent with Panofsky's expression for $\bar{\nabla} \times \bar{B}$ (see Supplementary Reading list) we find:

$$\begin{aligned}\bar{\nabla} \times \bar{\nabla} \times \bar{E} &= -\bar{\nabla} \times \frac{\partial \bar{B}}{\partial t} = -\frac{\partial}{\partial t}\left[\mu_0\left(\bar{J} + \frac{\partial \bar{D}}{\partial t}\right) + \mu_0\left(\bar{\nabla} \times \bar{M}\right)\right] \\ &= -\mu_0\frac{\partial \bar{J}}{\partial t} - \mu_0\varepsilon_0\frac{\partial^2 E}{\partial t^2} - \mu_0\frac{\partial^2 \bar{P}}{\partial t^2} - \mu_0\bar{\nabla} \times \frac{\partial \bar{M}}{\partial t}.\end{aligned} \tag{1.2.8}$$

If we restrict ourselves to non-magnetic, insulating materials ($\bar{M} = 0; \bar{J} = 0$), then we can use a vector identity in Eq. (1.2.8) to find

$$\bar{\nabla} \times \bar{\nabla} \times \bar{E} = -\nabla^2 \bar{E} + \bar{\nabla} \left(\bar{\nabla} \cdot \bar{E} \right) = -\nabla^2 E + \frac{\bar{\nabla}}{\varepsilon} \left[\bar{\nabla} \cdot \bar{D} - \bar{\nabla} \cdot \bar{P} \right]. \tag{1.2.9}$$

Provided that the local free charge density ρ_v is zero and there are no spatial variations of polarization due to charge migration or field gradients, we find $\bar{\nabla} \cdot \bar{D} = \bar{\nabla} \cdot \bar{P} = 0$ and in free space the wave equation becomes

$$\nabla^2 \bar{E} - \frac{1}{c^2} \frac{\partial^2 \bar{E}}{\partial t^2} = \mu_0 \frac{\partial^2 \bar{P}}{\partial t^2}, \tag{1.2.10}$$

where $c \equiv (\mu_0 \varepsilon_0)^{-1/2}$ is the speed of light in vacuum.

1.2.4 Absorption and Dispersion

Solutions of Eq. (1.2.10) are useful in describing many important classical phenomena in linear dielectrics ($P = P^{(1)}$), such as absorption and dispersion. As an example of such a solution, consider a linearly polarized plane wave $E(z, t)$ propagating along $\hat{z}$. If the wave is polarized along $\hat{x}$, meaning that the vector E points along the direction x, this wave can be written in the form

$$\bar{E}(z, t) = \frac{1}{2} E_{0x}(z) \hat{x} \exp[i(kz - \omega t)] + c.c., \tag{1.2.11}$$

where *c.c.* stands for complex conjugate. Other possible polarization states are considered in Section 1.2.6. Here however we show, by substituting Eq. (1.2.11) into Eq. (1.2.10), that the wave equation can often be written in a scalar form that ignores the vector character of light. After substitution, the orientation of the field emerges as a common factor, which may therefore simply be dropped from the equation. In this way, we find

$$\frac{\partial^2 E_{0x}(z)}{\partial z^2} + 2ik \frac{\partial E_{0x}(z)}{\partial z} - k^2 E_{0x}(z) + \frac{\omega^2}{c^2} E_{0x}(z) = -\frac{\omega^2}{c^2} \chi_e E_{0x}(z). \tag{1.2.12}$$

An important mathematical simplification can be introduced at this point, called the slowly varying envelope approximation (SVEA), by assuming that the amplitude $E_{0x}(z)$ varies slowly over distance scales comparable to the wavelength. This permits neglect of small terms like $\partial^2 E_0(z)/\partial z^2$ in Eq. (1.2.12), whereupon the wave equation reduces to

$$2ik \frac{\partial E_{0x}(z)}{\partial z} - \left(k^2 - \frac{\omega^2}{c^2} \right) E_{0x}(z) = -\frac{\omega^2}{c^2} \chi_e E_{0x}(z). \tag{1.2.13}$$

Recognizing that the electric susceptibility χ_e can be complex ($\chi_e = \chi' + i\chi''$), we can equate the real parts of Eq. (1.2.13) to find

$$k^2 = \frac{\omega^2}{c^2}\left(1 + \chi'\right). \tag{1.2.14}$$

This relationship between frequency ω and wavenumber k is known as the linear dispersion relation. It is a defining relationship for light, usually written as

$$\omega = c_n k, \tag{1.2.15}$$

where $c_n(\omega) = c/n(\omega)$ is the phase velocity of the wave in which c is the speed of light in vacuum and n is the refractive index given by $n^2 = 1 + \chi'$. Equating the imaginary parts of Eq. (1.2.13), one finds

$$2k\frac{\partial E_{0x}(z)}{\partial z} = -\frac{\omega^2}{c^2}\chi'' E_{0x}(z). \tag{1.2.16}$$

Multiplication of both sides of Eq. (1.2.16) by the conjugate field amplitude $E^*_{0x}(z)$ yields

$$\frac{\partial |E_0|^2}{\partial z} = -\frac{\omega^2}{c^2}\frac{\chi''}{k}|E_0|^2, \tag{1.2.17}$$

and thus

$$\frac{\partial I}{\partial z} = -\alpha I, \tag{1.2.18}$$

where I is the optical intensity and α is the absorption coefficient, defined by

$$\alpha \equiv \frac{\omega^2}{c^2}\frac{\chi''}{k}. \tag{1.2.19}$$

In vacuum this result is just $\alpha = k\chi''$. The solution of Eq. (1.2.18) can be written as

$$I(z) = I_0 \exp[-\alpha z], \tag{1.2.20}$$

which is Beer's law. When $\alpha > 0$, the electromagnetic wave undergoes exponential absorption (loss). When $\alpha < 0$, there is exponential amplification (gain). Beer's law applies to optical propagation in many systems. Exceptions include systems where radiation trapping, saturation, or multiple scattering take place.

1.2.5 Resonant Response

So far, we have characterized the electric response by introducing a polarization P that depends on the incident field. The proportionality constant is the classical macroscopic susceptibility χ, and we find that we can describe some well-known propagation effects such as the dependence of the phase velocity c_n on refractive index and the exponential nature of absorption/gain in simple materials. However, to this point we have no model or fundamental theory for the electric susceptibility χ itself.

For this purpose we turn to a classical harmonic oscillator model based on linear response (and Hooke's law). The electron and proton comprising a fictitious atom are joined by a mechanical spring with a restoring force constant k_0. It is assumed the proton does not move significantly in response to the applied field (Born approximation), whereas the displacement $x(t)$ about equilibrium of the bound electron depends on the applied field according to $\bar{F} = -e\bar{E} = -k_0\bar{x}$. If $\bar{E}$ is harmonic in time, solutions will have the form $x(t) = x_0 \exp(-i\omega t)$. Upon substitution into Newton's second law

$$m_e \frac{\partial^2 x(t)}{\partial t^2} - m_e \gamma \frac{\partial x(t)}{\partial t} + k_0 x(t) = -eE_{0x} \exp(-i\omega t), \tag{1.2.21}$$

one can solve for the amplitude $x(t)$ of driven charge motion, which is given by

$$x_0 = \frac{eE_{0x}/m_e}{\omega^2 - i\omega\gamma - \omega_0^2}, \tag{1.2.22}$$

where $\omega_0 \equiv \sqrt{k_0/m_e}$ is the resonant frequency of the oscillator and γ is an empirical damping constant. The total polarization is

$$P(t) = -\sum_1^N ex(t) = -Nex(t), \tag{1.2.23}$$

for N identically prepared atoms. A comparison with our former expression $P = \varepsilon_0 \chi E$ determines the frequency-dependent susceptibility components, and through them the absorption and dispersion curves shown in Figure 1.1.

Although this Lorentz model is strictly empirical, it provides a qualitatively useful picture of the frequency dependence of system response near electronic resonances. Regrettably it does not provide a way to derive the spring and damping constants from first principles or to explain the stability of atoms. We turn to quantum mechanics to remedy such deficiencies.

1.2.6 The Vectorial Character of Light

The wave equation (Eq. (1.2.10)) is a vector relation. Often, as shown in Section 1.2.4, it can be reduced to a scalar relation because the vector character of light plays no role in the particular atom–field interaction of interest. However, this is certainly not always true, and it is helpful to identify what circumstances require full vector analysis. In later chapters, a few topics such as Zeeman coherence and transverse optical magnetism are covered that are strongly governed by vectorial aspects of light. In anticipation of these subjects, we close this introductory chapter by describing the basic polarization states of light that are possible and the angular momentum carried by them.

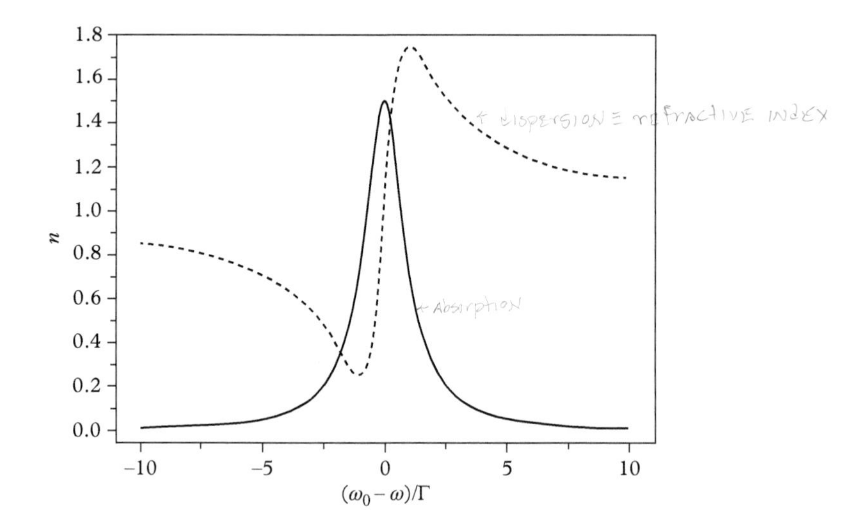

Figure 1.1 *Frequency dependence of absorption (solid curve) and dispersion (refractive index; dashed curve) near a resonance at frequency ω_0 in the classical model.*

In Section 1.2.4, the polarization of light was introduced as a property of light specifying the direction in which the optical electric field is oriented. Since the field orientation may vary in time, this definition still admits three fundamentally different states of polarization which are called linear, right circular, and left circular. Linear polarization describes light waves in which the electric field points in a fixed direction. For example, x-polarized light propagating along $\hat{z}$ has an E field oriented along the unit vector $\hat{x}$. Hence it may be written in the form $\bar{E}(z,t) = E_x(z,t)\hat{x}$ or $\bar{E}(z,t) = E(z,t)\hat{e}$ if we let $\hat{e}$ represent a more general polarization unit vector. Circular polarization describes light with an electric field vector that executes a circle around the propagation axis. The electric field rotates once per cycle, either clockwise or counterclockwise, as the wave propagates forward. The circular basis vectors for these polarizations are

$$\hat{\varepsilon}_{\pm} = \mp\left(1/\sqrt{2}\right)(\hat{x} \pm i\hat{y}). \tag{1.2.24}$$

In the case of circular polarization, the vector $\hat{e} = \hat{\varepsilon}_{\pm}$ is an axial rather than a polar vector. The 90° difference in phase of oscillations along $\hat{x}$ and $\hat{y}$ in Eq. (1.2.24) results in fields of the form $\bar{E}(z,t) = E_x(z,t)\hat{\varepsilon}_{\pm}$ that do not point in a fixed direction. Instead, they rotate as the wave propagates. Circular and linear states of polarization are illustrated in Figure 1.2.

Circular polarizations carry spin angular momentum of $\pm\hbar$, as indicated in Figure 1.2. The field direction and the energy density associated with the field twists as it moves, undergoing a helical rotation with respect to $\hat{z}$. Linear polarization carries no angular momentum, since the electric field vector does not rotate about $\hat{z}$ at all. This latter state of polarization is a special combination or superposition state, consisting of the sum of two equal-amplitude, circularly polarized fields of opposite helicity.

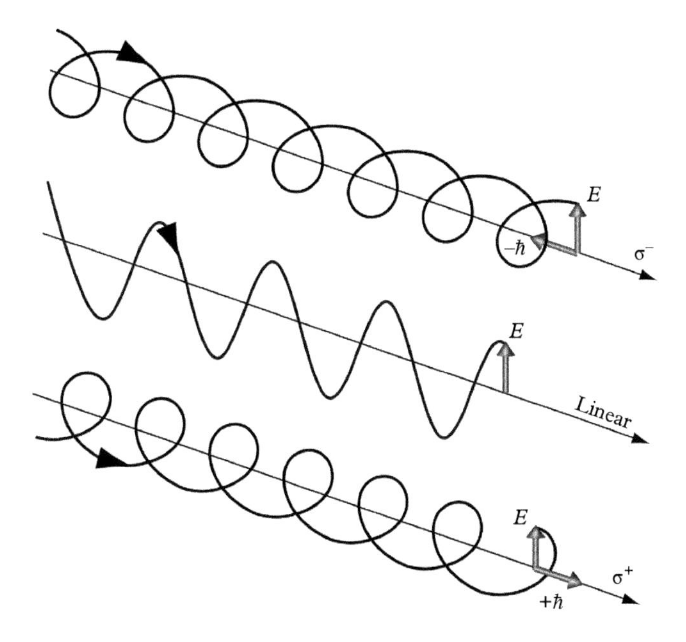

Figure 1.2 *Fundamental polarizations of light, illustrating light fields that carry spin angular momenta of $-\hbar$, 0, and $+\hbar$ (top to bottom, respectively). Angular momentum carried by a light wave affects the way it interacts with matter.*

Exercise: (a) Show that light which is linearly polarized along $\hat{x}$ can be written as the sum of two opposite circularly polarized fields. That is, show from Eq. (1.2.24) that x-polarized light is a phased, superposition state of two equal amplitude right and left circularly polarized components:

$$\hat{x} = -\left(1/\sqrt{2}\right)(\hat{\varepsilon}_+ - \hat{\varepsilon}_-). \tag{1.2.25}$$

(b) Conversely, show that circular polarization can be expressed as a superposition of two orthogonal, linearly polarized waves of equal amplitude.

Since momentum must be conserved in optical interactions, just as energy must be conserved, we can anticipate that any angular momentum carried by light may affect the way it interacts with materials. This point is addressed in later chapters.

SUPPLEMENTARY READING

G.R. Fowles, *Introduction to Modern Optics*. New York: Dover, 1987.

W.K.H. Panofsky and M. Phillips, *Classical Electricity and Magnetism*, 2nd ed. London: Addison-Wesley Publishing Co., 1962.

2

Basic Quantum Mechanics

2.1 Particles and Waves

Quantum mechanics postulates a concept known as wave–particle duality in the description of dynamical systems by associating a de Broglie wavelength λ_B with each particle. This wavelength depends on the linear momentum p of the particle, and is given by

$$\lambda_B = h/p, \tag{2.1.1}$$

where h is Planck's constant. The de Broglie wave causes particles to exhibit wave interference effects whenever the wavelength becomes comparable to or larger than the space occupied by the wave. In fact de Broglie derived the Bohr–Sommerfeld quantization rules from Eq. (2.1.1). Remarkably, this is the only essentially new idea of "wave" mechanics that is missing in "classical" mechanics, and curiously, although matter can exhibit particle or wave-like properties, its particle and wave characteristics are never observed together.

It follows specifically from Eq. (2.1.1) that electron waves can form self-consistent standing wave patterns ("stationary" states, "orbitals," or "eigenstates") that have constant angular momentum and energy. Charge distributions can be simultaneously localized and stable. The forces from Coulomb attraction between an electron and a nucleus, together with nuclear repulsion, must balance within atoms at equilibrium. This balance and the requirement that the standing wave be self-consistent uniquely determine the ground state of atoms. Excited states must similarly satisfy the boundary conditions imposed by the de Broglie wavelength, but tend to be short-lived because the forces on the electron are no longer balanced. Consequently excitation of such a state is quickly followed by a transition to the ground state, accompanied by emission of the energy difference between the ground and excited state as electromagnetic energy.

Conversely, light of frequency ν_0 has particle-like properties and may cause resonant transitions of atoms between well-defined initial and final states of an atom with energies E_i and E_f. To effect a transition between two specific states, light must supply the energy difference, according to the formula

$$h\nu_0 = E_f - E_i. \tag{2.1.2}$$

Lectures on Light. Second Edition. Stephen C. Rand.
© Stephen C. Rand 2016. Published in 2016 by Oxford University Press.

where ν_0 is called the transition frequency. Individual transitions are rarely observable because even small particles of bulk matter contain a great many atoms, and light of even modest intensity $I \propto |E|^2$ consists of a high density $(N/V = I/ch\nu)$ of energy-bearing particles (photons) which cause transitions at random times. So individual interactions generally go unnoticed. Particle-like interactions of light waves are rarely observed unless intensities are extremely low. If a light field is pictured as a collection of quantum particles of energy $h\nu_0$, then at low intensity the particles must be widely separated in time, presuming they are all the same. We then require a description of light in terms of mathematical operators that produce discrete changes of the field when they interact. The particle nature of light is therefore an important factor determining the noise and statistical correlation properties of weak fields. However, a great many important aspects of light–matter interactions can be described adequately by treating only the atomic variables quantum mechanically as operators and the light field classically, with an amplitude and phase that are continuous scalar functions. This approach, called the semi-classical approach, is the one adopted throughout the first few chapters of this book. Aspects of light–matter interactions that depend explicitly on the discrete character of the field are reserved for the final chapters.

Charge motion caused by light is typically a dipolar response such as that determined in our classical model (Eq. (1.2.21)). However, in systems of bound charges excited near internal resonant frequencies, the dipole polarization that develops must clearly depend on the stationary quantum states and the corresponding resonant frequencies. Hence under these conditions, polarization of the medium becomes a quantum mechanical observable determined to first order by non-zero values of the first moment of the microscopic polarization operator $\hat{p}$. Hence we shall need a procedure for determining the expected values of the dipole moment and other operators weighted by available states of the atomic system (specified by the probability amplitude ψ of finding the system in a given state at a given time and location) to predict the outcome of experimental measurements. This predictor of repeated measurements is called the expectation value, and it resembles an average weighted by the probability amplitude of the state of the system. The most important physical observable considered throughout this book is the electric dipole $\hat{p} = -e\bar{r}$, which has an expectation value given by

$$\langle \hat{p} \rangle = -\langle \psi | e\bar{r} | \psi \rangle \equiv -e \int \psi * \bar{r} \psi \, dV. \tag{2.1.3}$$

Here the wavefunction $\psi(r,t)$ accounts for the proportion of eigenstates contributing to the actual state of the system. (See Section 2.2 and Appendix A for a more complete discussion of the definition of expectation value.) The bra-ket symbols used in Eq. (2.1.3) were introduced by Dirac to represent not only the conjugate state functions $\psi \leftrightarrow |\psi\rangle$ and $\psi^* \leftrightarrow \langle\psi|$ respectively but also to simplify spatial integrals where $|\psi\rangle$and $\langle\psi|$ appear in combination, as in the dipole moment $\langle \psi | e\bar{r} | \psi \rangle$ of Eq. (2.1.3). This notation provides a convenient shorthand and will be used to shorten calculations throughout the book.

The total polarization for N identical atomic dipoles in the quantum mechanical limit is obtained by summing individual contributions with an expression similar to

Eq. (1.2.21). However, the notation now implicitly includes functions (wavefunctions) on which the operator of interest must operate for the expression to have meaning:

$$\langle \bar{P} \rangle = -\left\langle \sum_{i=1}^{N} e\bar{r}_i \right\rangle. \tag{2.1.4}$$

Dipoles may be static or time varying. When $\langle \bar{P} \rangle$ is time varying, it may have observable effects even if its time average is zero, because the charge distribution oscillates and may cause a change of state of the atom or lead to radiation. Indeed, the magnitude of the dipole operator $e\bar{r}$ evaluated between *different* states of a single atom, molecule, or optical center will be shown to be the dominant factor determining the rate of optically induced transitions between quantum states. For a collection of atoms however, we shall find that the relative phase variation of $e\bar{r}_i$ from one atom to the next also plays an important role in determining the magnitude and temporal development of the ensemble-averaged or *macroscopic* polarization $\langle \bar{P} \rangle$ in Eq. (2.1.4). The behavior of $\langle \bar{P} \rangle$ due to initial phasing, dephasing, and even rephasing is important in coherent optical processes, and the density matrix is especially well-suited to keep track of important phase information.

2.2 Quantum Observables

2.2.1 Calculation of Quantum Observables

Operators generate eigenvalues corresponding to specific energy states by their action on components of a wavefunction ψ. However, all systems, particularly systems undergoing transition, contain components with a spread in energy and momentum. So for a system described by a wavefunction $\psi(\bar{r}, t)$ that is in general not an eigenstate, measurement of an observable O yields a simple average or first moment of O. This moment is weighted by the states the system is in or passing through. The only quantum mechanical aspect of this calculation is that O must be replaced by the corresponding operator $\hat{O}$:

$$\langle \hat{O} \rangle = \langle \psi | \hat{O} | \psi \rangle = \int d^3 r \psi * \hat{O} \psi. \tag{2.2.1}$$

This expression will then incorporate commutation properties, as discussed in Appendix A.

Because quantum theory deals with wave-like properties of matter, its predictions reflect the delocalization of waves and include a certain amount of "inexactitude." This is not a defect of the theory, but merely reflects the difficulty inherent in answering the question "Where is a wave?" Once a wave-based, probabilistic description of matter is introduced, there is a minimum uncertainty associated with measurements. To be more precise, conjugate variables cannot be determined simultaneously with infinite precision. For example, linear momentum p_x and position x are conjugate variables that have uncertainties Δp_x and Δx that are mutually related.

$$\Delta p_x \Delta x \geq \hbar/2 \tag{2.2.2}$$

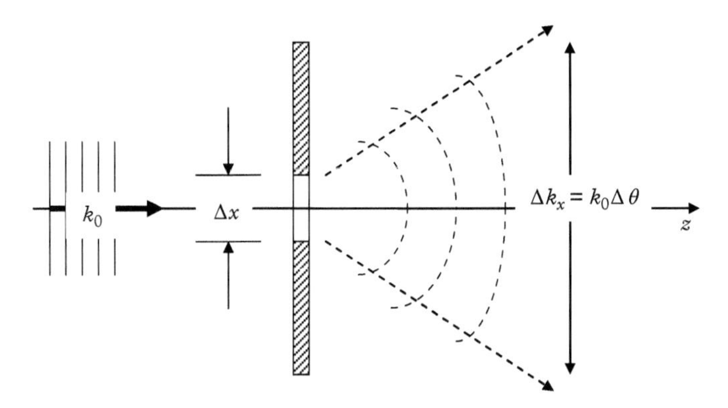

Figure 2.1 *Illustration of a typical uncertainty relationship between spatial resolution Δx and the corresponding angular spread that accompanies optical diffraction at an aperture according to the relation $\Delta p_x \Delta x = \hbar k_0 \Delta\theta \Delta x \geq \hbar/2$.*

The exact form of this Heisenberg uncertainty relation is justified and generalized in Appendix B for all pairs of conjugate variables. A simple example of the reciprocal relationship between spatial and angular resolution governed by Eq. (2.2.2) is illustrated by diffraction from an aperture (Figure 2.1), when a plane wave develops a spread of propagation directions as it passes through a small aperture.

Measurements acquire minimum uncertainty when the product in Eq. (2.2.2) takes on its minimum value. For the variables in the previous paragraph, the minimum product is $\Delta p_x \Delta x = \hbar/2$. When systems are prepared in minimum uncertainty or coherent states they obey this equality, as discussed in Chapter 6. However, it will be shown that despite the inescapable uncertainty associated with wave mechanics, the Heisenberg limit expressed by Eq. (2.2.2) can be partly circumvented by preparing systems in special states called coherent states.

2.2.2 Time Development

The wavefunction $\psi(\bar{r}, t)$ that we use to describe a material system gives the probability amplitude for finding the particle at $(\bar{r}, t)$, and its square modulus $|\psi(\bar{r}, t)|^2$ yields the probability density. Typically, experiments measure $|\psi(\bar{r}, t)|^2$, but we note in passing that it is possible to measure the eigenstates forming the basis of $\psi(\bar{r}, t)$ directly [2.1]. We expect to find the particle (with unit probability) if we search through all space. Hence the wavefunction is assumed to be normalized according to

$$\int \psi * (\bar{r}, t)\psi(\bar{r}, t) d^3 r = 1. \tag{2.2.3}$$

Evolution of the wavefunction in time is described by Schrödinger's wave equation

$$i\hbar \frac{\partial \psi}{\partial t} = H\psi. \tag{2.2.4}$$

In classical mechanics, the Hamiltonian H describing system energy is expressible in terms of canonically conjugate variables q_i and p_i of the motion. The operator form of the Hamiltonian, designated by $\hat{H}$, is obtained by a procedure that replaces the Poisson bracket $\{q_i, p_j\}$ in Hamilton's classical equations of motion by a similar bracket that keeps track of operator commutation properties, namely the commutator $[\hat{q}_i, \hat{p}_j] \equiv \hat{q}_i\hat{p}_j - \hat{p}_j\hat{q}_i = i\hbar\delta_{ij}$. An analogous procedure is used in a later chapter to quantize the electromagnetic field, and the essential role played by commutation in optical interactions will be discussed there further.

Solutions to the Schrödinger equation can be constructed from the spatial energy eigenfunctions U_n of the Hamiltonian, together with harmonic functions of time. For example,

$$\psi_n(r,t) = U_n(r)\exp(-i\omega_n t) \tag{2.2.5}$$

is a particular solution which satisfies Schrödinger's equation when the system energy is constant (i.e., $\hat{H}\psi_n = i\hbar\frac{\partial}{\partial t}\psi_n = \hbar\omega_n\psi_n$ and $\hbar\omega_n$ must be the energy eigenvalue E of the spatial eigenstate):

$$\hat{H}U_n = \hbar\omega_n U_n.$$

The energy of real, non-decaying systems is constant and observable, so the eigenvalues of such systems are real. It is the adjoint nature of $\hat{H}$ that assures us mathematically that the eigenvalues are real, that the corresponding eigenfunctions U_n can be orthonormalized, and that they collectively furnish a complete description of the system. That is,

$$\int U_n^* U_m d^3r = \delta_{nm}, \tag{2.2.6}$$

$$U_n^*(\bar{r})U_n(\bar{r}') = \delta\left(\bar{r} - \bar{r}'\right). \tag{2.2.7}$$

Formally, the U_n form a complete basis set that can represent an arbitrary state of the system. The most general solution for the wavefunction is formed from linear combinations of solutions like Eq. (2.2.5), such as

$$\psi(\bar{r},t) = \sum_n C_n U_n(\bar{r})e^{-i\omega_n t}. \tag{2.2.8}$$

If the total probability $|\psi|^2$ for the system to occupy the space defined by the basis functions U_n is to be normalized to unity, one can easily show from Eq. (2.2.8) that the constant coefficients must satisfy

$$\sum_n |C_n|^2 = 1. \tag{2.2.9}$$

In Eq. (2.2.9), $|C_n|^2$ gives the occupation probability associated with a particular eigenstate labeled by index n. This closure relation is exploited in Section 2.4 as the main tool for changing between alternative descriptions of a system, expressed in terms of various basis sets.

2.2.3 Symmetry

When researchers are confronted by a new problem, the greatest simplifications arise from considering system symmetries. Symmetry can be exploited to reduce complex problems to simpler ones and to derive general rules for atom–field interactions. The optimal approach to using symmetry in this way is the domain of group theory, which is too extensive a topic to be included here. However, we review one symmetry here, inversion symmetry, because it has important implications for many problems in optical science. It affects both linear and nonlinear interactions of light with matter. In a later chapter we shall encounter the Wigner–Eckart theorem (see also Appendix H) which can be used to implement other symmetry considerations, without group theory. For a broader discussion of symmetry, see Ref. [2.2].

When a system has inversion symmetry, it is energetically indistinguishable in inverted and non-inverted states (see Figure 2.2). Then, one can conclude from symmetry alone that (i) the wavefunctions for different stationary states may be classified into odd or even "parity," (ii) the system cannot sustain a permanent dipole moment, and (iii) "parity" must change when an electromagnetic transition occurs. Hence the presence or absence of this symmetry often determines whether electromagnetic transitions are allowed or not. To confirm these results we introduce parity operator $\hat{I}$, defined by its inverting action on all three spatial coordinates, and allow it to operate on the energy eigenvalue equation:

$$\hat{I}\left(\hat{H}(\bar{r})\psi(\bar{r})\right) = \hat{I}E\psi(\bar{r}) = E\psi(-\bar{r}). \tag{2.2.10}$$

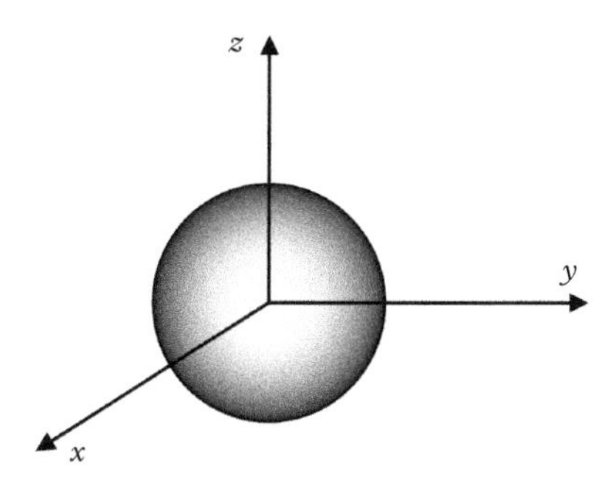

Figure 2.2 *Example of a three-dimensional object with inversion symmetry. A featureless sphere looks the same if the coordinates are all reversed.*

The left side can be rewritten as

$$\hat{I}\left(\hat{H}(\bar{r})\psi(\bar{r})\right) = \hat{H}(-\bar{r})\psi(-\bar{r}) = \hat{H}(\bar{r})\hat{I}\psi(\bar{r}), \tag{2.2.11}$$

since $\hat{H}(\bar{r}) = \hat{H}(-\bar{r})$. This last point is the result of the system being physically indistinguishable in inverted or non-inverted states. Eq. (2.2.11) shows that the commutator $[\hat{H}, \hat{I}] = 0$. Since $\hat{I}$ commutes with $\hat{H}$, the energy eigenfunctions of $\hat{H}$ are also eigenfunctions of $\hat{I}$. The eigenvalues I of the parity operator are easily found by applying inversion twice to return the original wavefunction. Thus

$$\hat{I}^2\psi = I^2\psi = \psi, \tag{2.2.12}$$

and wavefunctions of the system separate into two groups distinguished by $I = \pm 1$.

In the presence of inversion symmetry we reach three conclusions:

1. Energy eigenfunctions in systems with inversion symmetry have a definite parity. That is, with respect to inversion of the coordinates they are either even (e) or odd (o).

$$\psi^{(e)} = \psi(r) + \psi(-r), \ \text{for } I = +1, \tag{2.2.13}$$

$$\psi^{(o)} = \psi(r) - \psi(-r)), \ \text{for } I = -1. \tag{2.2.14}$$

2. The system has no permanent dipole moment.

$$\langle\psi(r)|er|\psi(r)\rangle = \hat{I}\,\langle\psi(r)|er|\psi(r)\rangle = \langle\pm\psi|-er|\pm\psi\rangle = -\langle\psi(r)|er|\psi(r)\rangle\,.$$

Since the dipole moment must equal its negative, the only solution is

$$\langle\psi|er|\psi\rangle = 0, \tag{2.2.15}$$

regardless of the state ψ of the system.

3. Transition dipole moments (where the initial and final states are different) only exist between states of opposite parity.

$$\begin{aligned}\langle\psi^{(o)}(r)|er|\psi^{(e)}(r)\rangle &= \hat{I}\langle\psi^{(o)}(r)|er|\psi^{(e)}(r)\rangle = \langle-\psi^{(o)}(r)|-er|\psi^{(e)}(r)\rangle \\ &= \langle\psi^{(o)}(r)|er|\psi^{(e)}(r)\rangle \neq 0,\end{aligned} \tag{2.2.16}$$

$$\left\langle\psi_1^{(o)}\middle|er\middle|\psi_2^{(o)}\right\rangle = \hat{I}\left\langle\psi_1^{(o)}\middle|er\middle|\psi_2^{(o)}\right\rangle = \left\langle-\psi_1^{(o)}\middle|-er\middle|-\psi_2^{(o)}\right\rangle = -\left\langle\psi_1^{(o)}\middle|er\middle|\psi_2^{(o)}\right\rangle = 0, \tag{2.2.17}$$

$$\left\langle\psi_1^{(e)}\middle|er\middle|\psi_2^{(e)}\right\rangle = \hat{I}\left\langle\psi_1^{(e)}\middle|er\middle|\psi_2^{(e)}\right\rangle = \left\langle\psi_1^{(e)}\middle|-er\middle|\psi_2^{(e)}\right\rangle = -\left\langle\psi_1^{(e)}\middle|er\middle|\psi_2^{(e)}\right\rangle = 0. \tag{2.2.18}$$

These selection rules based on inversion symmetry play an important role in some of the simple examples of quantum systems that follow.

2.2.4 Examples of Simple Quantum Systems

2.2.4.1 Simple Harmonic Oscillator

One of the most rudimentary dynamical systems with inversion symmetry consists of a simple harmonic oscillator, exemplified by a spring that obeys Hooke's law of motion on an immoveable support attached to a mass m. Its allowed energy levels are depicted in Figure 2.3.

In such a system, the restoring force of the spring is linear in the displacement x from equilibrium in accord with $\bar{F} = -k\bar{x}$ (where $k = m\omega_0^2$) and the system energy consists of potential and kinetic energy terms that appear explicitly in the Hamiltonian

$$\hat{H} = \hat{p}^2/2m + m\omega_0^2\hat{x}^2/2. \tag{2.2.19}$$

By solving Schrödinger's equation one finds the stationary eigenfunctions

$$U_n = \left(\frac{m\omega_0}{\hbar\sqrt{\pi}2^n n!}\right)^{1/2} H_n(\xi)\exp(-\xi^2/2), \tag{2.2.20}$$

where $\xi \equiv \sqrt{m\omega_0/\hbar}x$ and $H_n(\xi)$ is a Hermite polynomial. These will later be referred to as basis states and written simply as $|n\rangle \equiv |U_n\rangle$. The eigenvalues are

$$\hbar\omega_n = \left(n + \frac{1}{2}\right)\hbar\omega_0, n = 0, 1, 2, \ldots \tag{2.2.21}$$

Note that the eigenfunctions separate into even and odd sets, reflecting the inversion symmetry of the motion about the origin.

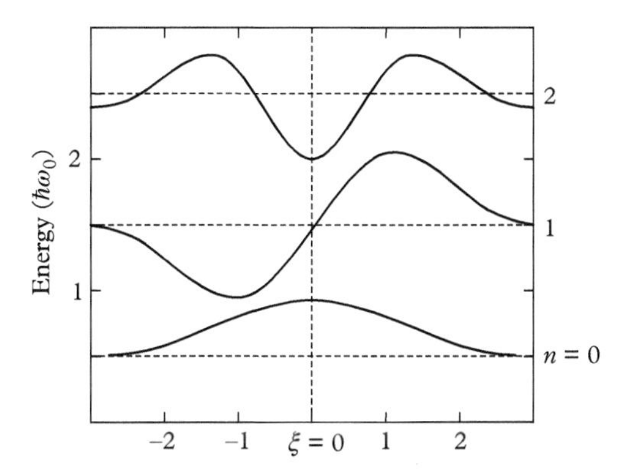

Figure 2.3 *The lowest three eigenfunctions of the simple harmonic oscillator.*

2.2.4.2 Particle in a One-Dimensional (Symmetric) Potential Well

The situation of a particle in an infinitely deep potential is illustrated in Figure 2.4:

$$\hat{H} = \frac{\hat{p}^2}{2m} + \hat{V}, \hat{V} = \begin{cases} 0, & 0 \leq z \leq L \\ \infty, & z < 0, z > L \end{cases}. \tag{2.2.22}$$

The momentum operator is $\hat{p} = -i\hbar\partial/\partial z$. Energy eigenfunctions are therefore

$$U_n(z) = \begin{cases} \sqrt{2/L}\sin k_n z, & 0 \leq z \leq L \\ 0, & z < 0, z > L \end{cases}, \tag{2.2.23}$$

where $k_n = n\pi/L$ and $n = 1, 2, 3, \ldots$

The energy eigenvalues are

$$\hbar\omega_n = \frac{\hbar^2 k_n^2}{2m}, \tag{2.2.24}$$

and a general wavefunction for this problem is

$$\psi = \left(\frac{2}{L}\right)^{1/2} \sum_n C_n \sin(k_n z)\, \mathrm{e}^{-i\omega_n t}. \tag{2.2.25}$$

Notice that the solutions in Eq. (2.2.25) again separate into contributions with n even or odd, reflecting the inversion symmetry of the potential, as in Section 2.2.3.

2.2.4.3 Particle in a One-Dimensional (Asymmetric) Potential Well

In modulation-doped field-effect transistor (MODFET) structures, energy bands near interfaces bend in such a way that carriers can be trapped in shallow potential wells that lack inversion symmetry, forming a two-dimensional (2-D) electron gas, as illustrated in Figure 2.5.

The potential $V(x)$ is approximately proportional to the distance x from the interface:

$$V(x) = V_0 x. \tag{2.2.26}$$

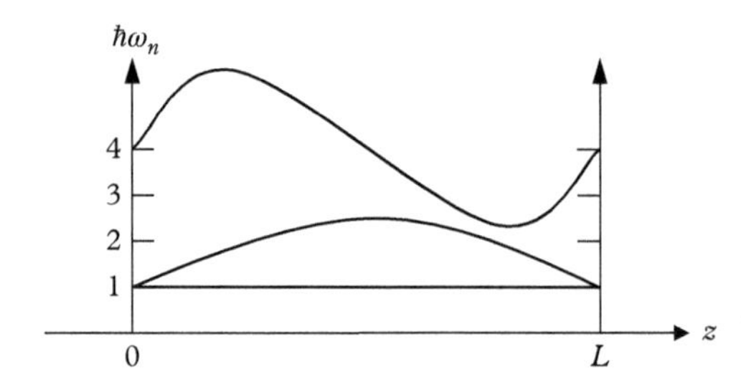

Figure 2.4 *The two lowest eigenfunctions of a one-dimensional infinite square well.*

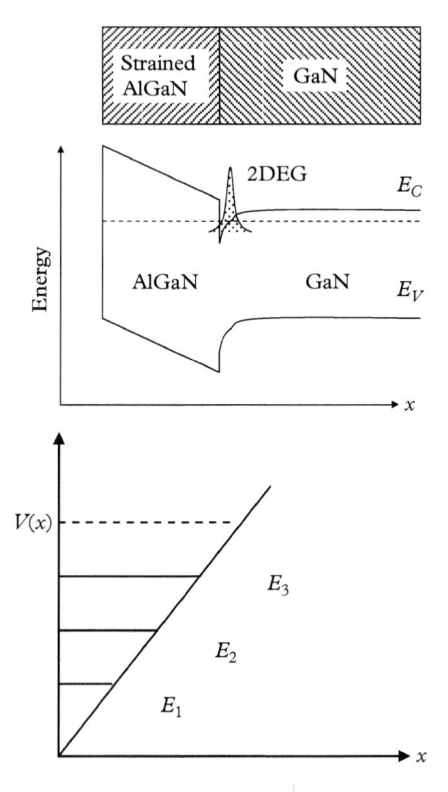

Figure 2.5 *(a) Schematic cross-section of a ternary nitride semiconductor structure giving rise to a two-dimensional electron gas* (2DEG) *at the* AlGaN/GaN *interface (in the middle). (b) The linear potential well and quantized energy levels at the* 2DEG *interface.*

Hence the time-independent Schrödinger equation is

$$\left[\frac{-\hbar^2}{2m_e}\frac{d^2}{dx^2} + (V_0 x - E)\right]U(x) = 0. \tag{2.2.27}$$

By introducing the dimensionless variable

$$\xi_n \equiv \left(x - \frac{E_n}{V_0}\right)\left(\frac{2m_e V_0}{\hbar^2}\right)^{1/3}, \tag{2.2.28}$$

where $n = 1, 2, 3, \ldots$ is an integer index for the various solutions, this equation simplifies to

$$\left[\frac{d^2}{d\xi_n^2} - \xi_n\right]U_n(\xi_n) = 0. \tag{2.2.29}$$

The solution of this equation is an Airy function, related to Bessel functions of fractional order. A suitable integral form of this function can be written as

$$\Phi_n(\xi_n) = \frac{1}{\pi}\int_0^{\infty} \cos\left(\frac{1}{3}u^3 + u\xi_n\right) du, \tag{2.2.30}$$

with the explicit eigenfunctions being

$$U_n(\xi_n) = A_n \Phi_n(-\xi_n), \tag{2.2.31}$$

where A_n is a constant amplitude for the nth solution. To find energy eigenvalues it is necessary to apply boundary conditions. For this we can approximate the well height at the origin as infinite, so that $U(\xi) = 0$ at $x = 0$. That is,

$$U_n\left(-\frac{E_n}{V_0}\left[\frac{2m_e V_0}{\hbar^2}\right]^{1/3}\right) = U_n(-R_n) = 0. \tag{2.2.32}$$

This shows that the eigenvalues are proportional to roots (R_n) of the Airy function. The first few roots are $R_1 = 2.34$, $R_2 = 4.09$, $R_3 = 5.52$, and $R_4 = 6.78$. Hence the eigenvalues, obtained from Eq. (2.2.32), are

$$E_n = \left(\frac{V_0^2\hbar^2}{2m_e}\right)^{1/3} R_n. \tag{2.2.33}$$

The wavefunctions must be normalized to unity. This requires that

$$\int_{-R_n}^{\infty} U_n^*(\xi) U_n(\xi) d\xi = |A_n|^2, \tag{2.2.34}$$

showing that the undetermined coefficients are given by

$$A_n = \left(\int_{-R_n}^{\infty} |U_n(\xi_n)|^2 \, d\xi_n\right)^{-1/2}. \tag{2.2.35}$$

In this example, no reduction of the wavefunction into sets of even and odd energy levels takes place. Because of the lack of inversion symmetry, the wavefunctions do not separate into different parity classes. (See *Handbook of Mathematical Functions*, NBS, 1964, pp.446–447, Eqs. 10.4.1 and 10.4.32 and Ref. [2.3]).

2.2.4.4 *Particle in an Adjustable Three-Dimensional Box*

F centers in alkali halides are excellent examples of real particles in real cubic boxes of adjustable size. These color centers consist of an electron in an anion vacancy of an ionic crystal (a defect of the crystal structure), and give rise to dramatic coloration of otherwise transparent crystals [2.3]. One does not need to know much in the way of specifics about them to make quite detailed, verifiable predictions about them. We picture an electron in a cubic space left by a missing anion, a space somewhat larger than a hydrogen atom (Figure 2.6), with a positive charge and an infinitely high potential at the walls. Because it is caged in by the otherwise perfect crystal, the wavefunction must vanish at the "walls" comprised of neighboring cations. Hence, borrowing from the one-dimensional symmetric case in Section 2.2.4.3, we would guess that possible wavefunctions have the form

$$\psi(x, y, z) = \sin(k_x x)\sin(k_y y)\sin(k_z z) \tag{2.2.36}$$

where $k_x = \frac{\pi}{a} n_x$, $k_y = \frac{\pi}{a} n_y$, and $k_z = \frac{\pi}{a} n_z$.

The energy eigenvalues are

$$E(n_x, n_y, n_z) = \frac{\hbar^2 k^2}{2m} = \frac{\hbar^2\left(k_x^2 + k_y^2 + k_z^2\right)}{2m} = \frac{\hbar^2 \pi^2}{2ma^2}\left(n_x^2 + n_y^2 + n_z^2\right). \tag{2.2.37}$$

The ground state energy is

$$E_0 = \frac{3h^2}{8ma^2}, \tag{2.2.38}$$

since $n_x = n_y = n_z = 1$. The energy of the first excited state is

$$E_1 = \frac{6h^2}{8ma^2}, \tag{2.2.39}$$

since $n_x = 2$ and $n_y = n_z = 1$. The transition frequency of the first resonance of the F center is therefore given by

$$h\nu = E_1 - E_0 = \frac{3n^2}{8ma^2}, \tag{2.2.40}$$

from which one obtains directly the Mollwo–Ivey relation [2.4] for the relationship between the edge length of the crystal unit cell and the frequency of resonant absorption:

$$\ln(a) = \frac{1}{2}\ln(\nu) + c. \tag{2.2.41}$$

As shown in Figure 2.7, this relation has been confirmed by measurements on various alkali halide crystals.

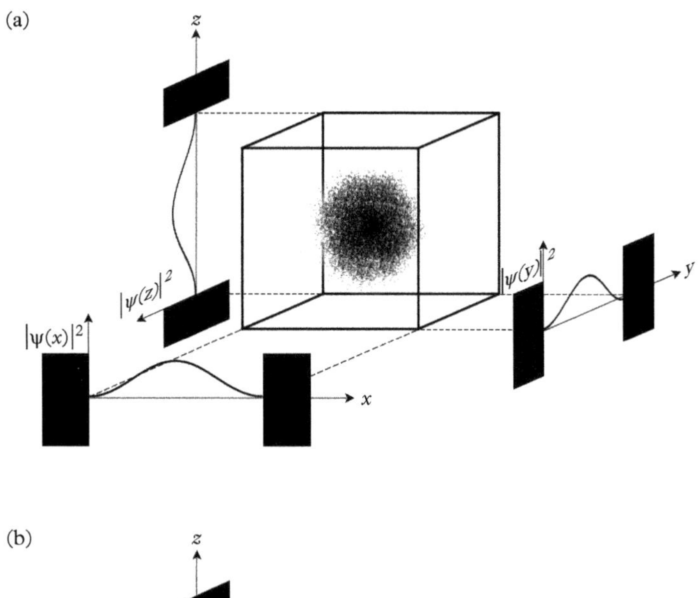

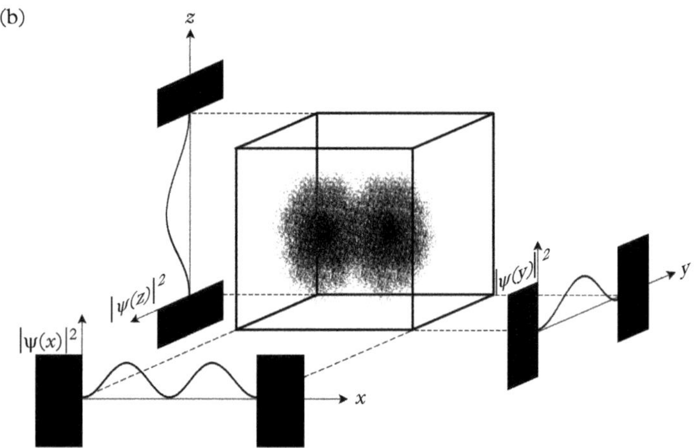

Figure 2.6 *Pictorial representation of the wavefunctions and charge distributions of an F center in the (a) ground state and (b) the first excited state. (After Ref. [2.4]).*

Multi-electron color centers are not only good examples of particle-in-a-box analysis, but have been widely used in high-gain tunable laser technology [2.5]. Another important application relies on their mobility under applied fields. Oxygen vacancies in semiconducting rutile (TiO_2) are thermally more stable than the alkali F centers and can be made to migrate by the application of voltage across the crystal, causing a change in resistance of the medium proportional to the integral of voltage over time. This understanding led to the first realization of a fundamental circuit element called the memristor [2.6].

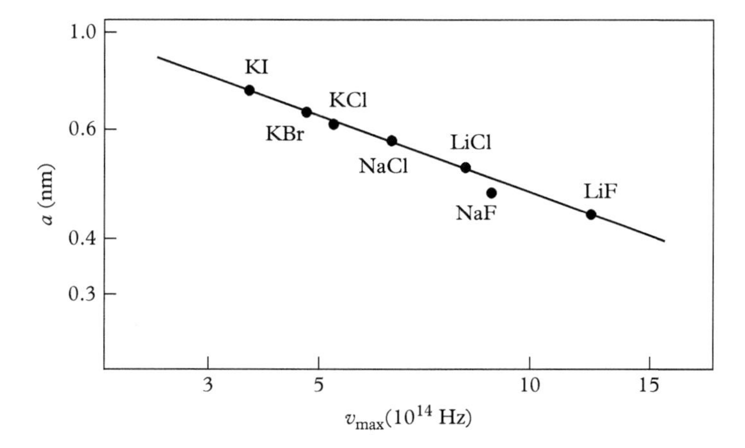

Figure 2.7 *Plot of experimental measurements of the first absorption resonance of F centers in several alkali halides, compared with the prediction of the Mollwo–Ivey relation (solid curve). (After Ref. [2.4].)*

2.3 Dynamics of Two-Level Systems

Knowledge of system symmetries and eigenenergies is often enough to predict dynamics in systems where the wavefunction and Hamiltonian H_0 are unknown. This is fortunate, because systems encountered in research often have unknown properties and structure that is complicated enough to prevent determination of the wavefunction. The limited number of models available for which ψ can be found analytically might even suggest that these case studies are irrelevant to new problems in which ψ is essentially unknowable. However, there are at least three reasons why this is not so. The first is that many aspects of dynamics are dominated by the interaction of light with only two energy levels coupled by the light. Other levels in the system often are not strongly coupled by the electromagnetic field and play a secondary role. So, two-level models capture dominant aspects of the dynamics even in complex systems. The second reason is that complicated systems can sometimes be converted into truly two-level systems by a judicious choice of experimental conditions. Thirdly, as subsequently illustrated, the development that a system undergoes as time progresses depends on its resonant frequencies and state symmetries, not on any particular mathematical representation of H_0 or ψ.

Consider a solid known to have two levels separated by an energy in the range of an available light source. When light propagates along the z-axis of the crystal, the structure absorbs strongly regardless of polarization. What, if anything, can be said about such a system, its dynamics and transitions? Presuming the absorption process involves a single electron as in the vast majority of cases, and recalling that the classical energy of an electric dipole formed by the electron in a field $\bar{E}$ is given by the projection of the moment on the external field, the interaction energy should be

$$\hat{H}_I = -\bar{\mu} \cdot \bar{E}(z, t). \tag{2.3.1}$$

Since the light field is under the experimentalist's control, assume it is a plane wave traveling along z with no variation of its amplitude in x or y. The dipole moment could be oriented in an arbitrary, fixed direction described by

$$\bar{\mu} = \mu_x \hat{x} + \mu_y \hat{y} + \mu_z \hat{z}. \tag{2.3.2}$$

Perpendicular to z, only the x and y components of the moment are relevant for interaction with the wave. In free space the orientation of $\bar{E}$ is perpendicular to the direction of propagation, so $\mu_z \hat{z} \cdot \bar{E}(z,t) = 0$. Since we are told the absorption is isotropic, we may also assume $\mu_x = \mu_y = \mu_0$, and without loss of generality reduce Eq. (2.3.1) to

$$\hat{H}_{\text{int}} = -\mu_0 E_0, \tag{2.3.3}$$

where E_0 is the amplitude of the electromagnetic wave. The charge oscillations excited by the field are restricted to the x–y plane. So the interaction spans a two-dimensional Hilbert space. Equation (2.3.3) therefore describes a 2-D interaction operator, and can be represented using 2×2 matrices—the Pauli spin matrices which (together with the unit matrix σ_I) form a complete basis set:

$$\hat{\sigma}_x = \begin{bmatrix} 0 & 1 \\ 1 & 0 \end{bmatrix}; \quad \hat{\sigma}_y = \begin{bmatrix} 0 & -i \\ i & 0 \end{bmatrix}; \quad \hat{\sigma}_z = \begin{bmatrix} 1 & 0 \\ 0 & -1 \end{bmatrix}. \tag{2.3.4}$$

This set of matrices is called complete because any complex two-component vector can be written as a linear combination of σ_x, σ_y, and σ_z together with the unit matrix σ_I using appropriate coefficients. Also, the product of any two Pauli matrices is a Pauli matrix, showing that they form a complete group. Since light can at most transform the state of the atom among its available states in Hilbert space, the interaction Hamiltonian $\hat{H}_I$ must be an operator that can convert a 2-vector into another 2-vector in the same Hilbert space. So its matrix representation must also be 2×2 and we choose the Pauli matrices as basis operators. The operator representation of Eq. (2.3.3) must change state $\begin{bmatrix} 0 \\ 1 \end{bmatrix}$ into $\begin{bmatrix} 1 \\ 0 \end{bmatrix}$ and vice versa. Hence it is given by

$$\hat{H}_{\text{int}} = -\mu_0 E_0 \hat{\sigma}_x. \tag{2.3.5}$$

Exercise: Confirm that the form of $\hat{H}_I$ in Eq. (2.3.5) is correct by applying it to either eigenstate of the two-level atom and showing that it causes a transition to the other.

Now $\hat{\sigma}_x$ can be written in terms of raising and lowering operators for atomic states to simplify interpretation of dynamics. Consider the following two matrices.

$$\hat{\sigma}^+ \equiv \frac{1}{2}\left(\hat{\sigma}_x + i\hat{\sigma}_y\right) = \begin{bmatrix} 0 & 1 \\ 0 & 0 \end{bmatrix} \rightarrow \hat{\sigma}^+ \begin{bmatrix} 0 \\ 1 \end{bmatrix} = \begin{bmatrix} 1 \\ 0 \end{bmatrix}, \tag{2.3.6}$$

$$\hat{\sigma}^- \equiv \frac{1}{2}\left(\hat{\sigma}_x - i\hat{\sigma}_y\right) = \begin{bmatrix} 0 & 0 \\ 1 & 0 \end{bmatrix} \rightarrow \hat{\sigma}^- \begin{bmatrix} 1 \\ 0 \end{bmatrix} = \begin{bmatrix} 0 \\ 1 \end{bmatrix}. \tag{2.3.7}$$

Writing the interaction Hamiltonian in terms of these operators, which raise or lower the state of the atom, Eq. (2.3.5) becomes

$$\hat{H}_{\text{int}} = -\mu_0 E_0 \begin{bmatrix} 0 & 1 \\ 1 & 0 \end{bmatrix} = -\mu_0 E_0 \left(\hat{\sigma}^+ + \hat{\sigma}^-\right). \tag{2.3.8}$$

It may now be deduced that a continuous field–atom interaction causes the atom to make periodic transitions up and down between states 1 and 2 by stimulated absorption and emission, respectively, at a frequency related to the absorption coefficient (since it depends on the transition moment μ_0).

As shown by the exercise, discrete applications of the interaction Hamiltonian show the individual transitions caused by optical irradiation:

$$\hat{H}_{\text{int}} \begin{bmatrix} 0 \\ 1 \end{bmatrix} = -\mu_0 E_0 \begin{bmatrix} 1 \\ 0 \end{bmatrix} \tag{2.3.9}$$

and

$$\hat{H}_{\text{int}} \begin{bmatrix} 1 \\ 0 \end{bmatrix} = -\mu_0 E_0 \begin{bmatrix} 0 \\ 1 \end{bmatrix}. \tag{2.3.10}$$

A more detailed picture of system dynamics can only emerge by solving explicitly for temporal evolution of the state of the system. In the present problem there is no information on the stationary states of the system. So we do not know the wavefunction itself, but we do know there are at least two states in the system. There is a ground state which might be represented as ψ^- and an excited state ψ^+. The challenge of predicting what the system will do in response to irradiation requires the determination of the most general wavefunction we can write based on this information. This is $\psi(t) = c^+\psi^+ + c^-\psi^-$, where c^+ and c^- are time-dependent coefficients. So the time-dependent Schrödinger equation yields

$$\begin{aligned} \frac{d}{dt}\left(c^+(t)\psi^+ + c^-(t)\psi^-\right) &= i\Omega\left(\hat{\sigma}^+ + \hat{\sigma}^-\right)\left(c^+(t)\psi^+ + c^-(t)\psi^-\right) \\ &= i\Omega\left(c^+(t)\psi^- + c^-(t)\psi^+\right), \end{aligned} \tag{2.3.11}$$

where $\Omega \equiv \mu_0 E_0/\hbar$. Hence

$$\left[\frac{\partial c^+(t)}{\partial t}\psi^+ + \frac{\partial c^-(t)}{\partial t}\psi^-\right] = i\Omega\left[c^+(t)\psi^- + c^-(t)\psi^+\right], \tag{2.3.12}$$

and because ψ^+ and ψ^- are independent vector states, Eq. (2.3.12) may be decomposed into the two equations

$$\frac{\partial c^+}{\partial t} = i\Omega c^-, \tag{2.3.13}$$

$$\frac{\partial c^-}{\partial t} = i\Omega c^+. \tag{2.3.14}$$

These two coupled equations are readily solved by differentiation and cross substitution.

$$\frac{\partial^2 c^\pm}{\partial t^2} = -\Omega^2 c^\pm \tag{2.3.15}$$

By taking into account the normalization condition $\left|c^+\right|^2 + \left|c^-\right|^2 = 1$ and initial condition $c^-(0) = 1$, the overall wavefunction can therefore be written

$$\psi(t) = c^+\psi^+ + c^-\psi^- = \psi^+ \sin\Omega t + \psi^- \cos\Omega t. \tag{2.3.16}$$

The system oscillates between the ground state ψ^- and the excited state ψ^+ at a frequency Ω called the resonant Rabi frequency. This oscillatory behavior is called Rabi flopping. What is interesting here is that we have predicted some system dynamics—an oscillation at a very specific rate in a poorly characterized system—without knowing the functional form of the wavefunction ψ or even the unperturbed system Hamiltonian H_0. It was sufficient to know a little about system symmetry, the energy eigenvalues, and to assume the optical interaction was dominated by the electric dipole Hamiltonian. It is particularly important to realize that one can calculate optical properties and dynamic response to light without knowing the wavefunction, since the wavefunction itself is rarely known in systems of interest.

2.4 Representations

2.4.1 Representations of Vector States and Operators

The Hamiltonian describing the interaction of matter with a perturbing influence is often written as the sum of a static portion $\hat{H}_0$ and an interaction term $\hat{V}$ (the electric dipole interaction between light and matter is considered in detail in Appendix C):

$$\hat{H} = \hat{H}_0 + \hat{V} \tag{2.4.1}$$

The Schrödinger equation then becomes

$$\frac{d}{dt}|\psi(t)\rangle = -\frac{i}{\hbar}\left(\hat{H}_0 + \hat{V}\right)|\psi(t)\rangle . \tag{2.4.2}$$

If the entire Hamiltonian $\hat{H}$ itself is time independent, direct integration of Eq. (2.4.2) yields

$$|\psi(t)\rangle = \exp\left(-\frac{i}{\hbar}\hat{H}t\right)|\psi(0)\rangle , \tag{2.4.3}$$

showing that the system wavefunction can still evolve by changing the admixture of its eigenstates. Hence an observable O acquires a time dependence which, if not too rapid, is reflected in the measured matrix element

$$\left\langle \hat{O}(t) \right\rangle = \langle\psi|\hat{O}|\psi\rangle . \tag{2.4.4}$$

There are three common ways to partition the time dependence in Eq. (2.4.4) between the state vector $|\psi\rangle$ and the operator $\hat{O}$. The most common divisions are the following:

Schrödinger picture:

$$\left\langle \hat{O} \right\rangle = \left\langle \psi^S(t) \middle| \hat{O}^S(0) \middle| \psi^S(t) \right\rangle , \tag{2.4.5}$$

where $\left|\psi^S(t)\right\rangle \equiv \exp(-i\hat{H}t/\hbar)\left|\psi^S(0)\right\rangle$.

Interaction picture:

$$\left\langle \hat{O} \right\rangle = \left\langle \psi^I(t) \middle| \hat{O}^I(t) \middle| \psi^I(t) \right\rangle , \tag{2.4.6}$$

where $\left|\psi^I(t)\right\rangle \equiv \exp(i\hat{H}_0 t/\hbar)\left|\psi^S(t)\right\rangle$ and $\hat{O}^I(t) \equiv \exp(i\hat{H}_0 t/\hbar)\hat{O}^S(0)\exp(-i\hat{H}_0 t/\hbar)$.

Heisenberg picture:

$$\left\langle \hat{O} \right\rangle = \left\langle \psi^H \middle| \hat{O}^H(t) \middle| \psi^H \right\rangle , \tag{2.4.7}$$

where $\hat{O}^H(t) \equiv \exp(i\hat{H}t/\hbar)\hat{O}^S(0)\exp(-i\hat{H}t/\hbar)$ and $\left|\psi^H\right\rangle \equiv \left|\psi^S(0)\right\rangle$.

These options for representing problems are explored further in the next section. One of the three pictures is generally preferable to another for a given problem. The choice of representation is determined by what aspect of system dynamics is to be emphasized in calculations.

2.4.2 Equations of Motion in Different Representations

The basic equation of motion is Schrödinger's equation, given by Eq. (2.4.2). However, it is a matter of convenience as to whether one views the physical effects of system dynamics as being associated primarily with the atomic wavefunction, or the perturbing field operator, or as being divided between the two. For easy analysis, the choice of representation should reflect whether it is the state of the atom, or the state of the field, or the interaction between them (i.e., coherent atom–field coupling) that is of most interest respectively. Why is this a concern? This matters because we generally have in mind experiments of a specific type with which we plan to probe a system. It may be just the atomic state that we wish to monitor. Or, alterations in the spectrum and intensity of the perturbing field itself may be the chief interest. Or, the chief concern might be measurement of the amplitude of a transient coherent polarization that simultaneously reflects the harmonic oscillation of a superposition state of the atom and its phase-dependent coupling with the applied field. For ease of interpretation, our calculations should reflect the way we choose to view the problem.

2.4.2.1 *Time-Independent $\hat{H}$*

Direct integration of the Schrödinger equation when $\hat{H}$ is time independent gives

$$|\psi(t)\rangle = \exp\left(-i\hat{H}t/\hbar\right)|\psi(0)\rangle, \tag{2.4.8}$$

where

$$\exp\left(-i\hat{H}t/\hbar\right) \equiv 1 - i\left(\hat{H}t/\hbar\right) - \frac{1}{2}\left(\hat{H}t/\hbar\right)^2 + \ldots \tag{2.4.9}$$

This result can then be used to develop slightly different equations of motion for the various pictures of system dynamics.

(i) *Schrödinger picture:*

$$i\hbar\frac{d}{dt}\left|\psi^S(t)\right\rangle = \hat{H}\left|\psi^S(t)\right\rangle, \tag{2.4.10}$$

$$\left\langle\hat{O}(t)\right\rangle = \left\langle\psi^S(t)\middle|\hat{O}(0)\middle|\psi^S(t)\right\rangle, \tag{2.4.11}$$

where

$$\left|\psi^S(t)\right\rangle = \exp\left(-i\hat{H}t/\hbar\right)|\psi(0)\rangle. \tag{2.4.12}$$

In Eq. (2.4.11) the expectation value of the operator has been written in terms of its initial value $\hat{O}(0)$ to emphasize that it does not change with time. All time dependence is associated with the wavefunctions in the Schrödinger picture.

(ii) *Interaction picture:*
To extract the purely sinusoidal dynamics associated with the static Hamiltonian H_0 in our description of the time development of the state vector, the contribution of H_0 can be removed with the transformation

$$\left|\psi^S(t)\right\rangle = \exp\left(-i\hat{H}_0 t/\hbar\right)\left|\psi^I(t)\right\rangle. \tag{2.4.13}$$

Then

$$i\hbar\frac{d}{dt}\exp\left(-i\hat{H}_0 t/\hbar\right)\left|\psi^I(t)\right\rangle = \left(\hat{H}_0^S + \hat{V}^S\right)\exp\left(-i\hat{H}_0 t/\hbar\right)\left|\psi^I(t)\right\rangle$$

$$i\hbar\frac{d}{dt}\left|\psi^I(t)\right\rangle = \exp\left(i\hat{H}_0 t/\hbar\right)\hat{V}^S\exp\left(-i\hat{H}_0 t/\hbar\right)\left|\psi^I(t)\right\rangle.$$

This result may be rewritten in the form

$$i\hbar\frac{d}{dt}\left|\psi^I(t)\right\rangle = \hat{V}^I(t)\left|\psi^I(t)\right\rangle, \tag{2.4.14}$$

where

$$\hat{V}^I(t) \equiv \exp\left(i\hat{H}_0 t/\hbar\right)\hat{V}^S\exp\left(-i\hat{H}_0 t/\hbar\right) \tag{2.4.15}$$

is an effective Hamiltonian for the transformed state vector $\left|\psi^I(t)\right\rangle$. This representation is called the interaction picture, because according to Eq. (2.4.14) the development of $\left|\psi^I(t)\right\rangle$ is determined by the interaction $\hat{V}^I(t)$ alone, rather than $\hat{H}_0$ or $\hat{H}$.

On the other hand the time development of *operators* in this picture is determined by $\hat{H}_0$. To make a comparison with the Schrödinger picture we note from Eq. (2.4.15) that an arbitrary operator in the interaction picture relates to one in the Schrödinger picture according to

$$\hat{O}^I = \exp\left(i\hat{H}_0 t/\hbar\right)\hat{O}^S(0)\exp\left(-i\hat{H}_0 t/\hbar\right). \tag{2.4.16}$$

Consequently its expectation value may be evaluated by making use of Eq. (2.4.13):

$$\begin{aligned}\left\langle\hat{O}^I(t)\right\rangle &= \left\langle\psi^I(t)\right|\exp\left(i\hat{H}_0 t/\hbar\right)\hat{O}^S(0)\exp\left(-i\hat{H}_0 t/\hbar\right)\left|\psi^I(t)\right\rangle \\ &= \left\langle\psi^S(t)\exp\left(-i\hat{H}_0 t/\hbar\right)\right|\exp\left(i\hat{H}_0 t/\hbar\right)\hat{O}^S(0)\exp\left(-i\hat{H}_0 t/\hbar\right)\left|\exp\left(i\hat{H}_0 t/\hbar\right)\psi^S(t)\right\rangle \\ &= \left\langle\hat{O}^S(t)\right\rangle.\end{aligned} \tag{2.4.17}$$

This shows that expectation values of operators are *independent of representation.* However, in the interaction picture operators are assigned part of the overall time dependence, unlike the Schrödinger picture. Hence we shall need an equation of motion for the operator $\hat{O}^I(t)$. Direct differentiation of Eq. (2.4.16) yields:

$$\frac{d}{dt}\hat{O}^I(t) = \frac{i}{\hbar}\left[\hat{H}_0, \hat{O}^I\right]. \tag{2.4.18}$$

Note that the expansion coefficients in the Schrödinger and interaction pictures are different. The complete time dependence is described by Schrödinger coefficients $c_n(t)$ in the expansion $|\psi\rangle = \sum_n c_n(t)|n\rangle$, whereas the time dependence due only to the interaction is described by interaction picture coefficients $C_n(t)$ in $|\psi\rangle = \sum_n C_n(t)\exp(-i\omega_n t)|n\rangle$. For each eigenfrequency ω_n, the coefficients are therefore related by

$$c_n(t) = C_n(t)\exp(-i\omega_n t). \tag{2.4.19}$$

(iii) *Heisenberg picture:*

$$\begin{aligned}\langle\hat{O}(t)\rangle &= \langle\psi^S(0)|\exp(i\hat{H}t/\hbar)\hat{O}\exp(-i\hat{H}t/\hbar)|\psi^S(0)\rangle \\ &= \langle\psi^H|\hat{O}^H(t)|\psi^H\rangle,\end{aligned} \tag{2.4.20}$$

where

$$\hat{O}^H(t) \equiv \exp\left(i\hat{H}t/\hbar\right)\hat{O}\exp\left(-i\hat{H}t/\hbar\right) \tag{2.4.21}$$

and

$$|\psi^H\rangle = |\psi^S(0)\rangle. \tag{2.4.22}$$

By direct differentiation of Eq. (2.4.21) we obtain

$$\frac{d}{dt}\psi^H = \frac{d}{dt}\psi(0) = 0, \tag{2.4.23}$$

$$\frac{d}{dt}\hat{O}^H = \frac{i}{\hbar}\left[\hat{H}^H, \hat{O}^H\right]. \tag{2.4.24}$$

Great care must be exercised when manipulating functions of operators, for the simple reason that operators do not generally commute with one another. Some decompositions of the time dependence which might suggest themselves intuitively are invalid.

For example,

$$|\psi(t)\rangle = \exp\left(-i\hat{H}t/\hbar\right)|\psi(0)\rangle = \exp\left(-i\hat{H}_0 t/\hbar\right)\exp\left(-i\hat{V}t/\hbar\right)|\psi(0)\rangle \tag{2.4.25}$$

is incorrect unless $\left[\hat{H}_0, \hat{V}\right] = 0$, because factorization of exponential functions of operators is prohibited when the operators in the argument do not commute. The Weyl formula (see Problem 6.1) prescribes $\exp[\hat{A}+\hat{B}] = \exp[\hat{A}]\exp[\hat{B}]\exp(-[\hat{A},\hat{B}]/2)]$. While the physical meaning of commutation in quantum mechanics is not obvious or intuitive at first, it may be physically understood by simply considering the different effects that reversed sequences of interactions have on optical systems. In the next section we discuss a sequence of light pulses as an example of complex time-varying interactions where commutation keeps track of the order of events.

As a final note on representations, although different choices of representation are available, the Schrödinger representation will be used for most of the analysis presented in this book. This choice helps maintain consistency from one topic to another to the maximum extent possible. Also, for Chapters 3–5, the optical interactions do not change the energy level structure of the atom itself, so occupation of the various levels and transition rates between them are the main concern. For this purpose, calculations of the probability amplitudes in the Schrödinger picture are generally adequate. An exception is the analysis of photon echoes covered in Chapter 4, where a combination of Schrödinger and interaction pictures makes the analysis tractable. Other exceptions appear in Chapters 6 and 7, where the state of the field rather than the state of the atom is of primary interest. Such cases are best treated in the Heisenberg picture which assigns the time development to the field operator as we have already seen.

2.4.2.2 *Time-Dependent* $\hat{H}$

A useful approach when the Hamiltonian is time dependent is to introduce an evolution operator $U_R(t)$. This operator is defined by

$$|\psi(t)\rangle = U_R(t)|\psi(0)\rangle. \tag{2.4.26}$$

Hence the equation of motion may be obtained directly as

$$\frac{d}{dt}|\psi(t)\rangle = \frac{d}{dt}U_R(t)|\psi(0)\rangle. \tag{2.4.27}$$

But the Schrödinger equation yields

$$\frac{d}{dt}|\psi(t)\rangle = -\frac{i}{\hbar}\hat{H}(t)|\psi(t)\rangle = -\frac{i}{\hbar}\hat{H}(t)U_R(t)|\psi(0)\rangle, \tag{2.4.28}$$

$$\frac{d}{dt}U_R(t) = -\frac{i}{\hbar}\hat{H}(t)U_R(t). \tag{2.4.29}$$

Solutions to Eq. (2.4.29) can be obtained by an iterative approach over small time intervals, and should reduce to $U_R(t, t_0) = \exp(-i\hat{H}(t - t_0)/\hbar)$ in the limit that $\hat{H}$ is time independent. For the time interval t_0 to t, formal integration yields

$$\int_{t_0}^{t} dU_R(t', t_0)dt' = -\frac{i}{\hbar}\int_{t_0}^{t} \hat{H}(t')U_R(t', t_0)dt'. \quad (2.4.30)$$

This result permits the first iterative contribution to the evolution operator to be identified as

$$U_R(t, t_0) = 1 - \frac{i}{\hbar}\int_{t_0}^{t} \hat{H}(t')U_R(t', t_0)dt', \quad (2.4.31)$$

since the zeroth order term is $U_R(t_0, t_0) = 1$. By successive substitutions for $U_R(t', t_0)$ into Eq. (2.4.31), a series expansion may be found for $U(t_1, t_0)$:

$$U(t, t_0) = 1 - \frac{i}{\hbar}\int_{t_0}^{t} \hat{H}(t_1)dt_1 + \left(\frac{-i}{\hbar}\right)^2 \int_{t_0}^{t}\int_{t_0}^{t} dt_1 dt_2 \hat{H}(t_1)\hat{H}(t_2)\ldots \quad (2.4.32)$$

Standard perturbative approaches to static and dynamic interactions are considered further in the review section of Appendix D. Note that Eq. (2.4.32) is a *time-ordered* expression, because the operators $\hat{H}(t_1)$, $\hat{H}(t_2)$, etc. do not commute in general.

What is the meaning of the restriction on the time order of operators? Among other things, it reflects the difference in dynamics caused by different sequences of interactions. If three pulses are incident on a system at separate times t_1, t_2, and t_3, it may easily be anticipated that the outcome depends on which pulse arrives first and whether its frequency is close to resonance. This situation is depicted in Figure 2.8 where three different sequences are compared. In the 123 sequence on the left the ground state absorption transition is driven first, so the upper state can be reached if pulses 2 and 3 arrive before the atom decays back to ground. In the 321 sequence in the middle of the figure however, the pulse connecting excited state 3 to the intended final state 4 is applied first. Since state 3 is presumably unoccupied prior to the arrival of the pulses, there is no population in this state to absorb the first arriving pulse and nothing happens—no atoms reach state 4 in this case. This illustrates the fact that permutation of the time of arrival is important in pulsed optical interactions. The third sequence in Figure 2.8 illustrates that the permutation of frequencies leads to still other excitation channels. Different time-domain and frequency-domain sequences of pulses lead to physically distinct processes and outcomes, and commutation algebra automatically accounts for this.

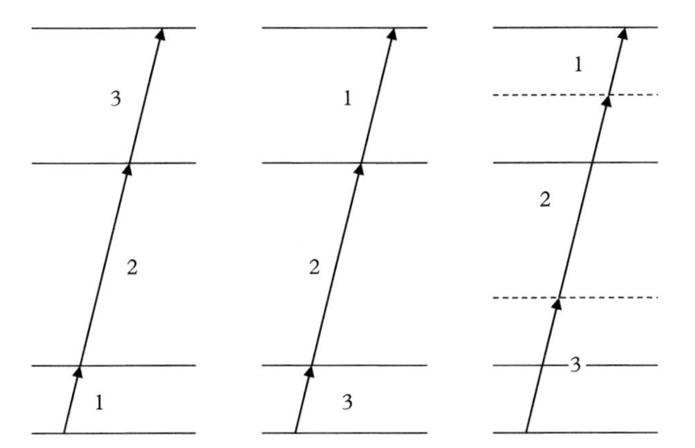

Figure 2.8 *Schematic diagram showing three different sequences of pulses with a fixed set of frequencies applied to excite the upper state of a four-level system. The sequence on the left begins from a populated state. All transitions are resonant. Because the transitions progress sequentially toward the upper level, this sequence is the most effective. The one in the middle illustrates a time-reversed sequence initiated from an excited state. It is ineffective if the excited states are unpopulated, but constitutes an independent excitation channel otherwise. The final sequence on the right permutes the frequencies of the first, by imagining them to excite unintended transitions in an off-resonant fashion. This illustrates a third independent excitation channel that could be effective if the detunings are not too large. What should be concluded from the figure is that the result of each sequence is different.*

2.4.3 Matrix Representations of Operators

To this point, we have described how one can divide up the overall time dependence of changes effected by operators, but we have not discussed the notation with which we plan to represent the operators themselves. There are several options, if we wish to avoid approaches that presume we know the wavefunction itself. For example, in a system with a finite number of energy levels, we know in advance that there is a finite number of expectation values and matrix elements between the available states. For the operator $\hat{O}$ in a two-level system these quantities are $\langle 1|\hat{O}|1\rangle$, $\langle 1|\hat{O}|2\rangle$, $\langle 2|\hat{O}|1\rangle$, and $\langle 2|\hat{O}|2\rangle$. These four quantities represent all the reproducible, measurable properties of the system. In this respect these eigenvalues provide a complete description of the role of the operator $\hat{O}$ on the system. Thus, a complete representation of the operator can be written as a collection of matrix entries:

$$\begin{bmatrix} \langle 2|\hat{O}|2\rangle & \langle 2|\hat{O}|1\rangle \\ \langle 1|\hat{O}|2\rangle & \langle 1|\hat{O}|1\rangle \end{bmatrix} = \begin{bmatrix} O_{22} & O_{21} \\ O_{12} & O_{11} \end{bmatrix}.$$

Alternatively, in Dirac notation, the representation of $\hat{O}$ can be written as $\hat{O} = O_{11}|1\rangle\langle 1| + O_{12}|1\rangle\langle 2| + O_{21}|2\rangle\langle 1| + O_{22}|2\rangle\langle 2|$. Any matrix element O_{ij} in the basis of states $|1\rangle$

and $|2\rangle$ can be retrieved by calculating $\langle i|\hat{O}|j\rangle = \langle i|\{O_{11}|1\rangle\langle 1| + O_{12}|1\rangle\langle 2| + O_{21}|2\rangle\langle 1| + O_{22}|2\rangle\langle 2|\}|j\rangle$ with $i, j = 1, 2$. For example, one finds $\langle 1|\hat{O}|2\rangle = \langle 1|\{O_{11}|1\rangle\langle 1| + O_{12}|1\rangle\langle 2| + O_{21}|2\rangle\langle 1| + O_{22}|2\rangle\langle 2|\}|2\rangle = O_{12}$.

In a system with an arbitrary number of states, use can therefore be made of the expansion

$$\hat{O} = \sum_{m',n'} O_{m',n'}|n'\rangle\langle m'|, \tag{2.4.33}$$

which represents the operator as a matrix of its coefficients between various states, as in the example. It may readily be checked whether Eq. (2.4.33) yields the appropriate matrix element in the basis $|n\rangle$ when evaluated between states n and m. The n,m entry in the matrix should be given by

$$\langle n|\hat{O}|m\rangle = \langle n|\sum_{n'm'} O_{m'n'}\,|n'\rangle\langle m'|m\rangle. \tag{2.4.34}$$

Since the $O_{m'n'}$ on the right of Eq. (2.4.34) are merely scalar coefficients and the sum is over primed variables only, this can be re-ordered as follows.

$$\langle n|\hat{O}|m\rangle = \sum_{n'm'} O_{m'n'}\langle n|n'\rangle\langle m'|m\rangle = \sum_{n'm'} O_{m'n'}\delta_{nn'}\delta_{mm'} = O_{mn}. \tag{2.4.35}$$

In obtaining Eq. (2.4.35), use has been made of the Kronecker delta δ_{ij} which is equal to one or zero, depending on whether the indices i,j are the same or different. The point is that the representation given by Eq. (2.4.33) reproduces all possible matrix elements for an operator in the space spanned by $|n\rangle$.

Matrix representations are very useful for calculations in systems with only a few levels. For example, the Hamiltonian of a two-level system is

$$\hat{H} = \sum_{n_1 m}^{2} H_{mn}|n\rangle\langle m| = H_{11}|1\rangle\langle 1| + H_{21}|1\rangle\langle 2| + H_{12}|2\rangle\langle 1| + H_{22}|2\rangle\langle 2|$$
$$= \begin{pmatrix} H_{11} & H_{12} \\ H_{21} & H_{22} \end{pmatrix}. \tag{2.4.36}$$

Using this kind of representation of an operator in terms of its matrix elements instead of its explicit mathematical form makes it easy to write a full equation of motion for a system based on minimal information about it. For example, in the presence of an electric dipole perturbation $\hat{V}$ the equation of motion (Schrödinger's equation) can be written

$$i\hbar\frac{d}{dt}\begin{pmatrix} C_1\exp(-i\omega_1 t) \\ C_2\exp(-i\omega_2 t)\end{pmatrix} = \begin{pmatrix} \hbar\omega_1 & V_{12} \\ V_{21} & \hbar\omega_2 \end{pmatrix}\begin{pmatrix} C_1\exp(-i\omega_1 t) \\ C_2\exp(-i\omega_2 t)\end{pmatrix}. \tag{2.4.37}$$

2.4.4 Changing Representations

Any continuous regular function ψ can be expanded in terms of a complete set of discrete basis functions U_n. Provided the functions U_n are mutually orthogonal and normalized, the expansion has the form

$$\psi = \sum_n C_n U_n, \tag{2.4.38}$$

where each expansion coefficient C_n gives the relative contribution of a single basis function. If we consider the functions U_n to constitute a set of basis vectors $|n\rangle$ which span the space, then ψ becomes a state vector, denoted in the Dirac notation introduced earlier by

$$|\psi\rangle = \sum_n C_n |n\rangle, \tag{2.4.39}$$

and the coefficient C_i is the result of vectorial projection of $|\psi\rangle$ on the basis vector $|i\rangle$. The projection procedure thus projects out of the full state $|\psi\rangle$ the probability amplitude of finding it in a particular eigenstate $|i\rangle$ in the basis of states $|n\rangle$:

$$\langle i|\psi\rangle = \langle i|\sum_n C_n |n\rangle = \sum_n C_n \langle i|n\rangle = \sum_n C_n\delta_{in} = C_i. \tag{2.4.40}$$

The set of all coefficients C_i constitutes a representation of ψ in the basis n.

It often happens that initial information about a system is available in a representation that is inconvenient for solving dynamics. Sometimes too, solutions are needed in the laboratory reference frame, whereas the basis states that make a problem mathematically tractable (perhaps by better reflecting the symmetry of the problem) are in a different frame. This situation can be rectified by changing representations. Provided the basis states form complete sets, the procedure for transforming between two arbitrary basis sets $|\xi\rangle$ and $|n\rangle$ is straightforward, as described next.

If we think of the set of coefficients C as the representation itself, then the problem of changing representations becomes one of relating an initial set of expansion coefficients C_n from a description like Eq. (2.4.39) to another set C_ξ in the desired expansion

$$|\psi\rangle = \sum_\xi C_\xi |\xi\rangle. \tag{2.4.41}$$

The tool for finding the relationship between C_n and C_ξ is obtained by multiplying Eq. (2.4.41) from the left by ψ^* and integrating over all space. Since $\langle\psi|\psi\rangle = 1$, the result equals unity:

$$\langle\psi| \sum_n C_n |n\rangle = \sum_{m,n} \langle m|C_m^* C_n |n\rangle = 1. \tag{2.4.42}$$

With the help of Eq. (2.4.40) one can substitute for the coefficients in Eq. (2.4.42) to obtain

$$1 = \sum_{m,n} \langle m|\langle\psi|m\rangle\langle n|\psi\rangle|n\rangle , \tag{2.4.43}$$

and then rearrangement of the right side yields

$$\begin{aligned} 1 &= \sum_{m,n} \langle m|n\rangle\langle\psi|m\rangle\langle n|\psi\rangle \\ &= \sum_{m,n} \delta_{mn} \langle\psi|m\rangle\langle n|\psi\rangle . \end{aligned} \tag{2.4.44}$$

Since ψ does not depend on m, n it can be removed from the summation. Hence

$$1 = \langle\psi|\left\{\sum_n |n\rangle\langle n|\right\}|\psi\rangle . \tag{2.4.45}$$

For this equation to be valid the sum in parentheses must equal 1. Hence this quantity furnishes a representation of the identity operator $\hat{I}$, namely

$$\hat{I} = \sum_n |n\rangle\langle n| . \tag{2.4.46}$$

Equation (2.4.46) is the most convenient mathematical tool for changing representations. With it, one can change from the $|n\rangle$ basis to the $|\xi\rangle$ basis or from (x, y, z) configuration space coordinates to (r, θ, ϕ) coordinates by simply inserting the identity operator for the desired basis into the original expansion of the wavefunction. That is,

$$\begin{aligned} |\psi\rangle &= \sum_n C_n |n\rangle = \sum_{\xi,n} |\xi\rangle\langle\xi|C_n|n\rangle \\ &= \sum_{\xi,n} [C_n\langle\xi|n\rangle]|\xi\rangle. \end{aligned} \tag{2.4.47}$$

The square bracket in Eq. (2.4.47) encloses all but the desired basis vectors $|\xi\rangle$. Hence Eq. (2.4.47) is a representation of $|\psi\rangle$ in terms of basis vectors $|\xi\rangle$ with new coefficients

$C_\xi \equiv \sum_n C_n \langle \xi \, | n \rangle$. Since the new basis was chosen for convenience, it is helpful to know that the states $|\xi\rangle$ need not be orthogonal to $|n\rangle$ for this procedure to work. The new basis needs only to be complete.

One final point about representations bears mentioning. Throughout this book, wavefunctions for collections of particles are written without imposing additional restrictions due to the Pauli principle. Although atoms are typically bosons, actually assuming them to be bosons represents a significant assumption. Stable atoms consist of protons, neutrons, and electrons, all of which are fermions. But they contain an even total number of such particles and are found to obey Bose statistics. Consequently we usually assume any number of them may occupy the same state. Nevertheless in semiconductors, as one counter-example, the occupation of valence and conduction band spin states is governed by the Pauli exclusion principle, so spin statistics must be reflected in wavefunctions for the energy bands. As another example, bosonic atoms can sometimes be converted into fermions near Feshbach resonances [2.6], so in principle the occupation of spin states should reflect the Pauli principle in dense phases of such matter. Because the wavefunctions of fermionic systems must be written differently from Eq. (2.4.47) to account for spin exclusion, Appendix F provides an elementary introduction to the anti-commutation properties of spin systems, for the sake of completeness.

REFERENCES

[2.1] J. Itatani, J. Levesque, D. Zeidler, H. Niikura, H. Pepin, J.C. Kieffer, P.B. Corkum, and D.M. Villeneuve, *Nature* 2004, **432**, 867.

[2.2] M. Tinkham, *Group Theory and Quantum Mechanics*. New York: Courier Dover Publications, 2003.

[2.3] P.W. Langhoff, *Am. J. Phys.* 1971, **39**, 954.

[2.4] W. Kuhn, *Eur. J. Phys.* 1980, **1**, 65; see also W.B. Fowler, *The Physics of Color Centers*. New York: Academic Press, 1968.

[2.5] T.T. Basiev, S.B. Mirov, and V.V. Osiko, *IEEE J. Quantum Electron.* 1988, **24**, 1052.

[2.6] D.B. Strukov, G.S. Snider, D.R. Stewart, and R.S. Williams, *Nature* 2008, **453**, 80.

[2.7] C.A. Regal, M. Greiner, and D.S. Jin, *Phys. Rev. Lett.* 2004, **92**, 040403.

PROBLEMS

2.1. A particle in a one-dimensional well has the Hamiltonian

$$H = \begin{cases} \dfrac{p^2}{2m} = -\dfrac{\hbar^2}{2m}\dfrac{d^2}{dz^2}, & 0 \le z \le L \\ \infty, & z < 0, z > \mathrm{L} \end{cases}$$

and is in a superposition state described by

$$\Psi = \left(\frac{2}{L}\right)^{1/2} \sum_n C_n \sin(k_n z) e^{-i\omega_n t}.$$

(a) What is the expectation value of its energy?

(b) Under what condition might we expect the expectation value of energy to equal the energy of an individual state, for example, the mth state?

(c) Taking into account restrictions on the allowed values of k_n, what is the average result of repeated measurements of the momentum of particles in state Ψ?

2.2. A two-level atom has a Hamiltonian $H = \begin{pmatrix} H_{11} & H_{12} \\ H_{21} & H_{22} \end{pmatrix}$. Find the appropriate expansion coefficients to write this completely in terms of the three Pauli spin matrices plus the unit matrix.

2.3. An electron with spin angular momentum $S = \frac{\hbar}{2}\sigma$ where σ is a Pauli spin matrix experiences a static magnetic field $\bar{B}$ oriented along the z-axis. Assume that the spin is initially oriented perpendicular to the field, in state $\Psi(0) = \frac{1}{\sqrt{2}}\begin{pmatrix} 1 \\ 1 \end{pmatrix}$. Find $\Psi(t)$ and show that the expectation value of the spin in the x-direction precesses according to $\langle S_x \rangle = \frac{\hbar}{2}\cos\omega t$. (The magnetic moment of the spin is related to the spin angular momentum by $\mu_s = g\mu_B S/\hbar$, where g is the spin g factor, μ_B is the Bohr magneton, and the Hamiltonian is $H = -\bar{\mu}_s \cdot \bar{B}$).

2.4. Show that the Pauli spin matrices have the properties:

(a) $\sigma_i\sigma_j = i\sigma_k$ for cyclic (and $\sigma_i\sigma_j = -i\sigma_k$ for anti-cyclic) permutations, and

(b) $[\sigma_i, \sigma_j] = 2i\sigma_k$ for cyclic (and $[\sigma_i, \sigma_j] = -2i\sigma_k$ for anti-cyclic) permutations of the indices $i, j, k = x, y, z$.

2.5. Pauli spin matrices, and transition operators $\sigma^{\pm}$ based on them, are commonly encountered in calculations of photon echoes and resonance fluorescence. Products of these matrices must be simplified on the basis of their commutation relations. Show that

(a) $[\sigma^{\pm}, \sigma_z] = \mp 2\sigma^{\pm}$,

(b) $[\sigma^{+}, \sigma^{-}] = \sigma_z$,

where $\sigma^{\pm} \equiv \frac{1}{2}(\sigma_x \pm i\sigma_y)$.

2.6. Pauli matrices anti-commute.

(a) Show that $[\sigma_i, \sigma_j]_+ \equiv \sigma_i\sigma_j + \sigma_j\sigma_i = 2\delta_{ij}$, where δ_{ij} is a Kronecker delta whose value is zero unless the subscripts are equal.

(b) Use an inductive argument and part (a) to show that $\sigma_i(\sigma_j)^k = (-\sigma_j)^k\sigma_i$, when $i \neq j$ and k is an arbitrary positive integer.

(c) Prove that $\exp[-\frac{i}{2}\theta_2\sigma_1]\exp[-\frac{i}{2}\omega_0\tau\sigma_3]$

$$= \cos(\theta_2/2)e^{-(i/2)\omega_0\tau\sigma_3} - i\sin(\theta_2/2)e^{(i/2)\omega_2\tau\sigma_3}\sigma_1$$

2.7. The Pauli spin matrices are useful because as representations of angular momentum operators they are generators of important unitary transformations. For example, an exponential function of the three Pauli spin 1/2 operators generates rotations about the x, y, and z axes of a two-level atom as verified in part (c). Since the static and interaction Hamiltonians, and the transformation of arbitrary superposition states into other states of the atom, can all be represented by these matrices, the Pauli matrices can provide a complete representation of static and dynamic aspects of atoms in their "spin" space.

(a) Let ξ be a real number and suppose $\hat{A}$ is a matrix operator that satisfies $\hat{A}\hat{A} = 1$. Show that $\exp(i\xi\hat{A}) = \hat{I}\cos(\xi) + i\hat{A}\sin(\xi)$, where $\hat{I}$ is the unit operator.

(b) Show that if $\hat{A} = \hat{\sigma}_y$, the exponential operator in part (a) produces a rotation through an angle ξ about the $\hat{y}$-axis, by showing that it reduces to the usual 2×2 rotation matrix for transforming the two (x and z) components of a state vector projected onto the plane perpendicular to the rotation axis.

(c) Finally, if $\hat{n} = n_x\hat{x} + n_y\hat{y} + n_z\hat{z}$ is an arbitrary real unit vector in three dimensions, prove that $(\hat{n}\cdot\overleftrightarrow{\sigma})^2 = \hat{I}$ and use part (a) to find the explicit 2×2 matrix form of the operator $R_{\hat{n}}(\theta) \equiv \exp\left(-i\theta\hat{n}\cdot\overleftrightarrow{\sigma}/2\right)$ that produces a rotation by θ about the axis $\hat{n}$. The symbol $\overleftrightarrow{\sigma}$ in these expressions denotes the three component vector $\left(\sigma_x, \sigma_y, \sigma_z\right)$ of Pauli spin matrices. (Note: The factor of 2 difference between the angular arguments of R and the exponential operator merely accounts for the rotational degeneracy of real space with respect to rotations in spin space.)

(d) Write the interaction Hamiltonian for electric dipole response in a two-level atom in terms of Pauli spin matrices—both (i) as a single Hermitian matrix and (ii) as a sum of two non-Hermitian matrices (that are raising and lowering operators).

2.8. Use a symmetry argument to find the expectation value of the electric dipole moment $<e\bar{r}>$ of an atom in an eigenstate.

2.9. Show that the expectation value of the magnetic dipole moment $\bar{\mu}^{(m)} = e(\bar{r}\times\bar{p})/2m$ of a system with inversion symmetry is not necessarily zero in an eigenstate.

2.10. (a) Calculate the expectation value of the dipole moment $< e\bar{r} >$ of an atom with the specific wavefunction

$$\psi(\bar{r}, t) = C_{210}u_{210}(\bar{r})\exp(-i\omega_{210}t) + C_{100}u_{100}(\bar{r})\exp(-i\omega_{100}t),$$

where

$$u_{100}(r,\theta,\phi) = (\pi a_0^3)^{-1/2}\exp(-r/a_0),$$
$$u_{210}(r,\theta,\phi) = (32\pi a_0^3)^{-1/2}(r/a_0)\cos\theta\exp(-r/2a_0),$$

and a_0 is a constant (the Bohr radius). Use spherical coordinates and write the position vector as $\bar{r} = \frac{1}{2}r\sin\theta[(\hat{x}-i\hat{y})\exp(i\phi) + (\hat{x}+i\hat{y})\exp(-i\phi)] + \hat{z}r\cos\theta$.

(b) Give a physical reason why the result of part (a) is not zero.

2.11. A system is placed in a linear superposition state with the (unnormalized) wavefunction

$$\psi_1(t) = \exp(-i\pi t)\cos x + \exp(-i5\pi t)\cos 5x + \exp(-i11\pi t)\cos 11x.$$

(a) Plot the probability $|\psi_1(x,t)|^2$ of finding the system at $x = \pi/60$ as a function of time for $0 < t < 40$.

(b) Recalculate and plot the squared probability amplitude for finding the system at $x = \pi/60$ for $0 < t < 40$ when relaxation of the excited states is introduced, by modifying the wavefunction to

$$\psi_2(t) = e^{-i\pi t}\cos x + e^{-t/10}(e^{-i5\pi t}\cos 5x + e^{-i11\pi t}\cos 11x).$$

(c) On the basis of your plots, interpret the large amplitude features in the temporal evolution of the system with and without dissipation. Identify two fundamental differences in the basic dynamics of non-dissipative and dissipative systems.

2.12. A beam of light propagating in the z direction is in a state of polarization that is expressed in terms of linearly polarized basis states $|x\rangle = \begin{pmatrix}1\\0\end{pmatrix}$ and $|y\rangle = \begin{pmatrix}0\\1\end{pmatrix}$ according to $|\psi\rangle = \psi_x|x\rangle + \psi_y|y\rangle$.

(a) Find the expansion coefficients ψ_x and ψ_y of the wavefunction in Dirac notation.

(b) Change representations by inserting the identity operator to show that a completely equivalent description of $|\psi\rangle$ exists that is based on left- and right- circularly polarized states $|L\rangle = \frac{1}{\sqrt{2}}\begin{pmatrix}1\\-i\end{pmatrix}$ and $|R\rangle = \frac{1}{\sqrt{2}}\begin{pmatrix}1\\i\end{pmatrix}$. Find the coefficients c_1 and c_2 in the new representation $|\psi\rangle = c_1|R\rangle + c_2|L\rangle$.

(c) Configuration space representations can be given in terms of *rotated* basis vector states. This is done by noting that $|x\rangle = \cos\phi|x'\rangle - \sin\phi|y'\rangle$ and $|y\rangle = \sin\phi|x'\rangle + \cos\phi|y'\rangle$. Hence $\langle x'|x\rangle = \cos\phi$ and $\langle y'|y\rangle = \sin\phi$, where ϕ is the angle by which x' is rotated with respect to x about z. Show that a state vector in the transformed (primed) basis becomes

$$|\psi'\rangle = \begin{pmatrix}\psi_{x'}\\ \psi_{y'}\end{pmatrix} = \begin{pmatrix}\langle x'|\psi\rangle\\ \langle y'|\psi\rangle\end{pmatrix} = \begin{pmatrix}\cos\phi & \sin\phi\\ -\sin\phi & \cos\phi\end{pmatrix}\begin{pmatrix}\langle x|\psi\rangle\\ \langle y|\psi\rangle\end{pmatrix} = R(\phi)|\psi\rangle$$

where $R(\phi)$ is the rotation operator.

(d) Circularly polarized representations can be expressed in terms of rotated bases too. Separate the rotation operator into two parts—a diagonal matrix and the Pauli matrix σ_y, each with an appropriate coefficient. Then find the eigenvalues and eigenvectors of the Pauli matrix (the part of R which is traceless). Interpret the meaning of the eigenvalues in the context of photon polarization.

(e) Operate with $R(\phi)$, *after reduction into symmetric and anti-symmetric parts as in part (d)*, on the basis states $|L\rangle$ and $|R\rangle$ directly to show that a rotation multiplies each of them by a complex phase factor. (Note that the phase factor differs in sign in $|L'\rangle$ and $|R'\rangle$.) Is this consistent with part (d)?

(f) Show that an arbitrary linear superposition of rotated basis states is not an eigenvector of the rotation operator. (Hint: Apply the reduced operator to a superposition state like $|\psi\rangle = |R\rangle\langle R|\psi\rangle + |L\rangle\langle L|\psi\rangle$). What does this mean in general with regard to the angular momentum of linear superposition states of right- and left-circularly polarized light?

2.13. A two-level atom is subjected to a perturbation that only causes transitions between its two states. Thus the interaction Hamiltonian has the form $V = \begin{pmatrix} 0 & V_{12} \\ V_{21} & 0 \end{pmatrix}$. Find the appropriate expansion coefficients to write this completely in terms of the three Pauli spin matrices plus the unit matrix.

2.14. A single electron is subjected to an electric field E that is uniform throughout space and points along the z-axis.

(a) Find the potential energy $V(z)$ of the charge when its zero is defined to be at the origin ($z = 0$).

(b) Write down and simplify the time-independent Schrödinger equation including this potential energy term. (Since p_x and p_y are constants of the motion, the wavefunction can be written as $\psi(r) = e^{ik_x x}e^{ik_y y}Z(z)$ and the derivatives with respect to x, y performed to proceed with this.)

(c) By introducing an appropriate dimensionless variable ξ to replace z, show that the Schrödinger equation can be cast in the standard form

$$\frac{d^2Z(\xi)}{d\xi^2} - \xi Z(\xi) = 0.$$

(d) Show that the Airy function

$$Ai(\xi) = \frac{1}{2\pi}\int_{-\infty}^{\infty} \exp\left[i\left(\frac{u^3}{3} + u\xi\right)\right]du$$

is a solution of this equation and plot *or* sketch it versus ξ by making use of asymptotic forms of $Ai(\xi)$ for large positive and negative values of the argument:

$$Ai(\xi) = \begin{cases} \dfrac{1}{\sqrt{\pi}(-\xi)^{1/4}} \sin\left(\dfrac{2}{3}(-\xi)^{3/2} + \dfrac{\pi}{4}\right), & \xi \ll -1 \\[2ex] \dfrac{1}{2\sqrt{\pi}\xi^{1/4}} \exp\left(-\dfrac{2}{3}\xi^{3/2}\right), & \xi \gg +1 \end{cases}.$$

(e) Comment on the qualitative difference between the wavefunction in the positive and negative potential regions.

3

Atom–Field Interactions

3.1 The Interaction Hamiltonian

In the description of atoms interacting with light, the full system Hamiltonian H accounts not only for the energetics of each atom but also the energy density of the light field itself and the modification of system energy due to the coupling of atoms to the field. Our ability to understand and predict behavior relies on an accurate determination of H and subsequent solution of Schrödinger's equation for the wavefunction, which is usually deemed essential. However, in the century or so that has passed since the discovery of quantum mechanics, analytic solutions for the wavefunctions of only a handful of simple, isolated atoms like hydrogen and helium have emerged. So it may be hard to imagine applying this approach successfully to modern research on new materials. However, its limitations can be overcome by realizing that the vast majority of optical problems involve the same interaction Hamiltonian, do not require knowledge of the wavefunction itself, and can be enormously simplified by experimental as well as theoretical procedures. Moreover, with the so-called reduced density matrix, mathematically rigorous ways can be found to ignore parts of complicated systems while focusing on specific portions of interest. In this chapter we prepare for this by identifying the one interaction Hamiltonian to be used in subsequent chapters, by introducing the density matrix to replace the wavefunction, and discussing experimental tricks that can be used to convert multi-level systems into simple two-level systems.

When it is chiefly the *dynamics* of an atom or some other system that are of interest, it is natural to separate the total Hamiltonian into a part H_0 that describes the isolated (static) system and an electromagnetic interaction Hamiltonian $V(t)$ which initiates dynamics:

$$H = H_0 + V(t). \tag{3.1.1}$$

H_0 determines the allowed energies and wavefunction of the unperturbed system, but knowledge of the exact wavefunction is fortunately not necessary to predict the effects of $V(t)$. For this, knowledge of the placement of energy levels and system symmetries is adequate, and these properties of the system can be determined experimentally without ever knowing H_0 or the wavefunction. $V(t)$ can be derived (see Appendix C) in a

Lectures on Light. Second Edition. Stephen C. Rand.
© Stephen C. Rand 2016. Published in 2016 by Oxford University Press.

general multipole form that is not only the same for all problems considered in this book but can be readily quantized and that displays dependences on (i) the electric and magnetic fields of light and (ii) the main charge motions induced by its passage. The largest induced moments are the electric dipole moment $\bar{\mu}^{(e)} = e\bar{r}$ and the magnetic dipole moment $\bar{\mu}^{(m)} = \frac{1}{2}\bar{r} \times e\bar{v}$, where $\bar{r}$ and $\bar{v}$ are the displacement and velocity vectors of a single positive charge. If we include both types of moment, we have

$$V(t) = -\bar{\mu}^{(e)} \cdot \bar{E} - \bar{\mu}^{(m)} \cdot \bar{B}. \tag{3.1.2}$$

At low intensities, the magnitudes of plane electric and magnetic fields are related by $B = E/c$, where c is the velocity of light. Because $\bar{\mu}^{(m)}$ is proportional to velocity $\bar{v}$ of the charge whereas $\bar{\mu}^{(e)}$ is not, the second term in Eq. (3.1.2) is almost always negligible compared to the first, by virtue of the factor $v/c \ll 1$. There are rare exceptions in dynamic systems where resonant enhancement of magnetic interactions takes place, but in subsequent chapters the interaction Hamiltonian will be taken to be

$$V(t) = -\bar{\mu}^{(e)} \cdot \bar{E}(t) = -e\bar{r} \cdot \bar{E}(t). \tag{3.1.3}$$

In this expression, although $\bar{\mu}^{(e)} = e\bar{r}$ is the electric dipole induced by a time-varying field and is therefore time-varying itself, it is a constant vector in the rotating frame of the optical field $E(t)$. Hence its dependence on time is dropped in Eq. (3.1.3) whenever it is possible to make the rotating wave approximation described in Section 3.2. We shall see later in Section 7.5 that the situation is quite different for transverse magnetic moments created by intense light.

3.2 Perturbation Theory

We now introduce an approximate method for calculating time development of a simple system in which the atomic properties are quantized but the radiation field is left in classical form (as a sinusoidal wave). This is called semi-classical perturbation theory. Results will be compared in the next section with an exact method, setting the stage for later combinations of exact and approximate techniques.

Consider an atom (Figure 3.1) with only two levels $|1\rangle$ and $|2\rangle$ that is subjected to excitation by a standing wave:

$$\bar{E}(z,t) = \begin{Bmatrix} \hat{x}E_0(z)\cos\omega t, & t \geq 0 \\ 0, & t < 0 \end{Bmatrix}. \tag{3.2.1}$$

Provided the light wave does not cause significant shifts in the energy levels of the atom, a solution to Eq. (3.2.1) can be written as a linear superposition of eigenfunctions of H_0. We take the wavefunction to be

$$\psi(\bar{r},t) = \sum_n C_n U_n(\bar{r}) \exp(-i\omega_n t), \tag{3.2.2}$$

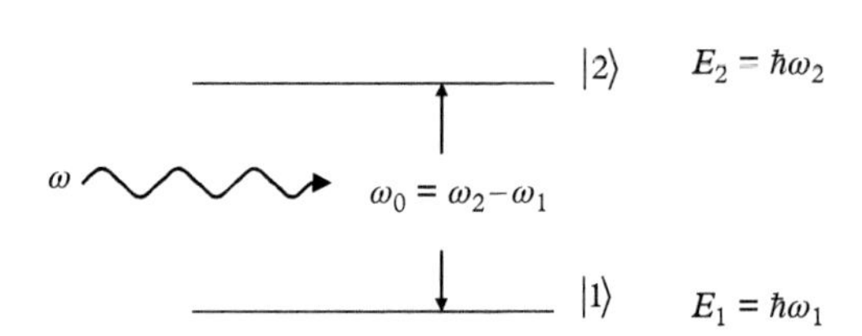

Figure 3.1 *Light of frequency ω interacting with a two-level atom which has a resonant frequency of ω_0.*

where the basis functions U_n are eigenfunctions of H_0 determined by

$$H_0 U_n = E_n U_n, \tag{3.2.3}$$

and eigenfrequencies ω_n are related to the energy eigenvalues according to

$$E_n = \hbar\omega_n. \tag{3.2.4}$$

Energies of the upper and lower levels are $\hbar\omega_2$ and $\hbar\omega_1$, respectively.

Assume that the light frequency ω is near the resonant transition frequency between levels 2 and 1, defined by

$$\omega_0 = \omega_2 - \omega_1, \tag{3.2.5}$$

so that $\omega \cong \omega_0$. Equation (3.2.2) can then be written out explicitly as

$$\psi(\bar{r}, t) = C_2(t)\exp(-i\omega_2 t)U_2(\bar{r}) + C_1(t)\exp(-i\omega_1 t)U_1(\bar{r}) \tag{3.2.6}$$

and substituted into Schrödinger's equation to find the temporal evolution of the system. This procedure yields

$$i\hbar\frac{\partial}{\partial t}\sum_i C_i \exp(-i\omega_i t)U_i = (H_0 + V)\sum_i C_i \exp(-i\omega_i t)U_i,$$

$$i\hbar\sum_i \left(\dot{C}_i - i\omega_i C_i\right)\exp(-i\omega_i t)U_i = \sum_i C_i \exp(-i\omega_i t)\,(E_i + V)\,U_i. \tag{3.2.7}$$

Multiplying from the left by U_j^* and integrating over all space, one obtains

$$i\hbar\left(\dot{C}_j - i\omega_j C_j\right)\exp(-i\omega_j t) = C_j \exp(-i\omega_j t)E_j + \sum_{i\neq j} V_{ji}C_i \exp(-i\omega_i t), \tag{3.2.8}$$

where the matrix elements V_{ji} are

$$V_{ji} = -e\bar{E} \cdot \int dV \psi_j^*(\bar{r}, t)\bar{r}\psi_i(\bar{r}, t) = \langle U_j | -e\bar{r} \cdot \bar{E} | U_i \rangle = -e\langle U_j | \bar{r} | U_i \rangle \cdot \bar{E}$$

$$= \begin{Bmatrix} -e\bar{r}_{ji} \cdot \bar{E}, & j \neq i \\ 0, & j = i \end{Bmatrix}.$$

Equation (3.2.8) simplifies to

$$\dot{C}_j = \frac{1}{i\hbar} \sum_{i \neq j} V_{ji} C_i \exp[-i(\omega_i - \omega_j)t]. \tag{3.2.9}$$

In the present case the sum over i runs over only the two values $i = 1, 2$ giving

$$\dot{C}_2 = \frac{1}{i\hbar} \exp[-i(\omega_1 - \omega_2)t] V_{21} C_1(t), \tag{3.2.10}$$

$$\dot{C}_1 = \frac{1}{i\hbar} \exp[-i(\omega_2 - \omega_1)t] V_{12} C_2(t), \tag{3.2.11}$$

and the off-diagonal, interaction matrix element is

$$V_{21} = -\bar{\mu}_{21} \cdot \bar{E}_0 \cos \omega t = -\frac{ex_{21}E_0}{2} [\exp(i\omega t) + \exp(-i\omega t)]. \tag{3.2.12}$$

Note that if the induced dipole moment is not parallel to $\bar{E}_0 = E_0\hat{x}$ as assumed previously, then the interaction strength is reduced by its projection on $\hat{x}$. This is one example of polarization effects to be covered later.

The time development of the atom is completely described by the two equations

$$\dot{C}_2 = \frac{i}{2} \frac{\bar{\mu}_{21} \cdot \bar{E}_0}{\hbar} \{\exp[i(\omega_0 + \omega)t] + \exp[i(\omega_0 - \omega)t]\} C_1 \tag{3.2.13}$$

and

$$\dot{C}_1 = \frac{i}{2} \frac{\bar{\mu}_{12} \cdot \bar{E}_0}{\hbar} \{\exp[-i(\omega_0 - \omega)t] + \exp[-i(\omega_0 + \omega)t]\} C_2. \tag{3.2.14}$$

All that remains is to solve these coupled partial differential equations.

Consider an iterative approach for a ground state atom that is weakly excited by light of frequency $\omega \neq \omega_0$, beginning at $t = 0$. The initial (zeroth order) conditions are $C_2^{(0)} = 0$

and $C_1^{(0)} = 1$. Substituting the initial C_2 and C_1 into Eq. (3.2.14), we obtain equations for changes of the coefficients at short times.

$$\dot{C}_2^{(1)} \cong \frac{i}{2}\frac{\bar{\mu}_{21} \cdot \bar{E}_0}{\hbar}(\exp[i(\omega_0 + \omega)t] + \exp[i(\omega_0 - \omega)t]), \tag{3.2.15a}$$

$$\dot{C}_1^{(1)} \cong 0. \tag{3.2.15b}$$

These first-order equations may be integrated directly. The results are

$$C_1^{(1)}(t) \cong C_1(0) = 1 \tag{3.2.16a}$$

and

$$C_2^{(1)}(t) \cong \frac{1}{2}\frac{\bar{\mu}_{21} \cdot \bar{E}_0}{\hbar}\left\{\frac{\exp[i(\omega_0 + \omega)t] - 1}{\omega_0 + \omega} + \frac{\exp[i(\omega_0 - \omega)t] - 1}{\omega_0 - \omega}\right\}. \tag{3.2.16b}$$

Since $\omega \sim 10^{15}$ rad/s at optical frequencies, it is an excellent approximation to drop the term proportional to $(\omega_0 + \omega)^{-1}$. This is called the rotating wave approximation (RWA). We retain only the term in which the atomic and field phasors rotate together, rather than in opposite senses. (A useful geometric model that pictures time development as rotation in a fictitious space is outlined in Section 3.8). Hence the first-order predictions for the time dependence of the probability amplitudes in this problem are

$$C_1^{(1)}(t) \cong C_1(0) = 1 \tag{3.2.17}$$

and

$$C_2^{(1)}(t) \cong \frac{1}{2}\frac{\bar{\mu}_{21} \cdot \bar{E}_0}{\hbar}\left[\frac{\exp[i(\omega_0 - \omega)t] - 1}{(\omega_0 - \omega)}\right]. \tag{3.2.18}$$

Equation (3.2.18) can be rewritten as

$$C_2(t) = \frac{i}{2}\frac{\bar{\mu}_{21} \cdot \bar{E}_0}{\hbar}\exp[i(\omega_0 - \omega)t/2]\frac{\sin([\omega_0 - \omega]t/2)}{([\omega_0 - \omega]/2)}, \tag{3.2.19}$$

yielding a time-dependent probability of excitation to state 2 which is

$$|C_2(t)|^2 = \left(\frac{\bar{\mu}_{21} \cdot \bar{E}_0}{2\hbar}\right)^2 \frac{\sin^2([\omega_0 - \omega]t/2)}{([\omega_0 - \omega]/2)^2}. \tag{3.2.20}$$

Atomic states have a natural lifetime τ which is the inverse of the total decay rate γ out of that state. So far, we have ignored spontaneous decay of this kind in our analysis. In

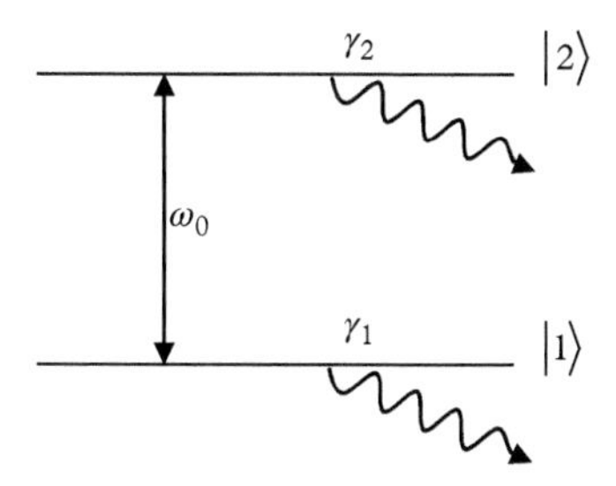

Figure 3.2 *Energy level diagram of an atom with levels 2 and 1 undergoing decay at rates γ_2 and γ_1.*

an attempt to include exponential relaxation processes like those depicted in Figure 3.2, let us try augmenting the equations of motion, Eq. (3.2.9), in an ad hoc manner to introduce decay:

$$\dot{C}_2 = -\frac{1}{2}\gamma_2 C_2 + \frac{i}{2}\frac{\bar{\mu}_{21}\cdot\bar{E}_0}{\hbar}\exp[i(\omega_0-\omega)t]C_1, \tag{3.2.21}$$

$$\dot{C}_1 = -\frac{1}{2}\gamma_1 C_1 + \frac{i}{2}\frac{\bar{\mu}_{12}\cdot\bar{E}_0}{\hbar}\exp[-i(\omega_0-\omega)t]C_2. \tag{3.2.22}$$

Note that these equations correctly predict exponential decay in the absence of light. That is, if we ignore the second term on the right of Eq. (3.2.21) by setting $C_1 = 0$ (or $E_0 = 0$) one can readily integrate the equation to find $C_2(t) = C_2(0)\exp(-\gamma_2 t/2)$. Thus spontaneous, exponential decay of the amplitude takes place at the expected rate. Moreover if we treat level 1 as the true ground state by choosing $\gamma_1 = -\gamma_2$, then in the absence of light note that occupation probability for the system as a whole should be conserved, since all the atoms that decay out of the excited state arrive in the ground state. The factor of 1/2 in the first term on the right of both Eq. (3.2.21) and Eq. (3.2.22) at least correctly accounts for the difference between probability amplitude and probability density.

Presuming the additions to Eqs. (3.2.21) and (3.2.22) are correct despite their ad hoc nature, a perturbative approach similar to that developed earlier could be used to solve them. For initial conditions $C_2(0) = 0$ and $C_1(0) = 1$ one finds

$$C_1(t) = C_1(0)\exp(-\gamma_1 t/2) = \exp(-\gamma_1 t/2)\ . \tag{3.2.23}$$

Substituting Eq. (3.2.23) into Eq. (3.2.21) with an integrating factor of $\exp(\gamma_2\, t/2)$, one finds

$$\frac{d}{dt}\exp(\gamma_2 t/2)C_2 = \frac{i}{2}\frac{\bar{\mu}_{21}\cdot\bar{E}_0}{\hbar}\exp[i(\omega_0-\omega)t]\exp[(-\gamma_1+\gamma_2)t/2].$$

Hence the perturbation theory solution is

$$C_2^{(1)}(t) = \exp(-\gamma_2 t/2)\left(\frac{i\Omega}{2}\right)\left[\frac{\exp[i(\omega_0-\omega)t + (\gamma_2-\gamma_1)t/2] - 1}{i(\omega_0-\omega) + (\gamma_2-\gamma_1/2)}\right], \tag{3.2.24}$$

where $\Omega \equiv \bar{\mu}_{21}\cdot\bar{E}_0/\hbar$. The probability of excitation to state 2 is therefore

$$|C_2(t)|^2 = \left(\frac{\Omega}{2}\right)^2\left\{\frac{\exp(-\gamma_1 t) - 2\exp[-(\gamma_2+\gamma_1)t/2]\cos\Delta t + \exp(-\gamma_2 t)}{\Delta^2 + [(\gamma_2-\gamma_1)/2]^2}\right\} \tag{3.2.25}$$

where we have introduced the frequency detuning factor $\Delta \equiv \omega_0 - \omega$.

Exercise: Show that in this perturbative limit Eq. (3.2.25) reduces to a probability that is given at very short times $(t \ll \Delta^{-1}, \gamma_2^{-1})$ by

$$|C_2(t)|^2 \cong \left(\frac{\Omega t}{2}\right)^2\frac{\sin^2(\Delta t/2)}{[\Delta t/2]^2}. \tag{3.2.26}$$

At short times we find the same transition probability as in Eq. (3.2.20). At first glance, the time dependence in Eq. (3.2.26) appears to be incompatible with a constant transition rate, which is the expected result. Despite the apparent quadratic dependence of the excitation probability on time t in the first factor on the right of Eq. (3.2.26), the transition rate is correctly predicted to be proportional to t itself, however. This is due to the fact that the area under the $(\sin x/x)^2$ curve diminishes with time as $1/t$. Hence a perturbation calculation based on the squares of C coefficients seems to incorporate decay in a reasonable way. In Figure 3.3 the perturbation result even gives qualitatively

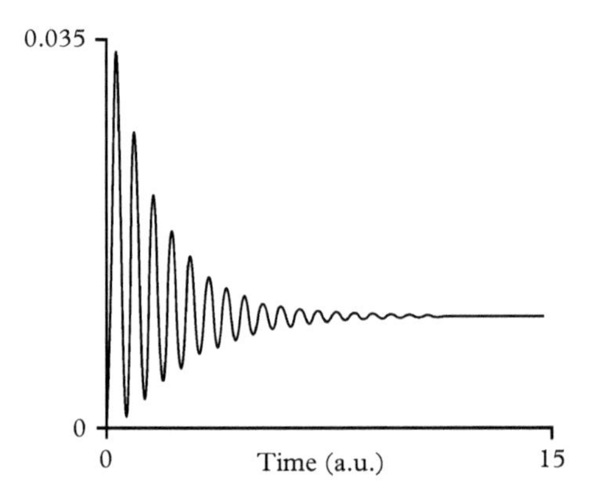

Figure 3.3 *Excited state occupation probability versus time for a two-level atom initially in the ground state, obtained from perturbation theory (plotted beyond its range of validity). This calculation (for Eq. (3.2.25) parameter values $\Delta = 10$; $\gamma_1 = 0$; $\gamma_2 = 1$) correctly foreshadows the transient phenomenon of nutation (Chapter 4) even though a perturbation approach is not justified at long times.*

correct results beyond its range of applicability. Oscillations caused by competition between stimulated absorption and stimulated emission are found to gradually give way to a small but finite steady-state population in the upper state. (Transients of this kind are considered in more detail in Chapter 4). We shall see shortly however that there are serious problems with this description at anything other than the very shortest timescales. Not only will this analytic approach be found to fail on resonance, but it also fails utterly off resonance ($\omega \neq \omega_0$) where we *do expect* perturbation theory to work well.

3.3 Exact Analysis

To gain perspective on the weaknesses and strengths of the perturbation approach, we now reconsider Eq. (3.2.18) in the RWA using an exact approach developed by I.I. Rabi. The Rabi approach ignores decay, in favor of finding a solution for atomic dynamics in a strong field, to compare with perturbation results.

$$\dot{C}_2(t) = (i\Omega/2)\exp[i(\omega_0 - \omega)t]C_1, \tag{3.3.1}$$

$$\dot{C}_1(t) = (i\Omega^*/2)\exp[-i(\omega_0 - \omega)t]C_2. \tag{3.3.2}$$

Without losing generality, we shall assume that Ω is not complex and try a solution of the form

$$C_1(t) = \exp[i\mu t], \tag{3.3.3}$$

and substitute this into Eq. (3.3.2) to solve for C_2. This yields

$$C_2(t) = \frac{2\mu}{\Omega}\exp[i(\Delta + \mu)t]. \tag{3.3.4}$$

Using both Eq. (3.3.3) and Eq. (3.3.4) in Eq. (3.3.2) we find that solutions only exist if

$$\mu_{1,2} = -\frac{1}{2}\Delta \pm \frac{1}{2}\left[\Delta^2 + \Omega^2\right]^{1/2}. \tag{3.3.5}$$

The most general solutions for C_2 and C_1 are therefore

$$C_1(t) = A\exp(i\mu_1 t) + B\exp(i\mu_2 t), \tag{3.3.6}$$

$$C_2(t) = (2/\Omega)\exp[i\Delta t]\,[A\mu_1\exp(i\mu_1 t) + B\mu_2\exp(i\mu_2 t)]. \tag{3.3.7}$$

Making use of initial conditions $C_2(0) = 0$ and $C_1(0) = 1$ to determine unknown coefficients A and B, we find

$$A = \left(\frac{\mu_2}{\mu_2 - \mu_1}\right)$$

and

$$B = \left(\frac{\mu_1}{\mu_1 - \mu_2} \right).$$

It is convenient to define the difference between the eigenfrequencies in Eq. (3.3.5) as the generalized Rabi flopping frequency:

$$\Omega_R \equiv \mu_1 - \mu_2 = \left[\Delta^2 + \Omega^2 \right]^{1/2}. \tag{3.3.8}$$

The product of μ_1 and μ_2 is related to the resonant Rabi frequency $\Omega \equiv \mu E_0/\hbar$ by

$$\mu_1 \mu_2 = - \left(\frac{\Omega}{2} \right)^2, \tag{3.3.9}$$

so we can write a solution for the coefficient $C_2(t)$ of the wavefunction which is

$$C_2(t) = i(\Omega/\Omega_R) \exp[i\Delta t/2] \sin(\Omega_R t/2), \tag{3.3.10}$$

with the result that the occupation probability becomes

$$|C_2(t)|^2 = \left(\frac{\Omega}{2} \right)^2 \left[\frac{\sin(\Omega_R t/2)}{(\Omega_R/2)} \right]^2. \tag{3.3.11}$$

For large detuning (far off resonance) we find

$$|C_2(t)|^2 = \left(\frac{\Omega}{2} \right)^2 \left[\frac{\sin(\Delta t/2)}{\Delta/2} \right]^2, \tag{3.3.12}$$

in agreement with the result, Eq. (3.3.23), from perturbation theory, ignoring decay. For zero detuning (on resonance) the exact procedure yields

$$|C_2(t)|^2 = \left(\frac{\Omega}{2} \right)^2 \left[\frac{\sin(\Omega t/2)}{\Omega/2} \right]^2 = \sin^2(\Omega t/2). \tag{3.3.13}$$

The predicted dynamics for this case are illustrated in Figure 3.4.

The predictions of Figures 3.3 and 3.4 are obviously quite different. The perturbation approach considered in the last section therefore fails under conditions of resonant excitation—even if incident radiation is weak. The reason is that on resonance, the perturbation expansion no longer converges. High-order interactions become important. Probability amplitudes like the second term in Eq. (3.2.16b) diverge. The exact solution

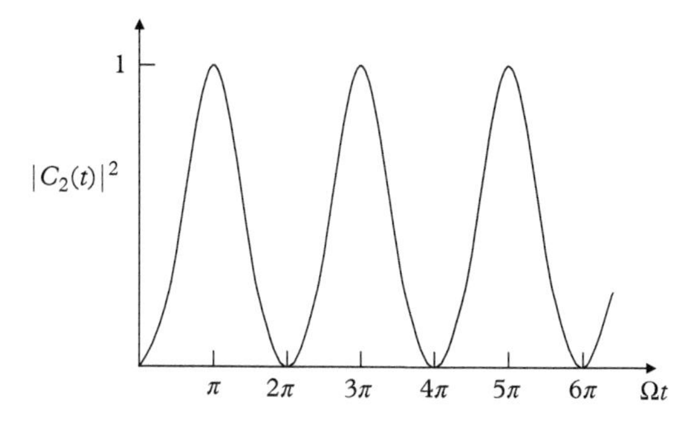

Figure 3.4 *Excited state occupation probability versus time at resonance* ($\Delta = 0$).

plotted in Figure 3.4 contains no such divergences, and indicates that on resonance undamped pulsations of the excited state population occur. The atom alternates regularly between the excited state and the ground state, undergoing a process known as Rabi flopping. Unlike the dynamics of Figure 3.3, undamped oscillations can take place with resonant excitation and the ground state can be emptied periodically, giving rise to a transient population inversion.

3.4 Preliminary Consideration of AC Stark or Rabi Splitting

On resonance ($\omega = \omega_0$), the eigenfrequencies in the Rabi solution reduce to

$$\mu_{1,2} = \pm\Omega/2,$$

where the subscripts 1, 2 refer to + or − solutions, respectively. The probability amplitudes for states 2 and 1 from Eqs. (3.3.6) and (3.3.7) in the RWA are

$$C_2 = \frac{1}{2}\left[\exp(i\Omega t/2) - \exp(-i\Omega t/2)\right] \tag{3.4.1}$$

and

$$C_1 = \frac{1}{2}\left(\exp(i\Omega t/2) + \exp(-i\Omega t/2)\right). \tag{3.4.2}$$

Hence, on the basis of Eq. (3.2.2), the complete wavefunction should be

$$\psi = \frac{1}{2}\left\{e^{-i\left(\omega_2-\frac{\Omega}{2}\right)t}\psi_2 - e^{-i\left(\omega_2+\frac{\Omega}{2}\right)t}\psi_2 + e^{-i\left(\omega_1-\frac{\Omega}{2}\right)t}\psi_1 + e^{-i\left(\omega_1+\frac{\Omega}{2}\right)}\psi_1\right\}. \tag{3.4.3}$$

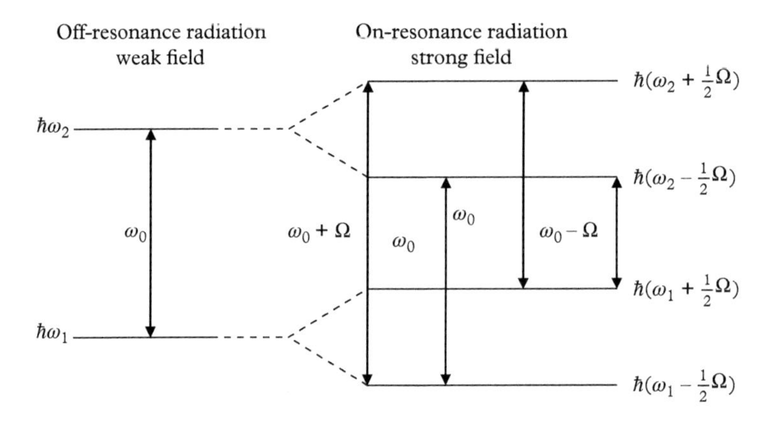

Figure 3.5 *A preliminary picture of strong-field effects in a two-level atom.*

Notice that, according to Eq. (3.4.3), the total wavefunction incorporates four oscillation frequencies. That is, Eq. (3.4.3) implies the existence of four distinct eigenfrequencies whereas there are only two eigenstates in the system. The energy level structure suggested by this result is shown in Figure 3.5. Because the number of eigenfrequencies exceeds the number of eigenstates, it is evident that a serious inconsistency has somehow entered into the calculation.

Ironically, the improper procedure used in arriving at Eq. (3.4.3) correctly foreshadows the response of a two-level atom driven by an optical field exactly on resonance, in which the atomic levels shift dynamically at the optical frequency through the Stark effect of the electric field. In this way the atom undergoes what is known as AC Stark splitting or Rabi splitting (see Figure 3.6). In later chapters we shall approach time-dependent problems near resonances more cautiously, and not only resolve these discrepancies using strong-field analysis that incorporates decay consistently but achieve agreement with experiment [3.1, 3.2].

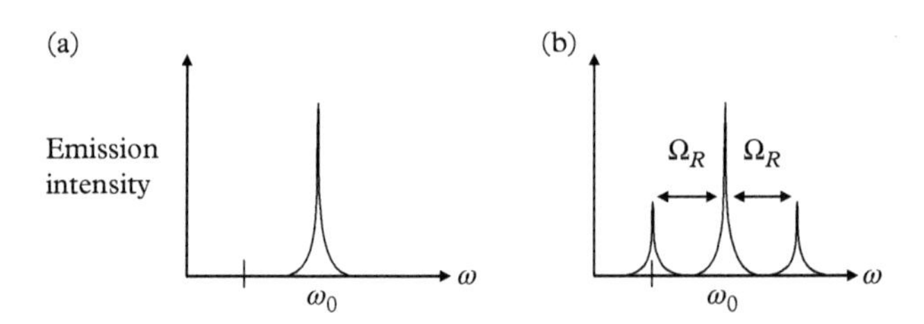

Figure 3.6 *Anticipated differences in the fluorescence spectrum of a two-level atom excited (a) off resonance, and (b) on resonance, based on Eq. (3.4.3) and Figure 3.5.*

3.5 Transition Rates

Common experience reveals three basic radiative processes: spontaneous decay (via emission) from a high-energy unstable state to a lower energy stable state, and transitions induced by an external light field upward or downward between states. From analysis of Sections 3.2 and 3.3, we already know the probability of finding a two-level atom in state 2 is $|C_2(t)|^2$. Hence, if the atom is prepared in the lower state 1, the initial probability of atoms leaving state 1 (by absorption) is also given by

$$P = |C_2(t)|^2 . \tag{3.5.1}$$

Consequently, the transition rate for absorption is

$$\frac{dP}{dt} = \frac{d}{dt}|C_2(t)|^2 . \tag{3.5.2}$$

For a two-level atom, which by assumption has no energy level degeneracies, Eq. (3.5.2) also gives the rate of stimulated emission from state 2. To extend its applicability, we should account for the fact that both the incident light field and the resonant response of atoms have finite bandwidths. As a first step, let us ignore the atomic bandwidth, proportional to the spontaneous emission rate, and replace the radiation field factor E_0^2 in $|C_2(t)|^2$ by its spectral energy density $U(\omega)$. That is, we make the replacement

$$\frac{1}{2}\varepsilon_0 E_0^2(\omega)\delta\left(\omega-\omega_0\right) \Rightarrow U(\omega)d\omega. \tag{3.5.3}$$

The field E_0 might be that of a blackbody radiator or a single-mode laser field, with spectral densities that are very different, but the net probability P can always be found by using Eq. (3.5.3) to replace E_0 with $U(\omega)$ and integrating over frequency.

What is the absorption rate in the case of broadband radiation, for example? Using the expression, Eq. (3.3.11), for $|C_2(t)|^2$, we find an absorption probability P_2 of

$$P_2 = \frac{1}{2}\left\langle\cos^2\theta\right\rangle\frac{\mu_{21}^2}{\hbar^2\varepsilon_0}\int d\omega t^2 U(\omega)\frac{\sin^2\left[\left(\omega_0-\omega\right)t/2\right]}{\left[\left(\omega_0-\omega\right)t/2\right]^2},$$

where $\left\langle\cos^2\theta\right\rangle$ is the average of the square of the geometric cosine factor from the inner product of $\bar{\mu}$ and $\bar{E}$. If the spectral density $U(\omega)$ varies slowly compared to the factor $\sin^2 x/x^2$ in the integral, we can approximate it with the resonance value $U(\omega_0)$. In this way we find

$$P_2 = \left\langle\cos^2\theta\right\rangle\frac{\mu_{21}{}^2}{\hbar^2\varepsilon_0}U(\omega_0)t\int_{-\infty}^{\infty}\frac{\sin^2 x}{x^2}dx = \left(\frac{\pi\mu_{21}{}^2}{\hbar^2\varepsilon_0}\right)U(\omega_0)t\left\langle\cos^2\theta\right\rangle . \tag{3.5.4}$$

For linearly polarized light, the angular average yields $\langle\cos^2\theta\rangle = 1$. Hence

$$\frac{dP_2}{dt} = \pi\left(\frac{\mu_{21}{}^2}{\hbar^2\varepsilon_0}\right)U(\omega_0). \tag{3.5.5}$$

Exercise: Show that for light with spectral density $U(\omega) = \frac{1}{2}\varepsilon_0 E_0^2\delta(\omega-\omega_0)$ that matches the lineshape of a transition between discrete states at frequency ω_0, the transition rate given by Eq. (3.5.5) equals that from Fermi's golden rule (Appendix D). Assume the RWA applies, so that the interaction Hamiltonian can be written simply as $V_{21} = \frac{1}{2}(\mu_{21}\cdot E_0)\exp(i\omega t)$.

For unpolarized, isotropic radiation (like blackbody radiation) the orientational average yields $\langle\cos^2\theta\rangle = \frac{1}{3}$, and we instead find

$$\frac{dP_2}{dt} = \frac{\pi}{3}\left(\frac{\mu_{21}^2}{\hbar^2\varepsilon_0}\right)U(\omega_0)\,. \tag{3.5.6}$$

The Einstein B coefficient for blackbody radiation can be obtained by comparing Eq. (3.5.6) with the defining relation for the stimulated emission rate, $dP_2/dt = BU(\omega_0)$, to obtain

$$B = \frac{\pi}{3}\left(\frac{\mu_{21}^2}{\hbar^2\varepsilon_0}\right). \tag{3.5.7}$$

To find A from B an assumption must be made regarding the spectral density. Blackbody sources have a spectral density function of the form derived by Planck:

$$U(\omega) = \frac{\hbar\omega^3/\pi^2c^3}{\exp(\hbar\omega/k_bT)-1}. \tag{3.5.8}$$

The spontaneous emission rate, the Einstein A coefficient, may then be obtained on the basis of "detailed balance" in the system (i.e., "What goes up must come down!"):

$$\dot{n}_2 = -An_2 - BU(\omega_0)(n_2 - n_1),$$

$$\dot{n}_1 = An_2 + BU(\omega_0)(n_2 - n_1).$$

In the steady state ($\dot{n} = 0$), both population equations give the same result, namely Einstein's detailed balance relation:

$$[A + BU(\omega_0)]\,n_2 = BU(\omega_0)n_1. \tag{3.5.9}$$

In order for the populations to satisfy the Boltzmann distribution

$$n_2 = n_1\exp[-\hbar\omega/k_BT],$$

Table 3.1 *Two-level system transition rates.*

	Unpolarized Isotropic	Linearly Polarized	Circularly Polarized
B_{21}	$\frac{\pi}{3}\frac{\mu_{21}^2}{\hbar^2\varepsilon_0}$	$\pi\frac{\mu_{21}^2}{\hbar^2\varepsilon_0}$	$\frac{\pi}{3}\frac{\mu_{21}^2}{\hbar^2\varepsilon_0}$
B_{12}	$\frac{g_2}{g_1}\cdot\frac{\pi}{3}\frac{\mu_{21}^2}{\hbar^2\varepsilon_0}$	$\frac{g_2}{g_1}\cdot\pi\frac{\mu_{21}^2}{\hbar^2\varepsilon_0}$	$\frac{g_2}{g_1}\cdot\frac{\pi}{3}\frac{\mu_{21}^2}{\hbar^2\varepsilon_0}$
A_{21}	$\frac{\pi}{3}\cdot\left(\frac{\hbar\omega^3}{\pi^2c^3}\right)\frac{\mu_{21}^2}{\hbar^2\varepsilon_0}$	$\pi\left(\frac{\hbar\omega^3}{\pi^2c^3}\right)\frac{\mu_{21}^2}{\hbar^2\varepsilon_0}$	$\frac{\pi}{3}\cdot\left(\frac{\hbar\omega^3}{\pi^2c^3}\right)\frac{\mu_{21}^2}{\hbar^2\varepsilon_0}$

the energy density must be given by Eq. (3.5.8), so the ratio of A and B must be

$$\frac{A}{B} = \frac{\hbar\omega^3}{\pi^2c^3}. \tag{3.5.10}$$

Exercise: For a quasi-two-level system with electronic degeneracies g_1 and g_2 in levels 1 and 2, respectively, find the A and B coefficients for all possible polarizations (as given in Table 3.1).

Spontaneous transition probabilities (Einstein A coefficients) are tabulated for most atoms and ions in spectroscopic handbooks. From these tabulations, the dipole moments μ_{ab} may be readily calculated for simple systems, as shown by the correspondences in Table 3.1. However, the determination of matrix elements or transition rates that are optically induced via the interaction Hamiltonian $\bar{\mu}\cdot\bar{E}$ may involve more complicated geometric considerations. If the directions of $\bar{\mu}$ and $\bar{E}$ are fixed and known, as in the case of dopants or defects in crystals of high symmetry, then the calculation of $\langle\psi|\,\bar{\mu}\cdot\bar{E}\,|\psi\rangle$ is straightforward. More generally however, this requires the use of the Wigner–Eckart theorem, as discussed in Chapter 4 in the context of Zeeman coherence and other phenomena.

3.6 The Density Matrix

Results for the probability of a transition or the value of some observable like an induced dipole moment are invariably expressed in terms of bilinear combinations of probability amplitudes like $C_2C_2^*$ or $C_2C_1^*$, as shown in Section 3.6.1. This is because the expectation value of any observable involves bilinear products of this type. As a consequence it is not surprising to find that, even in simple systems, one must carefully keep track of all possible bilinear combinations of probability amplitudes to describe dynamics consistently and accurately.

3.6.1 Electric Dipole Transition Moments

The dipole moment in a two-level system has the following matrix elements.

$$e\bar{r}_{22} = \langle 2\,|e\bar{r}|\,2\,\rangle = e\int d^3rU_2^*(\bar{r})\,\bar{r}U_2(\bar{r}),$$

$$e\bar{r}_{21} = e\bar{r}_{12}^* = \langle 2\,|e\bar{r}|\,1\rangle = e\int d^3rU_2^*(\bar{r})\,\bar{r}U_1(\bar{r}),$$

$$e\bar{r}_{11} = \langle 1\,|e\bar{r}|\,1\rangle = e\int d^3rU_1^*(\bar{r})\,\bar{r}U_1(\bar{r}).$$

With a time-dependent state vector

$$|\psi(t)\rangle = C_2\exp(-i\omega_2 t)|2\rangle + C_1\exp(-i\omega_1 t)|1\rangle, \tag{3.6.1}$$

the expectation value of *any* operator $\hat{O}$ is

$$\langle\hat{O}\rangle = \langle\psi|\hat{O}|\psi\rangle = C_2C_2^*O_{22} + C_1C_1^*O_{11} + \left\{C_2C_1^*\exp\left[-i(\omega_2-\omega_1)t\right]O_{12} + c.c.\right\} \tag{3.6.2}$$

according to Eq. (2.2.1). In a system with inversion symmetry, as we found in Section 2.2.3, all diagonal moments are zero (i.e., $\langle 2|e\bar{r}|2\rangle = \langle 1|e\bar{r}|1\rangle = 0$). Hence the expectation value of the dipole moment operator for an electron is given by

$$\langle\bar{\mu}^{(e)}\rangle = \langle\psi\,|e\bar{r}|\,\psi\rangle = eC_2C_1^*\exp\left[-i(\omega_2-\omega_1)t\right]\bar{r}_{12} + c.c. \tag{3.6.3}$$

This highlights the importance of the off-diagonal bilinear combination $C_2C_1^*$.

3.6.2 Pure Case Density Matrix

One way of dealing systematically with the bilinear probability amplitudes like those in Eq. (3.6.3) is to organize them into a so-called density matrix ρ constructed from the expansion coefficients $c_2 = C_2\exp(-i\omega_2 t)$ and $c_1 = C_1\exp(-i\omega_1 t)$ using the matrix element definitions

$$\begin{aligned}\rho_{22} &\equiv c_2c_2^*, & \rho_{21} &\equiv c_2c_1^*,\\ \rho_{12} &\equiv c_1c_2^* = \rho_{21}^*, & \rho_{11} &\equiv c_1c_1^*.\end{aligned}$$

The explicit form of the density matrix is

$$\hat{\rho} = \begin{pmatrix} c_2c_2^* & c_2c_1^* \\ c_1c_2^* & c_1c_1^* \end{pmatrix} = \begin{pmatrix} \rho_{22} & \rho_{21} \\ \rho_{12} & \rho_{11} \end{pmatrix}. \tag{3.6.4}$$

In terms of the two-level density matrix elements ρ_{ij}, the expectation value of an operator $\hat{O}$ becomes

$$\begin{aligned}\langle\hat{O}\rangle &= (\rho_{22}O_{22} + \rho_{21}O_{12}) + (\rho_{12}O_{21} + \rho_{11}O_{11}) \\ &= \sum_i \sum_j \rho_{ij}O_{ji} = \sum_i \left(\hat{\rho}\hat{O}\right)_{ii} \\ &= Tr(\hat{\rho}\hat{O}).\end{aligned} \tag{3.6.5}$$

This trace of the product of the density matrix $\hat{\rho}$ and operator $\hat{O}$, designated $Tr(\hat{\rho}\hat{O})$, is the sum of diagonal matrix elements of $\hat{\rho}\hat{O}$. Provided there are no diagonal (static) moments in the system, the expectation value of the electric dipole moment operator is

$$\langle e\bar{r}\rangle = \mu_{21}\rho_{12} + \mu_{12}\rho_{21}. \tag{3.6.6}$$

In virtually all atom–light interactions, we need to calculate system observables as a function of time. To follow temporal evolution, we therefore start by finding equations of motion for the amplitudes

$$c_1(t) = C_1(t)\exp(-i\omega_1 t),$$
$$c_2(t) = C_2(t)\exp(-i\omega_2 t),$$

and from these we find the equations for ρ_{ij}. In the Schrödinger picture, all the time dependence is associated with the state vector and the lower case c numbers incorporate all the time dependence, fast or slow. The wavefunction

$$|\psi(\bar{r},t)\rangle = c_2(t)|U_2(\bar{r})\rangle + c_1(t)|U_1(\bar{r})\rangle \tag{3.6.7}$$

must satisfy Schrödinger's equation

$$\frac{\partial}{\partial t}|\psi(t)\rangle = -\frac{i}{\hbar}H|\psi(t)\rangle. \tag{3.6.8}$$

From Eqs. (3.6.7) and (3.6.8) it is easily shown that

$$\dot{c}_2 = -i\omega_2 c_2 - \frac{i}{\hbar}V_{21}c_1, \tag{3.6.9}$$

$$\dot{c}_1 = -i\omega_1 c_1 - \frac{i}{\hbar}V_{12}c_2. \tag{3.6.10}$$

Hence equations of motions for individual elements of the density matrix are

$$\begin{aligned}\dot{\rho}_{22} &= \dot{c}_2 c_2^* + c_2 \dot{c}_2^* \\ &= \left(-i\omega_2 c_2 - \frac{i}{\hbar} V_{21} c_1\right) c_2^* + c_2 \left(i\omega_2 c_2^* + \frac{i}{\hbar} V_{12} c_1^*\right) \\ &= -\frac{i}{\hbar} V_{21} \rho_{12} + c.c., \end{aligned} \tag{3.6.11}$$

$$\dot{\rho}_{11} = \frac{i}{\hbar} V_{21} \rho_{12} + c.c., \tag{3.6.12}$$

$$\begin{aligned}\dot{\rho}_{21} &= \dot{c}_2 c_1^* + c_2 \dot{c}_1^* \\ &= \left(-i\omega_2 c_2 - \frac{i}{\hbar} V_{21} c_1\right) c_1^* + c_2 \left(i\omega_1 c_1^* + \frac{i}{\hbar} V_{21} c_2^*\right) \\ &= -i\omega_0 \rho_{21} + \frac{i}{\hbar} V_{21} \left(\rho_{22} - \rho_{11}\right), \end{aligned} \tag{3.6.13}$$

and

$$\rho_{12} = \rho_{21}^*, \tag{3.6.14}$$

where $\omega_0 \equiv \omega_2 - \omega_1$.

3.6.3 Mixed Case Density Matrix

The pure case density matrix describes a single atom or molecule. However in most real problems we have an ensemble of many similar systems to track. Typically, there are N identical systems all obeying the same equation of motion but exhibiting different phases or stages of evolution.

In the pure case we could have written

$$\hat{\rho} = |\psi\rangle\langle\psi|, \tag{3.6.15}$$

since this operator expression projects out the right probability amplitudes to agree with our original definition, Eq. (3.6.4), of ρ. By substituting Eq. (3.6.1) into Eq. (3.6.15), for example, the off-diagonal matrix element is found to be

$$\rho_{21} = \langle 2|\,\hat{\rho}\,|1\rangle = \langle 2\,|\psi\,\rangle\langle\psi\,|1\,\rangle = c_2 c_1^*.$$

If only a certain fraction P_j of the atoms is in state ψ_j at time t, then for the entire ensemble we would write

$$\hat{\rho} = \sum_{j=1}^{N} P_j \left|\psi_j\right\rangle\left\langle\psi_j\right|, \tag{3.6.16}$$

to describe a "mixed" case. That is, the system is in a mixture of pure case states. The summation can be either discrete or continuous. For example, the various ψ_j might differ only by a continuous distribution of phase angles developing in time (as in dephasing).

The important thing is that even after introducing this "statistical" aspect of an ensemble, expectation values and time dependence of the system can still be calculated the same way as before. In the case of the expectation value we find

$$\begin{aligned}\langle\hat{O}\rangle &= \sum_j P_j \langle\psi_j|\hat{O}|\psi_j\rangle \\ &= \sum_j P_j \sum_k \langle\psi_j|\hat{O}|k\rangle\langle k|\Psi_j\rangle \\ &= \sum_k \sum_j P_j \langle k|\Psi_j\rangle\langle\psi_j|\hat{O}|k\rangle \\ &= \sum_k \langle k| \sum_j P_j |\psi_j\rangle\langle\psi_j|\hat{O}|k\rangle \\ &= \sum_k \left(\hat{\rho}\hat{O}\right)_{kk} = Tr\left(\hat{\rho}\hat{O}\right).\end{aligned} \tag{3.6.17}$$

Similarly, the equation of motion is

$$\begin{aligned}\dot{\hat{\rho}} &= \sum_j P_j \left\{|\dot{\psi}_j\rangle\langle\psi_j| + |\psi_j\rangle\langle\dot{\psi}_j|\right\} \\ &= -\frac{i}{\hbar}\sum_j P_j \left\{\hat{H}|\psi_j\rangle\langle\psi_j| - |\psi_j\rangle\langle\psi_j|\hat{H}\right\} \\ &= -\frac{i}{\hbar}\left[\hat{H},\hat{\rho}\right]\end{aligned} \tag{3.6.18}$$

Individual matrix elements of the density matrix are therefore calculated from

$$\begin{aligned}\dot{\rho}_{ij} &= -\frac{i}{\hbar}\langle i|\hat{H}\hat{\rho} - \hat{\rho}\hat{H}|j\rangle \\ &= -\frac{i}{\hbar}\sum_k \left\{\langle i|\hat{H}|k\rangle\langle k|\hat{\rho}|j\rangle - \langle i|\hat{\rho}|k\rangle\langle k|\hat{H}|j\rangle\right\} \\ &= -\frac{i}{\hbar}\sum_k \left\{H_{ik}\rho_{kj} - \rho_{ik}H_{kj}\right\}.\end{aligned} \tag{3.6.19}$$

Since decay processes are not included in the Hamiltonian, Eq. (3.6.19) is the key formula for calculating $\rho_{ij}(t)$ to reveal population and polarization dynamics in the absence of losses. In the next section relaxation terms are added to obtain a more general equation of motion. The significance of pure versus mixed cases is explored further in Chapter 3 problems. Further discussion of the density matrix in the context of statistical mechanics may be found in Ref. [3.3].

3.7 Decay Phenomena

Since the Hamiltonian in Eq. (3.6.19) is Hermitian, its eigenvalues are all real. Consequently only lossless time-harmonic or constant temporal behavior can be described by Eq. (3.6.19). In real systems, daily experience teaches us that losses occur. To account for finite atomic lifetimes and other spontaneous relaxation processes in atoms that affect level populations we must add relaxation terms to Eq. (3.6.19), just as we did in Eqs. (3.2.21) and (3.2.22). Guided for the moment by the expectation that population of level i decays at a total rate γ_i which is twice the rate of decay of probability amplitude, we shall augment the density matrix equations to read

$$\dot{\rho}_{11} = -\gamma_1 \rho_{11} + \left[\frac{i}{\hbar} V_{21} \rho_{12} + c.c. \right], \tag{3.7.1}$$

$$\dot{\rho}_{22} = -\gamma_2 \rho_{22} - \left[\frac{i}{\hbar} V_{21} \rho_{12} + c.c. \right], \tag{3.7.2}$$

$$\dot{\rho}_{21} = -\frac{1}{2} (\gamma_2 + \gamma_1) \rho_{21} - i\omega_0 \rho_{21} + \frac{i}{\hbar} V_{21} (\rho_{22} - \rho_{11}). \tag{3.7.3}$$

This phenomenological approach to relaxation dynamics will be justified later in Section 6.4 on quantized reservoir theory.

Equations (3.7.1)–(3.7.3) still do not account for all the interactions that strongly affect coherent optical dynamics, such as the effects of elastic collisions between atoms in a gas, or equivalently between atoms and phonons in solids. Collisions change the phase of individual probability amplitudes and therefore potentially affect diagonal and off-diagonal elements of the density matrix quite differently. This is easily confirmed by assigning time-dependent phase angles to the probability amplitudes $c_i \propto c_i(0) \exp(i\phi_i(t))$ and $c_j \propto c_j(0) \exp\big(i\phi_j(t)\big)$. Then one can compare diagonal elements such as $\rho_{ii} = c_i c_i^* \propto |c_i(0)|^2$ with off-diagonal elements like $\rho_{ij} = c_i c_j^* \propto c_i(0) c_j^*(0) \exp\big(i\left(\phi_i(t) - \phi_j(t)\right)\big)$ to recognize that a time-dependent phase term is retained only in the latter case. Since all elements of the density matrix correspond to measurable quantities, unlike the probability amplitudes themselves, phase evolution evidently plays a significant role in coherent dynamics. In Chapter 6 we shall justify the ad hoc relaxation terms introduced here, from first principles.

To understand that the gradual loss of phase coherence between atoms (called dephasing) is a process distinct from population decay, one can consider the electromagnetic environment of Pr^{3+} ions in LaF_3 [3.4]. In isolation, these ions absorb light at a precise frequency ω_0, but in ensembles they experience random energy shifts due to crystal

Figure 3.7 *Temporal fluctuations in the transition frequency of atoms due to dynamic changes in their environment.*

vibrations and other dynamic processes, such as nearby spin flips that modulate their instantaneous transition frequency (Figure 3.7).

Let the time-varying transition frequency be $\omega(t)$ and its small, random frequency shift from the isolated resonance frequency be $\delta\omega(t)$. Ignoring other perturbations for the moment (i.e., setting $V_{21} = 0$ in Eq. (3.7.3)), we therefore write

$$\dot{\rho}_{21} = -\left[i\omega_0 + i\delta\omega(t) + \Gamma'_{21}\right]\rho_{21}, \tag{3.7.4}$$

where

$$\Gamma'_{21} = \frac{1}{2}(\gamma_2 + \gamma_1). \tag{3.7.5}$$

Formal integration of Eq. (3.7.4) yields

$$\rho_{21}(t) = \rho_{21}(0)\exp\left[-\left(i\omega_0 + \Gamma'_{21}\right)t - i\int_0^t dt'\delta\omega\left(t'\right)\right]. \tag{3.7.6}$$

Since all parameters except $\delta\omega(t)$ are fixed, an ensemble average only affects the final term in the argument of the exponential function in Eq. (3.7.6):

$$\begin{aligned}
&\left\langle \exp\left[-i\int_0^t \delta\omega\left(t'\right)dt'\right]\right\rangle \\
&= \left\langle 1 - i\int_0^t dt_1\delta\omega\left(t_1\right) - \frac{1}{2}\int_0^t dt_1\int dt_2\delta\omega\left(t_1\right)\delta\omega\left(t_2\right) + \ldots \right. \\
&\left. \ldots + \frac{(-i)}{(2n)!}^{2n}\int_0^t dt_1\ldots\int_0^t dt_{2n}\delta\omega\left(t_1\right)\ldots\delta\omega\left(t_{2n}\right) + \ldots\right\rangle.
\end{aligned} \tag{3.7.7}$$

Recognizing that

$$\langle \delta\omega(t) \rangle = 0, \tag{3.7.8}$$

and using a Markoff approximation for the pair correlation function, whereby frequency fluctuations at one time do not depend on prior dynamics, so that

$$\left\langle \delta\omega(t)\delta\omega\left(t'\right)\right\rangle = 2\gamma_{deph}\delta\left(t-t'\right), \tag{3.7.9}$$

in agreement with the fluctuation-dissipation theorem, we obtain

$$\left\langle \exp\left[-i\int_0^t \delta\omega\left(t'\right) dt'\right]\right\rangle = \exp\left(-\gamma_{deph}t\right). \tag{3.7.10}$$

Using Eq. (3.7.10) in Eq. (3.7.6), we find the result

$$\rho_{21}(t) = \rho_{21}(0)\exp\left[-\left(i\omega_0 + \Gamma'_{21} + \gamma_{deph}\right)t\right]. \tag{3.7.11}$$

By defining a total decay rate for dephasing according to

$$\Gamma \equiv \Gamma'_{21} + \gamma_{deph} = \frac{1}{2}(\gamma_2 + \gamma_1) + \gamma_{deph}, \tag{3.7.12}$$

the equation of motion for the coherence becomes

$$\dot{\rho}_{21} = -(i\omega_0 + \Gamma)\rho_{21} + \frac{i}{\hbar}V_{21}(\rho_{22} - \rho_{11}). \tag{3.7.13}$$

Exercise: Verify that a random frequency shift $\delta\omega \ll \omega$ has no effect on the level populations determined by ρ_{22} and ρ_{11}.

The phenomenological decay terms treated previously include two types, differing by whether they affect the diagonal or off-diagonal elements of the density matrix. The two types are associated with population decay and polarization dephasing, respectively, and are important not only because they extend the applicability of our analysis, but together they provide overall conservation of occupation probabilities, unlike the squared amplitudes $|C_2|^2$ and $|C_1|^2$ alone. In the next section a pictorial analogy is developed, called the vector model, between evolving populations and polarizations of a two-level atom and the motion of a gyroscope. The model is useful not only for understanding all the aspects of dynamics covered so far—off-resonant excitation, Rabi flopping, population decay, and dephasing—but permits the outcome of coherent interactions involving multiple pulses to be anticipated in a simple way. The importance of even weak dephasing is illustrated by Problem 3.3.

3.8 Bloch Equations

Having discussed a couple of important categories of relaxation process, and anticipating how to handle them, we now modify Eq. (3.6.19) using a phenomenological procedure that preserves the Hermitian character of H while adding decay terms to the equation of motion. The assumed form of these terms will be fully justified later in Chapter 6.

$$\begin{aligned} i\hbar\dot{\hat{\rho}} &= \left[\hat{H}, \hat{\rho}\right] + \textit{relaxation terms} \\ &= \left[\hat{H}, \hat{\rho}\right] \pm i\hbar\Lambda \pm i\hbar\gamma\hat{\rho} - i\hbar\Gamma\hat{\rho} \end{aligned} \tag{3.8.1}$$

In Eq. (3.8.1) we have separated decay contributions into separate categories for incoherent pumping (Λ), radiative and non-radiative population relaxation (γ), and dephasing (Γ). Population changes due to incoherent pumping processes are particularly challenging to write down at this stage, even if energy pathways of the system are thought to be fairly well understood. They may contribute population relaxation terms at fixed rates of the form $\dot{\rho}_{ii} \propto \pm i\hbar\Lambda$ in open systems where external mechanisms can add or remove population without preserving the number density in the levels of interest. On the other hand, incoherent pumping may contribute terms of the form $\dot{\rho}_{ii} \propto \pm i\hbar\Lambda_{ij}\rho_{jj}$ in closed systems subject to thermal excitation from one level to another. Although relaxation has been introduced in a semi-empirical way in Eq. (3.8.1), we shall make extensive use of this equation of motion in Chapters 4 and 5 to get started with the analysis of optical dynamics. Its form will be justified rigorously later.

Under certain conditions the density matrix equations of motion can be cast into a convenient form known as the optical Bloch equations, resembling equations for gyroscopic precession. The simplest version of these equations is called the vector model and illustrates that two-level atoms are analogous to spin 1/2 particles. This model provides a geometrical picture of the development of optical polarization in time, and simplifies calculations of coherent aspects of multiple pulse interactions. However, the Bloch equations originated from studies of spin magnetism and assume that diagonal and off-diagonal elements decay at similar rates. Hence, despite the visual appeal of the vector model, as described subsequently, solutions of the Bloch equations should not be confused with full solutions of the density matrix.

Using the RWA, the atom–field interaction is

$$V_{21} = -\frac{1}{2}\mu_{21}E_0 e^{-i\omega t}, \tag{3.8.2}$$

and we introduce the slowly varying envelope approximation (SVEA) by writing

$$\rho_{21} = \tilde{\rho}_{21} e^{-i\omega t}, \tag{3.8.3}$$

where $\tilde{\rho}_{21}$ denotes the slowly varying amplitude of the matrix element ρ_{21} proportional to the amplitude of charge oscillation at the optical frequency. With the use of Eqs. (3.8.2) and (3.8.3), the equation of motion (Eq. (3.7.3)) becomes

$$\dot{\tilde{\rho}}_{21} = -\left[i\left(\omega_0 - \omega\right) + \Gamma\right]\tilde{\rho}_{21} - \frac{i}{2}\frac{\mu_{21}E_0}{\hbar}\left(\rho_{22} - \rho_{11}\right).$$

By defining $\Delta \equiv \omega_0 - \omega$ as the detuning of the optical frequency ω from the resonance at ω_0, one obtains

$$\left(\frac{d}{dt} + i\Delta + \Gamma\right)\tilde{\rho}_{21} = -\frac{i}{2}\Omega\left(\rho_{22} - \rho_{11}\right). \tag{3.8.4}$$

A simple, geometrical picture of system evolution emerges if we define components of a 3-space vector by the real quantities

$$R_1 = \tilde{\rho}_{21} + \tilde{\rho}_{12}, \tag{3.8.5}$$

$$R_2 = i\left(\tilde{\rho}_{21} - \tilde{\rho}_{12}\right), \tag{3.8.6}$$

$$R_3 = \rho_{22} - \rho_{11}. \tag{3.8.7}$$

These quantities vary slowly in an optical period, and because they are mutually independent can be viewed as components of the vector

$$\bar{R} = R_1\hat{e}_1 + R_2\hat{e}_2 + R_3\hat{e}_3 \tag{3.8.8}$$

in a fictitious space spanned by the orthogonal basis set $\left(\hat{e}_1, \hat{e}_2, \hat{e}_3\right)$. The time development of each component is given by

$$\dot{R}_1 = -\Delta R_2 - \Gamma R_1, \tag{3.8.9}$$

$$\dot{R}_2 = \Delta R_1 - \Gamma R_2 + \Omega R_3, \tag{3.8.10}$$

$$\dot{R}_3 = -\left[\gamma_2\rho_{22} - \gamma_1\rho_{11}\right] - \Omega R_2. \tag{3.8.11}$$

The rate Γ is the reciprocal of the decay time T_2:

$$\Gamma = T_2^{-1}. \tag{3.8.12}$$

If we consider the limiting case $\Gamma = \gamma_2 = \gamma_1 = T_1^{-1} = T_2^{-1} = T^{-1}$ in Eqs. (3.8.9)–(3.8.11), we obtain

$$\dot{R}_1 = -\Delta R_2 - \frac{1}{T}R_1, \tag{3.8.13}$$

$$\dot{R}_2 = \Delta R_1 - \frac{1}{T}R_2 + \Omega R_3, \tag{3.8.14}$$

$$\dot{R}_3 = -\Omega R_2 - \frac{1}{T}R_3. \tag{3.8.15}$$

Equations (3.8.13)–(3.8.15) are the optical Bloch equations. They can be written more compactly as

$$\dot{\bar{R}} = -\frac{1}{T}\bar{R} + \bar{\beta} \times \bar{R}, \tag{3.8.16}$$

where

$$\bar{\beta} = -\Omega\hat{e}_1 + \Delta\hat{e}_3. \tag{3.8.17}$$

Because Eq. (3.8.16) is the equation of motion of a gyroscope, it shows that elements of the density matrix can be chosen (in the limiting case) as components of a Bloch vector $\bar{R}$ which executes simple precession about an effective field $\bar{\beta}$ (Figure 3.8). The length of $\bar{R}$ shrinks with time as $\exp[-t/T]$ due to decay processes.

Off resonance, the effective field or torque vector $\bar{\beta}$ is a constant vector in the $\hat{e}_1 - \hat{e}_3$ plane in the absence of decay. In this case a solution to Eq. (3.8.16) without the relaxation term is obtained by finding a unitary transformation which turns the Bloch vector into a constant vector. To freeze the action of the Bloch vector, we first rotate about $\hat{e}_2$ by angle α. Coordinates transform according to

$$\begin{pmatrix} x' \\ y' \\ z' \end{pmatrix} = \begin{pmatrix} \cos\alpha & 0 & \sin\alpha \\ 0 & 1 & 0 \\ -\sin\alpha & 0 & \cos\alpha \end{pmatrix} \begin{pmatrix} x \\ y \\ z \end{pmatrix},$$

so the Bloch vector transforms according to

$$\begin{pmatrix} R_1' \\ R_2' \\ R_3' \end{pmatrix} = \begin{pmatrix} \cos\alpha & 0 & -\sin\alpha \\ 0 & 1 & 0 \\ \sin\alpha & 0 & \cos\alpha \end{pmatrix} \begin{pmatrix} R_1 \\ R_2 \\ R_3 \end{pmatrix} = U_1\bar{R}. \tag{3.8.18}$$

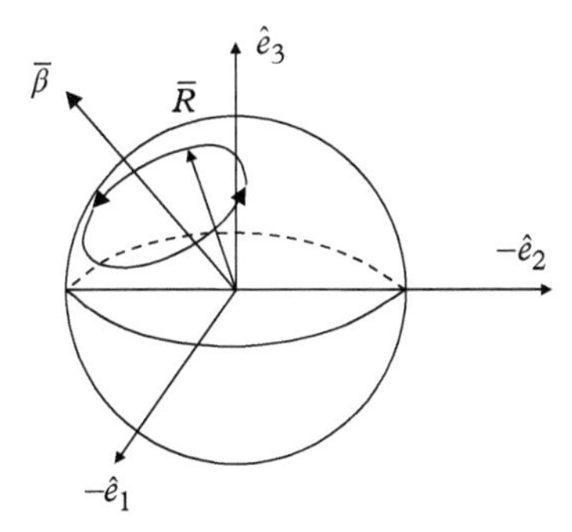

Figure 3.8 *Processional motion of the Bloch vector $\bar{R}$ around the effective field $\bar{\beta}$. For the orientation of $\bar{\beta}$ that is shown, the detuning is positive.*

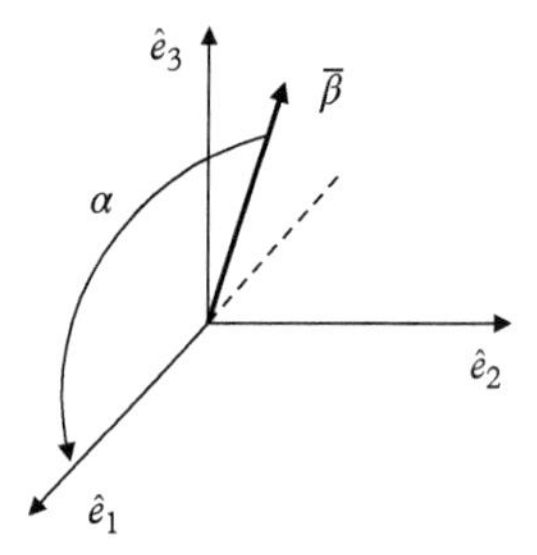

Figure 3.9 *The effective field $\bar{\beta}$ in the Bloch vector model. $\bar{\beta}$ makes an angle α with respect to the $\hat{e}_1$ axis in the plane of basis vectors $\hat{e}_1, \hat{e}_3$ such that* $\tan\alpha = \Delta/\Omega$ *for positive detuning.*

Exercise: Draw the direction of the effective field $\bar{\beta}$ (see Figure 3.9) for a resonant interaction and sketch the motion of the Bloch vector $\bar{R}$ as a function of time, ignoring decay ($T = \infty$). What is the significance of $\bar{R}$ passing periodically through the $\pm\hat{e}_3$ directions? What is the atomic state when $\bar{R}$ points along $\hat{e}_2$?

In this process, since the effective field $\bar{\beta}$ now lies along the new (primed) $\hat{e}_1$ axis, the Bloch vector merely executes a precession counter-clockwise about $\hat{e}'_1$. A second rotation about the $\hat{e}'_1$ axis therefore leads to a frame in which $\bar{R}$ is stationary. The rotation angle must be $\Omega_R(\Delta)t$.

$$\begin{pmatrix} R''_1 \\ R''_2 \\ R''_3 \end{pmatrix} = \begin{pmatrix} 1 & 0 & 0 \\ 0 & \cos\Omega_R t & -\sin\Omega_R t \\ 0 & \sin\Omega_R t & \cos\Omega_R t \end{pmatrix} \begin{pmatrix} R'_1 \\ R'_2 \\ R'_3 \end{pmatrix} = U_2\bar{R}' \tag{3.8.19}$$

Now we can relate $\bar{R}(t)$ to $\bar{R}''$ at any time. But at $t = 0$ the Bloch vector $\bar{R}''$ is the same as the initial Bloch vector $\bar{R}(0)$ apart from a single coordinate rotation through angle α which we must "undo:"

$$\begin{pmatrix} R_x(0) \\ R_y(0) \\ R_z(0) \end{pmatrix} = \begin{pmatrix} \cos\alpha & 0 & \sin\alpha \\ 0 & 1 & 0 \\ -\sin\alpha & 0 & \cos\alpha \end{pmatrix} \begin{pmatrix} R''_1 \\ R''_2 \\ R''_3 \end{pmatrix} = U_3\bar{R}''. \tag{3.8.20}$$

Hence $\bar{R}(0) = U_3U_2U_1\bar{R}(t)$ or

$$\bar{R}(t) = U_1U_2U_3\bar{R}(0) = \bar{\bar{U}}\bar{R}(0), \tag{3.8.21}$$

where the evolution operator U is given by

$$U = \begin{pmatrix} \cos\alpha & 0 & -\sin\alpha \\ 0 & 1 & 0 \\ \sin\alpha & 0 & \cos\alpha \end{pmatrix} \begin{pmatrix} 1 & 0 & 0 \\ 0 & \cos\Omega_R t & -\sin\Omega_R t \\ 0 & \sin\Omega_R t & \cos\Omega_R t \end{pmatrix} \begin{pmatrix} \cos\alpha & 0 & \sin\alpha \\ 0 & 1 & 0 \\ -\sin\alpha & 0 & \cos\alpha \end{pmatrix}.$$

Making the assignments $\sin\alpha \equiv \Delta/\Omega_R$ and $\cos\alpha \equiv \Omega/\Omega_R$, based on the definitions $\Omega \equiv \mu E/\hbar$ and $\Omega_R \equiv \sqrt{\Delta^2 + \Omega^2}$ and Figure 3.9, the final result is

$$U = \begin{bmatrix} \dfrac{\Omega^2 + \Delta^2 \cos\Omega_R t}{\Omega_R^2} & \dfrac{-\Delta}{\Omega_R}\sin\Omega_R t & \dfrac{\Delta\Omega}{\Omega_R^2}(1 - \cos\Omega_R t) \\ \dfrac{\Delta}{\Omega_R}\sin\Omega_R t & \cos\Omega_R t & -\dfrac{\Omega}{\Omega_R}\sin\Omega_R t \\ \dfrac{\Delta\Omega}{\Omega_R^2}(1 - \cos\Omega_R t) & \dfrac{\Omega}{\Omega_R}\sin\Omega_R t & \dfrac{\Delta^2 + \Omega^2 \cos\Omega_R t}{\Omega_R^2} \end{bmatrix}. \tag{3.8.22}$$

In the case of nuclear magnetic resonance, spins precess in real space (with an $\hat{e}_3$ axis determined by the direction of the static field) about an effective magnetic field $\bar{\beta}$ consisting of the vector sum of applied static and oscillating magnetic fields. In the electric dipole or optical case considered here there is no static field present, so the precession is not in real space. There is no fixed direction associated with the population difference $R_3 = \rho_{22} - \rho_{11}$. Nevertheless, there is a polarization axis, and in keeping with the spin analogy T_2 is called the "transverse" decay time since it describes relaxation of transverse components R_1 and R_2 of the Bloch vector. T_2 is also referred to as the dephasing time. T_1 is called the "longitudinal" relaxation time since it applies to the $\hat{e}_3$ or $\hat{z}$ component of the Bloch vector R_3. It is the characteristic time for population decay. In the next chapter the distinction between T_1 and T_2 will be restored in the Bloch model to permit an appreciation of the different roles these times play in coherent dynamics.

3.9 Inhomogeneous Broadening, Polarization, and Signal Fields

The microscopic polarization induced by light in a medium is given by the expectation value of the atomic electric dipole moment in accord with Eq. (3.6.6):

$$<\hat{\mu}> = Tr\left(\hat{\rho}\hat{\mu}\right) = \rho_{21}\mu_{12} + \rho_{12}\mu_{21}. \tag{3.9.1}$$

If we assume that atoms in a system are not identical, but have a distribution of resonant frequencies, due perhaps to a Maxwellian distribution of velocities or some other

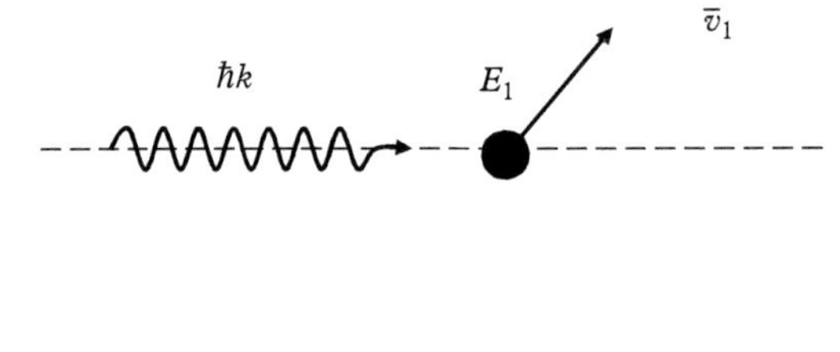

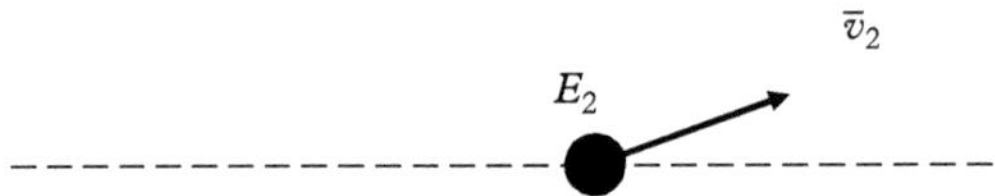

Figure 3.10 *Schematic diagram of an atom before (top) and after (bottom) interacting with a photon, illustrating the change in momentum of the atom that results from the inelastic scattering event.*

inhomogeneity in the system, then an ensemble average must be performed to account for this.

The resonant frequency of atom is shifted by motion according to the Doppler effect. This is a simple kinematic effect that may be readily understood in the following way. In Figure 3.10, an atom of mass M in state E_1 moving at velocity $\bar{v}_1$ encounters a photon of momentum $\hbar\bar{k}$. After absorption of the photon, the atom is in a new internal energy state E_2 and has a new velocity $\bar{v}_2$. Conservation of energy and momentum for this interaction are described by the equations:

$$M\bar{v}_1 + \hbar\bar{k} = M\bar{v}_2, \tag{3.9.2}$$

$$E_1 + \frac{1}{2}Mv_1^2 + \hbar\omega = E_2 + \frac{1}{2}Mv_2^2. \tag{3.9.3}$$

According to Eq. (3.9.3), a stationary atom ($v_1 = 0$) absorbs very close to its resonance frequency

$$\omega_0 \equiv (E_2 - E_1)/\hbar, \tag{3.9.4}$$

because from Eq. (3.9.2) the recoil velocity v_2 of massive atoms is seen to be very small.

For moving atoms however, we find upon substitution of Eqs. (3.9.4) and (3.9.2) into Eq. (3.9.3) that

$$\frac{1}{2}M\left(v_1^2 - v_2^2\right) + \hbar\omega = \hbar\omega_0. \tag{3.9.5}$$

If this expression is rewritten using Eq. (3.9.2) both as

$$Mv_1^2 + \hbar\bar{k}\cdot\bar{v}_1 = M\bar{v}_1\cdot\bar{v}_2, \tag{3.9.6}$$

and

$$Mv_2^2 = M\bar{v}_1\cdot\bar{v}_2 + \hbar\bar{k}\cdot\bar{v}_2, \tag{3.9.7}$$

then the energy conservation equation becomes

$$\hbar\omega_0 = \hbar\omega + \frac{1}{2}M\left(\text{v}_1^2 - \text{v}_2^2\right) = \hbar\omega + \frac{1}{2}\left(M\bar{\text{v}}_1 \cdot \bar{\text{v}}_2 - \hbar\bar{k} \cdot \bar{\text{v}}_1\right) - \frac{1}{2}\left(M\bar{\text{v}}_1 \cdot \bar{\text{v}}_2 + \hbar\bar{k} \cdot \bar{\text{v}}_2\right)$$
$$= \hbar\omega - \hbar\bar{\text{v}}_1 \cdot \bar{k} - \left(\hbar^2 k^2/2M\right). \tag{3.9.8}$$

The second term on the right of Eq. (3.9.8) is the first-order Doppler shift, proportional to the projection of velocity on the line of sight. The third term gives the shift arising from recoil of the atom.

Taking only the Doppler shift from Eq. (3.9.8) into account, we find

$$<\hat{\mu}> = \frac{1}{2}\mu_{12}<\tilde{\rho}_{21}> \exp[-i(\omega t - kz)] + c.c. \tag{3.9.9}$$

for example, where

$$<\tilde{\rho}_{21}> \equiv \frac{1}{k\text{v}\sqrt{\pi}} \int_{-\infty}^{\infty} \tilde{\rho}_{21} \exp[-(\Delta/k\text{v})^2] d\Delta. \tag{3.9.10}$$

Here $<\tilde{\rho}_{21}>$ denotes an average over the Maxwellian molecular velocity distribution and v is the root-mean-square (rms) velocity. Hence the macroscopic polarization in a sample of molecular density N where each molecule develops an electric dipole moment $p \equiv <\hat{\mu}>$ is

$$P(z,t) = Np = \frac{1}{2}N\,\mu_{12}<\tilde{\rho}_{21}> \exp[-i(\omega t - kz)] + c.c., \tag{3.9.11}$$

and $P(z,t)$ is the source of a signal field

$$E_s(z,t) = \frac{1}{2}E_{s0}(z,t)\exp[-i(\omega t - kz)] + c.c. \tag{3.9.12}$$

which must be calculated in a manner consistent with Maxwell's equations. That is, a signal wave emerges from the sample which must satisfy the wave equation

$$\bar{\nabla} \times \bar{\nabla} \times \bar{E} + \mu\sigma\frac{\partial \bar{E}}{\partial t} + \mu\varepsilon\frac{\partial^2 \bar{E}}{\partial t^2} = -\mu\frac{\partial^2 \bar{P}}{\partial t^2}. \tag{3.9.13}$$

A typical interaction problem begins by determining the atomic polarization using the Bloch or density matrix equations. Then the effects of medium polarization and propagation must be taken into account using Eq. (3.9.13). In this way both microscopic dynamics and macroscopic propagation effects are incorporated consistently through a coupled set of equations known as the Maxwell–Bloch equations.

As long as signal amplitude is much smaller than the incident laser field ($E_{s0} \ll E_0$) we only need the lowest order terms of the wave equation. In most dielectrics $\bar{\nabla} \cdot \bar{P} = 0$, and so Eq. (3.9.13) reduces to

$$\frac{\partial^2 E}{\partial z^2} - \frac{1}{c^2}\frac{\partial^2 E}{\partial t^2} = \mu \frac{\partial^2 P}{\partial t^2}. \tag{3.9.14}$$

After substituting Eqs. (3.9.11) and (3.9.12) into this equation, using the RWA, and dropping second-order derivatives of slowly varying amplitudes, this yields

$$\left(\frac{\partial}{\partial z} + \frac{1}{c}\frac{\partial}{\partial t}\right) E_{s0} = \frac{ik}{2\varepsilon} N \mu_{12} <\tilde{\rho}_{21}>. \tag{3.9.15}$$

Here, as before, it has been assumed that $\tilde{\rho}_{21}$ varies little within an optical period or an optical wavelength. This is consistent with the SVEA and Eq. (3.9.15) is the semi-classical equivalent of the classical expression Eq. (1.2.16). In the simplest case, for optically thin samples of length L, the signal field that emerges from the sample is obtained by direct integration of Eq. (3.9.15), yielding

$$E_{s0} = \frac{ik}{2\varepsilon} NL \mu_{12} <\tilde{\rho}_{21}>. \tag{3.9.16}$$

In agreement with physical intuition, the signal field is proportional to sample length.

3.10 Homogeneous Line Broadening through Relaxation

The expression for the microscopic dipole moment on each atom is

$$p(t) = \rho_{12}\mu_{21} + \rho_{21}\mu_{12}, \tag{3.10.1}$$

and it may be compared term by term with the classical expression for the macroscopic polarization,

$$P(t) = \frac{1}{2}\varepsilon_0 E_0 \left[\chi^{(e)}(\omega)\exp(-i\omega t) + \chi^{(e)}(-\omega)\exp(i\omega t)\right], \tag{3.10.2}$$

to relate the positive frequency component of electric susceptibility $\chi^{(e)}(\omega)$ to the off-diagonal density matrix element ρ_{21}. Since Eq. (1.2.17) shows that the atomic absorption coefficient $\alpha(\omega)$ can be found from the imaginary part of the susceptibility $\chi^{(e)}(\omega) = \chi_R(\omega) + i\chi_I(\omega)$ by $\alpha(\omega) = (\omega/c)\,\chi_I(\omega)$, the absorption spectrum for the case in which all the atoms are equivalent (homogeneous) can be predicted on the basis of solutions of the density matrix. According to Eq. (1.2.14) the frequency dependence of the dispersion (or refractive index) $n(\omega)$ can also be obtained from χ_R, but this is not of immediate interest here.

Let us proceed to calculate the absorption lineshape of a two-level system by solving for the off-diagonal density matrix elements and taking radiative damping, dephasing, and power broadening into account. To do this we simply use Eq. (3.7.3) to solve for ρ_{21} which can then be substituted into Eq. (3.10.1). Having already discussed Doppler broadening, let us assume the atoms are at rest and solve the equation of motion to investigate some other spectral effects of relaxation processes.

Throughout the next chapter detailed solutions of the density matrix under various conditions are considered. Consequently, for the purpose of merely illustrating different broadening mechanisms, we anticipate the general solution of Eq. (3.7.3) that incorporates the steady-state value of ρ_{22} from Eq. (3.7.2) together with the population difference in a form consistent with the closure relation ($\rho_{22} - \rho_{11} = 1 - 2\rho_{22}$). This yields

$$\rho_{21} = \frac{(\Omega_{21}/2)(\Delta + i\Gamma)}{\Delta^2 + \Gamma^2 + (2\Gamma/\gamma)\,|\Omega_{12}|^2}\exp(-i\omega t), \tag{3.10.3}$$

and the macroscopic susceptibility is

$$\chi^{(e)}(\omega) = \frac{N}{\varepsilon_0 \hbar}\frac{\left|\mu_{21}^2\right|(\Delta + i\Gamma)}{\Delta^2 + \Gamma^2 + (2\Gamma/\gamma)\,|\Omega_{12}|^2}. \tag{3.10.4}$$

Notice the width of the resonance in the susceptibility in Eq. (3.10.4) is broadened by several factors. Power broadening or saturation broadening is caused by the term $\left|\bar{\mu}_{21} \cdot \bar{E}_0/\hbar\right|^2 2\Gamma/\gamma_2$ in the denominator, which depends on the incident field strength. The term Γ^2 describes broadening due to the total rate of dephasing. According to Eq. (3.7.12) this term therefore includes contributions due to population decay (called radiative or natural broadening) and contributions due to pure dephasing (arising from elastic collisions, phonon interactions, etc.). Power broadening, radiative decay, and pure dephasing are categorized as homogeneous broadening mechanisms, because on average each atom or optical center is affected in the same way.

The Doppler mechanism described in the last section has an effect that depends on a distinguishable property of each atom, namely its velocity. For this reason, Doppler broadening is referred to as inhomogeneous. Static crystal field broadening in solids is similar. The spectral shifts depend on "local" properties that vary from atom to atom, such as the resonant frequency. While the spectral shift arises from the atomic velocity in the case of Doppler broadening and from local variations in the internal electric field in solids, both situations cause inhomogeneous broadening.

Before leaving this topic, let us consider how a particular mechanism like Doppler broadening changes the spectral response of atoms mathematically. The Doppler effect arises when the component of velocity of an atom along the direction of propagation of light (v_z) results in a shift of its effective resonant frequency proportional to $\pm\mathrm{v}_z/c$. To first order the shifted resonance is at

$$\omega \cong \omega_0\left(1 \pm \frac{\mathrm{v}_z}{c}\right), \tag{3.10.5}$$

where the lower (upper) signs correspond to motion parallel (anti-parallel) to the wavevector of light propagating along z. To proceed we must assume a particular velocity distribution, say a Maxwellian velocity distribution, in order to determine the probability of finding an atom with absorption frequency between ω and $\omega + d\omega$. From statistical mechanics, this probability is known to be

$$\exp\left\{-Mc^2(\omega-\omega_0)^2/2\omega_0^2 k_B T\right\}(c/\omega_0)d\omega, \tag{3.10.6}$$

where M is the mass of the atom, k_B is the Boltzmann's constant, and T is temperature. The inhomogeneous distribution described by Eq. (3.10.6) is Gaussian in form. This lineshape contribution will therefore be quite different from the Lorentzian one in Eq. (3.10.4), which described homogeneous broadening. In general we can expect to encounter composite lineshapes which are convolutions of lineshapes due to several line-broadening mechanisms. For example, the composite lineshape $F(\omega)$ due to two line-broadening mechanisms described individually by $F_1(\omega')$ and $F_2(\omega-\omega')$ would be

$$F(\omega) = \int_{-\infty}^{\infty} F(\omega')F_2(\omega-\omega')d\omega'. \tag{3.10.7}$$

Any number of line-broadening mechanisms can be accounted for by repeated application of Eq. (3.10.7). The convolution of a Gaussian with a Lorentzian lineshape is called the Voigt profile (related to the error function), and unlike its two parent functions is unfortunately not integrable. Hence lineshape analysis can be quite complicated. It is for this reason that experimentalists have gone to great efforts to overcome the effects of broadening, by inventing techniques such as sub-Doppler linewidth spectroscopy. The extent of this inventiveness has been elegantly captured in Ref. [3.5].

3.11 Two-Level Atoms Versus Real Atoms

Real atoms are not two-level atoms. Real atoms generally have many more energy levels than the two considered in Section 3.2. Yet they are often remarkably well described by this simplest of all models.

The theory of absorption and dispersion based on classical harmonic oscillators, for example, accounts successfully for numerous optical phenomena, often by assuming the existence of a single resonance (analogous to a transition between two quantum levels). But how legitimate is such a simplified picture? Surely real atoms with complicated electronic structures need to be understood in great detail before there can be any hope of accounting for their "essential" features? This is a key concern in any new research project and an important theme of this book, which systematically steps through different aspects of optical interactions seeking to identify what approaches are indispensable in any given situation for predicting atomic and molecular dynamics when the wavefunction is not known. We therefore close this chapter by illustrating how *exact* correspondence can be achieved between experiments in a representative multi-level real atom and two-level theory of the optical interaction. This is done by considering how to convert a real sodium atom *experimentally* into a strictly two-level system (see also Ref. [3.6]).

Sodium is referred to as a "one-electron" atom, and yet it has a very complex energy level structure, as shown in Figure 3.11. Each of its many electrons is characterized by several quantum numbers n, l, m_l, and m_s. Electromagnetically inactive core electrons form "closed shells" with spectroscopic labels of the form $^{2S+1}L_J$ that specify the spin, orbit, and total angular momenta S, L, and J. These quantities are all zero for electrons that form the closed shells. Electrons that are left over, called valence electrons, lie

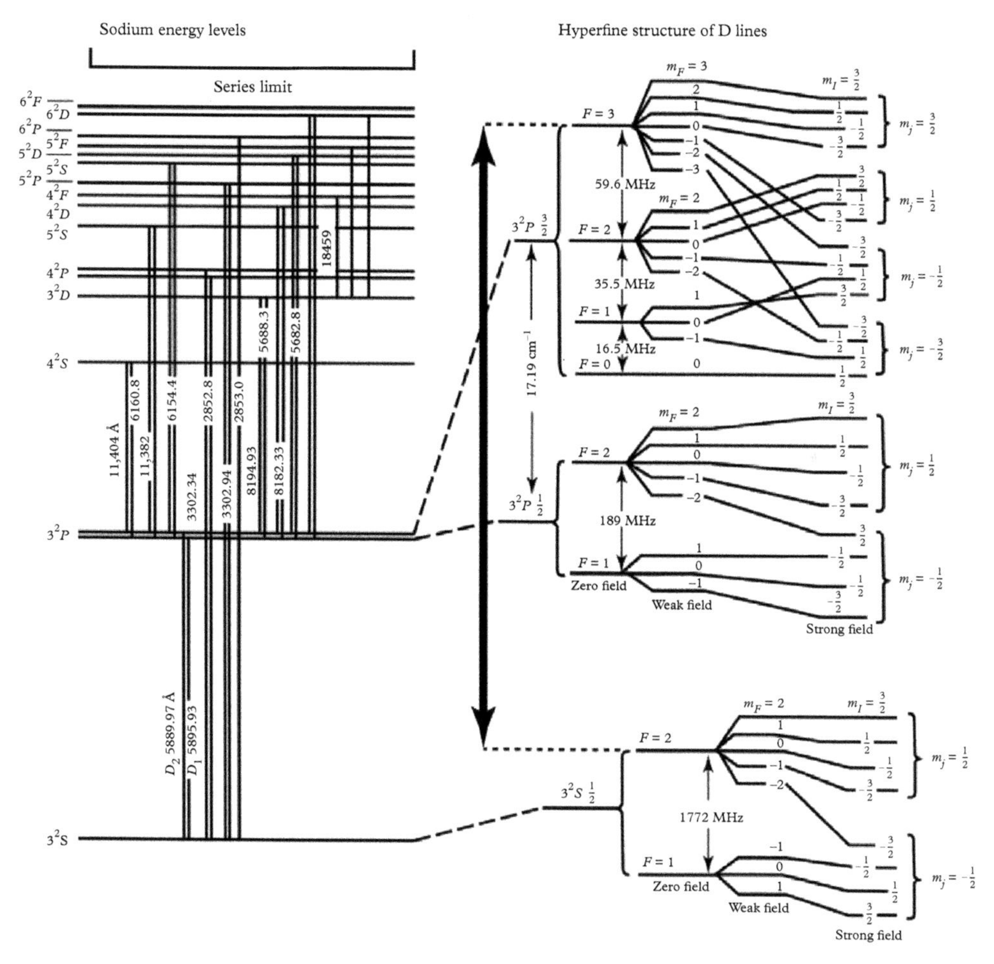

Figure 3.11 *Fine and hyperfine energy levels of the S, P, and D manifolds of sodium, showing optical transitions to the left. The thick vertical arrow illustrates a transition which allows interaction of light with a closed, truly two-level system as described in the text (after Ref. [3.7]).*

outside the closed shells and are responsible for electromagnetic transitions of Na in the optical range and its chemistry. So in addition to the closed shells, other states form by vector addition of the individual orbital and spin angular momenta (l_i and s_i) of valence electrons, yielding the key states for our purposes, of total angular momentum $\bar{J} = \bar{L} + \bar{S}$ in the weak-coupling, or Russell–Saunders, limit.

In sodium the closed shells comprise ten electrons in a configuration written $1s^2 2s^2 2p^6$. The superscripts identify how many electrons occupy each nl state. In addition, there is one valence electron which occupies a $3s$ orbital in the ground state, but which may be promoted in principle to $4s$, $5s\ldots$, $3p$, $4p\ldots$, or $3d$, $4d\ldots$, etc. by acquiring the appropriate energy. The electron spin may be "up" or "down", so the ground state has the spectroscopic designation $^2S_{1/2}$ (indicating $S = \frac{1}{2}, L = 0, J = \frac{1}{2}$). The first excited levels are $3p^2P_{3/2}$ and $3p^2P_{1/2}$, and the allowed electric dipole transitions between $3p^2P_{3/2}-3s^2S_{1/2}$ and $3p^2P_{1/2}-3S^2S_{1/2}$ are the sodium D lines at 589.0 nm and 589.6 nm, respectively.

The Na D lines are separated by 0.6 nm, but are not perfectly sharp. Collisions and the finite radiative lifetime of each level contribute to broadening. So one can ask whether it is possible to excite these transitions individually, even with perfectly monochromatic sources. That is, how do the linewidths and the separation of transitions affect experiments with real atoms? Is it possible to realize two-level atoms in practice by simply interrogating atoms with light of sufficiently narrow bandwidth such that only one transition is excited, or by pulsing the light so the interaction is over before other levels can become involved?

Consider some of the characteristic timescales and bandwidths in sodium vapor given in Table 3.2. From the entries in the table, it is evident that at ordinary temperatures coherent, quasi-monochromatic fields can indeed excite the D lines individually. We might be tempted to conclude that Na can therefore be treated as a two-level system. However, we have omitted important features of photon–atom interactions that invalidate this conclusion. Let us consider in more detail what is needed to make a two-level atom out of sodium by accounting for spectral width of the light source (which may cause overlap of neighboring transitions) and also by recognizing that sodium has additional (hyperfine) electronic structure.

Sodium has a nuclear spin of $I = 3/2$. The nuclear spin combines with each term of the total angular momentum J to produce additional levels labeled by quantum number F, which is the vector sum of J and I (hyperfine structure). Each hyperfine level is $(2F + 1)$-fold degenerate, as one can demonstrate by applying a magnetic field. Even in zero magnetic field, the width of optical excitation pulses must be carefully chosen if we are to think of sodium as a two-level atom. For example, to excite only the $3P^2P_{3/2}(F = 2) \leftrightarrow 3s^2S_{1/2}(F = 1)$ transition, the spectral width of the optical pulse must not embrace the $3s^2S_{1/2}(F = 2)$ or $3p^2P_{3/2}(F = 1)$ levels. The pulsewidth must therefore be less than 35.5 MHz $(\tau_p > 3 \times 10^{-8}s)$. On the other hand, it seems that the pulse duration has to be shorter than the collision time $(\tau_{coll} \sim 10^{-7}s)$ to avoid collisional redistribution of atoms to other states. So the required pulsewidth must lie in the range $3 \times 10^{-8}s < \tau_p < 10^{-7}s$, and even for pulse durations in this range the system could not remain a "two-level atom" for long.

Table 3.2 *Bandwidth associated with decay and line broadening in sodium atoms.*

Radiative lifetime:

$$\tau_{rad}\left({}^2P_{3/2}\right) = 16 \text{ ns}$$

$$\partial\omega(FWHM) = \frac{2}{\tau} \text{ (Homogeneous)}$$

$$\partial\lambda = \lambda(\partial\omega/\omega) \sim 10^{-3}\text{nm}$$

Collisional interval:

$$\tau_{coll} \sim 10^{-7}\text{–}10^{-8}\text{s}$$

$$\partial\lambda \sim 10^{-3}\text{nm (Homogeneous)}$$

Doppler broadening:

$$\partial\omega_{Doppler}(FWHM) = 2\omega(2\ln 2)^{1/2}\left(k_B T/Mc^2\right)^{1/2} \text{ (Inhomogeneous)}$$

$$\partial\nu \sim 10 \text{ GHz } \left(\sim 10^{10}\text{–}10^{13}\text{atoms/cm}^3\right)$$

$$\partial\lambda \sim 0.012 \text{ nm}$$

Transition interval:

$$\lambda_{D1} - \lambda_{D2} \sim 0.6 \text{ nm}$$

Strictly speaking, even when excited with an appropriate pulse width on the $3P^2P_{3/2}\ (F = 2) \leftrightarrow 3s^2S_{1/2}\ (F = 1)$ transition, Na cannot be considered a two-level atom. The reason for this is that spontaneous decay is allowed from $3p^2P_{3/2}\ (F = 2)$ to $3s^2S_{1/2}\ (F = 2)$. This process cannot be avoided and in this case removes atoms from the upper level and places them in a third state. Moreover we still have not taken the M_F sub-levels into account, which can influence the number of phased states participating in relaxation processes with or without applied magnetic fields in principle. This oversight might alter the outcome of coherent excitation of the atom, since the loss of phasing produces incoherent evolution.

In view of the previous discussion, the objective to convert an atom like sodium to a true two-level system would seem to be unattainable. Consider however what would happen if we worked instead on the transition $3^2S_{1/2}\ (F = 2) \leftrightarrow 3^2P_{3/2}\ (F = 3)$. This is the transition highlighted by the bold double-headed arrow in Figure 3.11. In this case, spontaneous emission from the upper state is only allowed back to the same initial state ($\Delta F = 0, \pm 1$), and to no other. Consequently our objective can be realized after all. This explains historically why this transition in sodium vapor was used in successful observations of the resonance fluorescence spectrum of a truly two-level atom [3.1, 3.2]. We shall return to the analysis of this important problem in Chapter 6 when resonance fluorescence is considered in more detail.

Exercise: Is there another transition between the hyperfine manifolds of sodium in Figure 3.11 that can yield a closed two-level system?

REFERENCES

[3.1] F. Schuda, C.R. Stroud, and M. Hercher, *J. Phys. B: At. Mol. Phys.* 1974, 7, L198.
[3.2] R.E. Grove, F.Y. Wu, and S. Ezekiel, *Phys. Rev. A* 1977, **15**, 227.
[3.3] R. Kubo, *Statistical Mechanics*. London: North-Holland, 1965.
[3.4] S.C. Rand, A. Wokaun, R.G. DeVoe, and R.G. Brewer, *Phys. Rev. Lett.* 1979, **43**, 1868.
[3.5] See, for example, M.D. Levenson and S. Kano, *Introduction to Nonlinear Laser Spectroscopy*. Amsterdam: Academic Press, 1988.
[3.6] L. Allen and J.H. Eberly, *Optical Resonance and Two-Level Atoms*. New York: Dover, 1987.
[3.7] F.J. Schuda, *High resolution spectroscopy using a cw stabilized dye laser*, PhD Dissertation, University of Rochester, 1974.

PROBLEMS

3.1. Unpolarized radiation presents an electric field vector $\bar{E}$ whose orientation is random in time. If the transition dipole $\bar{\mu}$ driven by $\bar{E}$ has a fixed direction because the atom is situated in a crystal where the electric fields of nearby (stationary) atoms impose a fixed axis of quantization, all values of the angle θ between $\bar{E}$ and $\bar{\mu}$ are equally probable. On the other hand, for circularly-polarized radiation, the argument of the $\cos\theta(t)$factor in the scalar product $\bar{\mu}\cdot\bar{E}$ is $\theta(t) = \omega t$. Show that the first moment or average value of $\cos^2\theta$ in both cases is equal to 1/3 in agreement with the transition rate prefactors in Table 3.1.

3.2. Just like the Hamiltonian operator of Problem 2.4, the density matrix operator for atoms in particular states may be represented by simply giving the entries for the matrix explicitly. For example, the ground-state density matrix for a two- level atom is $\rho_g = \begin{bmatrix} 0 & 0 \\ 0 & 1 \end{bmatrix}$ and the excited state matrix is $\rho_e = \begin{bmatrix} 1 & 0 \\ 0 & 0 \end{bmatrix}$.

(a) Verify that these two representations are correct by calculating the occupation probabilities $\rho_{11} = \langle 1|\,\rho_g\,|1\rangle$ and $\rho_{22} = \langle 2|\,\rho_e\,|2\rangle$ for states 1 and 2. Thus diagonal entries in ρ give occupation probabilities or populations of specific levels.

(b) Off-diagonal entries in ρ do not give populations. Show that $\rho_{12} = \langle 1|\,\psi\rangle\langle\psi\,|2\rangle = 0$, if $|\psi\rangle$ is an eigenstate and $\rho_{12} = \langle 1|\,\psi\rangle\langle\psi\,|2\rangle \neq 0$ if $|\psi\rangle$ is a superposition state like $|\psi\rangle = c_1\,|1\rangle + c_2\,|2\rangle$.

(c) To illustrate the difference between "pure" and "mixed" case density matrices, now consider both a single two-level atom in a superposition state described by

$$\rho_{pure} = \begin{bmatrix} \frac{1}{2} & \frac{1}{2} \\ \frac{1}{2} & \frac{1}{2} \end{bmatrix},$$

and an ensemble of such atoms at infinite temperature. Half the ensemble atoms are in the ground state and half are in the excited state, giving the matrix

$$\rho_{mixed} = \begin{bmatrix} \frac{1}{2} & 0 \\ 0 & \frac{1}{2} \end{bmatrix}.$$

Show that ρ_{pure} can be transformed using the rotation matrix $\begin{bmatrix} \cos\theta & \sin\theta \\ -\sin\theta & \cos\theta \end{bmatrix}$ to give ρ_g or ρ_e, but that it cannot be rotated to give ρ_{mixed}. (This illustrates the fact that a single interaction with light—which causes rotations in the Hilbert space of the atom—cannot transform an entire system of independent atoms into the same state at the same time. Only operations on coherently prepared ensembles provide total system control.)

3.3. Consider a collection of N two-level atoms in which each atom is labeled by index j. The wavefunction of each atom j has a random phase factor ϕ_j, such that $|\psi_j\rangle = C_1|1\rangle + \exp(i\phi_j)C_2|2\rangle$. The probability of atom j having any particular phase scales as $P_j = 1/N$ when ϕ_j is uniformly distributed between 0 and π.

(a) Show that $\rho_{12} = \left\{ \sum_{j=1}^{N} P_j |\psi_j\rangle\langle\psi_j| \right\}_{12} = 0$ in the limit $N \to \infty$.

(b) Calculate the macroscopic polarization by inserting the atomic dipole operator into Eq. (3.6.17) to show that while a system may contain many oscillating dipoles, their relative phasing plays an important role in determining observable properties of an ensemble.

3.4. Consider a two-level atom, subjected to radiation near its resonant frequency ω_0, in which both levels undergo incoherent pumping and spontaneous decay.

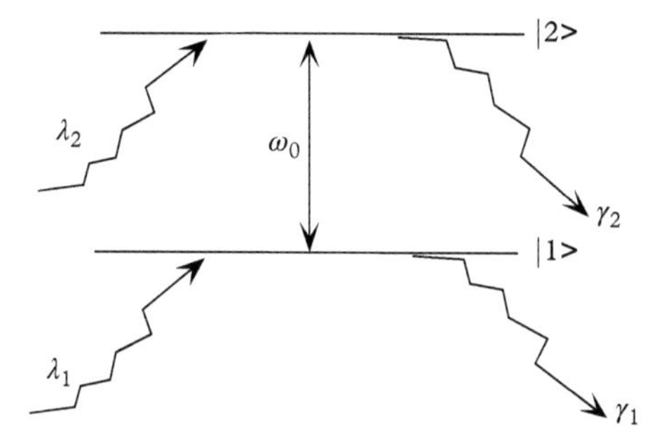

(a) Write down the equations of motion for all elements of the density matrix in the Schrödinger representation.

(b) Now suppose that state $|1>$ is the true ground state of the (closed) system and that state $|2>$ has contributions to its population due to thermal and optical excitation from $|1>$. Write down the equations of motion of ρ in this case ($\lambda_1 = 0$ and the magnitude of the thermal transition rate out of level 1 and into level 2 is $\lambda_2(\rho_{11} - \rho_{22})$.

(c) Use steady-state perturbation theory to solve the equations of part (b) for ρ_{11} and ρ_{22} to second order. What are the values of $\rho_{11} = \rho_{11}^{(0)} + \rho_{11}^{(1)} + \rho_{11}^{(2)}$ and $\rho_{22} = \rho_{22}^{(0)} + \rho_{22}^{(1)} + \rho_{22}^{(2)}$ at high intensities and do they make physical sense?

(d) Using the results of part (c), sketch absorption (proportional to the population difference) versus frequency at low intensity on an appropriate frequency scale, assuming collision broadening is negligible.

3.5. For a system with Hamiltonian $H = H_0 + V$, where V is a perturbation,

(a) show the equations of motion of the density matrix in the Schrödinger, interaction, and Heisenberg pictures are

$$i\hbar\dot{\rho}^S = \left[H_0 + V, \rho^S\right], i\hbar\dot{\rho}^I = \left[V^I, \rho^I\right], \quad \text{and} \quad i\hbar\dot{\rho}^H = 0.$$

(b) What can be said in general about eigenvalues and eigenfunctions of V if it is Hermitian?

(c) Not all interactions or perturbations are Hermitian. Write down the Schrödinger equation in matrix form for a four-level system subject to a Hermitian Hamiltonian H_0 and a perturbative, non-Hermitian interaction V which couples *only levels 3 and 4*. Find exact energy eigenvalues of the system in terms of matrix elements of H_0 and V using a diagonalization procedure, and compare their properties to eigenvalues of Hermitian operators.

3.6. Provide a proof that the largest magnitude to be expected for the off-diagonal density matrix element between two states is one-half.

3.7. Write down and solve (only) the equations of motion for *coherences* $\dot{\rho}_{ij}$ of a closed four-level system ($i, j = 1, 2, 3$, or 4) under steady-state conditions. The system is excited by a light field $E(t) = \frac{1}{2}E_0 \exp(i\omega t)$ + *c.c.* that couples only excited level $|2>$ to level $|4>$ via the dipole moment μ_{24} on the transition at frequency ω_0 (see figure). Assume that non-zero transition probabilities exist *between all levels* (no selection rules) and show explicitly from the equations of motion that all coherences other than that connecting levels 2 and 4 are negligible.

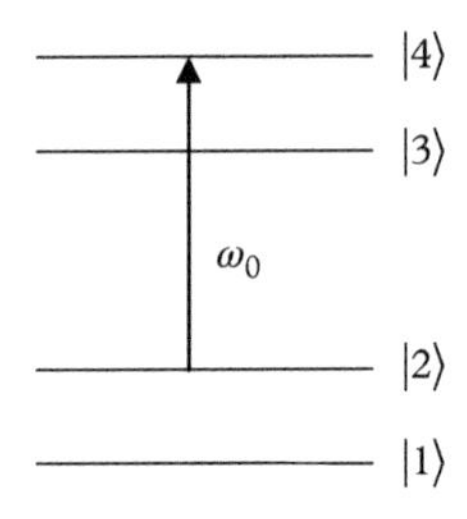

3.8. Prove that

(a) in a closed two-level system, no 2×2 matrix Γ exists which correctly describes relaxation of the density matrix in the presence of dephasing and natural decay.

(b) Determine the non-zero elements of the four-index quantity $\Gamma_{\alpha\alpha'\beta\beta'}$ which does have sufficient dimensionality to describe these relaxation processes correctly:

$$i\hbar\dot{\rho}_{\alpha\alpha'} = [H, \rho]_{\alpha\alpha'} - i\hbar \sum_{\beta,\beta'} \Gamma_{\alpha\alpha'\beta\beta'}\rho_{\beta\beta'}.$$

The indices α, α', β, and β' can take on values 1 or 2, corresponding to the two levels of the system.

3.9. Calculate general expressions for

(a) time-dependent elements of the density matrix ρ_{ij} of a two-level system in first-, second-, and third-order perturbation theory from the equation of motion

$$i\hbar\frac{d}{dt}\rho = [H_0, \rho] + [V, \rho] + i\hbar\left[\frac{d}{dt}\rho\right]_{relaxation}.$$

Assume V represents three weak applied light fields. In zeroth order no light field is present, but V is distinct and non-zero for each successive order of perturbation. That is, $V^{(0)} = 0$, but $V^{(1)}, V^{(2)}, V^{(3)} \neq 0$.

(b) Solve these equations in the steady-state limit, assuming the system is initially in the ground state. Show that when a single field ($V = V^{(1)}$) acts on a two-level system with one allowed transition ($\mu_{12} \neq 0$), new contributions to off-diagonal matrix elements ρ_{12} are generated only in odd orders of the perturbation sequence and diagonal matrix elements ρ_{11} and ρ_{22} (populations) change only in even orders. The "perturbation chain" may therefore be pictured as:

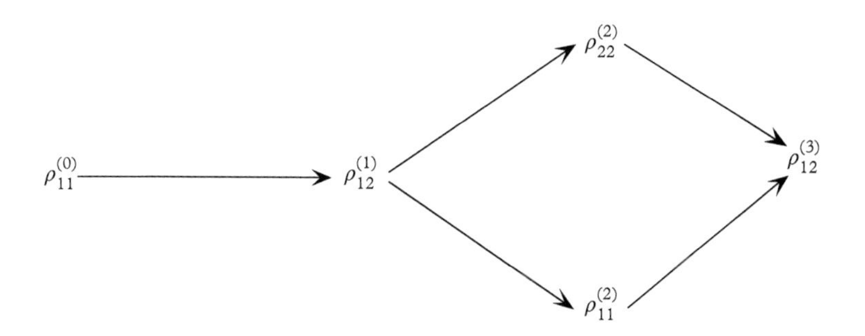

To do this, find $\rho_{12}^{(3)}$ and show that new contributions to coherence are generated in odd orders and population changes in even orders of perturbation. Note: This calculation provides valuable insight by showing that the response to a single light field proceeds first by the establishment of coherence, then a population correction which in turn modifies the coherence in third order, and so on. It tells us, for example, that to describe saturation effects (intensity-dependent absorption) we must include second order effects. To describe intensity-dependent refractive index changes, we must go to third order since this involves a second-order effect acting on a third wave.

3.10. (a) Solve the damped equations of motion for an "open" two-level atom system in the interaction picture using the Rabi method and the following equations for the coefficients in the wavefunction expansion:

$$\dot{C}_2 = -\frac{1}{2}\gamma_2 C_2 + \frac{1}{2}i\Omega e^{i(\omega_0-\omega)t}C_1,$$

$$\dot{C}_1 = -\frac{1}{2}\gamma_1 C_1 + \frac{1}{2}i\Omega^* e^{-i(\omega_0-\omega)t}C_2.$$

γ_2

γ_1

(b) Find $|C_1(t)|^2$ and $|C_2(t)|^2$ for initial conditions $C_2(0) = 1$ and $C_1(0) = 0$ and comment on their sum (Do you find what you expect for an "open" system?).

3.11. Show that if we consider a "closed" two-level system described by

$$\dot{C}_2 = -\frac{1}{2}\gamma_2 C_2 + \frac{1}{2}i\Omega e^{i(\omega_0-\omega)t}C_1,$$

$$\dot{C}_1 = +\frac{1}{2}\gamma_2 C_2 + \frac{1}{2}i\Omega^* e^{-i(\omega_0-\omega)t}C_2,$$

that the solution for $C_1(t)$ and $C_2(t)$ at arbitrary intensity, including the low intensity limit ($\Omega_R \to 0$), does not preserve trace. That is, despite the fact that the initial rate of decay of state 2 equals the rate of growth of 1 we do not obtain $|C_1(t)|^2 + |C_2(t)|^2 = 1$. (This failure of the rate equations is important motivation to retain bilinear coefficients like $C_2C_1^*$ and $C_1C_2^*$ in a density matrix description which does preserve trace in a consistent manner.)

3.12. To determine the spread expected in repeated measurements of the dipole moment, $e\bar{r} = \mu\begin{bmatrix} 0 & 1 \\ 1 & 0 \end{bmatrix}$, one can calculate the root-mean-square fluctuation given by

$\sigma \equiv \sqrt{\langle (e\bar{r})^2 \rangle - \langle e\bar{r} \rangle^2}$. Assuming μ is real, determine σ in terms of μ and elements of the density matrix for a two-level atom. (Hint: The angled brackets in the expression for σ indicate expectation values or means, calculated as traces of the operator with the density matrix.)

3.13. The time development of the coefficients of the wavefunction in a two-level system is given by the coupled equations

$$\dot{C}_2 = \frac{i}{2}\frac{\bar{\mu}_{21} \cdot \bar{E}_0}{\hbar}\{\exp[i(\omega_0 + \omega)t] + \exp[i(\omega_0 - \omega)t]\}C_1,$$

$$\dot{C}_1 = \frac{i}{2}\frac{\bar{\mu}_{12} \cdot \bar{E}_0}{\hbar}\{\exp[-i(\omega_0 - \omega)t] + \exp[-i(\omega_0 + \omega)t]\}C_2.$$

(a) Suppose the detuning $\Delta \equiv \omega_0 - \omega$ is much larger than any relaxation rates of $C_1(t)$or $C_2(t)$ so that both equations can be integrated directly (ignoring the temporal variation of C_1 and C_2 on the right-hand side). By substituting each result back into the right side of the second equation four terms are obtained. Retain only the largest time-invariant term, and represent the slow relaxation of the coefficients with the forms

$$C_1(t) = C_1(0)\exp(-\gamma_1 t); C_2(t) = C_2(0)\exp(-\gamma_2 t)$$

where γ_1 and γ_2 are complex. Then show that there are shifts of the energy levels given by

$$\gamma_1 = -\gamma_2 = -i\left(\frac{\Omega}{2}\right)^2 \frac{1}{\Delta}, \quad \Omega \equiv \frac{\bar{\mu}_{21} \cdot \bar{E}_0}{\hbar}.$$

(b) Based on part (a) and the origin of frequency shifts discussed in Appendix F, draw a diagram indicating how levels 1 and 2 are displaced (up or down) by the "light shifts" for positive and negative detuning.

3.14. Using the Bloch vector model,

(a) Calculate the expectation value of the electric dipole moment (i.e.,) in the state of a two-level system reached immediately following a resonant pulse of duration $\Delta t = \frac{\pi}{2\Omega}$.

(b) Does the sample radiate light?

(c) If the pulse length is simply doubled to $\Delta t = \frac{\pi}{\Omega}$, does the sample radiate all the more?

3.15. The probability of optical excitation to an excited energy level (i.e., Eq. (3.5.4)) contains a cosine squared factor in $\left|\langle \bar{\mu} \cdot \bar{E}(t) \rangle\right|^2$ whose average value accounts for the projection of the electric field $\bar{E}$ on the transition dipole moment $\bar{\mu}$. When high symmetry is present, $\bar{\mu}$ may point in a fixed direction specified by polar angles (θ, ϕ) defined with respect to the crystal z-axis. (a) Calculate the average of the angle cosine between $\bar{E}$ and $\bar{\mu}$ for circularly polarized light propagating along the y-axis. Does your answer agree with the entry in Table 3.1 when $\theta = \phi = 0$? (b) Why does the excitation probability go to zero when $\theta = \phi = \pi/2$?

3.16. Consider an atomic transition with a line-center frequency of v_0. Show that the ratio of the blackbody stimulated emission rate $W = BU$ to the spontaneous decay rate A is equal to the average number of photons $\langle n \rangle$ in the blackbody mode at frequency v_0. (Hint: The average number of photons in a mode equals the average energy $\langle E \rangle$ in the mode divided by the photon energy (hv_0).

4

Transient Optical Response

Although the principal focus of subsequent chapters will be detailed calculations of steady-state response to light under various conditions, it is helpful to have a mental picture of how polarization grows, decays, and mediates transient interactions using the Bloch equations of Chapter 3. Among other things, this helps one develop a sense for what determines when dynamics should be considered "fast" or "slow." A key part of the successful analysis of dynamics in new systems lies in categorizing processes as "fast" or "slow" and deciding which quantities should be retained as time-dependent variables once the timescale of the analysis is specified.

This chapter also provides physical insight into the meaning of "dephasing" or "decoherence," and provides an easy way to introduce some surprisingly interesting coherent phenomena induced by repeated applications of pulsed applied fields. Interesting connections also exist between irreversible transient phenomena and frequency shifts, but these are deferred to a discussion in Appendix F. Finally, some limitations associated with the use of the simple Bloch vector model are noted at the end of the chapter, as a prelude to focusing on the full term-by-term density matrix treatment used in Chapters 5, 6, and 7, which provides enough degrees of freedom for analytic solutions to all problems of interest.

4.1 Optical Nutation

4.1.1 Optical Nutation without Damping

Consider what happens when atoms are suddenly exposed to light that is resonant with a ground state transition. This gives rise to a coherent transient phenomenon called nutation, which is the driven response of atoms to light illustrated in Figure 4.1. Nutation describes the build-up of polarization caused by the application of an optical field.

Simple analytic solutions can be obtained, provided damping is ignored, even for atoms moving with a velocity v_z that modifies their detuning by the Doppler shift.

Lectures on Light. Second Edition. Stephen C. Rand.
© Stephen C. Rand 2016. Published in 2016 by Oxford University Press.

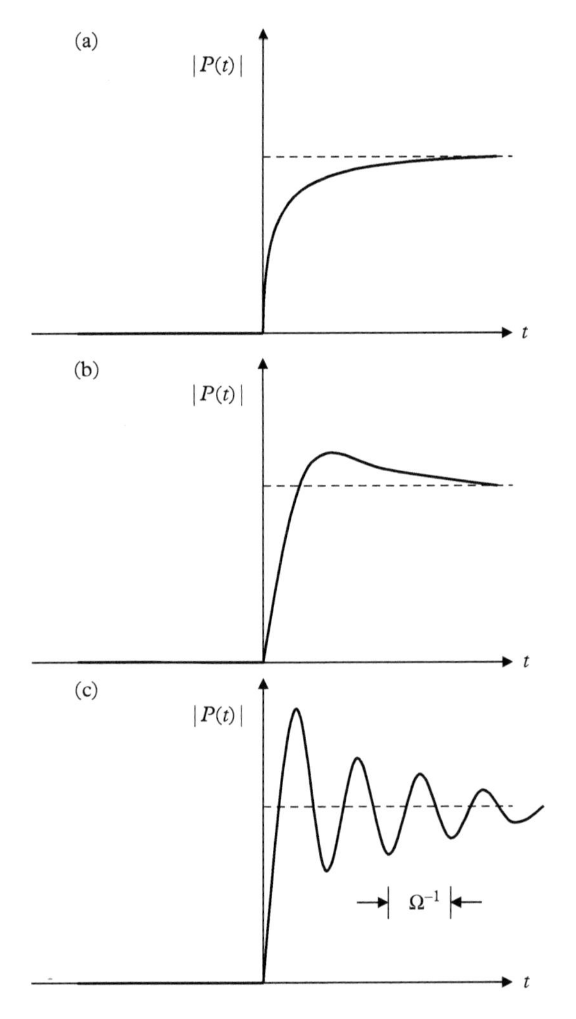

Figure 4.1 *Artist's concept of the onset of polarization due to nutation in (a) over-damped, (b) critically damped, and (c) under-damped conditions.*

The Bloch equations (Eqs. (3.8.13)–(3.8.15)) describing evolution of the Bloch vector reduce to

$$\dot{R}_1 + \Delta R_2 = 0, \tag{4.1.1}$$

$$\dot{R}_2 - \Delta R_1 - \Omega R_3 = 0, \tag{4.1.2}$$

$$\dot{R}_3 + \Omega R_2 = 0. \tag{4.1.3}$$

For a system whose initial population distribution is described by $R_3(0)$, these equations are readily solved for $t > 0$ and give

$$R_1(t) = \frac{\Omega\Delta}{\Omega_R^2} R_3(0)[\cos \Omega_R t - 1], \tag{4.1.4}$$

$$R_2(t) = \frac{\Omega}{\Omega_R} R_3(0) \sin \Omega_R t, \tag{4.1.5}$$

$$R_3(t) = R_3(0)\left[1 + \frac{\Omega^2}{\Omega_R^2}(\cos \Omega_R t - 1)\right], \tag{4.1.6}$$

where $\Omega_R^2 \equiv \Delta^2 + \Omega^2$, $\Delta \equiv \omega_0 - \omega - k\text{v}_z$, and $\Omega \equiv \mu_{21}E_{12}/\hbar$.

Exercise: Show that Eqs. (4.1.4)–(4.1.6) are reproduced by applying the evolution matrix in Eq. (3.8.22) to initial Bloch vector $\bar{R}(0) = (0, 0, R_3(0))$.

To calculate the signal field we need to evaluate the Doppler-averaged matrix element in Eq. (3.9.10). This yields

$$\begin{aligned}
<\tilde{\rho}_{21}> &= \frac{1}{k\text{v}\sqrt{\pi}} \int_{-\infty}^{\infty} \frac{1}{2}(R_1 - iR_2)\exp(-[\Delta/k\text{v}]^2)d\Delta \\
&= \frac{\Omega R_3(0)}{2k\text{v}\sqrt{\pi}} \int_{-\infty}^{\infty} \left\{\frac{\Delta}{\Omega_R^2}[\cos \Omega_R t - 1] - \frac{i}{\Omega_R}\sin \Omega_R t\right\}\exp(-[\Delta/k\text{v}]^2)d\Delta \\
&= \frac{\sqrt{\pi}}{2ik\text{v}} \Omega R_3(0)\exp(-[\Delta_0/k\text{v}]^2)J_0(\Omega t),
\end{aligned} \tag{4.1.7}$$

where we have assumed the excitation is tuned near (but not at) the Doppler peak at $\Delta = \Delta_0$ with the consequence that the R_1 contribution is approximately zero. According to Eqs. (3.9.12) and (3.9.16), the signal field is

$$E_s(t) \approx \frac{N\Omega L R_3(0)\sqrt{\pi}}{8\varepsilon\text{v}_0} \mu_{12}\exp(-[\Delta_0/k\text{v}]^2)J_0(\Omega t)\, e^{-i(\omega t - kz)} + c.c. \tag{4.1.8}$$

The signal at the detector will exhibit a slow oscillation described by the zero-order Bessel function J_0 at frequency Ω in Eq. (4.1.8), as illustrated in Figure 4.1(c).

Notice that the amplitude of the Bloch vector given by Eqs. (4.1.4)–(4.1.6) is constant:

$$|\bar{R}(t)| = \left(R_1^2(t) + R_2^2(t) + R_3^2(t)\right)^{1/2} = R_3(0). \tag{4.1.9}$$

Also, it precesses about the effective field vector $\bar{\beta}$ at the frequency

$$\Omega_R = \left(\Delta^2 + \Omega^2\right)^{1/2}. \tag{4.1.10}$$

The motion of the Bloch vector is particularly easy to visualize for the case of exact resonance ($\Delta = 0$). The tip of the Bloch vector sweeps around a circle in the $\hat{e}_2, \hat{e}_3$ plane perpendicular to the effective field $\bar{\beta} = \Omega\hat{e}_1$. Hence, at one instant of time it points "up" along $+\hat{e}_3$ and a half period later it points "down" along $-\hat{e}_3$. This corresponds to "Rabi flopping" behavior in which a collection of atoms alternately occupies the excited state or the ground state under the influence of a resonant driving field.

The frequency of driven population oscillations is the Rabi frequency, and experimentally this may be verified by increasing or decreasing the input field, as illustrated in Figure 4.2(a). When this is done, the Rabi oscillation frequency tracks the electric field linearly. On resonance, observations of the temporal oscillations (by any means that samples the excited versus ground state populations) provide a convenient way of measuring the transition dipole moment if the effective intensity of the wave is known [4.1].

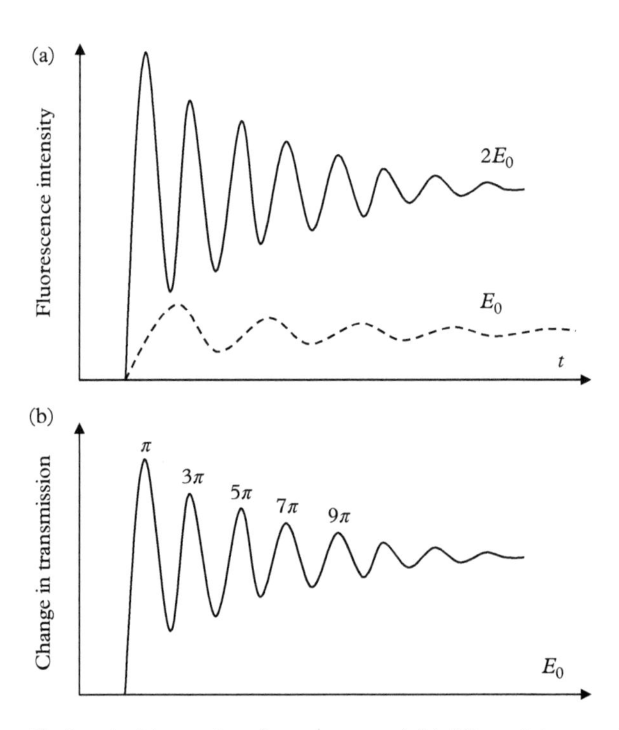

Figure 4.2 *Rabi oscillations in (a) transient fluorescence and (b) differential transmission experiments. The oscillation frequency in time-resolved experiments depends on the incident field strength. The oscillations versus power depend on the pulse area.*

With pulsed excitation, a precise number of Rabi oscillations may be induced by controlling the pulse area as illustrated in Figure 4.2(b). Measurement of the differential transmission (described in more detail in Chapter 5) versus input power can also serve to determine the resonant Rabi frequency. This method is commonly applied in the characterization of semiconductor quantum wells and quantum dots [4.2, 4.3].

4.1.2 Optical Nutation with Damping

Throughout the algebraic treatment of nutation given in Section 4.1.1, population and polarization decay (T_1 and T_2 processes, respectively) were ignored. For interactions with light that are longer than the characteristic decay times, this omission is obviously unacceptable. However, the Bloch equations do not yield to analysis when the decay terms are included, other than for a special case in which $T_1 = T_2$ [4.4]. This is due to the requirement that the Bloch model be based on the form of a simple gyroscopic equation of motion. Hence we shall await the density matrix methods of Chapters 5–7 which offer more degrees of freedom, and will not extend nutation analysis further here. Despite the limitations of the Bloch model, it works well for picturing the outcome of multiple-pulse interactions in which the pulse durations are much shorter than the characteristic times T_1 and T_2. In the next section we shall also find that decay processes that take place in the free precession periods between ultrafast pulses can be incorporated into this simplified analysis.

4.2 Free Induction Decay

Suppose a sample is resonantly excited by a laser beam until steady-state conditions are reached and then, at time $t = 0$, the excitation is suddenly switched off. After the switch, the polarization established in the sample prior to $t = 0$ continues to oscillate until contributions from different atoms get out of phase and the amplitude of the polarization drops to undetectable levels. During the time interval over which most of the atoms remain in phase with the original excitation (designated as the coherence time T_2 for historical reasons), an intense coherent beam is radiated in the forward direction. This is because the original excitation wave forced the array of atoms to oscillate in a fashion consistent with a forward-propagating coherent beam. After the atoms lose coherence, they are said to have undergone "dephasing." The time dependence of this process is illustrated in Figure 4.3.

Let us assume that the preparative stage ends by switching the transition frequency out of resonance with the laser frequency, in the manner of some of the earliest optical coherent transient experiments [4.5], and calculate the signal predicted by the Bloch equations. We can account for initial Boltzmann distributions of population in the absence of light by adding thermal source terms ρ_{11}^0 and ρ_{22}^0 to the Bloch equations of the last chapter:

$$\dot{\rho}_{22} - \dot{\rho}_{11} = -(\rho_{22} - \rho_{11})/T_1 - i\Omega\left(\tilde{\rho}_{21} - \tilde{\rho}_{12}\right) + \left(\rho_{22}^0 - \rho_{11}^0\right)/T_1. \tag{4.2.1}$$

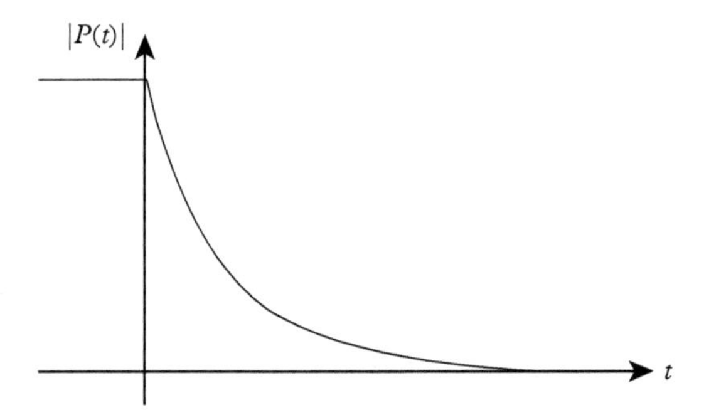

Figure 4.3 *Decay of the amplitude of optical polarization after the excitation is switched off at time t = 0.*

In this way we obtain modified Bloch equations from Eqs. (3.8.13)–(3.8.15):

$$\dot{R}_1 = -\Delta R_2 - R_1/T_2, \tag{4.2.2}$$

$$\dot{R}_2 = \Delta R_1 - R_2/T_2 + \Omega R_3, \tag{4.2.3}$$

$$\dot{R}_3 = -\Omega R_2 - \left(R_3 - R_3^0\right)/T_1, \tag{4.2.4}$$

where $R_3^0 \equiv \rho_{22}^0 - \rho_{11}^0$. Since the atoms are prepared by steady-state irradiation prior to the free precession period, we require steady-state solutions of the nutation equations to determine the initial conditions for the free precession which begins at $t = 0$. These are found by setting time derivatives in Eqs. (4.2.2)–(4.2.4) equal to zero:

$$R_1(0) = -\Delta\Omega R_3^0/\left(\Omega^2 T_1/T_2 + \Delta^2 + 1/T_2^2\right), \tag{4.2.5}$$

$$R_2(0) = \left(\Omega R_3^0\right)/\left(\Omega^2 T_1/T_2 + \Delta^2 + 1/T_2^2\right), \tag{4.2.6}$$

$$R_3(0) = R_3^0\left[\left(\Delta^2 + 1/T_2^2\right)/\left(\Omega^2 T_1/T_2 + \Delta^2 + 1/T_2^2\right)\right]. \tag{4.2.7}$$

At $t = 0$, the transition frequency is shifted instantaneously from ω by a small amount $\delta\omega_{21}$ to $\omega' \equiv \omega - \delta\omega_{21}$. Hence the detuning changes to the new value,

$$\Delta' = \Delta + \delta\omega, \tag{4.2.8}$$

and the field is suddenly far out of resonance with the original set of atoms. This is equivalent to setting the field amplitude to zero. That is, $\Omega = 0$ for $t \geq 0$. The Bloch equations therefore assume the form

$$\dot{R}_1 + \Delta R_2 + R_1/T_2 = 0, \tag{4.2.9}$$

$$\dot{R}_2 - \Delta R_1 + R_2/T_2 = 0, \tag{4.2.10}$$

$$\dot{R}_3 + \left(R_3 - R_3^0\right)/T_1 = 0. \tag{4.2.11}$$

The solution of Eq. (4.2.11) may be guessed immediately:

$$R_3(t) = R_3^0 + \left[R_3(0) - R_3^0\right]\exp(-t/T_1). \tag{4.2.12}$$

Solutions of Eqs. (4.2.9) and (4.2.10) may be readily obtained by Laplace transform techniques:

$$R_1(t) = [R_1(0)\cos\Delta t - R_2(0)\sin\Delta t]\exp(-t/T_2), \tag{4.2.13}$$

$$R_2(t) = [R_1(0)\sin\Delta t + R_2(0)\cos\Delta t]\exp(-t/T_2). \tag{4.2.14}$$

Equations (4.2.12)–(4.2.14) determine the motion of the Bloch vector $\bar{R}(t) = (R_1, R_2, R_3)$ in the absence of light, including "longitudinal" (population) and "transverse" (polarization) decay processes.

Exercise: Show that the evolution matrix for free precession that reproduces Eqs. (4.2.12)–(4.2.14) is

$$U_{FID}(t) = \begin{bmatrix} \exp(-t/T_2)\cos\Delta t & -\exp(-t/T_2)\sin\Delta t & 0 \\ \exp(-t/T_2)\sin\Delta t & \exp(-t/T_2)\cos\Delta t & 0 \\ 0 & 0 & \exp(-t/T_1) \end{bmatrix}, \tag{4.2.15}$$

and that it is identical (apart from the decay factors) to that derivable from Eq. (3.8.22).

To find the sample polarization and signal field we next use Eqs. (3.9.11) and (3.9.16), respectively:

$$\begin{aligned} <\tilde{\rho}_{21}> &= \frac{\sqrt{\pi}}{2ikv}\Omega R_3^0 \exp\left[-(\Delta_0/kv)^2\right]\left(\frac{1}{\sqrt{1+\Omega^2 T_1 T_2}} - 1\right) \\ &\quad \times \exp[-i\delta\omega t]\exp\left[-\left(1+\sqrt{1+\Omega^2 T_1 T_2}\right)t/T_2\right]. \end{aligned} \tag{4.2.16}$$

We have assumed that the excitation is centered at a detuning of $\Delta = \Delta_0$ near the Doppler peak and that the bandwidth of the excitation is narrow compared to the Doppler width $\sim kv$ so that the Gaussian factor may be removed from the integral. The amplitude of the signal field generated by this induced polarization is

$$E_{s0} = (ik/2\varepsilon)NL\mu_{12}<\tilde{\rho}_{21}>, \tag{4.2.17}$$

provided the sample is "optically thin." For a sample to be optically thin, the length L traversed by the incident beam must be less than the inverse absorption length ($\alpha L < 1$). The Doppler integral in Eq. (3.9.10) can also be evaluated exactly [4.6].

In the forward direction, the total field including the switched laser field is

$$E_T = \frac{1}{2}(E_{s0}\exp(-i[\omega t - kz]) + c.c.) + \frac{1}{2}E_0\exp(-i[\omega' t - kz])) + c.c. \tag{4.2.18}$$

The intensity, therefore, contains a cross term or beat signal contribution

$$|E_T|^2_{beat} = \frac{1}{4}E_0^* E_{s0}\exp(i\delta\omega t) + c.c., \tag{4.2.19}$$

where the signal field amplitude is

$$E_{s0} = \frac{\sqrt{\pi}NL\mu_{12}^2 E_0 R_3^{(0)}}{4\varepsilon v\hbar}\left(\frac{1}{\sqrt{1+\Omega^2 T_1 T_2}} - 1\right)\exp[-(\Delta_0/kv)^2]$$
$$\times \exp\left[-\left(1+\sqrt{1+\Omega^2 T_1 T_2}\right)t/T_2\right]. \tag{4.2.20}$$

The measured heterodyne signal intensity at the beat frequency is given by

$$I_s(t) \propto \mathrm{Re}\{|E_T|^2_{beat}\}, \tag{4.2.21}$$

and decays exponentially in time with an effective decay constant

$$\tau = T_2 \Big/ \left(1+\sqrt{1+\Omega^2 T_1 T_2}\right). \tag{4.2.22}$$

Exercise: At what intensity does power broadening result in effective decay constants of (i) $\tau = T_2/2$ and (ii) $\tau = T_2/4$?

Apparatus suitable for observing coherent transients by fast frequency switching of lasers is illustrated in Figure 4.4. In the absence of power broadening, observations of free induction decay (FID) behavior furnish the homogeneous decay time T_2 and

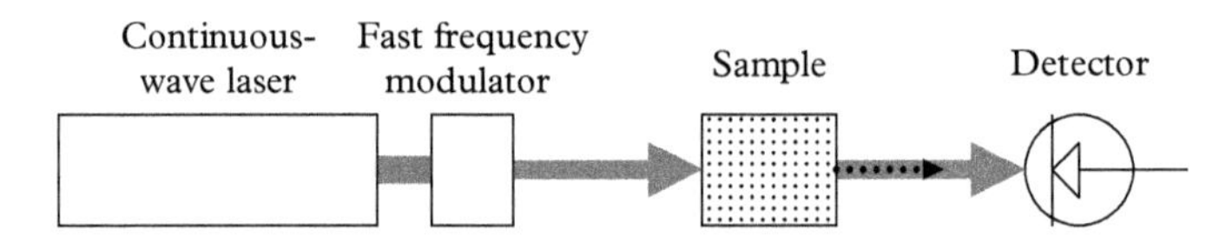

Figure 4.4 *A simple apparatus for coherent transient experimentation. The transmitted pump beam at the switched frequency (solid arrow) and the polarization established at the initial frequency prior to switching the frequency (dashed arrow) produce a heterodyne beat at the detector. The demodulated signal intensity is proportional to polarization amplitude.*

yield a measurement of the natural lifetime of the upper state in dielectric materials. In semiconductors, as a counter-example, the decay rate for recombination of electrons and holes is affected by the availability of final valence states, and consequently is not purely a measure of an excited state property. Also, at intensities approaching saturation it is not precisely the homogeneous linewidth that is measured, but a power-broadened value. In the time domain, decay time measurements then determine the effective decay constant τ given by Eq. (4.2.22). Since Ω^2 is proportional to optical intensity, the pure dephasing time T_2 must in general be inferred from extrapolations of decay time measurements to zero intensity. If the preparatory pulse period is shorter than T_2, a distribution of atoms with Doppler shifts covering a bandwidth in excess of T_2^{-1} will also be excited. Consequently the use of ultrashort preparation pulses results in apparent decay rates as high as $(T_2^*)^{-1}$, where $(T_2^*)^{-1}$ reflects the inhomogeneous rather than the homogeneous width of the transition.

A time-resolved FID signal observed by frequency switching [4.7] is illustrated in Figure 4.5. The dephasing time T_2 is obtained by fitting the exponential decay of the envelope of polarization signal oscillations. The time it takes the signal intensity to fall to $1/e$ of its initial value yields a measurement of $T_2/2$ in heterodyne-detected experiments of this type, when the preparation time exceeds T_2. To avoid inaccuracies due to power broadening, the measurements must be extrapolated to zero power ($\Omega = 0$), according to Eq. (4.2.22).

When frequency switching is used to observe coherent transients, it sometimes happens that the frequency shift is insufficient to tune the laser completely off resonance during the free precession period. In this case, the laser shifts into resonance with a new

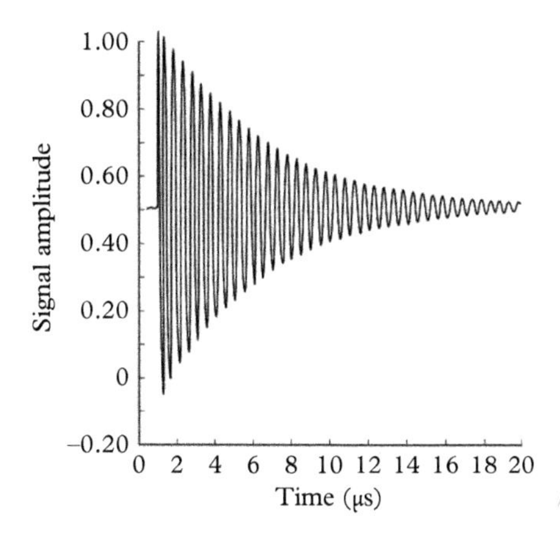

Figure 4.5 *An illustration of polarization decay observed in a heterodyne-detected FID experiment in* Pr^{3+}:LaF_3. *The oscillations in the figure occur at the difference frequency* $\nu_2 - \nu_1$ *between the optical polarization established by the excitation laser* (ν_2) *prior to the frequency switch and the heterodyne reference frequency* (ν_1). *(After Ref. [4.7].)*

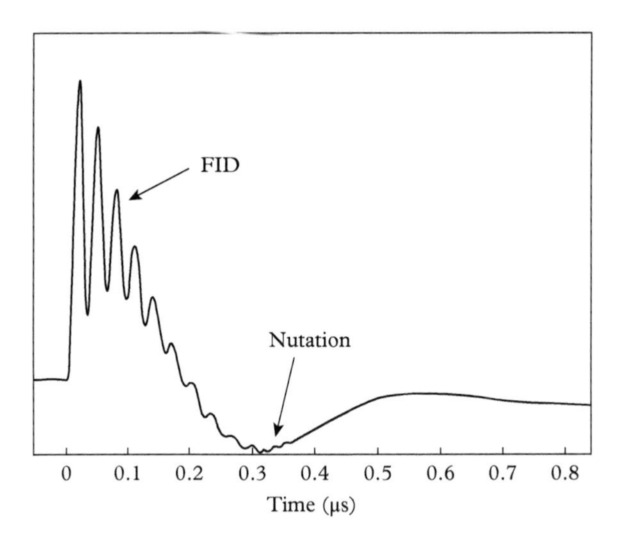

Figure 4.6 *FID and optical nutation signals observed simultaneously in iodine vapor by frequency switching. The FID oscillation frequency depends on the local oscillator frequency whereas the nutation oscillates at the resonant Rabi frequency which is intensity-dependent. (After Ref. [4.8].)*

packet of atoms that undergo nutation. Consequently FID of one group and nutation of another group of atoms may be observed simultaneously [4.8]. An example of this is shown in Figure 4.6.

4.3 Photon Echoes

Systems of atoms and molecules can be pulse-excited in at least three different ways. First, the excitation field may consist of output from a pulsed laser as in the original experiments [4.9], or it could be continuous laser output that is simply switched on and off. Alternatively, the energy levels of the system may be shifted in and out of resonance with a fixed frequency field (e.g., by applying Stark switching pulses to the sample). Finally, the frequency of the excitation field can be switched in and out of resonance while the optical field amplitude and the sample resonant frequencies are held constant. The net effect is similar in all cases.

In Figure 4.7, the optical intensity is assumed to be controlled with a fast modulator such as that shown in Figure 4.4, or by some other means, so that the optical field is constant during the desired pulse periods and zero otherwise. For the two-pulse sequence shown in the figure, let us proceed to calculate Bloch vector dynamics during the various time periods when conditions are fixed. For purposes of illustration, this will be done using three different approaches: algebraic analysis, rotation matrix analysis, and a more formal density matrix operator method.

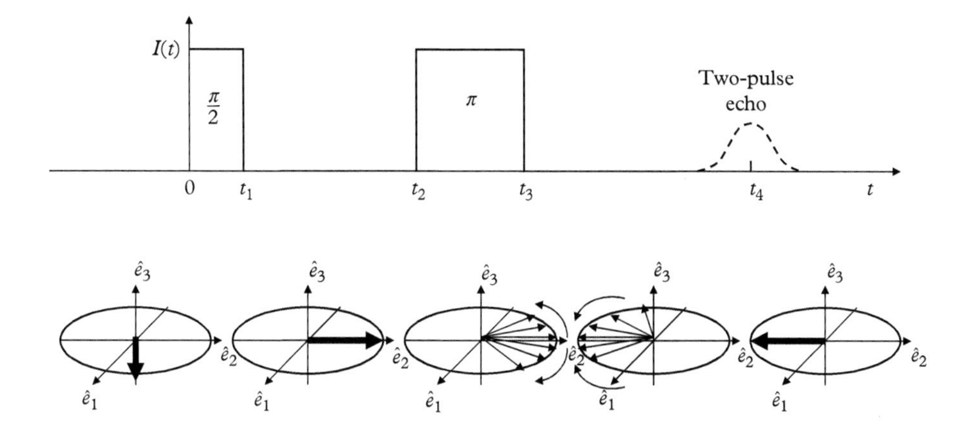

Figure 4.7 *Schematic diagram indicating how a photon echo polarization is formed. The echo results from rephasing of microscopic dipoles after a two-pulse sequence* (π/2 – π)*, which reverses dephasing through a* 180° *rotation in spin space.*

4.3.1 Algebraic Echo Analysis

4.3.1.1 $0 < t < t_1$

For sufficiently brief pulses ($t_1 \ll T_1, T_2$), no relaxation occurs during the applied pulse. As a consequence, relaxation terms in the optical Bloch equations may be dropped to obtain simplified equations during the first time period:

$$\dot{R}_1 + \Delta R_2 = 0, \tag{4.3.1}$$

$$\dot{R}_2 - \Delta R_1 - \Omega R_3 = 0, \tag{4.3.2}$$

$$\dot{R}_3 + \Omega R_2 = 0. \tag{4.3.3}$$

The solutions subject to the initial condition $R(0) = (0, 0, R_3(0))$ are:

$$R_1(t_1) = \frac{\Delta \Omega R_3(0)}{\Omega_R^2}(\cos \Omega_R t_1 - 1), \tag{4.3.4}$$

$$R_2(t_1) = \frac{\Omega R_3(0)}{\Omega_R} \sin \Omega_R t_1, \tag{4.3.5}$$

$$R_3(t_1) = R_3(0) + \frac{\Omega^2 R_3(0)}{\Omega_R^2}(\cos \Omega_R t_1 - 1). \tag{4.3.6}$$

Equations (4.3.4)–(4.3.6) are identical to the earlier results for nutation without damping.

4.3.1.2 $t_1 < t < t_2$

Between pulses the field is off (Figure 4.7). During this interval we set $\Omega = 0$ to solve the Bloch equations, simplified here by arbitrarily setting $R_3^0 = 0$. Note that if a different experimental approach is used, for example, if the sample were switched sufficiently far out of resonance that the optical interaction stops during this period, one can still account for this using the same substitution $\Omega = 0$. The optical Bloch equations then reduce to

$$\dot{R}_1 + \Delta R_2 + R_1/T_2 = 0, \tag{4.3.7}$$

$$\dot{R}_2 - \Delta R_1 + R_2/T_2 = 0, \tag{4.3.8}$$

$$\dot{R}_3 + R_3/T_1 = 0. \tag{4.3.9}$$

The solutions to these equations are of course the FID solutions given by Eqs. (4.2.12)–(4.2.14) with the time origin shifted to t_1:

$$R_1(t_2) = [R_1(t_1)\cos\Delta(t_2 - t_1) - R_2(t_1)\sin\Delta(t_2 - t_1)]\exp(-[t_2 - t_1]/T_2), \tag{4.3.10}$$

$$R_2(t_2) = [R_1(t_1)\sin\Delta(t_2 - t_1) + R_2(t_1)\cos\Delta(t_2 - t_1)]\exp(-[t_2 - t_1]/T_2), \tag{4.3.11}$$

$$R_3(t_2) = R_3(t_1)\exp(-[t_2 - t_1]/T_1). \tag{4.3.12}$$

The appearance of t_1 as the time origin in these expressions is just the consequence of the new initial condition:

$$R(t_1) = (R_1(t_1), R_2(t_1), R_3(t_1)). \tag{4.3.13}$$

During the second pulse the solutions are again given by the solutions for nutation without damping. However, the initial condition is

$$R(t_2) = (R_1(t_2), R_2(t_2), R_3(t_2)). \tag{4.3.14}$$

Therefore at $t = t_3$ we find

$$R_1(t_3) = R_1(t_2) - \frac{\Delta R_2(t_2)}{\Omega_R}\sin\Omega_R(t_3 - t_2) + \frac{\Delta^2 R_1(t_2)}{\Omega_R^2}[\cos\Omega_R(t_3 - t_2) - 1] + \frac{\Delta\Omega R_3(t_2)}{\Omega_R^2}[\cos\Omega_R(t_3 - t_2) - 1], \tag{4.3.15}$$

$$R_2(t_3) = R_2(t_2)\cos\Omega_R(t_3 - t_2) + \frac{1}{\Omega_R}[\Delta R_1(t_2) + \Omega R_3(t_2)]\sin\Omega_R(t_3 - t_2), \tag{4.3.16}$$

$$R_3(t_3) = R_3(t_2) - \frac{\Omega}{\Omega_R}R_2(t_2)\sin\Omega_R(t_3 - t_2) + \frac{\Omega}{\Omega_R^2}[\Delta R_1(t_2) + \Omega R_3(t_2)][\cos\Omega_R(t_3 - t_2) - 1]. \tag{4.3.17}$$

It is convenient to define a quantity at this point called the pulse "area," related to the time integral of the effective light field.

$$\theta(z) = \int_{-\infty}^{\infty} \Omega_R(z,t)dt, \quad \Delta \neq 0,$$

$$= \int_{-\infty}^{\infty} \frac{\mu_{21}E_{12}(z,t)}{\hbar} dt, \quad \Delta = 0. \tag{4.3.18}$$

Adopting definitions of pulse areas for specific time intervals as

$$\theta_{10} = \sqrt{\Omega^2 + \Delta^2}\,(t_1 - t_0) = \Omega_R t_1, \tag{4.3.19}$$

$$\theta_{32} = \sqrt{\Omega^2 + \Delta^2}\,(t_3 - t_2) = \Omega_R\,(t_3 - t_2), \tag{4.3.20}$$

the algebraic solutions become

$$R_1(t_3) = R_1(t_2) - \frac{\Delta}{\Omega_R} R_2(t_2)\sin\theta_{32} - \frac{2\Delta}{\Omega_R^2}\,[\Delta R_1(t_2) + \Omega R_3(t_2)]\sin^2\left(\frac{\theta_{32}}{2}\right), \tag{4.3.21}$$

$$R_2(t_3) = R_2(t_2)\cos\theta_{32} + \frac{1}{\Omega_R}\,[\Delta R_1(t_2) + \Omega R_3(t_2)]\sin\theta_{32} \tag{4.3.22}$$

$$R_3(t_3) = R_3(t_2) - \frac{\Omega}{\Omega_R} R_2(t_2)\sin\theta_{32} - \frac{2\Omega}{\Omega_R^2}\,[\Delta R_1(t_2) + \Omega R_3(t_2)]\sin^2\left(\frac{\theta_{32}}{2}\right), \tag{4.3.23}$$

by making use of the equality $(\cos\theta_{32} - 1) = -2\sin^2(\theta_{32}/2)$.

4.3.1.3 $t \geq t_3$

The period following application of the second pulse is described by the FID solutions once again. The appropriate expressions are

$$R_1(t) = [R_1(t_3)\cos\Delta(t - t_3) - R_2(t_3)\sin\Delta(t - t_3)]\exp(-[t - t_3]/T_2), \tag{4.3.24}$$

$$R_2(t) = [R_1(t_3)\sin\Delta(t - t_3) + R_2(t_3)\cos\Delta(t - t_3)]\exp(-[t - t_3]/T_2), \tag{4.3.25}$$

$$R_3(t) = R_3(t_3)\exp(-[t - t_3]/T_1). \tag{4.3.26}$$

By using the prior solutions for $R_1(t_3)$, $R_2(t_3)$, and $R_3(t_3)$ in Eqs. (4.3.24)–(4.3.26) we obtain the Bloch vector $R(t)$ for the time period $t > t_3$ after application of the pulses, in terms of the initial conditions. To calculate the radiant polarization during this period, we need only the transverse components of the Bloch vector $R_1(t)$ and $R_2(t)$. They contain the factor

$$\cos\Delta(t-t_3)\cos\Delta(t_2-t_1)\sin\Delta(t-t_3)\sin\Delta(t_2-t_1) = \cos\Delta(t-2\tau), \tag{4.3.27}$$

which is non-zero at time $t = 2\tau \equiv t_3 + t_2 - t_1$. These terms are responsible for the sudden appearance of radiative sample polarization at time $t = 2\tau$, since the Doppler phase vanishes. This rephasing of the polarizations of individual atoms at a time τ after the application of two pulses separated by an interval τ is called a photon echo. The polarization created by the first pulse is re-established by the second pulse after a short delay equal to interval τ.

Other terms in the Bloch vector such as $\sin\Delta(t-2\tau)$ are zero at $t = 2\tau$ or do not rephase and need not be considered further. Echo terms in $R_1(t)$ are odd functions of Δ. Hence if we assume the optical excitation is tuned near the Doppler peak, which is represented by an even function of Δ, the Doppler-averaged R_1 component will be $<R_1(t)> \sim 0$. Consequently, the polarization reduces to

$$<\tilde{\rho}_{12}(t)> = i < R_2(t)>/2, \tag{4.3.28}$$

and the photon echo signal amplitude will be

$$E_{s0}(L,t) = -kNL\mu_{21}R_2/4\varepsilon. \tag{4.3.29}$$

If the signal field is measured by heterodyne detection, as described in our earlier calculation of FID, the intensity contains a beat note.

$$\begin{aligned} |E_T|^2_{beat} &= \frac{1}{4}E_0E^*_{s0} + c.c. \\ &= \frac{\hbar kNL\Omega^4}{8\varepsilon\Omega_R^3}R_3(0)e^{-t/T_2}\cos[\delta\omega_{21}(t-2\tau)] < \sin\theta_{10}\sin^2\left(\frac{\theta_{32}}{2}\right)\cos[\Delta(t-2\tau)]>. \end{aligned} \tag{4.3.30}$$

The signal reaches a maximum value at $t = 2\tau$ given by

$$|E_T|^2_{beat} = \frac{\hbar kNL\Omega^4}{8\varepsilon\Omega_R^3}R_3(0)e^{-t/T_2} < \sin\theta_{10}\sin^2\left(\frac{\theta_{32}}{2}\right)>. \tag{4.3.31}$$

The echo envelope decays with a time constant T_2 and the signal will be largest for pulse areas $\theta_{10} = \pi/2$ and $\theta_{32} = \pi$. When optimized in this way, the input pulses are referred to as forming a $\frac{\pi}{2} - \pi$ sequence. An example of echo signals observed by heterodyne detection is given in Figure 4.8 (see Ref. [4.10]).

Exercise: Calculate the envelope decay times for FID and the photon echo when the signals are detected directly as $|E_{s0}|^2$ instead of by heterodyning as $E_0E^*_{s0} + c.c.$

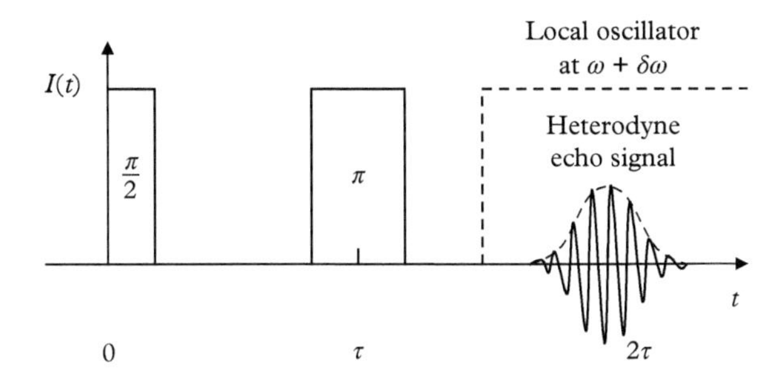

Figure 4.8 *Illustration of photon echoes detected by overlap of a local oscillator with the echo signal at the detector. The rapid oscillations within the echo pulse envelope reflect the heterodyne frequency difference between the oscillator* ($\omega + \delta\omega$) *and echo* (ω) *frequencies (see Ref. [4.10]).*

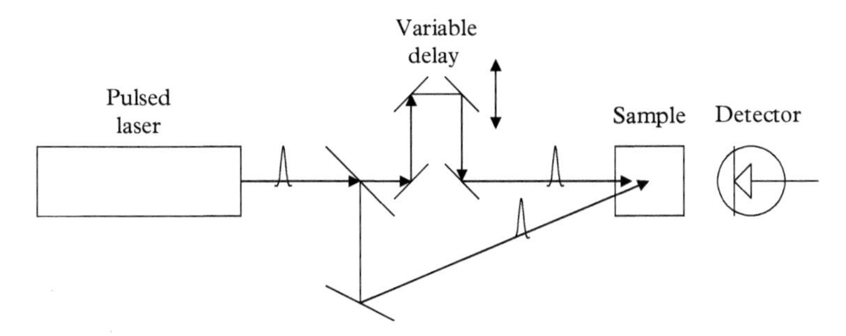

Figure 4.9 *Schematic diagram of a photon echo experiment based on a pulsed laser source. The detector is assumed to be placed in the direction of the phase-matched output determined by conservation of photon linear momentum.*

Photon echoes may also be excited using pulsed output obtained directly from appropriate laser sources. In this case beam splitters and adjustable mechanical delay lines are typically used to provide the desired two-pulse sequence as shown schematically in Figure 4.9. Pulse bandwidth may be sufficient to overlap more than one transition in such cases. The beating of multiple transitions then modulates the echo amplitude, furnishing a way to measure small differences in the transition frequencies. For example, by Fourier transforming the modulated echo envelope, energy level spacings can be inferred for the emitting centers [4.11, 4.12]. Whenever echo experiments are performed with angled beams, the direction of propagation of the signal beam must be determined by considering conservation of linear momentum in the coherent interaction, in order to know exactly where to place the detector. This consideration relates to the need for "phase-matching" in photon echo observation, a topic considered in a little more detail in the section on four-wave mixing in Chapter 5. Photon echoes are most commonly recorded as echo signal intensity versus inter-pulse delay. However, if frequency-switching techniques are used, heterodyne detection can be applied to advantage by providing a reference field at the expected time of the echo, as depicted in Figure 4.8.

4.3.2 Rotation Matrix Analysis

A rotation matrix approach to analyzing Bloch vector evolution was developed in Chapter 3. The general form for the overall rotation matrix applicable to a time interval characterized by constant (but arbitrary) conditions was given by Eq. (3.8.22). To describe multiple-pulse excitation of a system, we now need to combine the two specific forms of the general rotation matrix U that apply to nutation and FID periods.

Exercise:

(i) By ignoring decay during pulses and assuming strong, nearly resonant excitation for which $(\Delta/\Omega_R) \ll 1$, show that the rotation matrices $U_{pulse}(t_1, t_0)$ and $U_{FID}(t_2, t_1)$ for nutation and FID, respectively, over the specific time intervals t_0 to t_1 and t_1 to t_2 are:

$$U_{pulse}(t_1, t_0) = \begin{bmatrix} 1 & 0 & 0 \\ 0 & \cos[\Omega_R(t_1 - t_0)] & -\sin[\Omega_R(t_1 - t_0)] \\ 0 & \sin[\Omega_R(t_1 - t_0)] & \cos[\Omega_R(t_1 - t_0)] \end{bmatrix}, \tag{4.3.32}$$

$$U_{FID}(t_2, t_1) = \begin{bmatrix} \cos[\Delta(t_2 - t_1)] & -\sin[\Delta(t_2 - t_1)] & 0 \\ \sin[\Delta(t_2 - t_1)] & \cos[\Delta(t_2 - t_1)] & 0 \\ 0 & 0 & 1 \end{bmatrix}. \tag{4.3.33}$$

(ii) For a two-pulse sequence, show that evolution of the Bloch vector is described by

$$\bar{R}(t) = U(t, t_0)\bar{R}(0) = U_{FID}(t, t_3)U_{pulse}(t_3, t_2)U_{FID}(t_2, t_1)U_{pulse}(t_1, t_0)\bar{R}(0), \tag{4.3.34}$$

and show by matrix multiplication that the Bloch vector component $R_2(t)$, and consequently the echo polarization, exhibits the same dependence on pulse areas and inter-pulse delay as determined algebraically in Eq. (4.3.30).

In Figure 4.10, the results from three different pulsed optical experiments in methyl fluoride are compared. The upper trace shows that, contrary to the calculation of Eq. (4.3.31), echo amplitude obtained using a Carr–Purcell [4.13, 4.14] sequence of pulses $(\pi/2, \pi, \pi, \pi, \ldots)$ decays exponentially with a characteristic time that equals the population decay time T_1 rather than T_2. The results of delayed nutation shown in Figure 4.10 also have a characteristic time of T_1. In delayed nutation the recovery of probe absorption is monitored versus delay time, so that population decay is expected to govern this type of transient. However, the Carr–Purcell results present an unexpected example of how multiple-pulse sequences can measure decay processes other than pure dephasing. By repeatedly reversing the dephasing process using π-pulses, dephasing effects can be avoided altogether. This leaves only population decay to account for relaxation

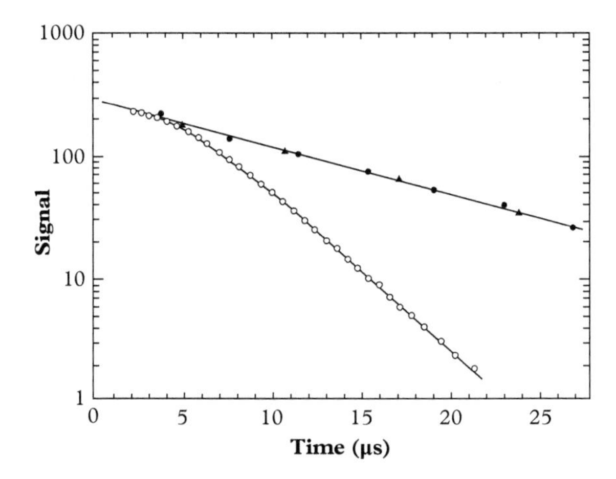

Figure 4.10 *Comparison of two-pulse echo signals with nutation and multiple-pulse (Carr–Purcell) echo results in* $^{13}CH_3F$. *Two-pulse echoes are sensitive to phase-changing collisions and measure* T_2. *Nutation and multiple-pulse sequences are limited by state-changing collisions, thereby reflecting population relaxation time* T_1. • *Carr–Purcell echoes;* ▲ *optical nutation;* ○ *two-pulse echoes. (After Ref. [4.13].)*

of the system. Further discussion of this point can be found in Ref. [4.15]. Application of this principle to inhibit intramolecular and intermolecular energy transfer processes is discussed in Ref. [4.16].

The lowest trace in Figure 4.10 gives the results of a simple two-pulse echo experiment, under conditions identical to those for the other two measurements. By its separation from the other curves, however, the data of this experiment can clearly be seen to reflect a decay rate $(T_2)^{-1}$ that is faster than $(T_1)^{-1}$. At long times, the dephasing rate is governed by velocity-changing collisions that are elastic, and that lead only to phase changes of the excited atoms rather than abrupt changes of state (population decay). Thus, coherent optical transients are sensitive to mild dynamic perturbations in the vicinity of luminescent centers, making them exquisite tools for the observation (and control) of extremely weak interactions. An example of observation and decoupling of low-energy nuclear spin interactions in optical spectroscopy is given in Ref. [4.17].

A simple experimental approach that may be applied to measure either T_1 or T_2 is provided by the stimulated echo, a three-pulse transient described in Ref. [4.18]. The dephasing and rephasing dynamics involved in this coherent transient are illustrated in Figure 4.11. An example of the use of stimulated echoes to characterize luminescent centers in a solid may be found in Refs. [4.11] and [4.12].

4.3.3 Density Matrix Operator Analysis

To conclude this chapter, a third approach to the calculation of echo signals in a two-level system is considered, as a prelude to choosing a methodology in the next chapter

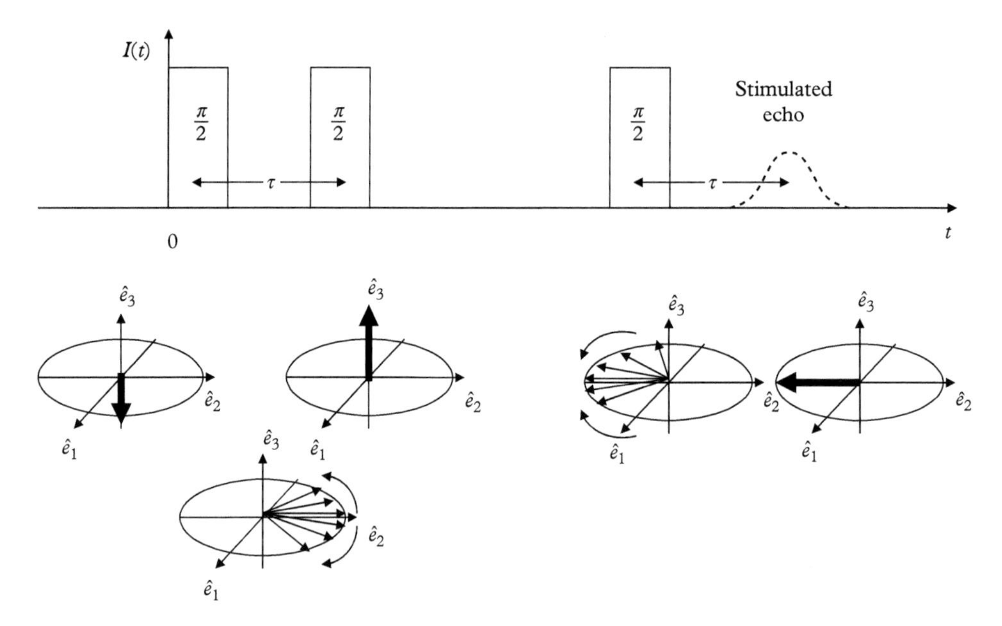

Figure 4.11 *Illustration of the pulse sequence and Bloch vector dynamics during three-pulse stimulated photon echoes.*

that will be applied throughout the remainder of the book. This one is based on formal integration of the equation of motion for the density matrix [4.19] and highlights exponential forms of the evolution operators based on the angular momentum operator. In quantum mechanics, angular momentum operators are the generators of rotations (see Appendix H). So their appearance here cements the idea that coherent interactions between light and matter in nonlinear and quantum optics are unitary transformations in Hilbert space. This accounts for their prominence in coherent laser spectroscopies of many types [4.18], as well as in the fields of coherent control and quantum information science [4.20], where precise manipulation of populations and superposition states is critical to accuracy.

A few introductory remarks are needed, since multiple representations are used in this section. The interaction picture is used in addition to the Schrödinger representation for reasons that will shortly be made clear. In the Schrödinger picture, the density matrix satisfies the equation

$$i\hbar\frac{d\rho^S}{dt} = [H, \rho^S]. \tag{4.3.35}$$

If H is time independent ($H = H_0$), this equation has the particularly simple solution

$$\rho^S(t - t_0) = U_0(t - t_0)\rho^S(t_0)U_0^+(t - t_0), \tag{4.3.36}$$

where

$$U_0(t - t_0) = \exp[-i(t - t_0)H_0/\hbar]. \tag{4.3.37}$$

Exercise: Verify that the transformed density matrix in Eq. (4.3.36) is a solution of the master equation Eq. (4.3.35) using direct differentiation of Eq. (4.3.37).

During periods when the Hamiltonian is not time independent, for example, during an optical pulse when $H(t) = H_0 + V(t)$, the density matrix equation, Eq. (4.3.35), becomes difficult to solve. The two parts of the Hamiltonian cause essentially different dynamics. This is the motivation behind the interaction picture covered in Chapter 2, which removes the evolution of the state vector due solely to the static Hamiltonian (see Eq. (2.4.13)). The equation of motion for the density matrix in the interaction picture is

$$i\hbar\frac{d\rho^I}{dt} = \left[V^I, \rho^I\right], \tag{4.3.38}$$

where $V^I(t - t_0) = U_0^+(t - t_0)VU(t - t_0)$. This is often easier to solve than Eq. (4.3.35). The formal solution to Eq. (4.3.38) is

$$\rho^I(t - t_0) = U_I(t - t_0)\rho^I(t_0)U_I^+(t - t_0), \tag{4.3.39}$$

where

$$U_I(t - t_0) = \exp\left[-\frac{i}{\hbar}\int_{t_0}^{t} V^I(t')dt'\right]. \tag{4.3.40}$$

The integral in the evolution operator is readily evaluated in the interaction picture. This is due to the fact that the integrand consists of the slowly varying amplitude of the optical interaction $V^I = \mu E_0\sigma_1/2$ in the rotating wave approximation, not the rapidly varying interaction Hamiltonian of the lab frame. The operator $U_I(t - t_0)$ describes the temporal evolution from t to t_0, but at the same time can be interpreted as a simple rotation through an angle θ about axis $\hat{n}$ in Hilbert space:

$$U_I(\theta) = \exp[-i\sigma_1\theta/2] = \exp\left[-i\hat{n}\cdot\hat{S}_1\theta/\hbar\right], \tag{4.3.41}$$

A square pulse of duration ε and area $\theta(\varepsilon) = \mu E_0\varepsilon/\hbar$ is assumed, and the spin angular momentum $\hat{n}\cdot\hat{S}_1 = \hbar\sigma_1/2$ has been introduced formally as the generator of rotations in two-level spin space. That the operator in Eq. (4.3.41) docs indeed cause rotations like the matrices of the last section is justified more thoroughly in Ref. [4.21] and Appendix H.

We are now ready to write down the solution for the density matrix of a two-level system subjected to an arbitrary pulse sequence in terms of evolution operators. In principle, the temporal evolution is given by the appealingly simple expression

$$\rho(t) = U(t-t_0)\rho(t_0)U^+(t-t_0), \tag{4.3.42}$$

provided there is no dissipation in the system and we can determine the evolution operator $U(t-t_0)$. The initial matrix is assumed to correspond to an equilibrium state, with populations of ρ_{11} and ρ_{22} in the ground and excited states, respectively. Hence

$$\rho^I(t_0) = \rho^S(t_0) = \begin{bmatrix} \rho_{22} & 0 \\ 0 & \rho_{11} \end{bmatrix} = \frac{1}{2}(\sigma_0 - \sigma_3). \tag{4.3.43}$$

We consider the same sequence of two square pulses separated by the time interval τ that was analyzed in previous sections. The pulses are applied from $t = t_0$ to $t_1 = \varepsilon_1$ and from $t_1 = \varepsilon_1$ to $t_2 = \tau$. Each is followed by a free precession period, first from $t_1 = \varepsilon_1$ to $t_2 = \tau$ and then from $t_2 = \tau$ to $t_3 = \tau + \varepsilon_2$, as in Figure 4.7. The system is strongly driven for brief moments and evolves freely at other times. The problem of determining the overall operator $U(t-t_0)$ is therefore rather complex, but can be subdivided conveniently on the basis of time periods when the Hamiltonian is H_0 versus when it is $H_0 + V(t)$. That is, the problem may be broken down into discrete periods during which the system is either perturbed or not.

During the free precession periods, evolution may be readily described in the Schrödinger picture using Eq. (4.3.36). During the pulses, however, when the Hamiltonian is time dependent, this expression is no longer valid. It is simpler to calculate the driven dynamics in the interaction picture and subsequently transform the result back to the Schrödinger lab frame to finish the calculation in a single frame. In this approach, time periods when radiation is present are handled in the interaction picture. Driven dynamics may be analyzed separately and then transformed to the Schrödinger picture using the result of the following exercise.

Exercise: Show directly from Eq. (2.4.13) that the density matrices in the Schrödinger and interaction pictures are related by

$$\rho^S(t) = \exp[-iH_0t/\hbar]\rho^I(t)\exp[iH_0t/\hbar] \tag{4.3.44}$$

Following this strategy, a decomposition of the evolution operator is made into four time periods, and also into suitable representations. This gives the following result.

$$\begin{aligned} U(t-t_0) &= U(t-t_3)U(t_3-t_2)U(t_2-t_1)U(t_1-t_0) \\ &= U_S(t-t_3)\exp[-i\omega_0 t_3]U_I(t_3-t_2)U_S(t_2-t_1)\exp[-i\omega_0 t_1]U_I(t_1-t_0), \end{aligned} \tag{4.3.45}$$

where

$$U_I(t_1 - t_0) = \exp[-i\sigma_1\theta_1/2], \tag{4.3.46}$$

$$U_S(t_2 - t_1) = \exp[-(i/2)\omega_0(\tau - \varepsilon_1)\sigma_3], \tag{4.3.47}$$

$$U_I(t_3 - t_2) = \exp[-i\sigma_1\theta_2/2], \tag{4.3.48}$$

$$U_S(t - t_3) = \exp[-(i/2)\omega_0(t - \tau - \varepsilon_2)\sigma_3]. \tag{4.3.49}$$

Substitution of Eqs. (4.3.46)–(4.3.49) into Eq. (4.3.45), followed by the combination of Eq. (4.3.45) with Eq. (4.3.42), yields an expression for the density matrix after the pulses have been applied:

$$\rho(t) = \exp\left[-\frac{i}{2}\omega_0(t-\tau)\sigma_3\right]\exp\left[-\frac{i}{2}\theta_2\sigma_1\right]\exp\left[-\frac{i}{2}\omega_0\tau\sigma_3\right]\exp\left[-\frac{i}{2}\theta_1\sigma_1\right]\rho(0)\cdot h.c. \tag{4.3.50}$$

Here *h.c.* stands for the Hermitian conjugate of the entire product of operators acting on $\rho(0)$ from the left.

To simplify Eq. (4.3.50) and compare it with earlier results, it is necessary to commute the second and third terms in $\rho(t)$. For this purpose, commutation relations of the Pauli matrices must be taken into account, and the identity in Problem 4.9 is helpful. Using the result

$$\exp\left[-\frac{i}{2}\theta_2\sigma_1\right]\exp\left[-\frac{i}{2}\omega_0\tau\sigma_3\right] = \cos\left(\frac{1}{2}\theta_2\right)\exp\left[-\frac{i}{2}\omega_0\tau\sigma_3\right] - i\sin\left(\frac{1}{2}\theta_2\right)\exp\left[\frac{i}{2}\omega_0\tau\sigma_3\right]\cdot\sigma_1$$

in Eq. (4.3.50), one finds

$$\rho(t) = \left[\cos\left(\frac{1}{2}\theta_2\right)e^{-(i/2)\omega_0 t\sigma_3} - i\sin\left(\frac{1}{2}\theta_2\right)e^{-(i/2)\omega_0(t-2\tau)\sigma_3}\cdot\sigma_1\right]$$
$$\times\left[\cos\left(\frac{1}{2}\theta_1\right) - i\sigma_1\sin\left(\frac{1}{2}\theta_1\right)\right]\rho(0)\cdot h.c.$$

Only terms with the time argument $(t - 2\tau)$ can contribute to echo formation. Consequently, the density matrix reduces to

$$\rho(t) = \sin^2\left(\frac{1}{2}\theta_2\right)\left\{e^{-(i/2)\omega_0(t-2\tau)\sigma_3}\left[\cos\left(\frac{1}{2}\theta_1\right) - i\sigma_1\sin\left(\frac{1}{2}\theta_1\right)\right]\sigma_1\rho(0)\cdot h.c.\right\} \tag{4.3.51}$$

Now, only the σ_3 term in $\rho(0)$ leads to non-zero contributions to $\rho(t)$ in Eq. (4.3.51), since the unit matrix portion σ_0 allows conjugates to combine and cancel. Furthermore, terms containing $\sigma_1^2\rho(0)\sigma_1^2$ or $\sigma_1\rho(0)\sigma_1$ are proportional to σ_3. These terms are diagonal and cannot give rise to radiation. Diagonal terms of $\rho(t)$ do not appear in the calculation of polarization in Chapter 3. They do not correspond to oscillating dipole moments and cannot contribute to radiant emission by the sample.

Dropping the non-radiant terms, we therefore find

$$\rho(t) = \sin^2\left(\frac{1}{2}\theta_2\right)\left\{e^{-(i/2)\omega_0(t-2\tau)\sigma_3}\left[\frac{i}{2}\cos\left(\frac{1}{2}\theta_1\right)\sin\left(\frac{1}{2}\theta_1\right)\right][\sigma_1,\sigma_3]e^{(i/2)\omega_0(t-2\tau)\sigma_3}\right\}$$

$$= \frac{1}{2}\sin^2\left(\frac{1}{2}\theta_2\right)\sin(\theta_1)\left\{e^{-(i/2)\omega_0(t-2\tau)\sigma_3}\sigma_2 e^{(i/2)\omega_0(t-2\tau)\sigma_3}\right\} \quad (4.3.52)$$

In this result for $\rho(t)$, which determines the signal field (see Eq. (3.9.16)), all the basic features of two-pulse echoes established in earlier sections of this chapter are again evident. The echo appears at $t = 2\tau$ and the echo amplitude is maximum for pulse areas of $\theta_1 = \pi/2$ and $\theta_2 = \pi$.

Multiple-pulse photon echoes like the stimulated echo illustrated in Figure 4.11 provide a useful method of storing and retrieving information on demand and also of processing information. For example, information in the form of amplitude modulation of the input light can be stored as an index or population grating in the medium and recalled at a later time (on demand) by the third pulse in a three-pulse echo sequence. High fidelity applications of this kind have in fact been demonstrated by the retrieval of long pulse sequences constituting time-encoded bit strings [4.22]. High-speed, large-bandwidth spectrum analysis can also be performed using coherent transients [4.23].

This subsection has presented yet another way to calculate the signal amplitude of photon echoes, adding to the methods covered in Sections 4.3.1 and 4.3.2. This third method was based on the formally concise expression (Eq. (4.3.42)) which yields the density matrix $\rho(t)$ directly. While appealingly simple at the outset, this formula contained an exponential form of the evolution operator and matrix representations of the SU(2) Pauli spin operators which introduced inconvenient limitations. Explicit use had to be made, for example, of the SU(2) commutation relations, as well as expansions of the exponential functions in terms of the Pauli spin matrix operators.

A moment's pause makes one realize how awkward it would be to have to use complex functions of higher order spinors, for example, SU(n) spin matrices for an n-level system, to describe systems with more than two levels. Also, dissipation has been completely ignored in the treatment of this subsection. To include relaxation processes would greatly complicate the arguments of the temporal evolution operators even within a single time segment. This combination of drawbacks argues against adopting this approach as a general tool for analysis. Similar reservations can be leveled at the other approaches explored in Sections 4.3.1 and 4.3.2. Consequently, for the remainder of this book, we shall turn to solutions of the complete, differential master equation for the density matrix in component form. This will avoid restriction to a small number of energy levels or fields, and will include population and coherence decay processes which govern crucial aspects of the dynamics of real systems in a consistent manner.

REFERENCES

[4.1] R.L. Shoemaker and E.W. Van Stryland, *J. Chem. Phys.* 1976, **64**, 1733.
[4.2] D. Press, T.D. Ladd, B. Zhang, and Y. Yamamoto, *Nature* 2008, **456**, 218.
[4.3] T.H. Stievater, X. Li, D.G. Steel, D. Gammon, D.S. Katzer, D. Park, C. Piermarocchi, and L.J. Sham, *Phys. Rev. Lett.* 2001, **87**, 133603.
[4.4] F.A. Hopf, R.F. Shea, and M.O. Scully, *Phys. Rev. A* 1973, **7**, 2105.
[4.5] R.G. Brewer, in R. Balian, S. Haroche, and S. Liberman (eds.) *Frontiers in Laser Spectroscopy*, Vol. 1. Les Houches Lectures, Session XXVII. Amsterdam: North-Holland Publishing, 1977.
[4.6] R.G. DeVoe and R.G. Brewer, *IBM J. Res. Dev.* 1979, **23**, 527.
[4.7] R.G. DeVoe and R.G. Brewer, *Phys. Rev. Lett.* 1983, **50**, 1269.
[4.8] A.Z. Genack and R.G. Brewer, *Phys. Rev. A* 1978, **17**, 1463.
[4.9] N.A. Kurnit, I.D. Abella, and S.R. Hartmann, *Phys. Rev.* 1966, **141**, 391.
[4.10] R.M. Shelby and R.M. MacFarlane, *Phys. Rev. Lett.* 1980, **45**, 1098.
[4.11] Y.C. Chen, K. Chiang, and S.R. Hartmann, *Phys. Rev. B* 1980, **21**, 40.
[4.12] A. Lenef, S.W. Brown, D.A. Redman, S.C. Rand, J. Shigley, and E. Fritsch, *Phys. Rev. B* 1996, **53**, 427.
[4.13] J. Schmidt, P. Berman, and R.G. Brewer, *Phys. Rev. Lett.* 1973, **31**, 1103.
[4.14] H.Y. Carr and E.M. Purcell, *Phys. Rev.* 1954, **94**, 630.
[4.15] For further discussion, see C.P. Slichter, *Principles of Magnetic Resonance*. New York: Springer-Verlag, 1980, pp.252–4.
[4.16] E.T. Sleva, A.H. Zewail, and M. Glasbeek, *J. Phys. Chem.* 1986, **90**, 1232.
[4.17] S.C. Rand, A. Wokaun, R.G. DeVoe, and R.G. Brewer, *Phys. Rev. Lett.* 1979, **43**, 1868.
[4.18] M.D. Levenson and S.S. Kano, *Introduction to Nonlinear Laser Spectroscopy*. Academic Press, 1988.
[4.19] L.G. Rowan, E.L. Hahn, and W.B. Mims, *Phys. Rev.* 1965, **137**, A61.
[4.20] M.A. Nielsen and I.L. Chuang, *Quantum Computation and Quantum Information*. Cambridge: Cambridge University Press, 2000.
[4.21] E. Merzbacher, *Quantum Mechanics*, 2nd ed. New York: John Wiley & Sons, 1970, pp.266–74.
[4.22] Y.S. Bai, W.R. Babbitt, and T.W. Mossberg, *Opt. Lett.* 1986, **11**, 724.
[4.23] V. Crozatier, G. Gorju, J.-L. LeGouet, F. Bretenaker, and I. Lorgere, *J. Lumin.* 2007, **127**, 104.

PROBLEMS

4.1. Apply the evolution matrix $U(t)$ defined by $R(t) = U(t)R(0)$, where R is the Bloch vector, to find the radiant polarization P of a two-level system at an arbitrary time after applying a pulse, as described. Notice that the evolution period $R(t) = U(t)R(0)$ of interest consists of two parts, namely nutation and precession.

(a) Assume the system is initially in the ground state and calculate the off-diagonal matrix element $<\tilde{\rho}_{12}>$ and the Doppler-averaged macroscopic polarization at times after the application of a single rectangular pulse that are short compared to T_2 in the limit of small detuning $((\Delta/\Omega_R) < 1$ and $(\Delta^2/\Omega_R^2) \ll 1)$. (Ignore population and polarization relaxation of the elements of R themselves.)

(b) If the system is initially prepared in a state characterized by $R_1(0) = 1$, then what will the polarization be at short times after the pulse?

4.2. Using the evolution matrix approach, show that the application of a single resonant π-pulse to a ground state atom leaves the system in an inverted state, by removing all population from the initial state.

4.3. An atomic ensemble is subjected to a $\pi/2 - \pi$ pulse sequence. If the system is allowed to evolve freely, it produces a photon echo at time 2τ, where τ is the temporal delay between the first two pulses, because the Bloch vector equals $R = (0,-1,0)$ at this instant. However, if an additional pulse of area $\pi/2$ is applied at $t = 2\tau$ the medium polarization can be converted into a population change. Use the evolution matrix to show that by selecting the final pulse to be co- or counter-propagating with respect to the first two pulses, the population change can be directed to either the ground or the excited state.

4.4. The Maxwellian distribution of velocities of atoms in a gas gives rise to a Gaussian distribution of resonance frequencies $D(\omega)$ that Doppler broadens the atomic transition linewidth.

(a) Given the probability of finding atoms of resonant frequency ω in the interval $d\omega$ is $D(\omega) = C\exp\{-Mc^2(\omega_0 - \omega)^2/2\omega_0^2 k_B T\}d\omega$, find the full width at half maximum intensity (FWHM) of the Doppler-broadened line.

(b) Collisions provide a second source of line broadening on optical transitions that is inversely proportional to the time τ_0 between collisions. The full width at half maximum of the collisional contribution is just twice the collision rate given by

$$\gamma_{\text{coll}} = \frac{1}{\tau_0} = \frac{4d^2 N}{V}\left(\frac{\pi k_B T}{M}\right)^{1/2}.$$

Show that Doppler and collisional contributions to the transition linewidth are equal at a gas density for which the volume per atom is close to λd^2, where λ is the optical wavelength of the transition and d is the (average) distance between the atomic centers during collision.

4.5. A one-dimensional atomic medium consists of only four atoms located at irregular (non-periodic) positions $z = 0$, $3\lambda/5$, $8\lambda/9$, and $4\lambda/3$. A linearly polarized field $E(z,t) = \frac{1}{2}E_0\hat{x}\exp[i(\omega t - kz)] + c.c.$ is incident on the medium and excites a microscopic dipolar response $\bar{p} = \varepsilon_0\chi^{(e)}\bar{E}$ in each atom.

(a) Calculate the net (macroscopic) polarization $\bar{P}(z,t) = \sum_{i=1}^{4} \bar{p}_i(z_i, t_i)$ of the system, being careful to account for the delay time associated with propagation of a fixed phase point of the driving field from one atom to the next, as well as their positions.

(b) Compare the squared polarization $|\bar{P}|^2$, which is proportional to the intensity of scattered light from the system, to the value expected from four independent atoms. Show that the re-emitted energy from the system has an intensity that is not linear in the number of emitters.

(This problem illustrates the fact that the incident light can coherently phase randomly positioned atoms, and that phasing of constituent dipoles plays an important role in determining the strength of the resulting macroscopic polarization.)

4.6. A sequence of pulses is applied to a two-level, Doppler-broadened medium. Consider the pulses to be delta functions applied with an interval τ, as shown in the accompanying figure. The pulse areas of the first and second pulses are θ_1 and θ_2, respectively.

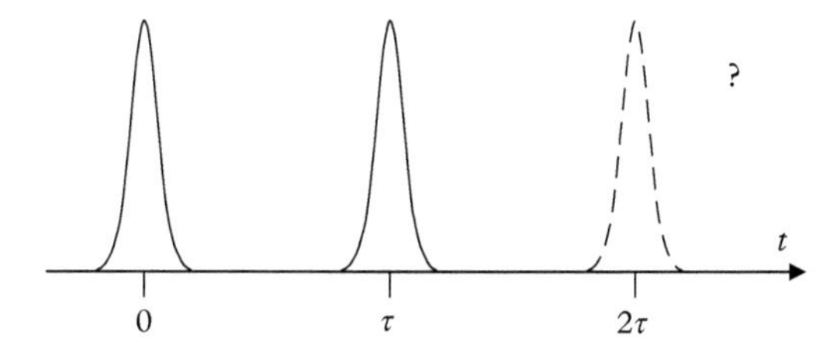

(a) Write out a simplified evolution matrix for nutation of the Bloch vector $R(t)$ in the limit that $\Omega_R \gg \Delta$ (i.e., $\Omega_R \cong \Omega$).

(b) Solve for the complete Bloch vector at $t \geq \tau$ using the matrix evolution operator method. (Hint: Do not set $\Delta = 0$ during free precession periods or you lose track of when various potential signal contributions peak in time.) Ignore decay between pulses and do not assume any special values for θ_1 and θ_2.

(c) When the Doppler width is broad compared to the homogeneous linewidth, averaging over the distribution yields a simplified polarization $\bar{\rho}_{12} \propto iR_2$. Then, only R_2 components contribute to the polarization. By examining your result for the final Bloch vector, determine whether an echo forms or not when $\theta_1 = \pi$ and $\theta_2 = \pi/2$. Justify.

(d) Is this pulse sequence equivalent to a stimulated echo sequence of three $\pi/2$ pulses? Draw Bloch *vector model pictures* to explain result (c) further and justify your conclusion.

(e) Finally, a third pulse of area $\theta_3 = \pi$ is applied at $t = 2\tau$ to complete a $\pi - \pi/2 - \pi$ sequence. Calculate the resulting behavior for $t \geq 2\tau$ and *explain* the result in words or pictures.

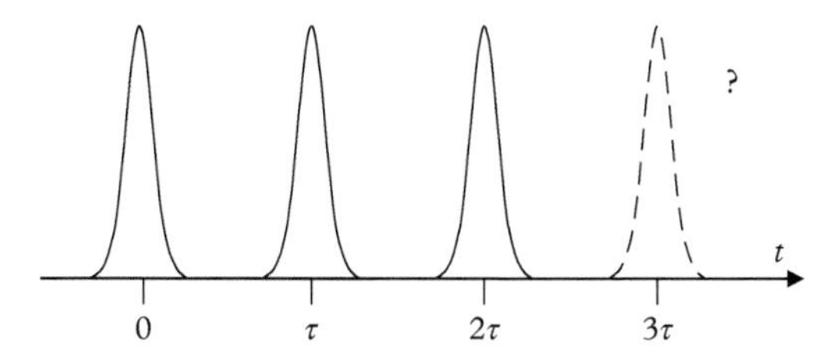

4.7. Assuming there is an allowed transition with a transition dipole moment μ between the two levels of a two-level system,

(a) find the Bloch vector of a two-level system, initially in the ground state, that is irradiated by a resonant π-pulse of short duration $\tau_p(\tau_p \ll T_1, T_2)$. Take the time of observation to be $t = \tau$ (where $\tau_p \ll \tau < T_1, T_2$).

(b) Calculate and explain the value of polarization P observed at time $t = \tau$?

(c) Draw the Bloch vector in a suitable coordinate system at time $t = \tau$.

(d) What is the excited state population at $t = \tau$?

(e) What is the sign of the absorption at $t = \tau$? Would you expect a probe pulse to experience loss (positive absorption) or gain (negative absorption) at this time?

(f) If a second pulse, a $\pi/2$-pulse, is applied at time $t = \tau$, what is the polarization $P(t)$ immediately following the pulse?

(g) Assuming the system is inhomogeneously broadened, would you expect an echo at time $t = 2\tau$? Why or why not?

4.8. Using Bloch vector diagrams similar to Figure 4.5, give an argument explaining why the Carr–Purcell sequence of pulses, with areas and timing described by $\pi/2(t = 0), \pi(\tau), \pi(3\tau), \pi(5\tau), \ldots$, produces echoes at times $2\tau, 4\tau, 6\tau \ldots$ whose envelope decays with a characteristic time T_1 instead of T_2.

4.9. By expanding the leftmost exponential function as an infinite sum, and using commutation properties of the Pauli matrices, show that

$$\exp\left[-\frac{i}{2}\theta_2\sigma_1\right]\exp\left[-\frac{i}{2}\omega_0\tau\sigma_3\right]$$
$$= \cos(\theta_2/2)\exp[-(i/2)\omega_0\tau\sigma_3] - i\sin(\theta_2/2)\exp[(i/2)\omega_0\tau\sigma_3]\cdot\sigma_1.$$

4.10. A two-level atom with an allowed ED transition is prepared in an initial Bloch vector state $R(0) = (0,1,0)$ and subjected to resonant excitation.

(a) Find $R(t)$ using the evolution matrix.

(b) On the assumption that the atom undergoes only internal decay processes, find the normalized excited state population $\rho_{22}(t)$.

4.11. Two resonant, ultrashort $\pi/2$ pulses separated by a variable delay time $\tau \approx T_2$ impinge on a transparent sample that should be considered a closed two-level system with $T_2 \ll T_1$. Using the Bloch vector model, and including population decay and dephasing only during the FID period, calculate

(a) the Bloch vector $R(\tau)$ at the trailing edge of the second pulse, and

(b) from $R(\tau)$ find the excited state occupation $\rho_{22}(\tau)$ at the trailing edge of the second pulse.

(c) State what characteristic time (T_1 or T_2) determines the decay of $\rho_{22}(\tau)$ at short times and describe in one sentence an experiment that could measure it.

4.12. Consider a three-level system like that in the figure, where the two dipole moments for transitions from the ground state are equal ($\mu_{12} = \mu_{13} = \mu$). An optical field of frequency ω is tuned to the exact mid-point between the transition frequencies ω_{01} and ω_{02} so that the respective detunings are equal but opposite ($\Delta_1 = -\Delta_2 = \Delta$). The interaction $V = -\frac{1}{2}\hbar\Omega e^{i\omega t} + c.c.$ is assumed to be sufficiently weak so that $\rho_{22} \approx \rho_{33} \approx 0$. Assume coherence between states 2 and 3 is negligible (far off resonance) and show that in the limit $\Gamma_{12}, \Gamma_{13} \ll \Delta$ the system polarization $P = (\mu_{12}\rho_{21} + \mu_{13}\rho_{31}) + c.c.$ is zero, indicating that radiant emission and elastic scattering are completely suppressed by interference.

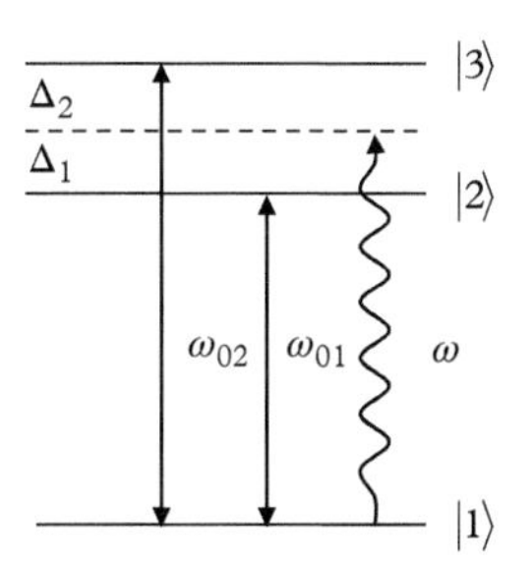

4.13. From the equations for nutation without damping, one can show that precise control of the applied field and pulse duration to form π pulses is not necessary to transfer population efficiently from one state to another. It is sufficient merely to sweep the frequency of the light rapidly across the resonance to cause *adiabatic rapid passage*.

To show this, consider a two-level system in an applied field with detunings of $\Delta = -\delta$ at negative times ($t < 0^-$) and $\Delta = +\delta$ at positive times ($t > 0^+$). Assume that $\Delta \gg \Omega$ and that $R_2(t)$ has the form $R_2(t) = \sin \Delta t$ in each time period ($t \neq 0$). By making the further assumption that Δ is constant except for the short interval $0^- < t > 0^+$ when the field sweeps through resonance, integrate the equation for $\dot{R}_3(t)$ separately over the negative and positive time periods to show that $R_3(0^+) = -R_3(0^-)$. (The sweep time is implicitly assumed to be faster than system decay yet slow enough that precessional motion follows the effective field adiabatically. Note also that using a convergence factor one can show that $\int_0^\infty \sin(\Delta t)dt = \frac{1}{\Delta}$).

4.14. Two ultrashort pulses with areas of $\pi/2$ are separated in time by a delay $\tau < T_2$ and detuned from resonance in a closed two-level system by an amount Δ (where $\Delta \gg 1/T_2$ and $\Delta \ll \Omega$). Use the Bloch vector model, including decay during the FID period between pulses, to determine the period of oscillations in the absorption (proportional to $\rho_{11} - \rho_{22}$) versus delay at the trailing edge of the second pulse. (These oscillations are manifestations of interference features in the excitation of atoms known as *Ramsey fringes*.)

Note: Spectroscopic measurements are sometimes made with Ramsey pulses. See for example B. Gross, A. Huber, M. Niering, M. Weitz, and T.W. Hansch, Europhysics Letters 44, pp. 186–191(1998).

4.15. Population inversion of a two-level system can be achieved in many ways using optical pulses with durations much shorter than coherence decay times. The most well-known method is to use a single resonant pulse whose area is π. A lesser-known method is the phase reversal method, which utilizes pairs of pulses that are phase reversed and non-resonant, as shown in the diagram. Use the Bloch vector evolution matrix for two sequential square pulses of duration $T_1 = \pi/\Omega_{R1}$ and $T_2 = \pi/\Omega_{R2}$ with equal field amplitudes and detunings ($\Delta_1 = \Delta_2$) but opposite phases ($\Omega_1 = -\Omega_2 \equiv \Omega$) to find a detuning that inverts a system initially in the ground state.

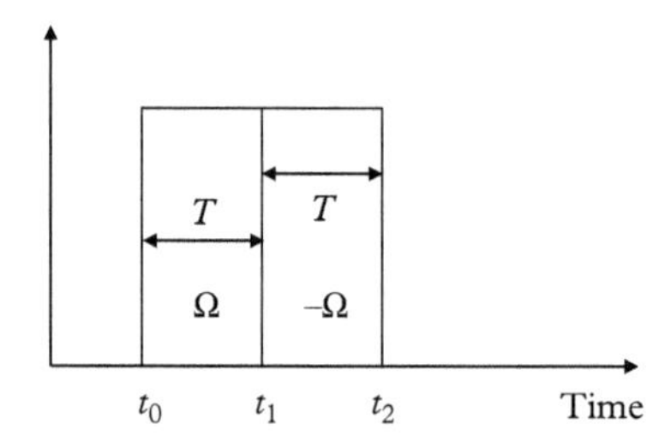

For further reading: N. Tsukuda, Opt. Commun. 24, 289(1978).

5

Coherent Interactions of Fields with Atoms

In this chapter, the tools and perspectives developed earlier are applied to predict the dynamics of many different systems, starting from the simplest and progressing to more complicated ones. Systems in each successive subsection are representative of a class of problem, and only one change is introduced at each step. In this way a veritable catalog of possibilities is assembled, all described with the density matrix and the same interaction Hamiltonian using the semi-classical approach. Situations in which it becomes necessary to consider the quantum structure of the electromagnetic field itself are deferred to Chapter 6, which covers quantized fields and coherent states.

5.1 Stationary Atoms

5.1.1 Stationary Two-Level Atoms in a Traveling Wave

Consider a system of stationary two-level atoms, as shown in Figure 5.1, characterized by a single resonance frequency $\omega_0 \equiv \omega_2 - \omega_1$. The system is subjected to traveling plane wave excitation of the form

$$V(t) = -\frac{1}{2}\bar{\mu} \cdot \bar{E}_0 \, \mathrm{e}^{i\omega t} + c.c. = -\frac{1}{2}\hbar\Omega \mathrm{e}^{i\omega t} + c.c. \tag{5.1.1}$$

According to Eqs. (3.6.19) and (3.8.1), the density matrix equations of motion are

$$i\hbar\dot{\rho}_{11} = V_{12}\rho_{21} - \rho_{12}V_{21} + i\hbar\gamma_{21}\rho_{22}, \tag{5.1.2}$$

$$i\hbar\dot{\rho}_{22} = -V_{12}\rho_{21} + \rho_{12}V_{21} - i\hbar\gamma_{21}\rho_{22}, \tag{5.1.3}$$

$$i\hbar\dot{\rho}_{12} = -\hbar\omega_0\rho_{12} + (V_{12}\rho_{22} - \rho_{11}V_{12}) - i\hbar\Gamma_{12}\rho_{12}, \tag{5.1.4}$$

$$\rho_{21} = \rho_{12}^{*}. \tag{5.1.5}$$

Lectures on Light. Second Edition. Stephen C. Rand.
© Stephen C. Rand 2016. Published in 2016 by Oxford University Press.

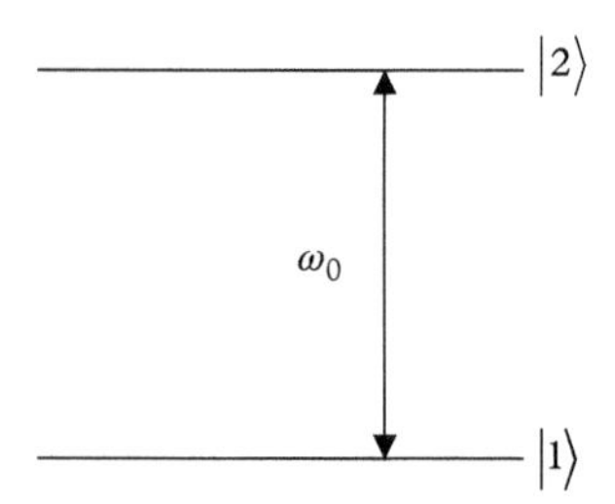

Figure 5.1 *Energy levels and resonant transition frequency in a two-level atom.*

Note that either a plus or a minus sign may apply to the phenomenological population decay terms on the far right of Eqs. (5.1.2) and (5.1.3). The choice of sign must reflect whether population is "arriving" (+) in a given level from more energetic states above it, or "leaving" (–). On the other hand, the sign for the decay of the off-diagonal matrix element (the "coherence") in Eq. (5.1.4) is always negative, reflecting the fact that coherence of a system never spontaneously increases. When there is no driving field, it always decreases. In addition the dephasing rate constant is real, so that $\Gamma_{12} = \Gamma_{21}$.

For steady-state behavior we assume that the response of the electron, described by off-diagonal element ρ_{12}, follows only one frequency component of the driving field, namely the positive one. This is called the rotating wave approximation (RWA). Rather than perform the time-integration of Eq. (5.1.4), which is necessary to find transient solutions, we begin by simply assuming the charge oscillation is at the optical frequency, so that the solution has the form

$$\rho_{12} = \tilde{\rho}_{12}\, e^{i\omega t}. \tag{5.1.6}$$

(This is justified by direct integration of the equation of motion for the off-diagonal density matrix element ρ_{12} in Appendix G). Upon substitution of Eq. (5.1.6) into Eq. (5.1.4), one immediately finds

$$[\Gamma_{12} + i(\omega - \omega_0)]\, \tilde{\rho}_{12} + \dot{\tilde{\rho}}_{12} = i\Omega_{12}\, (\rho_{11} - \rho_{22})\, /2. \tag{5.1.7}$$

We now make the further assumption that $\tilde{\rho}_{12}$ varies much more slowly than the optical period. This corresponds to the assumption that $d\tilde{V}_{12}(t')/dt' \cong 0$ in Appendix G and is called the slowly varying envelope approximation (SVEA). It is implemented by setting $\dot{\tilde{\rho}}_{12} = 0$, which yields

$$\tilde{\rho}_{12} = \left[\frac{\Omega_{12}/2}{\Delta + i\Gamma}\right](\rho_{11} - \rho_{22}). \tag{5.1.8}$$

Note that we have dropped the subscripts on the dephasing rate for simplicity, since $\Gamma_{12} = \Gamma_{21} \equiv \Gamma$ is a real decay rate and there are only two energy levels in the system, so there is no possible ambiguity as to the polarization decay to which it refers. The steady-state solution for $\rho_{12}(t)$ is therefore

$$\rho_{12}(t) = \left(\frac{\Omega_{12}/2}{\Delta + i\Gamma}\right)(\rho_{11} - \rho_{22})e^{i\omega t}. \tag{5.1.9}$$

For systems in which emission and coherence properties are not of interest, often it is only the temporal development of populations that is needed to describe basic system dynamics. For this purpose, population rate equations that do not require knowledge of the off-diagonal elements of the density matrix suffice. For example, absorption of the system, which is proportional to the population difference $\rho_{11} - \rho_{22}$, can be predicted without knowing ρ_{12}. This can be demonstrated by substituting Eq. (5.1.9) and its conjugate into the density matrix equations for $\dot{\rho}_{11}$ and $\dot{\rho}_{22}$.

Exercise: Show that substitution of Eqs. (5.1.9) and (5.1.5) into Eqs. (5.1.2) and (5.1.3) results in the population rate equations:

$$\dot{\rho}_{11} = -\frac{\Gamma/2}{\Delta^2 + \Gamma^2}|\Omega_{12}|^2(\rho_{11} - \rho_{22}) + \gamma_{21}\rho_{22}, \tag{5.1.10}$$

$$\dot{\rho}_{22} = \frac{\Gamma/2}{\Delta^2 + \Gamma^2}|\Omega_{12}|^2(\rho_{11} - \rho_{22}) - \gamma_{21}\rho_{22}. \tag{5.1.11}$$

These are coupled equations for the populations in levels 1 and 2 that can be solved exactly and include intensity-dependent dynamics. Without having to evaluate ρ_{12} explicitly, they can be used to describe the bleaching of absorbing systems like colored glass filters by intense light beams.

The next step is to solve for the steady-state populations. Let us start with ρ_{22}, using Eq. (5.1.9) in Eq. (5.1.3). Setting $\dot{\rho}_{22} = 0$ for steady-state response, this procedure yields

$$\begin{aligned}\gamma_{21}\rho_{22} &= \frac{i}{\hbar}(V_{12}\rho_{21} - \rho_{12}V_{21}) \\ &= -(i\Omega_{12}\tilde{\rho}_{21} - i\Omega_{21}\tilde{\rho}_{12})/2 \\ &= |\Omega_{12}/2|^2\left[L + L^*\right](\rho_{11} - \rho_{22}),\end{aligned} \tag{5.1.12}$$

where $L \equiv (i\Delta + \Gamma)^{-1}$. Solving for the excited state occupation in terms of that of the ground state one finds

$$\rho_{22} = \left[\frac{(L + L^*)|\Omega_{12}/2|^2/\gamma_{21}}{1 + (L + L^*)|\Omega_{12}/2|^2/\gamma_{21}}\right]\rho_{11}. \tag{5.1.13}$$

Since the total occupation probability must be unity ($\rho_{11} + \rho_{22} = 1$), one also obtains

$$\rho_{11} = \left[\frac{1 + (L + L^*) \left|\Omega_{12}/2\right|^2 / \gamma_{21}}{1 + 2(L + L^*) \left|\Omega_{12}/2\right|^2 / \gamma_{21}} \right] = \frac{\Delta^2 + \Gamma^2 + |\Omega_{12}/2|^2 \, 2\Gamma/\gamma_{21}}{\Delta^2 + \Gamma^2 + |\Omega_{12}/2|^2 \, 4\Gamma/\gamma_{21}}, \tag{5.1.14a}$$

$$\rho_{22} = \frac{|\Omega_{12}/2|^2 \, 2\Gamma/\gamma_{21}}{\Delta^2 + \Gamma^2 + |\Omega_{12}/2|^2 \, 4\Gamma/\gamma_{21}}. \tag{5.1.14b}$$

Exercise: Verify that ρ_{11}, ρ_{22} are purely real quantities that tend to appropriate limiting values as $\Omega \to \infty$ (i.e., at high intensity).

The absorption of light depends on the number of absorbers and their distribution among the available states. Hence it is proportional to $N(\rho_{11} - \rho_{22})$, where the population difference is given by Eqs. (5.1.13) and (5.1.14) as

$$\begin{aligned} \rho_{11} - \rho_{22} &= [1 + 2(L + L^*)|\Omega/2|^2/\gamma_{21}]^{-1} \\ &= \left[1 + \frac{\Gamma|\Omega|^2}{(\Delta^2 + \Gamma^2)\gamma_{21}} \right]^{-1}. \end{aligned} \tag{5.1.15}$$

Exercise: Find the limiting value of absorption as $\Omega \to \infty$ (high intensity limit)? Does the result in Eq. (5.1.15) make physical sense in this limit?

Equation (5.1.15) has the form $[1 + I/I_{sat}]^{-1}$, where the intensity I_{sat} at which absorption drops to half its maximum, a quantity known as the saturation intensity, is defined by

$$\begin{aligned} I_{sat} &\equiv \hbar^2 \gamma_{21} \mu_0 c / (|\mu_{12}|^2 \, [L + L^*]) \\ &= \left[\frac{\hbar^2 \gamma_{21} \mu_0 c (\Delta^2 + \Gamma^2)}{|\mu_{12}|^2 \, 2\Gamma} \right]. \end{aligned} \tag{5.1.16}$$

Exercise: At a positive detuning from resonance equal to the linewidth (i.e., $\Delta = \Gamma$), does the intensity required to saturate the system increase or decrease, and by what factor? Does the saturation intensity depend on whether the detuning Δ is negative or positive?

5.1.2 Stationary Three-Level Atoms in a Traveling Wave

We now consider the interaction of traveling wave excitation of the form in Eq. (5.1.1) with a system that has one additional level. Introduction of this one additional level can have significant implications for optical behavior. In the scheme of Figure 5.2, it

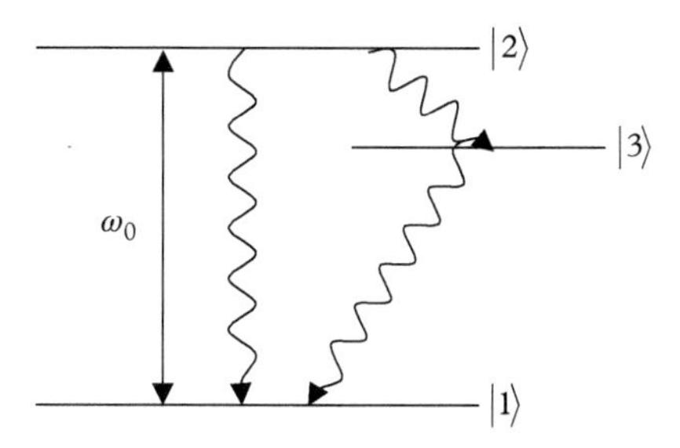

Figure 5.2 *Schematic of the excitation and relaxation processes of a three-level atom.*

is implicitly assumed that ω is near or on resonance for the $|1\rangle \leftrightarrow |2\rangle$ transition but is far off resonance for $|1\rangle \leftrightarrow |3\rangle$.

The equations of motion are

$$\dot{\rho}_{11} = \frac{1}{i\hbar}(V_{12}\rho_{21} - \rho_{12}V_{21}) + \gamma_{21}\rho_{22} + \gamma_{31}\rho_{33}, \tag{5.1.17}$$

$$\dot{\rho}_{22} = -\frac{1}{i\hbar}(V_{12}\rho_{21} - \rho_{12}V_{21}) - \gamma_{21}\rho_{22} - \gamma_{23}\rho_{22}, \tag{5.1.18}$$

$$\dot{\rho}_{33} = \gamma_{23}\rho_{22} - \gamma_{31}\rho_{33}, \tag{5.1.19}$$

$$\dot{\rho}_{12} = i\omega_0\rho_{12} - \frac{1}{i\hbar}V_{12}(\rho_{11} - \rho_{22}) - \Gamma_{21}\rho_{12}, \tag{5.1.20}$$

$$\rho_{21} = \rho_{12}^*. \tag{5.1.21}$$

In writing Eqs. (5.1.17)–(5.1.21), a very important simplifying assumption has been made regarding driven processes. Since no light fields were intentionally applied to couple state $|1\rangle$ to $|3\rangle$ or $|2\rangle$ to $|3\rangle$ directly, equations for the other possible coherences ρ_{13} and ρ_{23} were dropped from all equations. In point of fact, coherences do develop between other pairs of states, but provided the optical frequency ω has a small detuning with respect to $\omega_0 = \omega_{12}$, and large detunings with respect to ω_{13} and ω_{23}, these coherences are extremely small due to the detuning dependence evident in the denominator of Eq. (5.1.9).

Spontaneous processes in three-level systems lead to population transfer among the states. Population can build up in state 3 if it is long-lived, and significant changes in the distribution of population strongly alter the system saturation behavior. This can be shown starting from Eq. (5.1.19) by setting $\dot{\rho}_{33} = 0$ to examine steady-state behavior. We find

$$\rho_{33} = (\gamma_{23}/\gamma_{31})\rho_{22}. \tag{5.1.22}$$

To obtain ρ_{22} we first need to determine ρ_{12} from Eq. (5.1.20). Using the same procedure as in the last subsection, we find

$$\rho_{12} = \left[\frac{\Omega_{12}/2}{\Delta + i\Gamma}\right](\rho_{11} - \rho_{22})e^{i\omega t}. \tag{5.1.23}$$

This is identical to Eq. (5.1.9) for two-level systems. Using this in Eq. (5.1.18), and setting $\dot{\rho}_{22} = 0$ for the steady-state solution, we find

$$\rho_{22} = \left[\frac{(L + L^*)|\Omega_{12}/2|^2/\gamma_2}{1 + (L + L^*)|\Omega_{12}/2|^2/\gamma_2}\right]\rho_{11}, \tag{5.1.24}$$

where $\gamma_2 = \gamma_{21} + \gamma_{23}$ is the total relaxation rate from level 2. From Eq. (5.1.22) we then find

$$\rho_{33} = \left[\frac{(L + L^*)|\Omega_{12}/2|^2\,\gamma_{23}/\gamma_{31}\gamma_2}{1 + (L + L^*)|\Omega_{12}/2|^2/\gamma_2}\right]\rho_{11}. \tag{5.1.25}$$

Now by using the closure relation for three levels, namely

$$\rho_{11} + \rho_{22} + \rho_{33} = 1, \tag{5.1.26}$$

we can determine ρ_{11} explicitly by substituting Eqs. (5.1.24) and (5.1.25) into Eq. (5.1.26). The result is

$$\rho_{11} = \left[\frac{1 + (L + L^*)|\Omega_{12}/2|^2/\gamma_2}{1 + (L + L^*)(2 + \gamma_{23}/\gamma_{31})|\Omega_{12}/2|^2/\gamma_2}\right]. \tag{5.1.27}$$

Exercise: Do the solutions for $\rho_{11}, \rho_{22}, \rho_{33}$ present sensible values in the high-intensity limit ($|\Omega|^2 \to \infty$)? Do they agree with thermodynamic predictions for $T \to \infty$?

As before, absorption from the ground state is proportional to $\rho_{11} - \rho_{22}$. For the three-level system we find

$$\rho_{11} - \rho_{22} = \left[1 + \frac{|\Omega_{12}/2|^2}{\gamma_2}\left(2 + \frac{\gamma_{23}}{\gamma_{31}}\right)(L + L^*)\right]^{-1}. \tag{5.1.28}$$

This is of the form

$$\rho_{11} - \rho_{22} = \frac{1}{1 + I/I_{SAT}}, \tag{5.1.29}$$

where

$$I_{SAT} \equiv \frac{\hbar^2\gamma_2\mu_0 c}{|\mu_{12}|^2(1 + \gamma_{23}/2\gamma_{31})[L + L^*]} = \frac{\hbar^2\gamma_2\mu_0 c(\Delta^2 + \Gamma^2)}{2\Gamma|\mu_{12}|^2(1 + \gamma_{23}/2\gamma_{31})}. \tag{5.1.30}$$

Exercise: (i) If the rate of decay from level 2 to level 3 is much less than the rate from 2 to 1, the three-level system essentially reduces to a two-level system. Show that in this limit one obtains I_{SAT} (3 levels) = I_{SAT} (2 levels). (ii) Show by contrast that for finite γ_{23} the three-level saturation intensity can be significantly lower than that of the corresponding two-level system if level $|3\rangle$ is metastable (very long-lived).

An important thing to notice is that nonlinear behavior (i.e., saturable absorption) is more readily obtainable in three-level than two-level systems. However, one must then tolerate the losses and reduced speed of dynamics caused by population decay processes.

5.1.3 Stationary Two-Level Atoms in a Standing Wave

Consider the open system depicted in Figure 5.3 of stationary, two-level atoms acted upon by a standing wave field. Incoherent pumping is assumed to occur to both levels from unobserved external sources with rates that do not depend on the state of the system. This pumping is balanced by spontaneous decay from both levels. To simplify the calculation a little, spontaneous internal decay from level 2 to level 1 is ignored.

The interaction may be written

$$V(t) = -\frac{1}{2}\mu E_0 e^{i(\omega t - kz)} + c.c. - \frac{1}{2}\mu E_0 e^{i(\omega t + kz)} + c.c. = -2\hbar\Omega \cos kz \cos \omega t, \tag{5.1.31}$$

and the component equations of the density matrix are

$$\dot{\rho}_{11} = \lambda_1 - \gamma_1 \rho_{11} + 2i\Omega \cos kz \cos \omega t(\rho_{21} - \rho_{12}), \tag{5.1.32}$$

$$\dot{\rho}_{22} = \lambda_2 - \gamma_2 \rho_{22} - 2i\Omega \cos kz \cos \omega t(\rho_{21} - \rho_{12}), \tag{5.1.33}$$

$$\dot{\rho}_{12} = -(\Gamma_{12} - i\omega_0)\rho_{12} + 2i\Omega \cos kz \cos \omega t(\rho_{22} - \rho_{11}), \tag{5.1.34}$$

$$\rho_{21} = \rho_{12}^*. \tag{5.1.35}$$

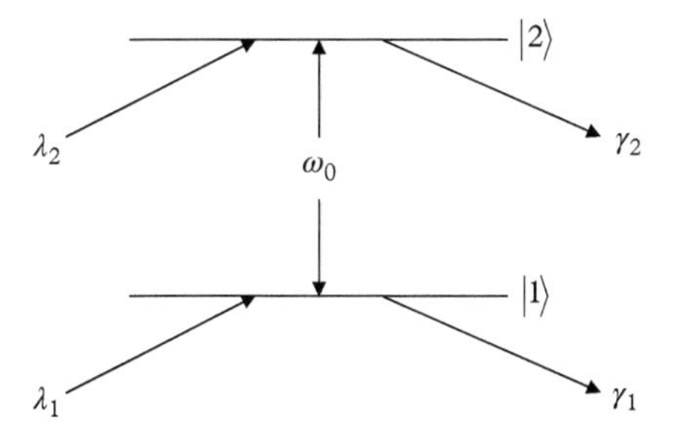

Figure 5.3 *Schematic diagram of a two-level atom interacting with a standing wave under conditions of incoherent pumping at rates λ_1 and λ_2.*

Letting

$$\rho_{12} = \tilde{\rho}_{12} e^{i\omega t} \tag{5.1.36}$$

as before, Eq. (5.1.34) yields

$$\dot{\tilde{\rho}}_{12} = -[\Gamma_{12} - i\Delta]\tilde{\rho}_{12} + 2i\Omega \cos kz\, (\rho_{22} - \rho_{11}). \tag{5.1.37}$$

In the steady state this gives

$$\tilde{\rho}_{12} = \left[\frac{2\Omega \cos kz}{\Delta + i\Gamma_{12}}\right] (\rho_{11} - \rho_{22}). \tag{5.1.38}$$

Using the RWA we can now evaluate the factor

$$\cos \omega t (\rho_{21} - \rho_{12}) = \frac{1}{2}\left(e^{i\omega t} + e^{-i\omega t}\right)\left(\tilde{\rho}_{21} e^{-i\omega t} - \tilde{\rho}_{12} e^{i\omega t}\right).$$

The result is

$$\cos \omega t (\rho_{21} - \rho_{12}) \cong \frac{1}{2}(\tilde{\rho}_{21} - \tilde{\rho}_{12}) = \frac{-i\Omega}{\Gamma_{21}} \cos kz (\rho_{22} - \rho_{11})\, L(\Delta). \tag{5.1.39}$$

Returning now to Eqs. (5.1.32) and (5.1.33), we can write

$$\dot{\rho}_{11} = \lambda_1 - \gamma_1 \rho_{11} + \frac{2\Omega^2}{\Gamma_{21}} \cos^2 kz (\rho_{22} - \rho_{11}) L(\Delta), \tag{5.1.40}$$

$$\dot{\rho}_{22} = \lambda_2 - \gamma_2 \rho_{22} - \frac{2\Omega^2}{\Gamma_{21}} \cos^2 kz (\rho_{22} - \rho_{11}) L(\Delta). \tag{5.1.41}$$

The Lorentzian factor $L(\Delta)$ is defined by $L(\Delta) \equiv \Gamma_{12}^2/(\Delta^2 + \Gamma_{12}^2)$. Steady-state solutions are:

$$\rho_{22} - \rho_{11} = \bar{N} / \left[1 + \left(\frac{2(\gamma_1 + \gamma_2)}{\gamma_1 \gamma_2 \Gamma_{21}}\right) \Omega^2 L(\Delta) \cos^2 kz\right], \tag{5.1.42}$$

$$\rho_{11} = \frac{\lambda_1}{\gamma_1} - \frac{2\Omega^2}{\gamma_1 \Gamma_{21}} L(\Delta) \cos^2 kz \cdot \bar{N} / \left[1 + 2\left(\frac{\gamma_1 + \gamma_2}{\gamma_1 \gamma_2 \Gamma_{21}}\right) \Omega^2 L(\Delta) \cos^2 kz\right], \tag{5.1.43}$$

$$\rho_{22} = \frac{\lambda_2}{\gamma_2} + \frac{2\Omega^2}{\gamma_2 \Gamma_{21}} L(\Delta) \cos^2 kz \cdot \bar{N} / \left[1 + 2\left(\frac{\gamma_1 + \gamma_2}{\gamma_1 \gamma_2 \Gamma_{21}}\right) \Omega^2 L(\Delta) \cos^2 kz\right]. \tag{5.1.44}$$

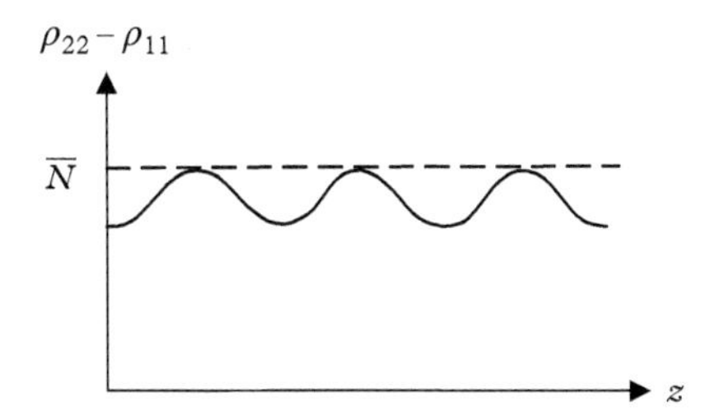

Figure 5.4 *Population difference versus position in a system of stationary two-level atoms subjected to standing wave excitation.*

In these expressions, following Ref. [5.1], we have introduced the average population difference $\bar{N} \equiv \left[\frac{\lambda_2}{\gamma_2} - \frac{\lambda_1}{\gamma_1}\right]$ that exists in the absence of light. Notice that atoms saturate in a way that depends on their location z with respect to the standing wave pattern. This is reflected not only in the population difference $\rho_{22} - \rho_{11}$, but in the fact that the power-broadened transition linewidth varies from point to point, as shown in Figure 5.4.

Exercise: Show explicitly that the population difference can be expressed in the form of a Lorentzian function of detuning Δ with a power-broadened linewidth of

$$\Gamma(z) = \Gamma_{21}\left[1 + \frac{2(\gamma_1+\gamma_2)}{\gamma_1\gamma_2\Gamma_{21}}\Omega^2\cos^2 kz\right]^{1/2}. \tag{5.1.45}$$

The off-diagonal density matrix element is found by using Eq. (5.1.42) in Eq. (5.1.38):

$$\begin{aligned}\tilde{\rho}_{12} &= \left[\frac{i\Omega_{12}\cos kz}{\Gamma_{21}-i\Delta}\right]\bar{N}/\left[1+\left(\frac{2(\gamma_1+\gamma_2)}{\gamma_1\gamma_2\Gamma_{21}}\right)\Omega^2 L(\Delta)\cos^2 kz\right] \\ &= \frac{2i\Omega_{12}\bar{N}\cos kz\,(\Gamma_{21}+i\Delta)}{\Delta^2+\Gamma_{21}^2\left[1+\frac{8(\gamma_1+\gamma_2)}{\gamma_1\gamma_2\Gamma_{21}}\Omega^2\cos^2 kz\right]}\end{aligned} \tag{5.1.46}$$

The total polarization of the sample is found in the usual way from

$$P(z,t) = N_0\langle\mu\rangle = N_0(\mu_{12}\rho_{21} + \rho_{12}\mu_{21}). \tag{5.1.47}$$

In this instance, however, a spatial average of Eq. (5.1.47) is needed to compute $P(z,t)$, since $\tilde{\rho}_{12}$ and $\tilde{\rho}_{21}$ depend on z. The atoms were assumed to be stationary, so they are distinguishable by position. The signal field radiated by the sample polarization must account for the spatial variations of polarization, requiring a less trivial integration of Eq. (3.9.15) than before.

Exercise: (i) Calculate the spatial average of $\tilde{\rho}_{12}$ given in (5.1.47). (ii) If the atoms were moving with the usual Doppler distribution of frequencies, the frequency average would involve a Lorentzian lineshape and a Gaussian distribution function. What lineshape would result?

The modulation of $\rho_{22} - \rho_{11}$ as a function of z is referred to as *spatial hole-burning*. In the presence of hole-burning the absorption spectrum shows dips or holes due to saturation. Holes may appear as a function of z as we have here, or as a function of frequency in inhomogeneously broadened systems. In the latter case we speak of *frequency-domain hole-burning*, a phenomenon which may be exploited for high-resolution saturation spectroscopy [5.2]. This is discussed in the subsequent subsection.

Exercise: Can frequency-domain hole-burning occur on a homogeneously broadened optical transition?

5.2 Moving Atoms

5.2.1 Moving Atoms in a Traveling Wave

We now consider moving two-level atoms subjected to traveling wave excitation in the open system of the last section. The interaction is written

$$V(t) = -(\mu E_0/2\hbar)\left(\exp[i(\omega t - kz)] + c.c.\right) = -\frac{1}{2}\hbar\Omega \exp[i(\omega t - kz)] + c.c. \quad (5.2.1)$$

Due to the atomic motion at velocity $\bar{\mathrm{v}}$, the full time derivative of ρ consists of its explicit time dependence plus a velocity-dependent contribution from the time dependence of the atomic position $z(t)$. The quantum mechanical transport equation is therefore modified to

$$i\hbar\left(\frac{\partial}{\partial t} + \bar{\mathrm{v}}\cdot\bar{\nabla}\right)\rho = [H, \rho] + \frac{d\rho}{dt}\Big|_{relax}. \quad (5.2.2)$$

$$\left(\frac{\partial}{\partial t} + \mathrm{v}_z\frac{\partial}{\partial z}\right)\rho_{11} + \gamma_1\rho_{11} = \lambda_1 - i\Omega\cos(kz - \omega t)\,(\rho_{12} - \rho_{21}), \quad (5.2.3)$$

$$\left(\frac{\partial}{\partial t} + \mathrm{v}_z\frac{\partial}{\partial t}\right)\rho_{22} + \gamma_2\rho_{22} = \lambda_2 + i\Omega\cos(kz - \omega t)\,(\rho_{12} - \rho_{21}), \quad (5.2.4)$$

$$\left(\frac{\partial}{\partial t} + \mathrm{v}_z\frac{\partial}{\partial z} - i\omega_0\right)\rho_{12} + \Gamma_{12}\rho_{12} = i\Omega\cos(kz - \omega t)\,(\rho_{22} - \rho_{11}). \quad (5.2.5)$$

As usual we set

$$\rho_{12} = \tilde{\rho}_{12} \exp[i(\omega t - kz)], \tag{5.2.6}$$

and solve for $\tilde{\rho}_{12}$ in steady state. If $\tilde{\rho}_{12}$ is sufficiently slowly varying in space and time, then the derivatives on the left side yield

$$\frac{d\rho_{12}}{dt} = \left(\frac{\partial}{\partial t} + \mathrm{v}\frac{\partial}{\partial z}\right)\rho_{12} = i(\omega - k\mathrm{v})\rho_{12}, \tag{5.2.7}$$

where the subscript z on velocity has been dropped. Then

$$\tilde{\rho}_{12} = i\left[\frac{(\Omega/2)}{\Gamma_{21} - i(\Delta + k\mathrm{v})}\right](\rho_{22} - \rho_{11}) \tag{5.2.8}$$

and

$$\rho_{22} - \rho_{11} = \bar{N}(\mathrm{v}) / \left[1 + \frac{2(\Omega/2)^2\,[\gamma_1 + \gamma_2]}{\Gamma_{21}\gamma_1\gamma_2} L(\Delta + k\mathrm{v})\right], \tag{5.2.9}$$

where the equilibrium occupation versus frequency, $\bar{N}(\mathrm{v}) \equiv \left(\frac{\lambda_2(\mathrm{v})}{\gamma_2} - \frac{\lambda_1(\mathrm{v})}{\gamma_1}\right)$, reflects the lineshape of the inhomogeneous velocity distribution. For a gas, this is given by the distribution of Doppler-shifted atomic resonance frequencies. $L(\Delta + k\mathrm{v})$ in (5.2.9) is the real Lorentzian factor defined in Eq. (5.1.39). The behavior of the population difference in Eq. (5.2.9) is illustrated in Figure 5.5.

Equation (5.2.9) can also be written as

$$\rho_{22} - \rho_{11} = \bar{N}(\mathrm{v})\left[1 - \frac{2I\eta\Gamma_{21}^2}{(\Delta + k\mathrm{v})^2 + \Gamma_{21}^2\,(1 + 2I\eta)}\right], \tag{5.2.10}$$

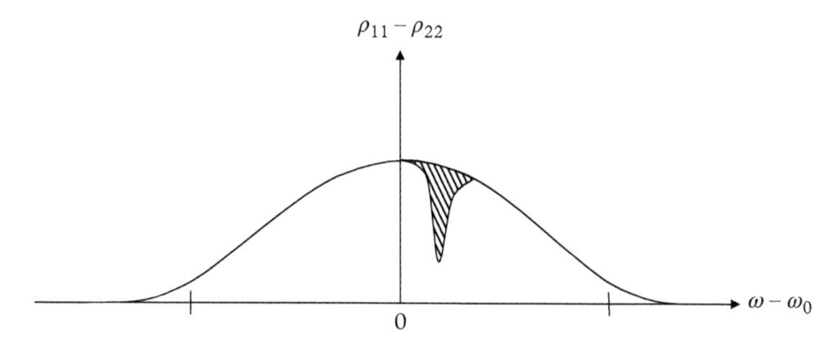

Figure 5.5 Spectral hole-burning *in an inhomogeneously broadened system of two-level atoms subjected to a single traveling wave. The shaded region shows the spectral hole that is burnt into the equilibrium distribution of atoms versus frequency (for negative $\bar{N}$).*

where $I \equiv 2(\Omega/2)^2 / \gamma_1\gamma_2$ and $\eta \equiv (\gamma_1 + \gamma_2)/2\Gamma_{12}$. From this expression it is more obvious that the deviation from the unperturbed value $-\bar{N}(\mathrm{v})$ is a power-broadened Lorentzian of width

$$\Gamma = \Gamma_{12}\,(1 + 2I\eta)^{1/2}. \tag{5.2.11}$$

Thus the atom–field interaction burns a hole in the population difference $\rho_{11} - \rho_{22}$. Figure 5.6 illustrates the population changes induced by the field in the individual states 1 and 2:

$$\rho_{11}(\mathrm{v}) = \frac{\lambda_1(\mathrm{v})}{\gamma_1} + \bar{N}(\mathrm{v})\frac{I\gamma_1\Gamma_{21}}{(\Delta + k\mathrm{v})^2 + \Gamma_{21}^2\,(1 + 2I\eta)}, \tag{5.2.12}$$

$$\rho_{22}(\mathrm{v}) = \frac{\lambda_2(\mathrm{v})}{\gamma_2} - \bar{N}(\mathrm{v})\frac{I\gamma_2\Gamma_{21}}{(\Delta + k\mathrm{v})^2 + \Gamma_{21}^2\,(1 + 2I\eta)} \tag{5.2.13}$$

Previously we obtained the response of the whole medium, the total polarization P, by calculating an average microscopic dipole and simply multiplying by N the number of absorbers per unit volume. Here the weighting function for atoms of a particular velocity is already included in the expression for ρ_{21} through the factor $\bar{N}(\mathrm{v})$. In the present situation, we must integrate over velocity in addition to multiplying by N:

$$P(z,t) = N\int_{-\infty}^{\infty} (\mu_{12}\rho_{21} + \mu_{21}\rho_{12})\,d\mathrm{v} \tag{5.2.14}$$

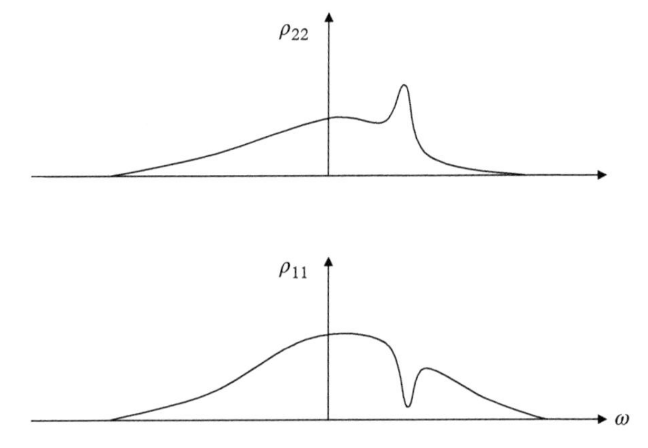

Figure 5.6 *Non-equilibrium features in the individual population distributions of states 1 and 2 of a system of moving two-level atoms subjected to traveling wave excitation.*

By comparing this with Eq. (3.10.2), which defines susceptibilities according to

$$P(z,t) = \frac{1}{2}\varepsilon_0 E_0 \left\{\chi(\omega)e^{-i(\omega t-kz)} + \chi(-\omega)e^{i(\omega t-kz)}\right\}, \tag{5.2.15}$$

we can look at the in-phase and out-of-phase response by evaluating real and imaginary parts of the linear susceptibility (quantities related by the Kramers–Kronig relations):

$$\chi = \chi' + i\chi'', \tag{5.2.16}$$

$$\chi(\omega) = \frac{N\mu}{\varepsilon_0 E_0}\int_{-\infty}^{\infty}\tilde{\rho}_{21}(\mathrm{v})d\mathrm{v} = -i\frac{N\mu^2}{\varepsilon_0\hbar}\int_{-\infty}^{\infty}\left[\frac{\Gamma_{21} - i(\Delta + k\mathrm{v})}{(\Delta + k\mathrm{v})^2 + \Gamma_{21}^2(1+2I\eta)}\right]\bar{N}(\mathrm{v})d\mathrm{v} \tag{5.2.17}$$

We assume $\bar{N}(\mathrm{v})$ has the form

$$\bar{N}(\mathrm{v}) = \frac{\Lambda_0}{\sqrt{\pi}\mathrm{v}_0}\exp\left(-\mathrm{v}^2/\mathrm{v}_0^2\right), \tag{5.2.18}$$

where v_0 is the 1/e width of the Maxwell–Boltzmann velocity distribution from statistical thermodynamics, and evaluate real and imaginary parts of Eq. (5.2.17) corresponding to dispersion and absorption of the sample, respectively. The imaginary or absorptive part is

$$\chi'' = \frac{N\Lambda_0\mu^2}{\sqrt{\pi}\varepsilon_0\hbar}\int_{-\infty}^{\infty}\left[\frac{\Gamma_{12}}{(\Delta + k\mathrm{v})^2 + \Gamma_{21}^2(1+2I\eta)}\right]\frac{\exp(-\mathrm{v}^2/\mathrm{v}_0^2)}{\mathrm{v}_0}d\mathrm{v}. \tag{5.2.19}$$

The integral in Eq. (5.2.19) is the convolution of a Gaussian and a Lorentzian. This is the plasma dispersion function for which tables and asymptotic forms are available in mathematical handbooks.

When the linewidth $k\mathrm{v}_0$ of the Doppler distribution greatly exceeds the homogeneous width Γ_{12} the plasma dispersion function has a simple asymptotic form, and

$$\chi'' = \frac{N\Lambda_0\mu^2 e^{-\Delta^2/k^2\mathrm{v}_0^2}}{\sqrt{\pi}\varepsilon_0\hbar h\mathrm{v}_0(1+2I\eta)^{1/2}}\int_{-\infty}^{\infty}\frac{\Gamma_p}{x^2+\Gamma_p^2}dx = \frac{\sqrt{\pi}N\Lambda_0\mu^2 e^{-\Delta^2/k^2\mathrm{v}_0^2}}{\varepsilon_0\hbar k\mathrm{v}_0(1+2I\eta)^{1/2}}. \tag{5.2.20}$$

This limit, in which $kv_0 \gg \Gamma_p \equiv (1 + 2I\eta)^{1/2}\,\Gamma_{12}$, is known as the Doppler limit and also permits asymptotic evaluation of the dispersion:

$$\begin{aligned}
\chi' &= -\frac{N\Lambda_0\mu^2}{\sqrt{\pi}\varepsilon_0\hbar v_0}\int_{-\infty}^{\infty}\frac{(\Delta + kv)\,e^{-v^2/v_0^2}}{(\Delta + kv)^2 + \Gamma_p^2}dv \\
&= -\frac{N\Lambda_0\mu^2}{\sqrt{\pi}\varepsilon_0\hbar v_0}\int_{-\infty}^{\infty}e^{-(x-\Delta)^2/k^2v_0^2}\frac{xdx}{x^2 + \Gamma_p^2} \\
&= -\frac{N\Lambda_0\mu^2}{\sqrt{\pi}\varepsilon_0\hbar v_0}e^{-\Delta^2/k^2v_0^2}\int_{-\infty}^{\infty}\frac{e^{2\Delta x/k^2v_0^2}x\,e^{-x^2/k^2v_0^2}}{x^2 + \Gamma_p^2}dx \\
&= -\frac{2\Delta N\Lambda_0\mu^2}{\sqrt{\pi}\varepsilon_0\hbar k^3v_0^3}e^{-\Delta^2/k^2v_0^2}\int_{-\infty}^{\infty}\frac{x^2}{x^2 + \Gamma_p^2}e^{-x^2/k^2v_0^2}dx \\
&= -\frac{2\Delta N\Lambda_0\mu^2}{\varepsilon_0\hbar\,(kv_0)^2}e^{-\Delta^2/k^2v_0^2}.
\end{aligned} \tag{5.2.21}$$

Notice that in the low-intensity limit, $I \to 0$, both χ' and χ'' are constants, so that we recover the usual relationship $P = \varepsilon_0\chi E$ for linear response with χ a constant. On the other hand, for high intensities $I \gg 1$ the absorption becomes nonlinear and $\chi'' \propto 1/\sqrt{I}$ in this inhomogeneously broadened system according to Eq. (5.2.20). This behavior is quite distinct from the absorption saturation in homogeneously broadened systems, as described by Eq. (5.1.15), for example (see Ref. [5.3]). There $\chi'' \propto 1/I$ for $I \gg 1$. Finally, notice that the dispersion, by which we mean χ', does not saturate. Physically this is because all atoms contribute to the dispersion, even in the low-intensity limit.

5.2.2 Moving Atoms in a Standing Wave

We consider the same standing wave interaction introduced previously in Eq. (5.1.29):

$$V(t) = -2\hbar\Omega\cos kz\cos\omega t \tag{5.2.22}$$

For moving atoms, Eqs. (5.1.30)–(5.1.33) become

$$\left(\frac{\partial}{\partial t} + v\frac{\partial}{\partial z}\right)\rho_{11} = \lambda_1 - \gamma_1\rho_{11} + 2i\Omega\cos kz\cos\omega t\,(\rho_{21} - \rho_{12}), \tag{5.2.23}$$

$$\left(\frac{\partial}{\partial t} + v\frac{\partial}{\partial z}\right)\rho_{22} = \lambda_2 - \gamma_2\rho_{22} - 2i\Omega\cos kz\cos\omega t\,(\rho_{21} - \rho_{12}), \tag{5.2.24}$$

$$\left(\frac{\partial}{\partial t} + v\frac{\partial}{\partial z}\right)\rho_{12} = -(\Gamma_{12} - i\omega_0)\rho_{12} + 2i\Omega\cos kz\cos\omega t(\rho_{22} - \rho_{11}), \tag{5.2.25}$$

$$\rho_{21} = \rho_{12}^*. \tag{5.2.26}$$

Since the standing wave corresponds to two oppositely directed traveling waves, we expect off-diagonal density matrix elements of the form

$$\rho_{21} = e^{-i\omega t}\left(\rho_+ e^{ikz} + \rho_- e^{-ikz}\right). \tag{5.2.27}$$

Populations will also display spatial variation, changing sinusoidally at twice the periodicity of either traveling wave. This is the same behavior we calculated earlier for stationary atoms in a standing wave field. Hence we expect populations of the form

$$\rho_{11} = \bar{\rho}_{11} + \left[\rho_1 e^{2ikz} + \rho_1^* e^{-2ikz}\right], \tag{5.2.28}$$

$$\rho_{22} = \bar{\rho}_{22} + \left[\rho_2 e^{2ikz} + \rho_2^* e^{-2ikz}\right], \tag{5.2.29}$$

where $\bar{\rho}_{11}, \bar{\rho}_{22}$ are constant, spatially averaged values of the populations.

Using the RWA we find

$$\cos kz \cos \omega t(\rho_{21} - \rho_{12}) = \frac{1}{4}\left\{(\rho_+ - \rho_-^*)\left[e^{2ikz} + 1\right] - (\rho_+^* - \rho_-)\ \left[e^{-2ikz} + 1\right]\right\} \tag{5.2.30}$$

$$\begin{aligned}\cos kz \cos \omega t(\rho_{22} - \rho_{11}) = \frac{1}{4}\Big\{&(\bar{\rho}_{22} - \bar{\rho}_{11})\left[e^{-i(\omega t+kz)} + e^{-i(\omega t-kz)}\right] \\ &+ (\rho_2 - \rho_1)e^{-i(\omega t-kz)} + \left(\rho_2^* - \rho_1^*\right)e^{-i(\omega t+kz)}\Big\}\end{aligned} \tag{5.2.31}$$

Substitution of Eqs. (5.2.28)–(5.2.31) into our starting equations gives us the main working equations of this subsection:

$$\begin{aligned}&(-i\omega + ikv)\rho_+ e^{-i(\omega t-kz)} + (-i\omega - ikv)\rho_- e^{-i(\omega t+kz)} = \\ &\quad - (\Gamma_{21} + i\omega_0)[\rho_+ e^{-i(\omega t-kz)} + \rho_- e^{-i(\omega t+kz)}] - i(\Omega/2)(\bar{\rho}_{22} - \bar{\rho}_{11})e^{-i(\omega t+kz)} \\ &\quad + [-i(\Omega/2)(\bar{\rho}_{22} - \bar{\rho}_{11}) - i(\Omega/2)(\rho_2 - \rho_1)]\, e^{-i(\omega t-kz)} - i(\Omega/2)(\rho_2^* - \rho_1^*)e^{-i(\omega t+kz)} \ldots,\end{aligned} \tag{5.2.32}$$

$$\begin{aligned}&(\gamma_1 + 2ikv)\rho_1 e^{2ikz} + (\gamma_1 - 2ikv)\rho_1^* e^{-2ikz} = \lambda_1 - \gamma_1\bar{\rho}_{11} + \frac{i\Omega}{4}\Big\{\rho_+\left[e^{2ikz} + 1\right], \\ &\quad + \rho_-\left[e^{-2ikz} + 1\right] - \rho_+^*\left[e^{-2ikz} + 1\right] - \rho_-^*\left[e^{2ikz} + 1\right]\Big\},\end{aligned} \tag{5.2.33}$$

$$\begin{aligned}&(\gamma_2 + 2ikv)\rho_2 e^{2ikz} + (\gamma_2 - 2ikv)\rho_2^* e^{-2ikz} = \lambda_2 - \gamma_2\bar{\rho}_{22} - \frac{i\Omega}{4}\Big\{\rho_+\left[e^{2ikz} + 1\right] \\ &\quad + \rho_-\left[e^{-2ikz} + 1\right] - \rho_+^*\left[e^{-2ikz} + 1\right] - \rho_-^*\left[e^{2ikz} + 1\right]\Big\}.\end{aligned} \tag{5.2.34}$$

Solution of these equations yields

$$\rho_2 - \rho_1 = -\frac{i\Omega}{4}\left[\frac{1}{\gamma_1 + 2ik\mathrm{v}} + \frac{1}{\gamma_2 + 2ik\mathrm{v}}\right](\rho_+ - \rho_-^*), \tag{5.2.35}$$

$$\bar{\rho}_{22} - \bar{\rho}_{11} = \bar{N} - \frac{i\Omega}{4}\left(\rho_+ + \rho_- - \rho_+^* - \rho_-^*\right)\left[\frac{1}{\gamma_1} + \frac{1}{\gamma_2}\right]. \tag{5.2.36}$$

Use of these two relations in Eq. (5.2.32), together with the complex conjugate of Eq. (5.2.32), permits an explicit solution for $(\rho_+ - \rho_-^*)$. One finds

$$(\rho_+ - \rho_-^*) = -i\Omega(L_+ + L_-^*)\bar{N}/\left[1 + \frac{\Omega^2}{4}\left(\frac{1}{\gamma_1} + \frac{1}{\gamma_2} + \frac{1}{\gamma_1 + 2ik\mathrm{v}} + \frac{1}{\gamma_2 + 2ik\mathrm{v}}\right)(L_+ + L_-^*)\right], \tag{5.2.37}$$

where

$$L_+ \equiv \frac{1}{i(\Delta + k\mathrm{v}) + \Gamma_{21}}; L_- \equiv \frac{1}{i(\Delta - k\mathrm{v}) + \Gamma_{21}}. \tag{5.2.38}$$

Substitution of Eq. (5.2.37) into Eq. (5.2.36) then yields the main result, namely an explicit expression for the average population difference in states 1 and 2 in the presence of light:

$$\bar{\rho}_{22} - \bar{\rho}_{11} = \bar{N}\left\{1 - \frac{\Omega^2}{4}\left(\frac{1}{\gamma_1} + \frac{1}{\gamma_2}\right)\left[\left((L_+ + L_-^*)^{-1} + \frac{\Omega^2}{4}\left(\frac{1}{\gamma_1} + \frac{1}{\gamma_2} + \frac{1}{\gamma_1 + 2ik\mathrm{v}} + \frac{1}{\gamma_2 + 2ik\mathrm{v}}\right)\right)^{-1} + c.c.\right]\right\}\ldots \tag{5.2.39}$$

Now let us examine this result in two special cases to understand its meaning.

(a) First, in the limit of no atomic motion ($\mathrm{v} \to 0$), evaluation of Eq. (5.2.39) simply gives

$$\bar{\rho}_{22} - \bar{\rho}_{11} = \bar{N} \tag{5.2.40}$$

and

$$\rho_{22} - \rho_{11} = \bar{N}\left\{1 - \frac{\Omega^2}{\Gamma}\left[\frac{(L_+ + L_-^*)}{1 + \Omega^2(L_+ + L_-^*)/\Gamma}\right]\cos 2kz\right\}, \tag{5.2.41}$$

where we have taken $\gamma_1 = \gamma_2 = \Gamma$ for simplicity. Notice we recover the simple spatial hole burning dependence of the population difference of Eq. (5.1.42) in this limit.

(b) Second, we can specialize to a closed two-level system by taking the following values for key rates in the system:

$$\lambda_2 = 0; \lambda_1 = \gamma_1 W(\mathrm{v}); \gamma_1 = \gamma_2 = \Gamma; \Gamma_{21} = \Gamma/2.$$

The quantity $W(\mathrm{v})$ normalizes the population at velocity v according to

$$\rho_{11}(\mathrm{v}) + \rho_{22}(\mathrm{v}) = W(\mathrm{v}) \tag{5.2.42}$$

and is therefore also given by the relation

$$W(\mathrm{v}) = -\bar{N}(\mathrm{v}) = \frac{\lambda_1(\mathrm{v})}{\Gamma(\mathrm{v})}, \tag{5.2.43}$$

when there is no pumping into level 2. These substitutions ensure that no time-dependent accumulation of probability takes place in the analysis. The ground state population of the ground state is constant in the absence of light. In this instance, the population difference in Eq. (5.2.36) is

$$\bar{\rho}_{22} - \bar{\rho}_{11} = \bar{N} - \frac{i\Omega}{2\Gamma}\left(\rho_+ + \rho_- - \rho_+^* - \rho_-^*\right). \tag{5.2.44}$$

By combining Eqs. (5.2.37) and (5.2.35) with Eqs. (5.2.30) and (5.2.31) we can obtain an expression for $\left(\rho_+ - \rho_-^*\right)$ in terms of $(\bar{\rho}_{22} - \bar{\rho}_{11})$ which is particularly amenable to analysis, showing that saturation of moving atoms exhibits features which are quite different from that of stationary atoms. The result is

$$\rho_+ - \rho_-^* = -i\Omega(L_+ + L_-^*)(\bar{\rho}_{22} - \bar{\rho}_{11})/\left[1 + 2(\Omega/2)^2(L_+ + L_-^*)/(\Gamma + 2ik\mathrm{v})\right]. \tag{5.2.45}$$

The difference between this quantity and its conjugate is what is needed to evaluate Eq. (5.2.44):

$$\left(\rho_+ - \rho_-^*\right) - \left(\rho_+^* - \rho_-\right) = \left\{\frac{-i\Omega(L_+ + L_-^*)(\bar{\rho}_{22} - \bar{\rho}_{11})}{1 + 2(\Omega/2)^2(L_+ + L_-^*)/(\Gamma + 2ik\mathrm{v})} + c.c.\right\}. \tag{5.2.46}$$

Now we can make use of Eq. (5.2.46) in Eq. (5.2.44) to determine the average population difference in the presence of light. This yields

$$\begin{aligned}\bar{\rho}_{22} - \bar{\rho}_{11} &= \bar{N} - \frac{\Omega^2}{2\Gamma}\left\{\frac{(\bar{\rho}_{22} - \bar{\rho}_{11})}{(L_+ + L_-^*)^{-1} + 2(\Omega/2)^2/(\Gamma + 2ik\mathrm{v})} + c.c.\right\} \\ &= \bar{N} - \frac{\Omega^2}{2\Gamma}\left\{\frac{(\Gamma + 2ik\mathrm{v})\,(\bar{\rho}_{22} - \bar{\rho}_{11})}{\Delta^2 - k^2\mathrm{v}^2 + i\Gamma k\mathrm{v} + (\Gamma/2)^2 + 2(\Omega/2)^2} + c.c.\right\}.\end{aligned} \tag{5.2.47}$$

The final result is

$$\bar{\rho}_{11} - \bar{\rho}_{22} = \frac{W(\mathrm{v})}{1 + 2(\Omega/2\Gamma)^2\left[L(k\mathrm{v} + \Omega') + L(k\mathrm{v} - \Omega')\right]}, \tag{5.2.48}$$

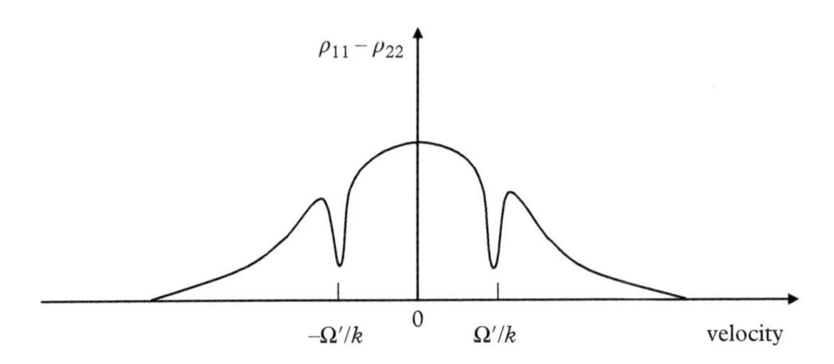

Figure 5.7 *Spectral hole-burning in a system of moving atoms subjected to standing wave excitation. In multi-level systems, hole-burning can be used for precision spectroscopy of energy level separations [5.2].*

where

$$\Omega' \equiv \left[\Delta^2 + (\Omega/2)^2/2\right]^{1/2} \tag{5.2.49}$$

and

$$L\,(x) \equiv (\Gamma\,/2)^2/\left[(\Gamma\,/2)^2 + x^2\right], \quad x \equiv k\mathrm{v} \pm \Omega'. \tag{5.2.50}$$

Notice in Figure 5.7 that two Lorentzian-shaped holes are burned in the population difference $\rho_{11} - \rho_{22}$ plotted as a function of velocity. These are frequency-domain holes since velocity is proportional to frequency shift across the Doppler-broadened lineshape. This makes obvious sense because the standing wave is composed of two traveling waves each of which burns its own hole.

Two remarkable consequences of the atomic motion under saturation conditions are contained in the population difference expression given by Eq. (5.2.48). First, the spectral holes generated by incident light are not located at the Doppler-shifted detuning values $k\mathrm{v} = \pm\Delta$. Instead holes are burned at slightly different values, given by

$$k\mathrm{v} = \pm\Omega' \cong \pm\left(\Delta + \Omega^2/16\Delta + \ldots\right). \tag{5.2.51}$$

This may be interpreted to mean that there is a light-induced shift of the atomic resonance (called the *light shift*) which is proportional to intensity. The magnitude of the shift is

$$\Delta\omega \cong \Omega^2/16\Delta. \tag{5.2.52}$$

The interaction between one of the waves and an atom is altered by the presence of the second wave in an unanticipated way. A second, related consequence is that if the pump frequency is tuned to exact resonance the absorption spectrum still appears to

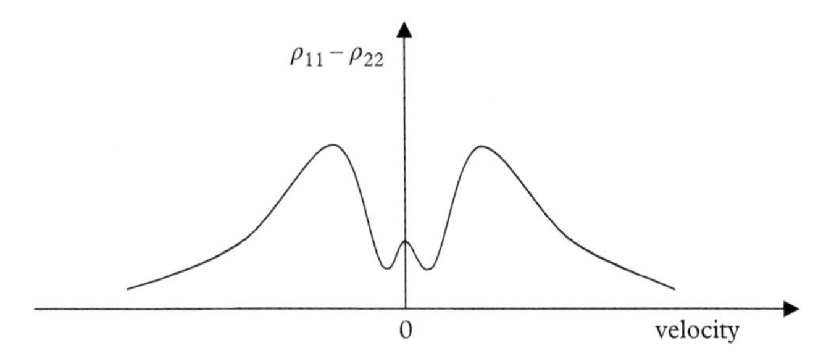

Figure 5.8 *Spectral hole-burning in a system of moving atoms subjected to standing wave excitation. The spectrum reveals that even at zero detuning* ($\Delta = 0$) *saturation produces a doublet due to the "light shift" effect.*

contain two features. The resonant hole-burning spectrum is a doublet. According to Eq. (5.2.48), when $\Delta = 0$ we still have two Lorentzians superimposed on the velocity distribution, centered at

$$kv = \pm\Omega/2\sqrt{2}. \tag{5.2.53}$$

The doublet predicted for zero detuning in Figure 5.8 is a reflection of AC Stark or Rabi splitting under resonant pumping conditions. However, while the analysis presented here anticipates Rabi splitting and the light shift effect, the analysis is not strictly valid when $\Delta = 0$, since we assumed $\Delta \gg \Gamma$ in the derivation. A proper analysis of strong-field effects on resonance will again have to await the analytic approaches of Chapter 6. Additional reading on the light shift may be found in Ref. [5.4] and suppression of the light shift for the purpose of improving measurement precision is discussed in Ref. [5.5].

5.3 Tri-Level Coherence

5.3.1 Two-Photon Coherence

We now turn to interactions involving two fields and three or more energy levels. Figure 5.9 shows three basic types of three-level systems. In discussing coherence involving three levels, we shall focus on only one of these, the cascade configuration shown in Figure 5.9(a), since only slight modifications of the analysis are needed to describe V- and Λ-configurations.

An important working assumption is that level 1 is optically connected to level 2 ($\mu_{12} \neq 0$), level 2 is connected to level 3 ($\mu_{23} \neq 0$), but level 1 is not connected to level 3 ($\mu_{13} = 0$). These properties are of course determined by the parity of the eigenstates, which must change for a transition to be electric dipole (ED) allowed. If the lowest level is assumed to be even parity, the intermediate state must be odd and the upper state even. Hence the transition directly from state 1 to state 3 has no change of parity and is not ED allowed (see Section 2.2.3). We shall also assume that the frequencies of two

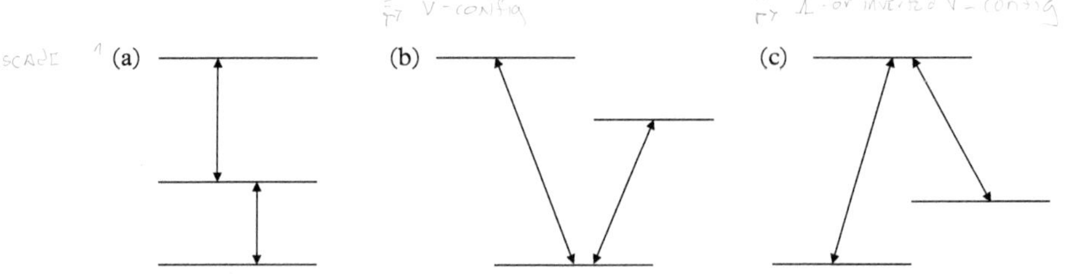

Figure 5.9 *Basic configurations of energy levels and allowed transitions in three-level systems. (a) Cascade, (b)* V*-configuration, and (c)* Λ*- or inverted-*V *configuration.*

incident light fields E_1 and E_2 are close enough to their intended transitions that they do not induce unintended transitions at any other frequency in the system:

$$E_1(z,t) = E_1 \cos(\omega_1 t + k_1 z), \tag{5.3.1}$$

$$E_2(z,t) = E_2 \cos(\omega_2 t + k_2 z). \tag{5.3.2}$$

Since we now have more than one transition to consider, the resonant Rabi frequency acquires a subscript to label the relevant transition, according to

$$\Omega_i \equiv \mu_i E_i/\hbar, \tag{5.3.3}$$

where $i = 1, 2$ and μ_i is the dipole moment associated with field amplitude E_i. In the remainder of this section, the resonant Rabi frequency will be treated as a real quantity.

The density matrix equation may then be written out in two groups as populations:

$$\dot{\rho}_{11} = \lambda_1 - \gamma_1 \rho_{11} + i\Omega_1 \cos(\omega_1 t + k_1 z)(\rho_{21} - \rho_{12}), \tag{5.3.4a}$$

$$\dot{\rho}_{22} = -\gamma_2 \rho_{22} - i\Omega_1 \cos(\omega_1 t + k_1 z)(\rho_{21} - \rho_{12}) + i\Omega_2 \cos(\omega_2 t + k_2 z)(\rho_{32} - \rho_{23}), \tag{5.3.4b}$$

$$\dot{\rho}_{33} = -\gamma_3 \rho_{33} - i\Omega_2 \cos(\omega_2 t + k_2 z)(\rho_{32} - \rho_{23}), \tag{5.3.4c}$$

and coherences:

$$\dot{\rho}_{21} = -(\Gamma_{21} + i\omega_{21})\rho_{21} - i\Omega_1 \cos(\omega_1 t + k_1 z)(\rho_{22} - \rho_{11}) + i\Omega_2 \cos(\omega_2 t + k_2 z)\rho_{31}, \tag{5.3.5a}$$

$$\dot{\rho}_{31} = -(\Gamma_{31} + i\omega_{31})\rho_{31} - i\Omega_1 \cos(\omega_1 t + k_1 z)\rho_{32} + i\Omega_2 \cos(\omega_2 t + k_2 z)\rho_{21}, \tag{5.3.5b}$$

$$\dot{\rho}_{32} = -(\Gamma_{32} + i\omega_{32})\rho_{32} - i\Omega_2 \cos(\omega_2 t + k_2 z)(\rho_{33} - \rho_{22}) - i\Omega_1 \cos(\omega_1 t + k_1 z)\rho_{31}. \tag{5.3.5c}$$

Since several resonant frequencies appear in Eqs. (5.3.5), subscripts are used to reflect the levels involved. For example, ω_{21} is the positive frequency difference between levels 1 and 2.

In the last section we considered values of the incoherent pumping rates that effectively closed the system. If we assume there is no pumping except to the lowest level, we can again use this approach:

$$\rho_{11}^{(0)} = -\bar{N} = \lambda_1/\gamma_1. \tag{5.3.6}$$

Now the RWA may be applied by writing the coherences as

$$\begin{aligned} \rho_{21} &= \tilde{\rho}_{21} \exp[-i(\omega_1 t - k_1 z)], \\ \rho_{32} &= \tilde{\rho}_{32} \exp[-i(\omega_2 t - k_2 z)], \\ \rho_{31} &= \tilde{\rho}_{31} \exp[-i(\omega_1 + \omega_2)t + i(k_1 + k_2)z]. \end{aligned} \tag{5.3.7}$$

The main purpose of this section is to illustrate that coherences in tri-level systems can be of two types, either radiative or non-radiative. To demonstrate this a perturbation approach may be used, in which we proceed to third order in the off-diagonal term ρ_{32}. Three detuning factors are encountered as the result of this, namely $\Delta_1 \equiv \omega_{21} - \omega_1 + k_1 \mathrm{v}$, $\Delta_2 \equiv \omega_{32} - \omega_2 + k_2 v$, and $\Delta_3 \equiv \omega_{31} - (\omega_1 + \omega_2) + (k_1 + k_2)\mathrm{v}$. Assuming the system starts in the lowest state, field E_1 alone is able to initiate dynamics in first order and field E_2 acts beginning in second order. So the first-order coherence in Eq. (5.3.5a) acquires the familiar form

$$\tilde{\rho}_{21}^{(1)} = \frac{(\Omega_1/2)}{(\Delta_1 - i\Gamma_{21})}(-\bar{N}). \tag{5.3.8}$$

Population changes make their first appearance in second order. Since $\rho_{32}^{(1)} = 0$, the solutions for populations in Eqs. (5.3.4a) and (5.3.4b) are

$$\rho_{11}^{(2)} = \left[1 - \frac{(\Omega_1/2)^2 2\Gamma_{21}/\gamma_2}{\Delta_1^2 + \Gamma_{21}^2}\right](-\bar{N}), \tag{5.3.9}$$

$$\rho_{22}^{(2)} = \left[\frac{(\Omega_1/2)^2 2\Gamma_{21}/\gamma_2}{\Delta_1^2 + \Gamma_{21}^2}\right](-\bar{N}) \tag{5.3.10}$$

Because a second field is applied beginning in second order of perturbation, a contribution to ρ_{31} also emerges in second order. This is the result of the structure of Eq. (5.3.5b). A non-zero amplitude $\tilde{\rho}_{31}^{(2)}$ can arise through the dependence on $\tilde{\rho}_{21}^{(1)}$ in the last term on the right side of Eq. (5.3.5b). Expressing the slowly varying amplitude in terms of $\tilde{\rho}_{21}^{(1)}$, one finds

$$\tilde{\rho}_{31}^{(2)} = \frac{(\Omega_2/2)}{(\Delta_3 - i\Gamma_{31})}\tilde{\rho}_{21}^{(1)}. \tag{5.3.11}$$

Proceeding to third order, contributions to ρ_{32} emerge from the last two terms on the right side of Eq. (5.3.5c). Since $\rho_{33}^{(2)} = 0$, and $\tilde{\rho}_{22}^{(2)}$ and $\tilde{\rho}_{31}^{(2)}$ are already given in Eqs. (5.3.10) and (5.3.11), this coherence can be written entirely in terms of $\tilde{\rho}_{21}^{(1)}$ as follows:

$$\begin{aligned}\rho_{32}^{(3)} &= (\Omega_2/2)\rho_{22}^{(2)} - (\Omega_1/2)\tilde{\rho}_{31}^{(2)} \\ &= -i\left(\frac{\Omega_1}{2\gamma_2}\right)\left(\frac{\Omega_2}{2}\right)\left(\tilde{\rho}_{21}^{(1)} - \tilde{\rho}_{21}^{(1)*}\right) - \frac{(\Omega_1/2)(\Omega_2/2)}{(\Delta_3 - i\Gamma_{31})}\tilde{\rho}_{21}^{(1)}.\end{aligned} \tag{5.3.12}$$

By substituting Eqs. (5.3.10) and (5.3.11) into Eq. (5.3.12) the solutions can also be written out explicitly in terms of $\bar{N}$ and the fields as

$$\tilde{\rho}_{31}^{(2)} = -\bar{N}\frac{(\Omega_1/2)(\Omega_2/2)}{(\Delta_1 - i\Gamma_{21})(\Delta_3 - i\Gamma_{31})}, \tag{5.3.13}$$

$$\tilde{\rho}_{32}^{(3)} = -\bar{N}\frac{(\Omega_1/2)^2(\Omega_2/2)}{(\Delta_2 - i\Gamma_{32})}\left\{\frac{2\Gamma_{21}/\gamma_2}{(\Delta_1^2 + \Gamma_{21}^2)} - \frac{1}{(\Delta_1 - i\Gamma_{21})(\Delta_3 - i\Gamma_{31})}\right\}. \tag{5.3.14}$$

Notice that according to Eqs. (5.3.11) and (5.3.12) both the tri-level coherences $\tilde{\rho}_{31}^{(2)}$ and $\tilde{\rho}_{32}^{(3)}$ are proportional to the primary coherence in the system, namely $\tilde{\rho}_{21}^{(1)}$. They share this common feature. A fundamental difference nevertheless exists between $\tilde{\rho}_{31}^{(2)}$ and $\tilde{\rho}_{32}^{(3)}$ in regard to their effect on system polarization. The coherence between levels 2 and 3 leads to polarization and radiation, whereas that between levels 1 and 3 does not. This is because the polarization in the former case, developed on the 3↔2 transition, is

$$P(\omega_2) = (\mu_{32}\rho_{23} + \mu_{23}\rho_{32}) \neq 0.$$

Both the dipole moment μ_{32} and the coherence $\tilde{\rho}_{32}^{(3)}$ itself are non-zero. By contrast, the coherence between levels 1 and 3 does not radiate, because

$$P(\omega_1 + \omega_2) = (\mu_{31}\rho_{13} + \mu_{13}\rho_{31}) = 0.$$

In this case, the coherence amplitude is not zero, but the dipole moment μ_{13} is zero. The excitation $\tilde{\rho}_{31}^{(2)}$ is therefore an example of a non-radiative, two-photon coherence.

5.3.2 Zeeman Coherence

We now consider another kind of tri-level coherence, involving magnetic sublevels of a nominally two-level system. So far, we have considered the light to be an essentially scalar field E, albeit one that is polarized in some particular direction. We have ignored the possibility that the field conveys angular momentum to the atoms or carries it away. It was assumed that the medium responds merely by charge motion along the polarization direction. Consequently we reduced the vector interaction $V = -\bar{\mu} \cdot \bar{E}$ to the scalar quantity μE_0. However, this is not always appropriate. Strictly speaking, the field $\bar{E}$ is vectorial and is capable of changing the angular momentum state of atoms, or the magnetic aspects of the wavefunction, when its axial rather than polar components exert themselves.

To investigate the coherences which can be established between selected magnetic sublevels using circularly polarized light we shall apply second-order perturbation theory to the density matrix and use the Wigner–Eckart (W-E) theorem (Appendix H) to keep track of vector properties of V. As usual we need to determine matrix elements $V_{ij} = \langle i | V | j \rangle$ of the atom–field interaction. Consider a two-level system with magnetic degeneracy like that in Figure 5.10. We write the field in irreducible form as

$$\bar{E}(z,t) = -\frac{1}{2}\left\{E_+\hat{\varepsilon}_- e^{-i(\omega t-kz)} + E_-\hat{\varepsilon}_+ e^{-i(\omega t-kz)}\right\} + c.c., \tag{5.3.15}$$

where

$$\hat{\varepsilon}_\pm = \mp\left(\hat{x} \pm i\hat{y}\right)/\sqrt{2} \tag{5.3.16}$$

are the transverse basis states of the rank one spherical tensor. By writing the field–atom interaction as an irreducible tensor, we can make use of the W-E theorem to assist with the evaluation of its matrix elements.

If $T_q^{(k)}$ is an irreducible spherical tensor operator, then according to the W-E theorem its matrix element can be decomposed as follows:

$$\left\langle \alpha jm \left| T_q^{(k)} \right| \alpha' j' m' \right\rangle = (-1)^{j-m} \left\langle \alpha j \left\| T^{(k)}\ \alpha' j' \right\rangle \right\| \alpha' j' \right\rangle \begin{pmatrix} -j & k & j' \\ -m & q & m' \end{pmatrix}, \tag{5.3.17}$$

where $\left\langle \alpha j \left\| T^{(k)} \right\| \alpha' j' \right\rangle$ is called the reduced matrix element and $\begin{pmatrix} -j & k & j' \\ -m & q & m'' \end{pmatrix}$ is a Wigner $3j$ symbol. The importance of the W-E theorem is that it separates the matrix element into two distinct parts. The $3j$ symbol contains all the geometrical aspects of the problem and is tabulated in many references [5.6]. The so-called "reduced" matrix element $\langle \| T \| \rangle$ is the "real" multipole transition moment, a quantity that is independent of coordinate

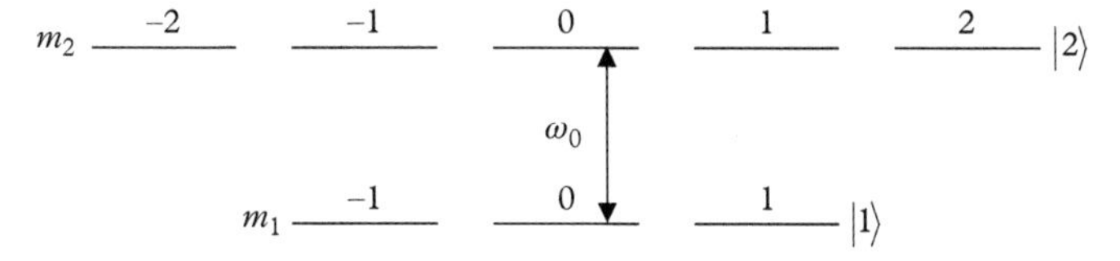

m_2: −2, −1, 0, 1, 2 $|2\rangle$

ω_0

m_1: −1, 0, 1 $|1\rangle$

Figure 5.10 *Magnetic sublevels of a quasi-two-level system with $J = 1$ ground state and a $J = 2$ excited state. Sublevels are distinguished by m_j, the quantum number specifying projections of J on the quantization axis for each state.*

system or geometry, which otherwise would be a rather ill-defined (origin-dependent) quantity in complex vector spaces.

The first step in applying Eq. (5.3.17) to the problem at hand is to write the interaction Hamiltonian V itself as an irreducible tensor. $V = -\bar{\mu} \cdot \bar{E}$ is a scalar product of the field vector and the dipole moment, which is given by

$$\begin{aligned}\bar{\mu} &= e\bar{r} = e\left[x\hat{x} + y\hat{y} + z\hat{z}\right] \\ &= \frac{1}{2}er\left[\left(\hat{x} - i\hat{y}\right)\exp i\phi + \left(\hat{x} + i\hat{y}\right)\exp\left(-i\phi\right)\right] + \hat{z}er\cos\theta \\ &= er\left[\sum_{q=-1}^{1}(-1)^q C_q^{(1)}\hat{\varepsilon}_{-q}\right].\end{aligned} \tag{5.3.18}$$

Here we have introduced the quantity $C_q^{(\ell)}$, which is the Racah tensor of rank l, related to the spherical harmonics by $C_q^{(\ell)} \equiv \left(\frac{4\pi}{2\ell+1}\right)^{1/2} Y_q^{(\ell)}$. On the basis of Eq. (5.3.18), we see that the general dipole moment μ has three irreducible tensor components:

$$\mu_\pm = -erC_{\pm 1}^{(1)},$$

$$\mu_0 = erC_0^{(1)}.$$

The field $\bar{E}$ in Eq. (5.3.15) is already in irreducible tensor form, since $\hat{\varepsilon}_\pm$ are basis vectors of the rank one spherical tensor. Ignoring time dependence, the interaction is therefore

$$\begin{aligned}V = -\bar{\mu} \cdot \bar{E} &= -\frac{1}{2}\sum_q (-1)^q \mu_q E_{-q} \\ &= +\frac{1}{2}er\left\{\sum_q (-1)^q\, C_q^{(\ell)} E_{-q}\right\} \\ &= -\frac{1}{2}\left\{\mu_+ E_- + \mu_- E_+\right\}.\end{aligned} \tag{5.3.19}$$

We need matrix elements of V starting from state j and ending up in state i. Adopting the notation $|j\rangle = |\alpha' j' m'\rangle$ and $|i\rangle = |\alpha\, j\, m\rangle$ for initial and final states we find

$$\left\langle V_{ij}\right\rangle_{E_-} = \langle i|\, V_-\, |j\rangle = \frac{1}{2}\left\langle \alpha\, j\, m\,|\mu_+ E_-|\, \alpha' j' m'\right\rangle = \frac{1}{2}eE_-\left\langle \alpha\, j\, m\left|rC_+^{(1)}\right| \alpha' j' m'\right\rangle,$$

$$\langle i|\, V_-\, |j\rangle = \frac{eE_-}{2}\left\langle \alpha j\left\|rC^{(1)}\right\| \alpha' j'\right\rangle (-1)^{j-m}\begin{pmatrix} j & l & j' \\ -m & l & m' \end{pmatrix}, \tag{5.3.20}$$

by using the W-E theorem. $\begin{pmatrix} j & l & j' \\ -m & l & m' \end{pmatrix}$ is a Wigner $3j$ symbol which renders the matrix element V_{ij} zero unless

$$\begin{aligned} \Delta j &= j - j' = 0, \pm 1, \\ \Delta m &= m - m' = +1, \\ j &= 0 \leftarrow \times \rightarrow j' = 0 \end{aligned} \tag{5.3.21}$$

These selection rules show that to first order in the perturbation, positive helicity light ($\bar{E}_- \propto E_- \hat{\varepsilon}_+$) excites only absorptive transitions in which m increases by one unit to $m = m' + 1$. That is, light with a positive projection of angular momentum on $\hat{z}$ adds one unit of momentum to the atom per absorbed quantum. Conversely in emission it carries away one unit of angular momentum. This behavior is illustrated in Figure 5.11 for a $J = 0 \rightarrow J = 1$ transition of a system that is initially in the ground state and is subjected to circularly polarized light of positive or negative helicity. Only one transition takes place, with an attendant change in the magnetic quantum number of the occupied state.

Now consider what happens in a second-order interaction with a single circularly polarized field, using perturbation theory (Appendix D):

$$i\hbar \dot{\rho}^{(2)} = \left[H_0, \rho^{(2)}\right] + \left[V^{(1)}, \rho^{(1)}\right] - i\hbar(\partial \rho^{(2)}/\partial t)_{rel} \tag{5.3.22}$$

With the RWA, the interaction matrix elements have the form

$$V_{m_{j2}m_{j1}} = -\frac{1}{2}\left\{\mu_-^{(m)} E_+ + \mu_+^{(m)} E_-\right\} \exp[-i(\omega t - kz)] \tag{5.3.23}$$

and

$$V_{m_{j1}m_{j2}} = -\frac{1}{2}\left\{\mu_-^{(m)} E_+ + \mu_+^{(m)} E_-\right\}^* \exp[i(\omega t - kz)]. \tag{5.3.24}$$

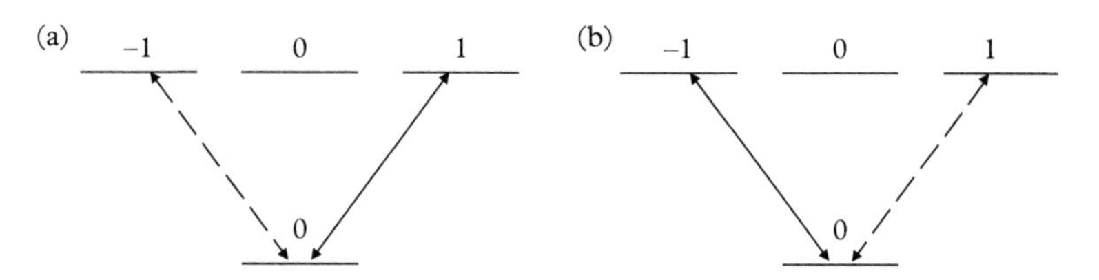

Figure 5.11 *Allowed absorption transitions (solid arrows) induced to different magnetic sublevels by circularly polarized light with (a) positive helicity (Δm = +1) and (b) negative helicity (Δm = −1). Forbidden transitions are dashed.*

For stationary two-level atoms we readily find the off-diagonal component

$$-i\omega\rho^{(1)}_{m_2m_1} = -i\omega_0\rho_{m_2m_1} - (i/\hbar)V_{m_2m_1}\left(\rho^{(0)}_{m_1m_1} - \rho^{(0)}_{m_2m_2}\right) - \Gamma_{m_2m_1}\rho^{(1)}_{m_2m_1}, \tag{5.3.25}$$

$$\tilde{\rho}^{(1)}_{m_2m_1} = \frac{\tilde{V}_{m_2m_1}}{\hbar[\Delta - i\Gamma_{m_2m_1}]}\left(\rho^{(0)}_{m_1m_1} - \rho^{(0)}_{m_2m_2}\right). \tag{5.3.26}$$

In second order the populations of states 1 and 2 change. Since there are sub-states within state 2, coherent relationships may arise between sub-state populations. Hence we encounter density matrix elements such as:

$$\begin{aligned}\rho^{(2)}_{m_2'm_2} &= -(i/\hbar\gamma_{m_2})\left[\tilde{V}_{m_2'm_1}\tilde{\rho}^{(1)}_{m_1m_2} - \tilde{\rho}^{(1)}_{m_2m_1}\tilde{V}_{m_1m_2'}\right] \\ &= \frac{-2\Gamma_{m_2m_1}}{\hbar^2\gamma_{m_2}}\left[\frac{\tilde{V}_{m_2'm_1}\tilde{V}^*_{m_2m_1}}{\Delta^2 + (\Gamma_{m_2m_1})^2}\right]\left(\rho^{(0)}_{m_2m_2} - \rho^{(0)}_{m_1m_1}\right).\end{aligned} \tag{5.3.27}$$

Upon substitution of the irreducible forms of the interaction given by Eqs. (5.3.23) and (5.3.24), the excited state population reduces to

$$\begin{aligned}\rho^{(2)}_{m_2'm_2} = \frac{\Gamma}{2\hbar^2\gamma_{m_2}[\Delta^2 + (\Gamma_{m_2m_1})^2]}&\left[\{<\mu_+^*>E_-^*<\mu_+>E_-\} + <\mu_+^*>E_-^*<\mu_->E_+\right. \\ &\left. + <\mu_-^*>E_+^*<\mu_+>E_- + \{<\mu_-^*>E_+^*<\mu_->E_+\}\right]\left(\rho^{(0)}_{m_1m_1} - \rho^{(0)}_{m_2m_2}\right).\end{aligned} \tag{5.3.28}$$

The first and last terms in Eq. (5.3.28), the ones in curly brackets, involve a sequence of two transitions that individually obey opposite selection rules. However, they are similar overall, because they involve the field combinations $E_-^*E_-$ and $E_+^*E_+$, and mediate two-photon transitions that result in no net angular momentum exchange with the atom. This process merely returns population to its original state, as shown in Figure 5.12, while modifying the occupation probability. This is the origin of common saturation, which is an example of nonlinear susceptibility. In this case the overall selection rule on the initial and final magnetic quantum numbers is

$$m_2 - m_2' = 0. \tag{5.3.29}$$

The other two terms in Eq. (5.3.28) are driven by the field combinations $E_-^*E_+$ and $E_+^*E_-$. These again mediate sequences of two transitions, as shown in Figure 5.13, but

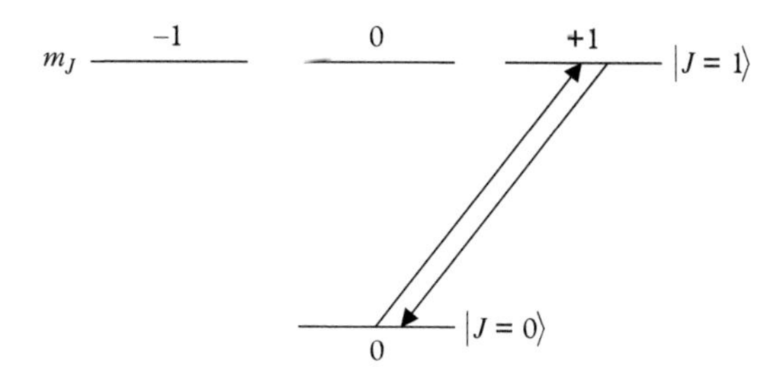

Figure 5.12 *A two-photon transition stimulated by circularly polarized light in a two-level system with magnetically degenerate level 2. The energy states are labeled by their total angular momentum quantum numbers $J = 1$ and $J = 0$ and their projections m_J on the quantization axis.*

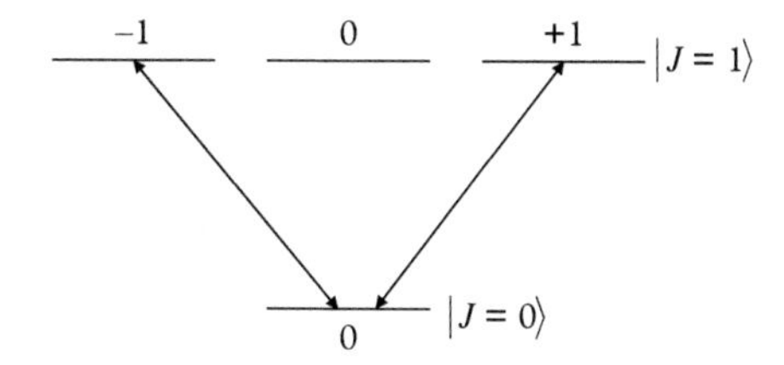

Figure 5.13 *A pair of allowed electric dipole transitions that connect two magnetic sublevels of the same state to establish Zeeman coherence via a two-photon process.*

in this case there is an overall exchange of angular momentum. These two-photon transitions obey the selection rule

$$m_2 - m_2' = \pm 2. \tag{5.3.30}$$

There is an important difference between the two types of two-photon transition present in the second-order perturbation calculation. The first type connects only two sublevels, whereas the latter connects three. This introduces a new degree of freedom in off-diagonal coherence terms $\rho^{(2)}_{m_2' m_2}$ established by interactions like $E_-^* E_-$ and $E_+^* E_+$ that makes them different from an ordinary saturation process. Figure 5.13 makes it clear that these interactions couple degenerate sublevels with the same J value. Thus it is possible to create phased charge oscillation between sublevels of the *same* state via this two-step process. This is another example of a non-radiative two-photon polarization called a Zeeman coherence.

Zeeman coherences first appear in second order, because the field must act at least twice to return the atom to a different sublevel of the original degenerate manifold. However, such polarizations are not proportional to diagonal elements of the density matrix, which would be population terms. $\rho_{m_1' m_1}$ is a diagonal element of the density matrix only if $m_j' = m_j$. If $m_j' \neq m_j$, the initial and final quantum numbers are different and the matrix element is off-diagonal. This is because there is no energy difference between the initial and final coupled levels that the coherence is non-radiative. In Chapter 7, additional important applications of non-radiative or "dark state" coherences will be covered.

The calculation of the Zeeman coherence for a case like the system shown in Figure 5.13 can be completed by setting $\rho_{22}^{(0)} \cong 0$ and assuming that thermal equilibrium among sublevels makes their populations equal. Then each of these sublevel populations has a value $\approx \rho_{11}^{(0)}/3$. Consequently the Zeeman coherence terms alone give the result

$$\rho_{1,-1}^{(2)} = \frac{\Gamma_{12}\rho_{11}^{(0)}}{6\hbar^2\gamma_2(\Delta^2+\Gamma_{12}^2)}\left\{\left(E_-^*E_+\right)\left\langle 1,-1\left|\mu_+^*\mu_-\right|11\right\rangle + \left(E_+^*E_-\right)\left\langle 11\left|\mu_-^*\mu_+\right|1,-1\right\rangle\right\}. \tag{5.3.31}$$

That this two-photon coherence is non-zero may readily be verified by including intermediate states in the two-photon matrix elements of Eq. (5.3.31). Angular momentum is conserved since the $\pm 2\hbar$ furnished by the field is acquired in two allowed steps during the atomic transitions:

$$\left\langle 11\left|\mu_-^*\right|00\right\rangle\langle 00\left|\mu_+\right|1,-1\rangle \neq 0, \qquad (\Delta m = +2) \tag{5.3.32}$$

or

$$\left\langle 1,-1\left|\mu_+^*\right|00\right\rangle\langle 00\left|\mu_-\right|11\rangle \neq 0, \qquad (\Delta m = -2). \tag{5.3.33}$$

5.4 Coherent Multiple Field Interactions

5.4.1 Four-Wave Mixing

Consider a three-level system with three collinear waves E_f, E_b, and E_p incident on it, as shown in Figure 5.14. Using perturbation theory, we wish to show in a simple way that a coherence appears *in third order* which radiates a phase conjugate, time-reversed replica of an input probe wave E_p.

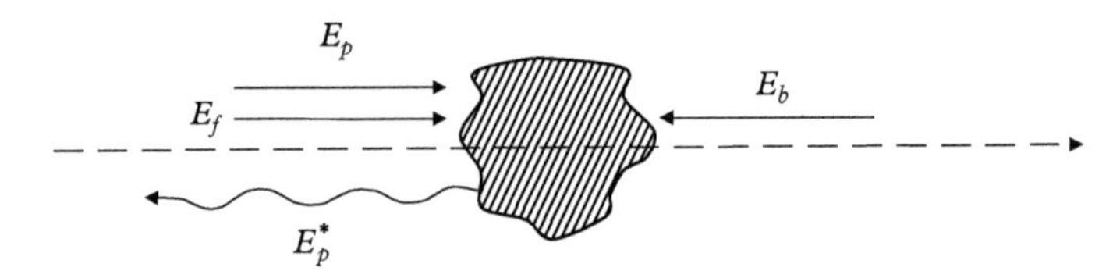

Figure 5.14 *Three incident fields impinging on a sample (shaded) and giving rise to a fourth wave, the output signal wave, via a third-order process called four-wave mixing.*

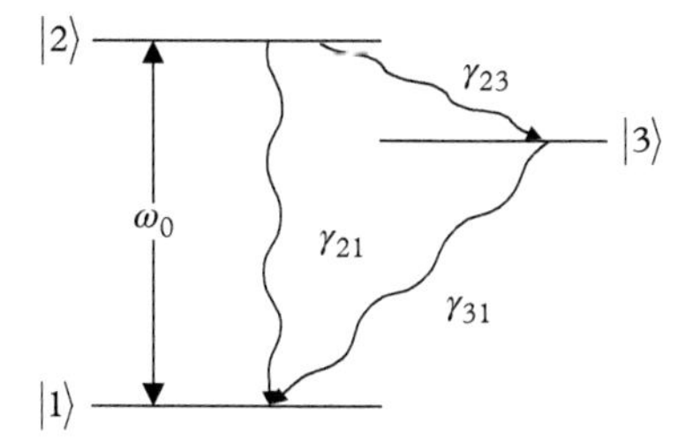

Figure 5.15 *Radiative and non-radiative transitions in a three-level system undergoing nearly degenerate four-wave mixing on the 1↔2 transition.*

We can also show that by tuning the probe frequency, information on relaxation pathways in the system can be obtained [5.2] which might otherwise be unmeasurable (Figure 5.15):

$$i\hbar\dot{\rho}_{11} = (V_{12}\rho_{21} - \rho_{12}V_{21}) + i\hbar\gamma_{31}\rho_{33} + i\hbar\gamma_{21}\rho_{22}, \tag{5.4.1}$$

$$i\hbar\dot{\rho}_{22} = -(V_{12}\rho_{21} - \rho_{12}V_{21}) - i\hbar(\gamma_{21} + \gamma_{23})\rho_{22}, \tag{5.4.2}$$

$$i\hbar\dot{\rho}_{33} = i\hbar\gamma_{23}\rho_{22} - i\hbar\gamma_{31}\rho_{33}, \tag{5.4.3}$$

$$i\hbar\dot{\rho}_{12} = -\hbar\omega_0\rho_{12} + (V_{12}\rho_{22} - \rho_{11}V_{12}) - i\hbar\Gamma_{21}\rho_{12}, \tag{5.4.4}$$

$$\rho_{21} = \rho_{12}^*. \tag{5.4.5}$$

The three perturbing waves yield traveling wave interactions given by

$$(V_{12})_j = -\frac{1}{2}\left[\mu_{12}E_j \exp[i(\omega_j t - k_j z)] + c.c.\right], \tag{5.4.6}$$

where the subscript $j = f, b, p$ denotes the forward pump, backward pump, and probe waves, respectively. Because there are multiple incident waves, we also apply the subscript j to earlier notations for the Rabi frequency, phase, and detuning:

$$\Omega_j \equiv \mu_{12}E_j/2\hbar,$$

$$\Phi_j \equiv \omega_j t - k_j z,$$

$$\Delta_j \equiv \omega_0 - \omega_j.$$

Now according to Eqs. (5.4.1)–(5.4.5) a third-order, off-diagonal matrix element $\rho_{12}^{(3)}$ can only be obtained from second-order population elements $\rho_{22}^{(2)}$ and $\rho_{11}^{(2)}$, which are in turn derived from a first-order element $\rho_{12}^{(1)}$. This perturbation chain is illustrated in Figure 5.16 and discussed in Appendix D.

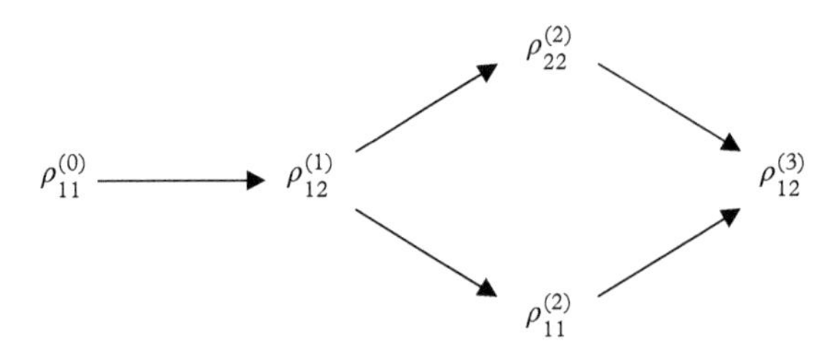

Figure 5.16 *Illustration of the perturbation changes expected when one field acts on a single transition in each new order.*

From Eq. (5.4.4) we therefore solve directly to find

$$\left(\rho_{12}^{(1)}\right)_j = \left[\frac{\Omega_j}{\Delta_j + i\Gamma_{21}}\right] e^{i\phi_j} \rho_{11}^{(0)}. \tag{5.4.7}$$

Since we are interested in tuning the frequency of the probe wave E_p we need to anticipate that population pulsations may appear in the solutions. First we write

$$i\hbar\dot{\rho}_{22}^{(2)} = -\hbar \sum_{j,k} \Omega_j^* e^{-i\phi_j} e^{i\phi_k} \left(\frac{\Omega_k}{\Delta_k + i\Gamma_{21}}\right) \rho_{11}^{(0)}$$

$$+ \hbar \sum_{j,k} \Omega_j e^{-i\phi_k} e^{i\phi_j} \left(\frac{\Omega_k^*}{\Delta_k - i\Gamma_{21}}\right) \rho_{11}^{(0)} - i\hbar\left(\gamma_{21} + \gamma_{23}\right) \rho_{22}^{(2)}. \tag{5.4.8}$$

Then set $\gamma_2 \equiv \gamma_{21} + \gamma_{23}$ and retain only phase-matched terms containing Ω_p^*. This means we shall ultimately ignore terms in which the spatial phase factor, or linear momentum, of the signal wave is different from the net phase factor or momentum of the three input waves:

$$i\hbar\left(\dot{\rho}_{22}^{(2)} + \gamma_2 \rho_{22}^{(2)}\right) = \rho_{11}^{(0)} \left[\Omega_p^* \Omega_f e^{-i(\phi_p - \phi_f)} \left(\frac{1}{\Delta_p - i\Gamma_{21}} - \frac{1}{\Delta_f + i\Gamma_{21}}\right)\right.$$

$$\left. + \Omega_p^* \Omega_b e^{-i(\phi_p - \phi_b)} \left(\frac{1}{\Delta_p - i\Gamma_{21}} - \frac{1}{\Delta_b + i\Gamma_{21}}\right)\right]. \tag{5.4.9}$$

Now we see that with a detuned probe frequency $\omega_p = \omega + \delta$, the terms on the right side oscillate as $\exp(-i\delta t)$. Hence, we use the substitution

$$\rho_{22} = \tilde{\rho}_{22} \exp(-i\delta t) \tag{5.4.10}$$

and find the following expression for $\tilde{\rho}_{22}$:

$$\tilde{\rho}_{22}^{(2)} = \frac{\rho_{11}^{(0)}}{\delta + i\gamma_2}\left[\Omega_p^*\Omega_f\left(\frac{1}{\Delta_p - i\Gamma_{21}} - \frac{1}{\Delta_f + i\Gamma_{21}}\right) + \right.$$
$$\left. + \Omega_p^*\Omega_b\left(\frac{1}{\Delta_p - i\Gamma_{21}} - \frac{1}{\Delta_b + i\Gamma_{21}}\right)\right]. \quad (5.4.11)$$

Similarly from Eq. (5.4.3) we may find ρ_{33},

$$\tilde{\rho}_{33} = \left(\frac{\gamma_{23}}{-i\delta + \gamma_2}\right)\tilde{\rho}_{22}, \quad (5.4.12)$$

and substitution in Eq. (5.4.1) gives the second-order ground state population:

$$\tilde{\rho}_{11}^{(2)} = \frac{-\rho_{11}^{(0)}}{\gamma_2 - \gamma_{31}}\left(\frac{\gamma_{23}}{\delta + i\gamma_{31}} + \frac{\gamma_{21} - \gamma_{31}}{\delta + i\gamma_2}\right)\left[\Omega_p^*\Omega_f\left(\frac{1}{\Delta_p - i\Gamma_{21}} - \frac{1}{\Delta_f + i\Gamma_{21}}\right)\right.$$
$$\left. + \Omega_p^*\Omega_b\left(\frac{1}{\Delta_p - i\Gamma_{21}} - \frac{1}{\Delta_b + i\Gamma_{21}}\right)\right]. \quad (5.4.13)$$

Notice that a resonant denominator $(\delta + i\gamma_{31})^{-1}$ appears in the ground state element $\rho_{11}^{(2)}$, but does not appear in $\rho_{22}^{(2)}$. Finally, with Eqs. (5.4.11) and (5.4.13) in Eq. (5.4.4) we find

$$\rho_{12}^{(3)} = \rho_{11}^{(0)}\left[\left(1 + \frac{\gamma_{21} - \gamma_{31}}{\gamma_2 - \gamma_{31}}\right)\left(\frac{1}{\delta + i\gamma_2}\right) + \left(\frac{1}{\gamma_2 - \gamma_{31}}\right)\left(\frac{1}{\delta + i\gamma_{31}}\right)\right]\cdot$$
$$\left[\frac{\Omega_p^*\Omega_f\Omega_b}{(\Delta_b + \delta + i\Gamma_{21})}\mathrm{e}^{i(\omega-\delta)t+ikz}\left(\frac{1}{\Delta_p - i\Gamma_{21}} - \frac{1}{\Delta_f + i\Gamma_{21}}\right)\right.$$
$$\left. + \frac{\Omega_p^*\Omega_b\Omega_f}{(-\Delta_f + \delta + i\Gamma_{21})}\mathrm{e}^{i(\omega-\delta)t+ikz}\left(\frac{1}{\Delta_p - i\Gamma_{21}} - \frac{1}{\Delta_b + i\Gamma_{21}}\right)\right]. \quad (5.4.14)$$

This third-order coherence between states $|1\rangle$ and $|2\rangle$ is radiative because $\mu_{12} \neq 0$. It gives rise to a time-reversed replica of the probe wave proportional to E_p^*. In this geometry, the fourth wave in the phase-matched interaction is thus a phase conjugate signal wave and its spectrum (Figure 5.17) versus detuning exhibits resonances whose widths reflect the rates of various decay processes possible within the atom.

In this derivation it was implicitly assumed that the sample was optically thin, since field amplitudes throughout the sample were taken to be constant. On the basis of Eq. (5.4.14) it is clear that on resonance ($\Delta \sim 0$), a system with large

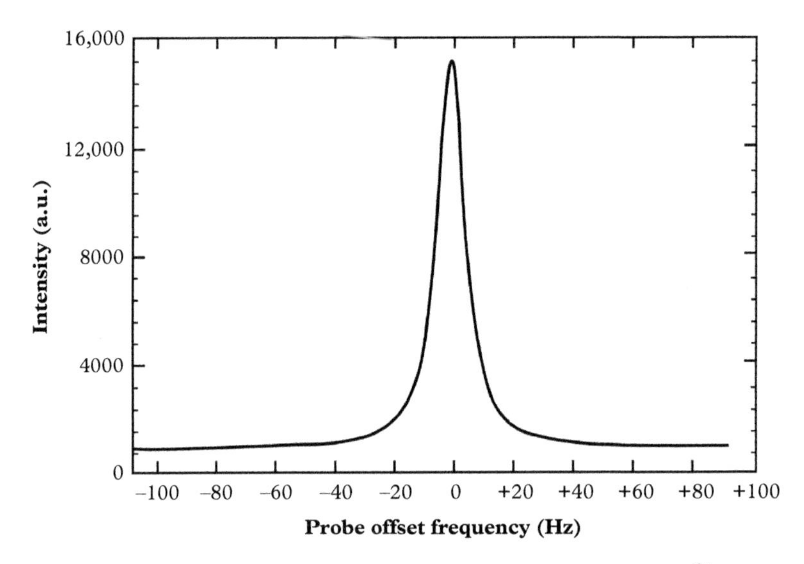

Figure 5.17 *The nearly degenerate four-wave mixing spectrum of three-level* Cr^{3+} *ions in sapphire, showing the Lorentzian-shaped resonance whose width is given by the slow decay rate* γ_{31} *from the metastable third level (after Ref. [5.3]). Underlying the main resonance in the figure is a second, much broader resonance with a width reflecting the decay rate* γ_{21} *of the second excited state in the system.*

homogeneous broadening ($\Gamma_2 \gg \gamma_{31}, \delta$) and a metastable third state ($\gamma_{31} \ll \gamma_{21}, \gamma_{23}$) exhibits a four-wave mixing spectrum which is just a single Lorentzian peak with a full width at half maximum (FWHM) equal to $2\gamma_{31}$. Thus an ultra-narrow resonance appears because of ground state saturation due to slow decay from level $|3\rangle$, as indicated in Figure 5.17. The simplicity of this result is a useful feature of nearly degenerate four-wave mixing (NDFWM). Probably the most remarkable aspect of it is that the decay time of the long-lived level can be measured accurately by optical means even if its decay to the ground state is completely non-radiative.

The phase conjugate character of the coherence in Eq. (5.4.14) is of great interest in its own right. Significant applications of phase conjugation exist in adaptive optics, imaging through obscuring media, and laser technology. The reader is directed to Ref. [5.7] for further discussion of this topic.

Variations of this nonlinear approach to laser spectroscopy are useful for extremely precise measurements of relative splittings in optical spectra. For example, sub-kilohertz resolution is readily obtained in measurements of hyperfine-split transitions of dopant ions in solids [5.8]. An extension of the theoretical treatment of nearly degenerate four-wave mixing to the regime of strong pump fields also reveals a spectrum with new features in it that are not present in Eq. (5.4.14). AC Stark resonances, with detunings determined by the transition dipole moment (and the optical field), appear when transition $1 \rightarrow 2$ saturates. This analysis is presented in the following section.

5.4.2 Pump–Probe Experiments

Third-order interactions are by no means restricted to the specialized geometry of NDFWM analyzed in the previous section. In fact third-order interactions may take place when only one or two beams are incident on a sample. Hence they are more common than might be expected, and researchers using optical characterization techniques not only need to be able to anticipate when saturation effects need to be taken into account, but also be aware of simple two-beam measurements in which the third-order susceptibility $\chi^{(3)}$ dominates the analysis, such as in pump–probe experiments.

To illustrate this, let us now discuss the *pump–probe* scenario of Figure 5.18. It is common practice to saturate systems with a strong pump wave and then monitor bleaching or recovery dynamics with a probe wave. The objective is usually to obtain information about time-dependent processes in completely unknown systems by forcing it slightly away from equilibrium. There are two basic approaches. The probe may be delayed in time or detuned in frequency. The first technique measures the dynamics directly in the temporal domain. The second measures dynamics in the corresponding Fourier space by frequency-domain spectroscopy.

What is not so obvious at first is that both types of experiment involve a third-order interaction. In this section we address the equations governing this general class of measurement and then analyze the frequency-domain spectrum in detail, to show that indeed three input fields determine the response just as in four-wave mixing [5.9]. To broaden the applicability of the analysis, we extend it to three-level systems however using an approach that combines exact and perturbative analysis. At the end we shall point out that pump–probe experiments measure the same basic quantities as NDFWM spectroscopy covered in the last section.

In a typical pump–probe experiment, a strong (saturating) pump wave and a weak probe wave intersect at a small angle within the sample. To investigate the probe transmission spectrum versus pump–probe detuning, we ignore the directions of the two input waves. This means that we ignore phase matching, and simply find the changes induced by the pump wave in the real and imaginary parts of the refractive index of the sample. These changes may be probed from any angle and depend on the relaxation rates among various levels in the system, so it may not seem too surprising that the

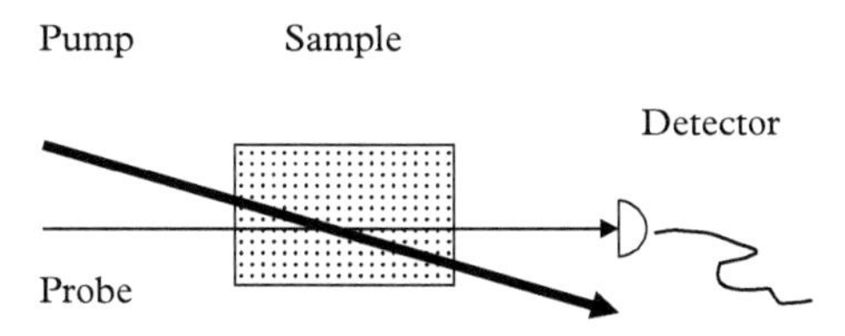

Figure 5.18 *Typical geometry of a pump–probe experiment in which a strong wave (dark arrow) partially saturates an absorptive transition or induces a parametric change in the birefringence or dispersion of a sample. The transmission of a probe wave is monitored with a detector.*

resulting probe spectrum contains the same basic information about the atomic system that was provided by NDFWM spectroscopy. However, the analysis of the last section was strictly perturbative, whereas in this section the pump wave is considered intense enough that an exact treatment that includes saturation effects is desired [5.10]. One consequence of this is that new high-field features appear (Rabi sidebands).

The component equations of the density matrix corresponding to the three-level system in Figure 5.19 are

$$\dot{\rho}_{11} = -(i/\hbar)(V_{12}\rho_{21} - \rho_{12}V_{21}) + \gamma_{31}\rho_{33} + \gamma_{21}\rho_{22}, \tag{5.4.15}$$

$$\dot{\rho}_{22} = +(i/\hbar)(V_{12}\rho_{21} - \rho_{12}V_{21}) - (\gamma_{21} + \gamma_{23})\rho_{22}, \tag{5.4.16}$$

$$\dot{\rho}_{33} = -\gamma_{31}\rho_{33} + \gamma_{23}\rho_{22}, \tag{5.4.17}$$

$$\dot{\rho}_{12} = -(i/\hbar)V_{12}(\rho_{22} - \rho_{11}) - (i/\hbar)(H_{11}\rho_{12} - \rho_{12}H_{22}) - \Gamma\rho_{12}, \tag{5.4.18}$$

$$\rho_{21} = \rho_{12}^*. \tag{5.4.19}$$

In this problem, the optical interaction is separated into two parts. The first is due to a strong pump wave E_1 and the second is due to a weak probe wave E_2. The pump wave is a traveling wave of the usual form

$$E_1(t) = \frac{1}{2}E_{10}\exp(i\omega_1 t) + c.c, \tag{5.4.20}$$

and the probe is similar,

$$E_2(t) = \frac{1}{2}E_{20}\exp(i\omega_2 t) + c.c., \tag{5.4.21}$$

except that $E_{20} \ll E_{10}$.

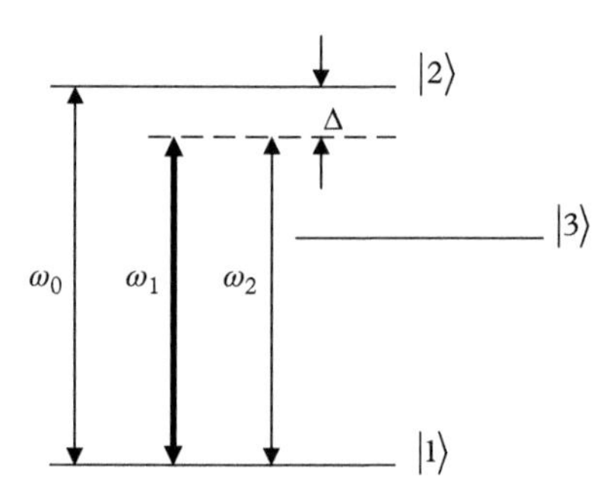

Figure 5.19 *Energy levels and detunings of a three-level system driven by two optical waves of similar frequency.*

The interaction Hamiltonian is written as the sum of a zeroth-order pump interaction and a first-order probe interaction. That is,

$$V = V^{(0)} + V^{(1)}, \tag{5.4.22}$$

where

$$V^{(0)} = -\frac{1}{2}\hbar\Omega_1 \exp(i\omega_1 t) + c.c., \tag{5.4.23}$$

$$V^{(1)} = -\frac{1}{2}\hbar\Omega_2 \exp(i\omega_2 t) + c.c., \tag{5.4.24}$$

and $\Omega_1 \equiv \mu E_{10}/\hbar$ and $\Omega_2 \equiv \mu E_{20}/\hbar$ are the corresponding resonant Rabi frequencies.

The effect of $V^{(0)}$ may be taken into account exactly, by the same procedure as in Section 5.1.2. This yields the results

$$\tilde{\rho}_{12}^{(0)} = \frac{(\Omega_1/2)}{\Delta + i\Gamma}(\rho_{11}^{(0)} - \rho_{22}^{(0)}), \tag{5.4.25}$$

$$\rho_{11}^{(0)} = \frac{1 + \Gamma[2\gamma_2(\Delta^2 + \Gamma^2)]^{-1}\,|\Omega_1|^2}{1 + \Gamma[2\gamma_2(\Delta^2 + \Gamma^2)]^{-1}(2 + \gamma_{23}/\gamma_{31})\,|\Omega_1|^2}, \tag{5.4.26}$$

$$\rho_{22}^{(0)} = \frac{\Gamma[2\gamma_2(\Delta^2 + \Gamma^2)]^{-1}\,|\Omega_1|^2}{1 + \Gamma[2\gamma_2(\Delta^2 + \Gamma^2)]^{-1}(2 + \gamma_{23}/\gamma_{31})\,|\Omega_1|^2}. \tag{5.4.27}$$

Although the population expressions in Eqs. (5.4.26) and (5.4.27) are exact, in the next step of the calculation they are taken as the zeroth-order solutions in a perturbation expansion (see Appendix D). The first-order equation of motion that includes the effect of the weak probe wave has the form

$$i\hbar\dot{\rho}^{(1)} = \left[H_0, \rho^{(1)}\right] + \left[V^{(0)}, \rho^{(1)}\right] + \left[V^{(1)}, \rho^{(0)}\right] + i\hbar\dot{\rho}_{relax}^{(1)}. \tag{5.4.28}$$

The various components are

$$\dot{\rho}_{11}^{(1)} = i(\Omega_1/2)\tilde{\rho}_{12}^{*(1)} + c.c. + i(\Omega_2/2)\tilde{\rho}_{12}^{*(0)} + c.c. + \gamma_{21}\rho_{22}^{(1)} + \gamma_{31}\rho_{33}^{(1)}, \tag{5.4.29}$$

$$\dot{\rho}_{22}^{(1)} = -i(\Omega_1/2)\tilde{\rho}_{12}^{*(1)} + c.c. - i(\Omega_2/2)\tilde{\rho}_{12}^{*(0)} + c.c. - (\gamma_{21} + \gamma_{23})\rho_{22}^{(1)}, \tag{5.4.30}$$

$$\dot{\rho}_{33}^{(1)} = \gamma_{31}\rho_{33}^{(1)} + \gamma_{32}\rho_{22}^{(1)}, \tag{5.4.31}$$

$$\begin{aligned}\dot{\rho}_{12}^{(1)} &= i(\Omega_1/2)\exp(i\omega_1 t)\left(\rho_{22}^{(1)} - \rho_{11}^{(1)}\right) \\ &\quad + i(\Omega_2/2)\exp(i\omega_2 t)\left(\rho_{22}^{(0)} - \rho_{11}^{(0)}\right) + i\omega_0\rho_{12}^{(1)} - \Gamma\rho_{12}^{(1)}.\end{aligned} \tag{5.4.32}$$

Because the pump and probe waves are detuned, their combined driving effect can produce pulsations in population and coherence at frequency $\pm\delta$. Consequently, the solutions acquire additional time dependences and are assumed to have the forms

$$\rho_{11} = \rho_{11}^{(0)} + \rho_{11}^{(1)} = \rho_{11}^{(0)} + \rho_{11}^{(+)} \exp(i\delta t) + \rho_{11}^{(-)} \exp(-i\delta t), \tag{5.4.33}$$

$$\rho_{22} = \rho_{22}^{(0)} + \rho_{22}^{(1)} = \rho_{22}^{(0)} + \rho_{22}^{(+)} \exp(i\delta t) + \rho_{22}^{(-)} \exp(-i\delta t), \tag{5.4.34}$$

$$\rho_{33} = \rho_{33}^{(0)} + \rho_{33}^{(1)} = \rho_{33}^{(0)} + \rho_{33}^{(+)} \exp(i\delta t) + \rho_{33}^{(-)} \exp(-i\delta t), \tag{5.4.35}$$

$$\tilde{\rho}_{12} = \tilde{\rho}_{12}^{(0)} + \tilde{\rho}_{12}^{(1)} = \tilde{\rho}_{12}^{(0)} + \tilde{\rho}_{12}^{(+)} \exp(i\delta t) + \tilde{\rho}_{12}^{(-)} \exp(-i\delta t). \tag{5.4.36}$$

The time derivatives on the left side of Eqs. (5.4.29)–(5.4.32) acquire new "beat" terms, even if we restrict our solutions to steady-state conditions by assuming that

$$\dot{\rho}_{11}^{(+)} = \dot{\rho}_{11}^{(-)} = \dot{\rho}_{22}^{(+)} = \dot{\rho}_{22}^{(-)} = \dot{\rho}_{33}^{(+)} = \dot{\rho}_{33}^{(-)} = 0 \tag{5.4.37}$$

and

$$\dot{\tilde{\rho}}_{12}^{(+)} = \dot{\tilde{\rho}}_{12}^{(-)} = 0. \tag{5.4.38}$$

For example, by substituting Eqs. (5.4.34) and (5.4.35) into Eq. (5.4.31) and setting the coefficients of individual frequency components equal to zero, we find

$$\rho_{33}^{(0)} = (\gamma_{23}/\gamma_{31})\rho_{22}^{(0)}, \tag{5.4.39}$$

$$\rho_{33}^{(+)} = [\gamma_{23}/(\gamma_{31} + i\delta)]\rho_{22}^{(+)}, \tag{5.4.40}$$

$$\rho_{33}^{(-)} = [\gamma_{23}/(\gamma_{31} - i\delta)]\rho_{22}^{(-)}. \tag{5.4.41}$$

Now, using the results for $\rho_{33}^{(\pm)}$ given in Eqs. (5.4.40) and (5.4.41), together with the first-order closure relation,

$$\rho_{11}^{(1)} + \rho_{22}^{(1)} + \rho_{33}^{(1)} = 0, \tag{5.4.42}$$

one finds

$$\rho_{11}^{(+)} = -\left\{1 + \left(\frac{\gamma_{23}}{\gamma_{31} + i\delta}\right)\right\} \rho_{22}^{(+)}, \tag{5.4.43}$$

$$\rho_{11}^{(-)} = -\left\{1 + \left(\frac{\gamma_{23}}{\gamma_{31} - i\delta}\right)\right\} \rho_{22}^{(-)}. \tag{5.4.44}$$

After substitution of these results in Eq. (5.4.29), expressions for the first-order excited state amplitudes are obtained:

$$\rho_{22}^{(+)} = i(\gamma_2 + i\delta)^{-1} \left\{(\Omega_1^*/2)\tilde{\rho}_{12}^{(+)} - (\Omega_1/2)\tilde{\rho}_{12}^{*(-)} - (\Omega_2/2)\tilde{\rho}_{12}^{*(0)}\right\}, \tag{5.4.45}$$

$$\rho_{22}^{(-)} = \rho_{22}^{(+)*}. \tag{5.4.46}$$

Next, by substituting Eq. (5.4.36) into Eq. (5.4.32), one can also find the coherence amplitudes

$$\tilde{\rho}_{12}^{(+)} = i[(\Delta - \delta) + i\Gamma]\left[(\Omega_1/2)\left(\rho_{11}^{(+)} - \rho_{22}^{(+)}\right) + (\Omega_2/2)\left(\rho_{11}^{(0)} - \rho_{22}^{(0)}\right)\right], \quad (5.4.47)$$

$$\tilde{\rho}_{12}^{(-)} = [(\Delta + \delta) + i\Gamma]^{-1}\left[(\Omega_1/2)\left(\rho_{11}^{(-)} - \rho_{22}^{(-)}\right)\right]. \quad (5.4.48)$$

Finally, substitution of these last two relations in Eq. (5.4.45) yields a general expression for $\rho_{22}^{(+)}$, from which all other quantities may be determined using Eqs. (5.4.43)–(5.4.48).

$$\rho_{22}^{(+)} = \frac{(\delta - 2i\Gamma)(\Omega_1^*\Omega_2/4)\left(\rho_{11}^{(0)} - \rho_{22}^{(0)}\right)}{(\delta - i\gamma_2)[(\Delta + \delta) - i\Gamma][(\Delta - \delta) + i\Gamma] + 2[2 + \gamma_{23}/(\gamma_{31} + i\delta)][\delta - i\Gamma]\,|\Omega_1/2|^2}. \quad (5.4.49)$$

With this result in hand, the coherence in Eq. (5.4.36) is fully determined to first order. The probe polarization and susceptibility can therefore be determined for any value of the pump intensity, resonance detuning Δ, and pump–probe detuning δ from the polarization $P(\omega + \delta) = \mu_{12}\rho_{21}(\omega + \delta) + \mu_{21}\rho_{12}(\omega + \delta)$.

Results are shown in Figure 5.20(a), (b), and (c) for the absorption spectrum of the probe at low, moderate, and high pump intensities, respectively. The pump detuning and decay parameters were fixed for these comparisons. At low intensities, weak dispersive beam coupling is seen at zero detuning and an absorption peak is present at a pump–probe detuning of $\delta/\Gamma = 10$. Just above saturation, the dispersive feature at zero detuning reveals two resonances, one for each decay process from the excited states, and the absorption peak at $\delta/\Gamma = 10$ decreases in strength. At 20 times saturation ($\Omega/\Gamma = 20$), power broadening smears out the narrowest central resonance and two Rabi sidebands appear at the generalized Rabi frequency splitting of $\delta/\Gamma = \pm 22.3$ in Figure 5.20(c). The central dispersive feature in each spectrum is a weak field response present at intensities below the saturation intensity. When magnified, the central feature shows two overlapping resonances whose widths are determined by the decay rates of levels 2 and 3. Consequently, experimental measurements of their widths provide direct determinations of the excited state decay rates, just as in NDFWM spectroscopy described in Section 5.4.1. Power broadening can obscure the central feature, as shown in Fig. 5.20(c) where the second resonance disappears at high intensity.

Pump–probe spectroscopy is frequently performed with short pulses to monitor changes in sample transmission versus time. System dynamics are then revealed in the time domain as "differential transmission" instead of in the frequency domain as calculated here and portrayed in Figure 5.20. Experimentally, measurements may be made by delaying the probe pulse with respect to the pump pulse using a mechanical delay line [5.11–5.13]. Theoretically, analysis requires direct integration of the time-dependent density matrix Eqs. (5.4.29)–(5.4.32). The outcome of analyzing the temporal profile of the transient transmission spectrum is however the same as that in frequency-domain measurements. The results reflect the decay rates of dynamic processes in excited states of the system.

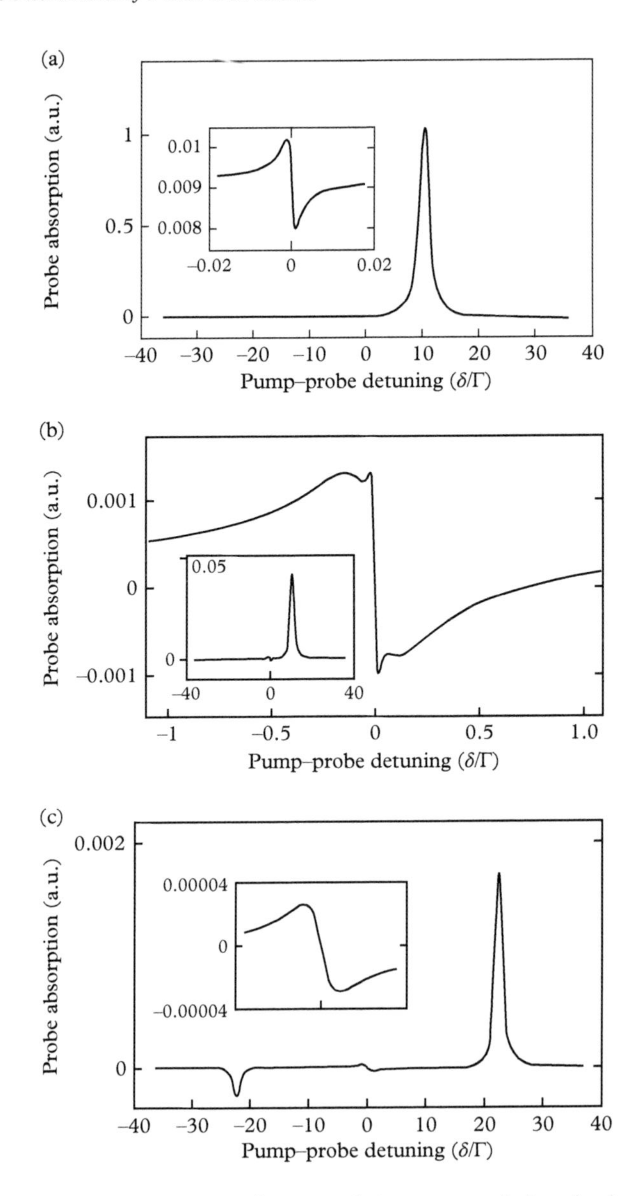

Figure 5.20 *Nearly degenerate pump–probe transmission spectra of three-level atoms with* $\gamma_{31} = 0.0001$, $\gamma_2 = 0.1$, $\gamma_{23} = 0.05$, *and* $\Delta = -10$, *at (a) low intensity* $(\Omega/\Gamma = 0.1)$, *(b) moderate intensity* $(\Omega/\Gamma = 3.0)$, *and (c) high intensity* $(\Omega/\Gamma = 20)$. *(After Ref. [5.10].)*

5.4.3 Quantum Interference

In optical processes multiple channels sometimes exist whereby an atom may be excited or de-excited. When two transitions can occur simultaneously between the same initial and final endpoints, interference takes place. Quantum interference generally manifests itself as a dispersively shaped resonance first investigated by Beutler [5.14] and Fano [5.15] in the context of auto-ionizing emission in gases. In this section a density matrix approach is developed to show that the detailed shape and spectral position of these asymmetric resonances depend not only on the wavelength and linewidth of both transitions contributing to the resonance, but on their coupling as well. If one of the two competing transitions is exceptionally weak (meaning the upper level is exceptionally long-lived), the interference features are generally visible enough to provide spectroscopic information on otherwise unobservable, metastable states. Another intriguing and potentially useful feature of quantum interference is that absorption sometimes vanishes at the resonance.

Consider the absorption in a three-level atom or center which has a localized or discrete excited state $|2\rangle$ "embedded" in a delocalized continuum or just a broad discrete level $|3\rangle$. Level $|3\rangle$ is assumed to be wide because the $|1\rangle \rightarrow |3\rangle$ transition is short-lived, whereas the $|1\rangle \rightarrow |2\rangle$ transition is narrow because state $|2\rangle$ is long-lived. The shaded region in Figure 5.21 depicts the full breadth of energy level $|3\rangle$. A dashed horizontal line indicates its center. The heavy solid horizontal line depicts level $|2\rangle$ in the lower half of the shaded region. Since level $|3\rangle$ directly overlaps the energy range of level $|2\rangle$ in our model, spin–orbit [5.16] or configuration interactions [5.15] between the two upper levels are possible. In the equations of motion the coupling frequency representing the strength of any such interaction will be denoted by $V_{23}/\hbar$.

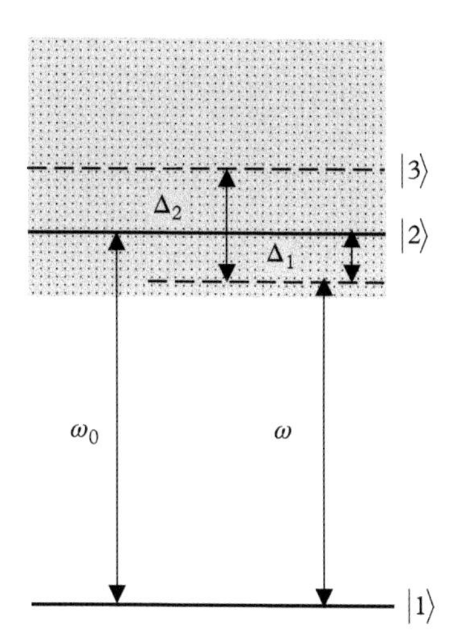

Figure 5.21 *Three-level model to illustrate quantum interference in absorption. The incident light at frequency ω is shown detuned from level $|2\rangle$ by Δ_1 and from level $|3\rangle$ by Δ_2.*

Level $|3\rangle$ is taken to lie at a higher energy than level $|2\rangle$, so that the detunings of incident light from the two resonances, denoted by $\Delta_1 \equiv \omega_0 - \omega$ and $\Delta_2 \equiv \omega_{13} - \omega$, are both positive (Figure 5.21). To calculate the system polarization under steady-state conditions, allowance needs to be made for the two coherences excited by a single incident light wave at frequency ω.

Consistent with the previous assumptions, the ED transition moments on the $|1\rangle \to |2\rangle$ and $|1\rangle \to |3\rangle$ transitions are taken to be non-zero ($\mu_{12} \neq 0$ and $\mu_{13} \neq 0$). The large linewidth of the $|1\rangle \to |3\rangle$ transition implies it is strongly ED-allowed whereas the $|1\rangle \to |2\rangle$ transition is comparatively weakly allowed. Since both transitions originate from the same ground state, the much weaker one must be allowed by virtue of a small admixture of angular momentum *different* from that of state $|1\rangle$. Such admixing often originates from spin–orbit or configuration interaction between states $|2\rangle$ and $|3\rangle$ and profoundly affects the shape and position of the overall $|1\rangle \to |2\rangle$ resonance.

Assume that angular momentum is a reasonably good quantum number and that the upper states of two absorption transitions effectively have the same L. Then the ED transition moment between states $|2\rangle$ and $|3\rangle$ is negligibly small ($\mu_{23} \cong 0$). This reduces the ED couplings among the states to those mentioned earlier ($\mu_{12} \neq 0$ and $\mu_{13} \neq 0$), and simplifies the equations of motion. However, we shall explicitly allow a magnetic dipole or other coupling between states $|2\rangle$ and $|3\rangle$ through an interaction denoted by V_{23}.

Interaction of the system with a traveling wave takes the familiar form $V(t) = -\frac{1}{2}\bar{\mu} \cdot \bar{E}_0 e^{i\omega t} + c.c. = -\frac{1}{2}\hbar\Omega e^{i\omega t} + c.c.$ The density matrix equations may then be written out explicitly:

$$i\hbar\dot{\rho}_{11} = (V_{12}\rho_{21} - \rho_{12}V_{21}) + (V_{13}\rho_{31} - \rho_{13}V_{31}) + i\hbar\gamma_{21}\rho_{22} + i\hbar\gamma_{31}\rho_{33}, \tag{5.4.50}$$

$$i\hbar\dot{\rho}_{22} = (V_{21}\rho_{12} - \rho_{21}V_{12}) - i\hbar\gamma_{21}\rho_{22}, \tag{5.4.51}$$

$$i\hbar\dot{\rho}_{33} = -(V_{31}\rho_{13} - \rho_{31}V_{13}) - i\hbar\gamma_{31}\rho_{33}, \tag{5.4.52}$$

$$i\hbar\dot{\rho}_{12} = -\hbar\omega_{12}\rho_{12} + V_{12}(\rho_{22} - \rho_{11}) + V_{13}\rho_{32} - \rho_{13}V_{32} - i\hbar\Gamma_{21}\rho_{12}, \tag{5.4.53}$$

$$i\hbar\dot{\rho}_{13} = -\hbar\omega_{13}\rho_{13} + V_{13}(\rho_{33} - \rho_{11}) + V_{12}\rho_{23} - \rho_{12}V_{23} - i\hbar\Gamma_{13}\rho_{13}, \tag{5.4.54}$$

$$i\hbar\dot{\rho}_{23} = -\hbar\omega_{23}\rho_{23} + (V_{21}\rho_{13} - \rho_{21}V_{13}) + V_{23}(\rho_{33} - \rho_{22}). \tag{5.4.55}$$

Relaxation terms between the excited states have been dropped ($\gamma_{23} = \Gamma_{23} = 0$) because the coupling between upper states is purely static.

Applying the SVEA, and recognizing that there is only one driving field, one can assume that the three coherences in the system have the forms:

$$\rho_{12} = \tilde{\rho}_{12}e^{i\omega t}; \quad \rho_{13} = \tilde{\rho}_{13}e^{i\omega t}; \quad \rho_{23} = \tilde{\rho}_{23}e^{i\omega t}.$$

Using a first-order perturbation approach, coupled equations for the slowly varying amplitudes of the optical coherences are readily obtained:

$$\tilde{\rho}_{12}^{(1)} = \frac{(\Omega_{12}/2)}{\Delta_1 + i\Gamma_{12}}\left(\rho_{11}^{(0)} - \rho_{22}^{(0)}\right) - \frac{(V_{32}/\hbar)}{\Delta_1 + i\Gamma_{12}}\tilde{\rho}_{13}^{(1)}, \tag{5.4.56}$$

$$\tilde{\rho}_{13}^{(1)} = \frac{(\Omega_{13}/2)}{\Delta_2 + i\Gamma_{13}}\left(\rho_{11}^{(0)} - \rho_{33}^{(0)}\right) - \frac{(V_{23}/\hbar)}{\Delta_2 + i\Gamma_{13}}\tilde{\rho}_{12}^{(1)}. \tag{5.4.57}$$

Note that terms in $\tilde{\rho}_{32}$ have been dropped. The amplitude for coherence between the upper states is vanishingly small since the coupling V_{23} is considered to be static. Since there is no charge oscillation there can be no coherence. Hence

$$\tilde{\rho}_{23}^{(1)} = 0. \tag{5.4.58}$$

Although Eqs. (5.4.56) and (5.4.57) are symmetric in structure, a strong asymmetry has already been established in the relationship between the two transitions through the assumptions of our model. When coupling between the upper states exceeds the transition rate on the weak transition, the $|1\rangle \rightarrow |2\rangle$ oscillation will be driven chiefly by its coupling to the strong $|1\rangle \rightarrow |3\rangle$ transition. By analogy, if one of two coupled mechanical oscillators is driven directly by an external force, motion of the second oscillator is initiated by virtue of its coupling to the first one. If the mechanical coupling is weak, the relative motion of the oscillators is out-of-phase. Similarly, for the case at hand, the oscillations are out-of-phase under weak coupling, resulting in destructive interference. As the coupling strength is increased, the motion evolves into the second eigenmode of coupled oscillators which is in-phase oscillation, resulting in constructive interference.

Assuming the system initially occupies ground state $|1\rangle$, our coupled expressions for the coherence amplitudes can be solved algebraically by substituting Eq. (5.4.56) into Eq. (5.4.57). The results are

$$\tilde{\rho}_{12} = \frac{(\Omega_{12}/2)}{\Delta_1 + i\Gamma_{12}} - \frac{(V_{32}/\hbar)}{[\Delta_1 + i\Gamma_{12}]}\left\{\frac{(\Omega_{13}/2)(\Delta_1 + i\Gamma_{12}) - (\Omega_{12}/2)(V_{23}/\hbar)}{[\Delta_1 + i\Gamma_{12}][\Delta_2 + i\Gamma_{13}] - |V_{23}/\hbar|^2}\right\}, \tag{5.4.59}$$

$$\tilde{\rho}_{13} = \frac{(\Omega_{13}/2)(\Delta_1 + i\Gamma_{12}) - (\Omega_{12}/2)(V_{23}/\hbar)}{[\Delta_1 + i\Gamma_{12}][\Delta_2 + i\Gamma_{13}] - |V_{23}/\hbar|^2}. \tag{5.4.60}$$

The absorption coefficient $\alpha(\omega)$ is determined by the imaginary part of the susceptibility (χ''), as described by Eq. (1.2.19). That is,

$$\alpha(\omega) = k\chi''. \tag{5.4.61}$$

χ'' in turn may be found from the positive frequency contribution to the macroscopic polarization $\tilde{P}(\omega)$. The relation between coherence, susceptibility, and absorption was examined previously in Section 3.10, where it was pointed out that by comparing two expressions for $\tilde{P}(\omega)$, namely

$$\tilde{P}(\omega) = \frac{1}{2}\varepsilon_0 E_0 \chi(\omega) \tag{5.4.62}$$

and

$$\tilde{P}(\omega) = N(\mu_{21}\tilde{\rho}_{12} + \mu_{31}\tilde{\rho}_{13}), \tag{5.4.63}$$

the susceptibility can be written in terms of the system coherences:

$$\chi(\omega) = \frac{2N}{\varepsilon_0 E_0}(\mu_{12}\tilde{\rho}_{21} + \mu_{13}\tilde{\rho}_{31}). \tag{5.4.64}$$

The imaginary part of the susceptibility, which governs absorption, is therefore

$$\begin{aligned}
\chi''(\omega) &= \mathrm{Im}\left\{\frac{2N}{\varepsilon_0 E_0}(\mu_{12}\tilde{\rho}_{21} + \mu_{13}\tilde{\rho}_{31})\right\} \\
&= \frac{N}{\varepsilon_0 \hbar}\left\{\frac{|\mu_{12}|^2\,\Gamma_{12}}{\Delta_1^2 + \Gamma_{12}^2} + \frac{|\mu_{13}|^2\,[\Gamma_{13}\left(\Delta_1^2 + \Gamma_{12}^2\right) + \Gamma_{12}\,|V_{23}/\hbar|^2]}{\left(\Delta_1\Delta_2 - \Gamma_{12}\Gamma_{13} - |V_{23}/\hbar|^2\right)^2 + (\Gamma_{12}\Delta_2 + \Gamma_{13}\Delta_1)^2}\right. \\
&\quad - \frac{2\mu_{12}\mu_{13}(V_{32}/\hbar)\,(\Gamma_{12}\Delta_2 + \Gamma_{13}\Delta_1)}{\left(\Delta_1\Delta_2 - \Gamma_{12}\Gamma_{13} - |V_{23}/\hbar|^2\right)^2 + (\Gamma_{12}\Delta_2 + \Gamma_{13}\Delta_1)^2} \\
&\quad \left. + \frac{|\mu_{12}|^2\,|V_{32}/\hbar|^2\left(\Delta_1^2\Gamma_{13} + 2\Delta_1\Delta_2\Gamma_{12} - \Gamma_{12}^2\Gamma_{13} - \Gamma_{12}\,|V_{32}/\hbar|^2\right)}{\left(\Delta_1^2 + \Gamma_{12}^2\right)\left[\left(\Delta_1\Delta_2 - \Gamma_{12}\Gamma_{13} - |V_{23}/\hbar|^2\right)^2 + (\Gamma_{12}\Delta_2 + \Gamma_{13}\Delta_1)^2\right]}\right\}.
\end{aligned} \tag{5.4.65}$$

If the excited state coupling is negligible ($V_{23} = 0$), this expression reduces to that of two distinguishable, non-interacting transitions. In this limit, the susceptibility simplifies to that of two independent, symmetric absorption resonances:

$$\chi''_{V=0}(\omega) = \frac{N}{\varepsilon_0 \hbar}\left\{\frac{|\mu_{12}|^2\,\Gamma_{12}}{\Delta_1^2 + \Gamma_{12}^2} + \frac{|\mu_{13}|^2\,\Gamma_{13}}{\Delta_2^2 + \Gamma_{13}^2}\right\}. \tag{5.4.66}$$

This limit is illustrated in Figure 5.22(a), where overlapping resonances are simply additive without excited state coupling ($V_{23} = 0$). When $V_{23} \neq 0$, however, interference from the last term in Eq. (5.4.65) causes significant modifications in the shape of the narrow resonance. In general the resonance becomes asymmetric (Figure 5.22(b)).

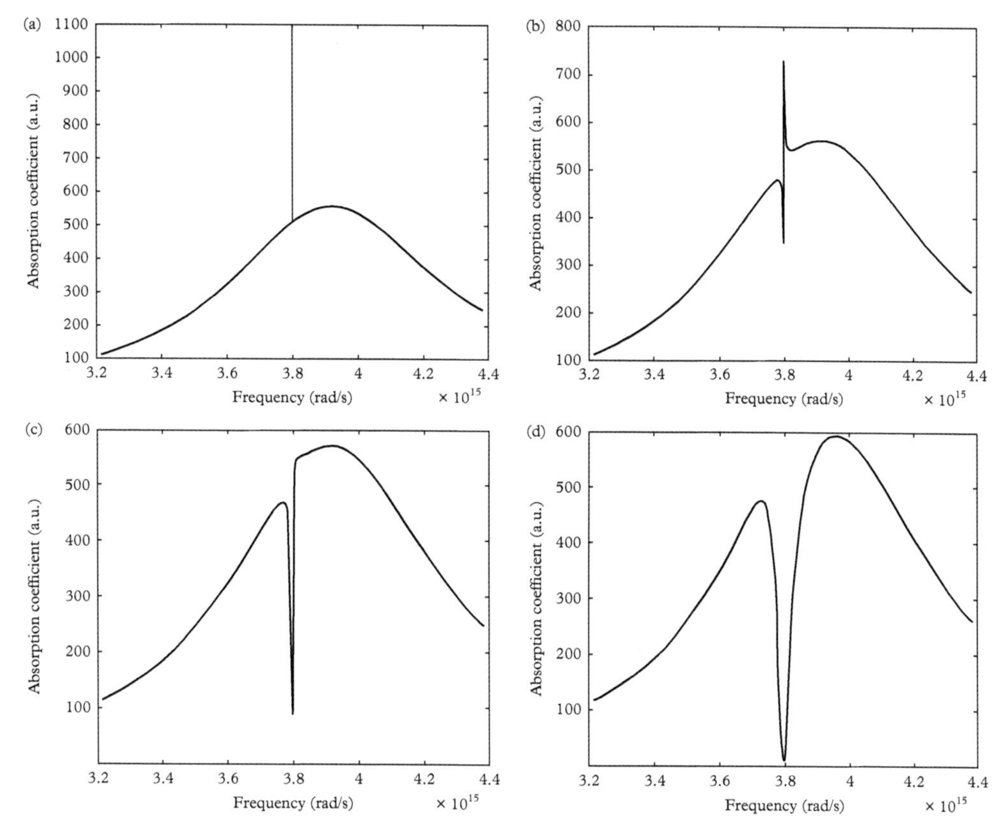

Figure 5.22 *Theoretical absorption spectra of (a) two uncoupled transitions* ($V_{23}/\hbar = 0$), *(b) weakly coupled transitions* ($\omega_{13} = 10\Gamma_{13} = 3.9 \times 10^{15}$ rad/s, $\Gamma_{12} = 5 \times 10^{11}$ rad/s, $\omega_0 = 3.8 \times 10^{15}$ rad/s, $V_{23}/\hbar = 10^{13}$ rad/s), *(c) anti-resonance with* $\Gamma_{12} = 5 \times 10^{11}$ rad/s, $\Gamma_{13} = 3.9 \times 10^{14}$ rad/s, $V_{23}/\hbar = 3.2 \times 10^{13}$ rad/s, *and (d) strongly coupled transitions, with the parameters of (b) except for* $V_{23}/\hbar = 1 \times 10^{14}$ rad/s.

For weak excited state coupling, anti-resonance is introduced (destructive interference), as illustrated in Figure 5.22(c) and (d). This behavior is analogous to that of coupled mechanical oscillators, as described previously.

If we consider a real system that can exhibit quantum interference, such as rare earth ions in crystals, it can easily happen that the number of impurity centers forming a complex with a Fano resonance in the absorption spectrum differs from the total number of impurity centers. This happens, for example, when an electron trapped in a shallow energy well acquires discrete energy levels determined by the well but also overlaps the electronic configuration of an adjacent impurity. The electron can thus undergo a configuration interaction with the center due to its proximity, leading to Fano interference. Centers with no trapped electron neighbor have the same underlying spectrum but without the Fano feature. Our analysis can be extended to encompass this common situation by defining the number of trapped electron complexes per unit volume to be N_1 and the number of impurities without trapped electrons to be N_2. The total number of impurities is $N = N_1 + N_2$ and the total susceptibility in Eq. (5.4.65) becomes

$$\chi''_{tot}(\omega) = \chi''(\omega) + \frac{N_2}{\varepsilon_0 \hbar} \frac{|\mu_{13}|^2 \Gamma_{13}}{\left(\Delta_2^2 + \Gamma_{13}^2\right)}. \tag{5.4.67}$$

The distinctive shapes of resonances exhibiting quantum interference not only provide accurate determinations of linewidths and resonant frequencies on weak transitions but improve their detectability. Weak narrow spectral lines are sometimes difficult to observe against strong backgrounds, whereas anti-resonances can produce a null in the background signal itself. In theory, high-contrast interference features can drive arbitrarily intense absorption peaks to zero (Figure 5.22(c)). This is borne out in experimental absorption spectra of Yb:SrF_2 crystals irradiated with gamma rays [5.17]. Immediately after exposure to the ionizing radiation, a pronounced null is observed at 365 nm (Figure 5.23) in the absorption spectrum of isolated Yb^{2+} ions. Analysis of the anti-resonance can determine the energy and lifetime of charges released by irradiation and trapped near Yb^{2+} sites. A second thing to note is that quantum interference

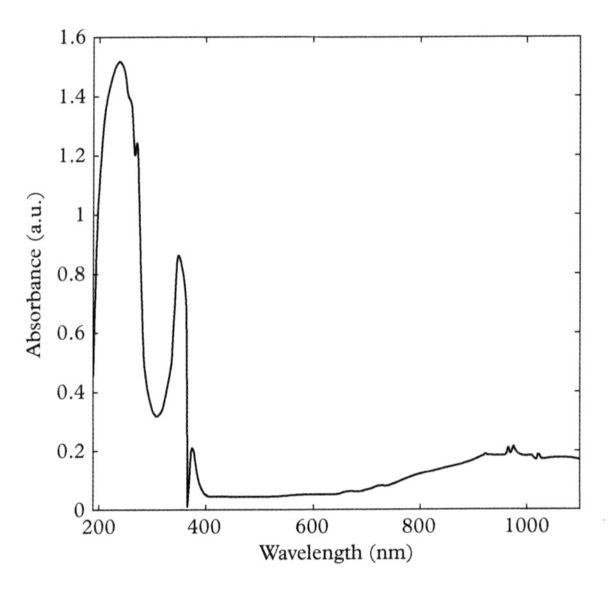

Figure 5.23 *The absorption spectrum of gamma-irradiated* Yb:SrF_2 *[5.17]. The anti-resonance near 360 nm is a Fano resonance due to transient centers composed of* Yb^{2+} *ions (responsible for the broad absorption features at 350 nm and at shorter wavelengths) and nearest-neighbor charges trapped in shallow potential wells. Spectral features between 900 and 1000 nm are* $^2F_{7/2} \rightarrow {}^2F_{5/2}$ *infrared absorption lines of trivalent* Yb^{3+} *impurities.*

cannot take place unless the wavefunctions of the impurity and the trapped charge actually overlap. This in itself provides useful structural information about complexes in the solid. In irradiated $Yb:SrF_2$ one may immediately conclude from the Fano interference in the Yb^{2+} spectrum that trapped charges are present in metastable energy levels where their wavefunctions overlap the rare earth impurity, presumably at interstitial sites. It is not possible to reach such specific conclusions in the case of independent resonances (Figure 5.22(a)).

Additional reading regarding experiments on Fano resonances in optical spectra may be found in Refs. [5.16–5.20]. Quantum interference is also encountered in nonlinear optics. An experimental observation of interference in harmonic generation is described, for example, in Ref. [5.21]. The achievement of induced transparency in the x-ray range for the purpose of generating coherent x-rays is described in Ref. [5.22]. The theory of electromagnetically induced transparency is covered in Section 7.5.

5.4.4 Higher Order Interactions and Feynman Diagrams

In Section 5.4.1, a four-wave mixing interaction with counter-propagating pump waves was analyzed to illustrate the appearance of coherence with phase-conjugate properties. In Section 5.4.3, beam coupling calculations revealed numerous resonances in the frequency domain spectrum. When as many as three input waves interact, there are of course many combinations of field amplitudes, when positive and negative frequencies are taken into account. All contribute to the total medium response. In Section 5.4.1 we simplified the analysis greatly by dropping all terms which did not contribute to the phase-matched interaction of interest. Similarly, in Section 5.4.3 we simplified the mathematical treatment by ignoring the spatial phases of interacting fields. Often these simplifications are difficult to foresee in new problems, and a diagrammatic approach can be helpful. The problem of writing down the appropriate density matrix elements or susceptibilities for a particular four-wave mixing interaction using perturbation theory is in itself not an easy task. This is obvious by simply evaluation the total number of terms possible in four-wave mixing: $\chi^{(3)}_{ijke}$ gives rise to $4 \times 3 \times 2 \times 1 = 24$ terms when fields are real or $24 \times 2 = 48$ terms when fields are taken to be complex (negative frequencies or phase conjugate amplitudes are included). To be useful a diagrammatic technique should therefore provide a simple picture of the dynamics and immediately yield the corresponding mathematical expression.

We proceed by drawing a "time line" for temporal development of a system. With the respect to its initial value, the density matrix can be written in terms of an evolution operator $U(t)$ as

$$\rho = |\psi\rangle\langle\psi| = U(t)\,|\psi_0\rangle\langle\psi_0|\,U^+(t). \tag{5.4.68}$$

Hence, to write down coherences like $\rho^{(n)} = P^{(n)}/\langle\mu\rangle$, which may be driven by positive or negative frequency components of incident fields, or susceptibilities such as

$\chi_{ijk}^{(n)} = P^{(n)}/E_i E_j E_k$, which depend on multiple field amplitudes and detunings, we shall need a double-sided Feynman diagram. In such a diagram, one side takes care of ket evolution $U(t)\,|\psi_0\rangle$ and the other follows bra development $\langle\psi_0|\,U(t)^+$. Each side of the diagram keeps track of one part of the system development in time, either the left (positive) or right (negative) side with the initial time and state of the system at the bottom and the final time and state at the top. The system moves from bottom to top by single photon interactions at each vertex and propagation from one vertex to the next.

Consider the simplest possible interaction, namely absorption of a single photon by a system initially in its ground state $\rho^0 = |g\rangle\langle g|$. Analytic results for first-order perturbation in a two-level system are:

$$\rho_{eg}^{(1)} = -\frac{i}{2\hbar}\left[\frac{\mu_{eg}E}{i\left(\omega-\omega_{eg}\right)+\Gamma_{eg}}\right]\rho_{gg}^{(0)}\mathrm{e}^{i\omega t} - \frac{i}{2\hbar}\left[\frac{\mu_{eg}^{*}E^{*}}{-i\left(\omega+\omega_{eg}\right)+\Gamma_{eg}}\right]\rho_{gg}^{(0)}\mathrm{e}^{-i\omega t}, \tag{5.4.69}$$

$$\chi_{ij}^{(1)} = \frac{P_i^{(1)}(\omega)}{E_j(\omega)} = \frac{1}{\hbar}\left[\frac{\left(\mu_j\right)_{eg}\left(\mu_i\right)_{eg}}{\left(\omega-\omega_{eg}\right)+i\Gamma_{eg}}\right]\rho_{gg}^{(0)} + \frac{1}{\hbar}\left[\frac{\left(\mu_j\right)_{ge}\left(\mu_i\right)_{ge}}{\left(\omega+\omega_{ge}\right)+i\Gamma_{ge}}\right]\rho_{gg}^{(0)}. \tag{5.4.70}$$

Based on Eq. (5.4.70), there should be two diagrams to describe the full susceptibility, corresponding to the rotating wave term and the counter-rotating wave term. In Figure 5.24, the diagram on the left shows a positive frequency absorptive interaction by the ket. The diagram on the right indicates absorption of a negative frequency (conjugate) field by the bra. Together, these diagrams comprise a diagrammatic equivalent to the two analytic terms in the first-order perturbation theory expression for $\rho_{eg}^{(1)}$ given in Eq. (5.4.69).

In first order, the final state of the system is $|e\rangle\langle g|$. Both the analytic expression and the diagram show this. To obtain the mathematical expression from the structure of

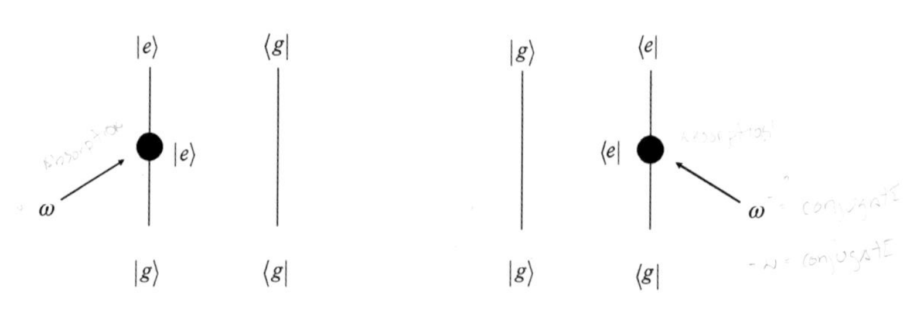

Figure 5.24 *Illustration of the two simplest double-sided Feynman diagrams representing evolution of the ket (left) and bra (right) components of the density matrix.*

the diagram, we choose to associate the matrix element with the vertex and the resonant denominator with propagation between vertices. Specifically, the interactions

$$I_j^L = \frac{1}{\hbar}\left(\mu_j\right)_{eg} E_j e^{i\omega_j t} \leftrightarrow \text{left vertex},$$

$$I_j^R = \frac{1}{\hbar}\left(\mu_j\right)_{eg}^* E_j^* e^{-i\omega_j t} \leftrightarrow \text{right vertex},$$

are vertex contributions and

$$\Pi_j^L = \left[\omega_{eg} - \omega + i\Gamma_{eg}\right]^{-1} \leftrightarrow \text{propagation from } |g\rangle \text{ to } |e\rangle \text{ on the left},$$

$$\Pi_j^R = \left[\omega_{eg} + \omega + i\Gamma_{eg}\right]^{-1} \leftrightarrow \text{propagation from } \langle g| \text{ to } \langle e| \text{ on the right},$$

are the propagators. With these substitutions we can write the density matrix as

$$\rho_{eg}^{(1)} = \frac{1}{2}\left[\Pi_j^L I_j^L + \Pi_j^R I_j^R\right]\rho_{gg}^{(0)}, \tag{5.4.71}$$

or we could write the susceptibility as

$$\chi_{ij}^{(1)}(\omega) = \frac{1}{2}\left[\Pi_j^L I_j^L \frac{(\mu_i)_{eg}}{E_j} + \Pi_j^R I_j^R \frac{(\mu_i)_{eg}^*}{E_j^*}\right]\rho_{gg}^{(0)}. \tag{5.4.72}$$

If the system has many levels, the net susceptibility $\chi^{(1)}$ taking contributions from all states into account requires summation over possible initial and final states g, e:

$$\chi_{ij}^{(1)}(\omega) = \frac{1}{2}\sum_{g,e}\left[\Pi_j^L I_j^L \frac{(\mu_i)_{eg}}{E_j} + \Pi_j^R I_j^R \frac{(\mu_i)_{eg}^*}{E_j^*}\right]\rho_{gg}^{(0)} \tag{5.4.73}$$

Notice that to calculate the correct density matrix elements from diagrams we multiply the vertex and propagator contributions and sum over states. For the correct susceptibility we have in addition to multiply by the dipole moment $(\mu_i)_{eg}$ between initial and final states before summing, because the diagrams strictly only show development of ρ.

For higher order processes (Figure 5.25) there are more numerous permutations of the diagrams. These involve reordering of the vertices vertically, and all left–right permutations to include conjugate processes, but the procedure is a straightforward extension of that already described. Spontaneous emission and other incoherent decay and non-parametric processes are omitted. We have the following rules:

1. Draw all possible diagrams, corresponding to all permutations of vertices ordered in time and distributed between the bra and ket lines.

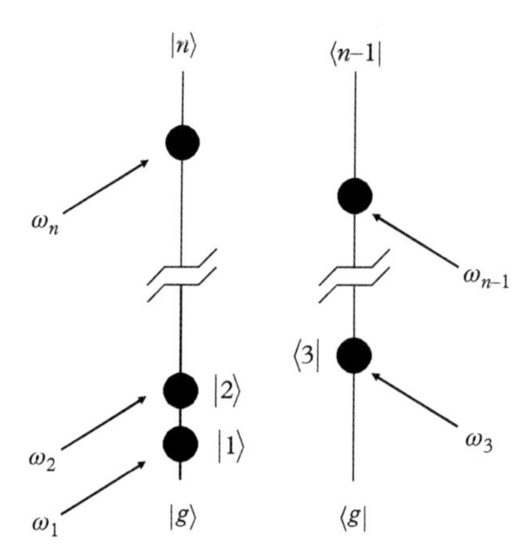

Figure 5.25 *Double-sided Feynman diagram illustrating a sequence of n optical interactions affecting the evolution of the ket and bra of the density matrix.*

2. For the jth vertex connecting state $|n\rangle$ to $|n'\rangle$, assign the following interaction factors:

$$I_j = \frac{1}{\hbar}\left(\mu_j\right)_{n'n} E_j e^{i\omega_j t} \leftrightarrow \text{absorption on the left,}$$

$$I_j = \frac{1}{\hbar}\left(\mu_j\right)_{n'n} E_j e^{i\omega_j t} \leftrightarrow \text{emission on the right,}$$

$$I_j = \frac{1}{\hbar}\left(\mu_j\right)^*_{n'n} E_j^* e^{-i\omega_j t} \leftrightarrow \text{absorption on the right,}$$

$$I_j = \frac{1}{\hbar}\left(\mu_j\right)^*_{n'n} E_j^* e^{-i\omega_j t} \leftrightarrow \text{emission on the left.}$$

3. Propagation from the jth to the $(j+1)$th vertex along ket–bra double timelines $|l\rangle\langle k|$ is described by propagator $\Pi_j = \left[\omega_{lk} + i\Gamma_{lk} + \sum_{i=1}^{j}(\mp\omega)_i\right]^{-1}$. The summation contains all frequencies up to and including ω_j, and each ω_i takes the negative or positive sign depending on whether the corresponding ith vertex shows absorption on the left or the right respectively. If the ith vertex indicates emission, the signs are reversed for a vertex on the right and left respectively.
4. Multiply all factors describing evolution from $|g\rangle\langle g|$ to $|n'\rangle\langle n|$ and sum over all possible states.
5. Find the contribution of one diagram to the nth-order density matrix $\rho^{(n)}$. The nth-order density matrix $\rho^{(n)}$ is the sum of all such diagrams.

The contribution to $\rho^{(n)}$ from the diagram in Figure 5.25 may be written down directly as

$$\sum_{g_1 1\ldots n} \frac{1}{(2\hbar)^n} \left\{ \frac{|n\rangle\, \mu_{(n)(n-1)} \ldots \mu_{1g}\rho_{gg}^{(0)}\mu_{g3} \ldots \mu_{(n-1)(n-2)}\, \langle n-1|}{\Pi_n^{-1}\Pi_{n-1}^{-1} \ldots \left(\omega_{2g} - (\omega_2 + \omega_1) + i\Gamma_{2g}\right)\left(\omega_{1g} - \omega_1 + i\Gamma_{1g}\right)} \right\}$$

$$\times E_1 E_2 E_3^* \ldots E_{n-1}^* E_n \exp(\omega_1 + \omega_2 - \omega_3 \ldots - \omega_{n-1} + \omega_n)t. \qquad (5.4.74)$$

It must be borne in mind that the diagrammatic technique illustrated here ignores all incoherent driving terms and assumes that population changes are negligible (all detunings are finite). That is, only terms driven by external fields are taken into account. This approach cannot be used to formulate solutions to problems in which dark (non-radiative) processes play an important role.

REFERENCES

[5.1] S. Stenholm, *Foundations of Laser Spectroscopy*. New York: John Wiley & Sons, 1984.

[5.2] See, for example, R.M. Macfarlane and R.M. Shelby, *Opt. Lett.* 1981, **6**, 96.

[5.3] S.C. Rand, in *Laser Spectroscopy and New Ideas*, Springer Series in Optical Sciences, Vol. 54. New York: Springer-Verlag, 1987.

[5.4] S.H. Autler and C.H. Townes, *Phys. Rev.* 1955, **100**, 703; P. Avan and C. Cohen-Tannoudji, *J. Phys. B: At. Mol. Phys.* 1977, **10**, 155.

[5.5] N. Huntermann, B. Lipphardt, M. Okhapkin, C. Tamm, E. Peik, A.V. Taichenachev, and V.I. Yudin, *Phys. Rev. Lett.* 2012, **109**, 213002; W.F. Holmgren, R. Trubko, I. Hromada, and A.D. Cronin, *Phys. Rev. Lett.* 2012, **109**, 243004; C.D. Herold, V.D. Vaidya, X. Li, S.L. Rolston, J.V. Porto, and M.S. Safronova, *Phys. Rev. Lett.* 2012, **109**, 243003.

[5.6] See, for example, I.I. Sobelman, *Atomic Spectra and Radiative Transitions*, Springer-Verlag Series in Chemical Physics. New York: Springer-Verlag, 1979.

[5.7] R.W. Fisher, *Optical Phase Conjugation*, New York: Academic Press, 1983.

[5.8] Y.S. Bai and R. Kachru, *Phys. Rev. Lett.* 1991, **67**, 1859.

[5.9] R.W. Boyd, M.G. Raymer, P. Narum, and D.J. Harter, *Phys. Rev. A* 1981, **24**, 411.

[5.10] Q. Shu, *Cooperative optical nonlinearities in Tm:LiYF$_4$*. PhD Dissertation, University of Michigan, 1996.

[5.11] R. Levy, B. Honerlage, and J.B. Grun, *Phys. Rev. B* 1979, **19**, 2326.

[5.12] Y. Aoyagi, Y. Segawa, and S. Namba, in R.R. Alfano (ed.), *Semiconductors Probed by Ultrafast Laser Spectroscopy*, Vol. 1. Academic Press, 1984, pp. 329–49.

[5.13] T.S. Sosnowski, T.B. Norris, H. Jiang, J. Singh, K. Kamath, and P. Bhattacharya, *Phys. Rev. B* 1991, **57**, R9423.

[5.14] H. Beutler, *Z. Phys.* 1935, **93**, 177.

[5.15] U. Fano, *Nuovo Cimento* 1935, **12**, 154; U. Fano, *Phys. Rev.* 1961, 124, 1866.

[5.16] M.D. Sturge, H.J. Guggenheim, and M.H.L. Pryce, *Phys. Rev. B* 1970, **2**, 2459.

[5.17] A. Tai, P. Machado, W. Meyer, D. Wehe, B. Roe, M. Reid, R. Reeves, and S.C. Rand, Valence conversion and Fano resonances in gamma-irradiated Tm:CaF$_2$, Yb:CaF$_2$, and Yb:SrF$_2$, unpublished.

[5.18] A. Lempicki, L. Andrews, S.J. Nettel, B.C. McCollum, and E.I. Solomon, *Phys. Rev. Lett.* 1980, **44**, 1234.
[5.19] U. Fano and J.W. Cooper, *Phys. Rev.* 1965, **137**, A1364.
[5.20] B. Luk'yanchuk, N.I. Zheludev, S.A. Maier, N.J. Halas, P. Nordlander, H. Giessen, and C.T. Chong, *Nat. Mater.* 2010, **9**, 707.
[5.21] D.J. Jackson and J.J. Wynne, *Phys. Rev. Lett.* 1982, **49**, 543.
[5.22] P. Ranitovic, X.M. Tong, C.W. Hogle, X. Zhou, Y. Liu, N. Toshima, M.M. Murnane, and H.C. Kapteyn, *Phys. Rev. Lett.* 2011, **106**, 193008.

PROBLEMS

5.1. The initial state of a pair of identical two-level atoms A and B located at the origin is $\psi = \frac{|1\rangle\,|0\rangle + |0\rangle|1\rangle}{\sqrt{2}}$, where the ket products denote the products of the wavefunctions for atoms A and B in the order $|A\rangle\,|B\rangle$. A, B may be 0 or 1. This state is the result of excitation by a single photon capable of providing enough energy to excite atom A or atom B of the pair but not both.

(a) Show that ensemble average measurements of the probability for finding the atoms in particular combination states, as given by the pair density matrix elements $\langle\psi|\,\rho_{AB}\,|\psi\rangle$ where $\rho_{AB} = |A\rangle\,|B\rangle\,\langle B|\,\langle A|$, confirm that half the time one finds atom A excited and half the time one finds B excited (i.e., find ρ_{01} and ρ_{10}).

(b) Next, separate the two atoms *of a single pair* by a distance on the order of a meter in the laboratory, without causing any phase or state change and make a sequential determination of the states of each separate atom. Because the pair is initially in a superposition state (and given the fundamental postulate of quantum mechanics) the state of atom A is unknowable until a measurement is made. Similarly the state of atom B is indeterminate until measured. Find the probability amplitude for ψ to yield each possible product state of A and B by projecting ψ directly onto each of the four possible eigenstate products in turn.

(c) In part (b), compare the probabilities of finding B in state 0 or 1 after A is found in 0. Suggest a resolution of the surprising fact that B seems to "know" the measured state of atom A and always conserves energy even though it has an overall 50% probability of being found in either state and is measured at a different location.

5.2. Assume that, in the case of resonant optical excitation of a closed two-level system, steady-state, oscillatory solutions for the level populations exist, of the form

$$\rho_{11} = \tilde{\rho}_{11}(1 - \cos\xi t),$$

$$\rho_{22} = \tilde{\rho}_{22}(1 + \cos\xi t).$$

(a) Use closure to find consistent (time-independent) values for $\tilde{\rho}_{11}$ and $\tilde{\rho}_{22}$.

(b) Find the slowly varying amplitude $\tilde{\rho}_{12}$ of the polarization $\rho_{12} = \tilde{\rho}_{12}e^{i\omega t}$ using the results of part (a).

(c) Based merely on what you know about resonant excitation, what would you expect the value of ξ to be?

(d) What is the physical process accounting for the time-dependence of $\tilde{\rho}_{12}$?

(e) The absorption of the system is proportional to $\rho_{11} - \rho_{22}$. Does it ever go to zero? Show and explain.

5.3. A short pulse of central frequency ω is incident on a closed two-level system at time $t = t_0$. The impulsive light-atom interaction is of the form

$$V(t) = V_0\delta(t - t_0)e^{i\omega t} + c.c.$$

Assume that although the interaction is represented mathematically by a delta function, it occurs over a time that is actually long compared to the optical period. Thus the slowly varying envelope approximation (SVEA) is still valid.

Solve for the density matrix elements, starting with ρ_{12}. Find the time-dependent solutions for the occupations of the two states of the system for times $t > t_0$. Ignore all considerations related to Fourier components of the pulse or pulse bandwidth.

5.4. In a two-level gas the polarization is proportional to the Doppler-averaged matrix element

$$\overline{\tilde{\rho}_{12}(t)} = \frac{1}{u\sqrt{\pi}} \int_{-\infty}^{\infty} \tilde{\rho}_{12} \exp\left(-(v_z/u)^2\right) dv_z.$$

In frequency-switched observations of free induction decay one has

$$\tilde{\rho}_{12} = \frac{i\Omega}{2} R_3(0) e^{-t/T_2} e^{i\delta\omega t} e^{i\Delta t} \left(\frac{i\Delta + \Gamma'}{\Delta^2 + \Gamma^2} \right),$$

where $\Gamma' \equiv \frac{1}{T_2}$, $\Gamma^2 \equiv \frac{1}{(T_2)^2} + \Omega^2 \frac{T_1}{T_2}$, $\Delta \equiv \omega_0 - \omega - kv_z$, and the frequency shift $\delta\omega$ is independent of Δ.

(a) Show that the integral above can be expressed analytically in terms of the error function $W(z)$ of complex argument, with different forms for cases $t < 2\Gamma/(ku)^2$ and $t > 2\Gamma/(ku)^2$.

(b) Show that for long times (greater than the inverse Doppler width) the result for $\overline{\tilde{\rho}_{12}}$ is the same as that obtained in lecture notes when one ignores a small imaginary component in the expression for the beat signal.

5.5. Verify that components $V_q^{(k)}$ of an irreducible spherical tensor operator of rank $k = 1$ are related to the rank 1 Cartesian components V_j of vector operator V by

$$V_1^{(1)} = -\frac{V_x + iV_y}{\sqrt{2}},$$

$$V_0^{(1)} = V_z,$$

$$V_{-1}^{(1)} = \frac{V_x - iV_y}{\sqrt{2}}.$$

Hint: One way to show this is to demonstrate that the given expressions satisfy the commutation relations of a rank $k = 1$ tensor.

5.6. By examining the conditions required for the Clebsch–Gordan coefficient or the $3j$ symbol to be non-zero (see Appendix H), show that the trace of any irreducible spherical tensor operator vanishes, except for rank $k = 0$ (scalar operators).

5.7. Prove that a state with angular momentum $J < k/2$ does not have a static 2^k-pole moment. For example, a state with $J = 1$ cannot have a 2^2-pole (quadrupole) moment. (Consider the expectation value of the tensor $\left\langle T_q^{(k)} \right\rangle \equiv \left\langle r^{(k)} Y_q^{(k)} \right\rangle$ and apply the Wigner–Eckart theorem or use tensor commutation relations of $T_q^{(k)}$ with the angular momentum operator to find any general restriction on q associated with static moments and to show that the 2^k-pole moment is zero for $J < k/2$).

5.8. Clebsch–Gordon coefficients are related to $3j$ symbols according to the formula

$$(j_i J; m_i m \,|\, j_i J j_f; m_f) = (-)^{-j_i + J - m_f} \sqrt{2j_f + 1} \begin{pmatrix} j_i & J & j_f \\ m_i & m & -m_f \end{pmatrix},$$

where j_i and j_f are the initial and final j values of the atomic transition and J is the angular momentum of the interaction. Their squares are proportional to the probabilities of transitions between states (Appendix H).

(a) Use a property of the $3j$ symbols to show that only initial and final sublevels differing in magnetic quantum number m by 0 or ± 1 have finite probabilities for electric dipole transitions.

(b) Calculate all seven of the Clebsch–Gordon coefficients for a transition from initial state $|j_i m_i\rangle$ with $j_i = 1$ to final state $|j_f m_f\rangle$ with $j_f = 1$ via the rank one electric dipole operator ($J = 1; m = 0, \pm 1$). Are any of the transitions forbidden?

(c) Repeat the calculation for all nine coefficients on a $j_i = 1$ to $j_f = 2$ transition. Are any of the transitions forbidden in this case?

5.9. A three-level atom has allowed single-photon electric dipole transitions between states 1 and 2 as well as 2 and 3. Hence, the two-photon absorption process depicted in the figure is allowed.

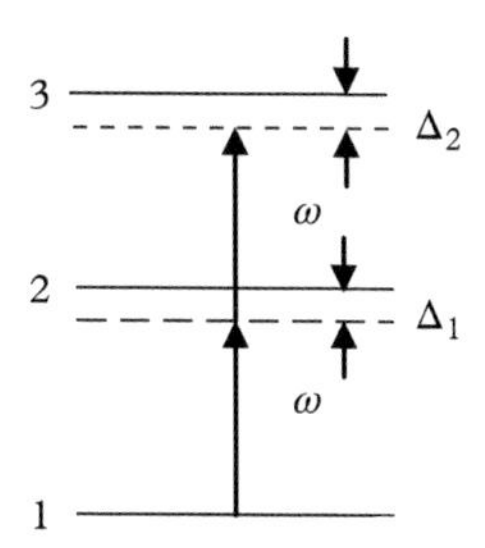

With only a single field incident on the atom, the interactions may be written $V_{12} = -(\hbar\Omega_{12}/2)\mathrm{e}^{i\omega t} + c.c.$ and $V_{23} = -(\hbar\Omega_{23}/2)\mathrm{e}^{i\omega t} + c.c.$, where $\Omega_{12} \equiv \mu_{12}E/\hbar$ and $\Omega_{23} \equiv \mu_{23}E/\hbar$. Assume the detunings Δ_1 and Δ_1 are small so that the rotating wave approximation holds for both transitions. Show that the third-order, steady-state population in state 3 depends quadratically on the incident intensity according to

$$\rho_{33}^{(3)} = \frac{|\Omega_{12}/2|^2 \, |\Omega_{23}/2|^2 \, (\Gamma_{12}/\gamma_{21})(4\Gamma_{23}/\gamma_3)}{\left(\Delta_1^2 + \Gamma_{12}^2 + |\Omega_{12}/2|^2 \, 4\Gamma_{12}/\gamma_{21}\right)\left(\Delta_2^2 + \Gamma_{23}^2\right)}$$

5.10. A three-level atom initially occupies the ground state. It undergoes resonant two-photon absorption to state 2, driven by an incident field at frequency ω as indicated in the accompanying diagram. Level 3 is very far off resonance for either one-photon or two-photon excitation, and the only non-zero transition dipole moments are μ_{13} and μ_{23}.

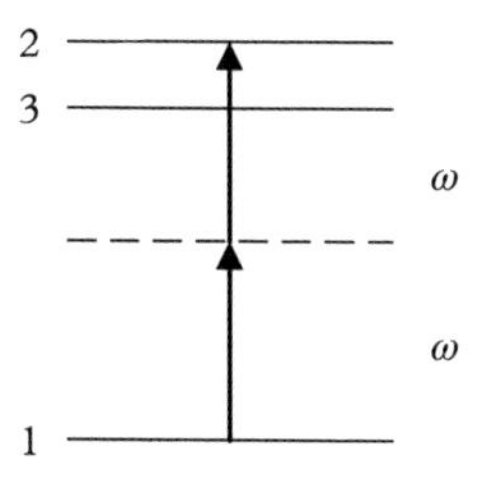

(a) Write down the density matrix equation of motion for the coherence ρ_{13}, omitting all decay terms. Assume the Hamiltonian is $H = H_0 + V$ and express your answer in terms of explicit matrix elements. (Do not solve the equation as yet.)

(b) At what frequencies do coherences ρ_{13} and ρ_{12} oscillate?

(c) *Without making any rotating wave approximation*, solve for a steady-state solution of ρ_{13} assuming that the perturbation

$$V = -\frac{1}{2}\mu_{13}E\mathrm{e}^{i\omega t} + c.c. - \frac{1}{2}\mu_{23}^* E^* e^{-i\omega t} + c.c.$$

does not cause a large change in ground state population. Leave your answer in terms of the slowly varying amplitude $\tilde{\rho}_{12}$ (i.e., it is not necessary to complete the problem by solving for the steady-state value of ρ_{12}).

5.11. In the three-level system shown in the figure, an incident light field composed of counter-propagating waves of the same frequency ω is tuned *between* the two closely spaced levels $|2>$ and $|3>$. These levels are close enough so that polarizations ρ_{21} and ρ_{31} can be created on both the $1\rightarrow 2$ and $1\rightarrow 3$ transitions of the moving atoms. Transitions to both excited states are dipole allowed and the Doppler-broadened linewidths of these transitions overlap. When effects of second order and higher are taken into account, one finds the resonance factor

$$\rho_{21} \propto \frac{1}{[i(\Delta_{21} + kv) + \Gamma_{21}]\,[-i(\Delta_{31} - kv) + \Gamma_{31}]}$$

in the expression for the polarization, where $\Delta_{21} \equiv \omega_{21} - \omega$, $\Delta_{31} \equiv \omega_{31} - \omega$, and v is the velocity.

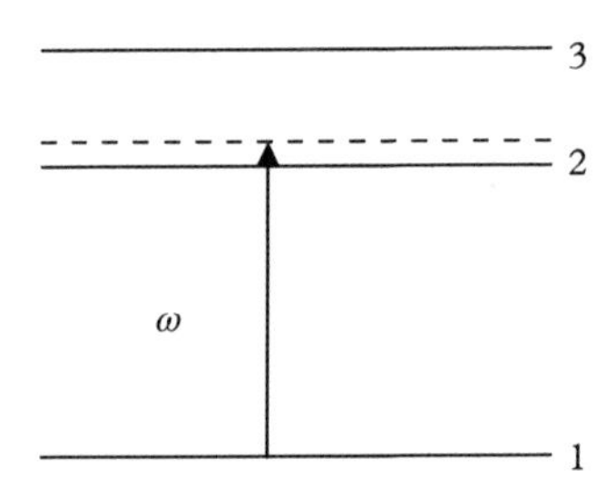

(a) Ignoring decay factors Γ (i.e., setting $\Gamma_{21} = \Gamma_{31} = 0$), identify a Doppler frequency shift kv at which resonant response is obtained that is *higher than first order*. (Because there are two free variables ω and v, proceed by eliminating ω to find kv.)

(b) Explain the physical origin of this "crossover" resonance by considering the Doppler shift from part (a) at which it occurs. (Note: ρ_{31} has a similar factor in it but need not be considered explicitly).

(c) What is the total macroscopic radiant polarization of the atoms? (Write down the formal expression and all terms contributing to the trace).

5.12. Prove that the general form of $\rho_{12}(t)$ *cannot be real.* (Hint: Instead of using Eq. (G.1), assume $\rho_{12}(t)$ is given by a real expression such as $\rho_{12}(t) = \frac{1}{2}\tilde{\rho}_{12}\exp(i\omega t) +$ *c.c.* Substitute this into the equation of motion (G.4). By equating coefficients for positive and negative frequency terms to separate Eq. (G.4) into one equation for the amplitude $\tilde{\rho}_{12}$ and one for $\tilde{\rho}^*_{12}$, show that steady-state solutions yield the result $\tilde{\rho}^*_{12} \neq (\tilde{\rho}_{12})^*$. Since $\tilde{\rho}^*_{12} \neq (\tilde{\rho}_{12})^*$ is necessary for the density matrix to provide consistent solutions of the equation of motion, this is sufficient to establish the fact that $\rho_{12}(t)$ is always complex.)

5.13. (a) Sketch all the double-sided Feynman diagrams for the state of a system undergoing two-photon emission from an initial state $|e><e|$.

(b) In how many ways can this occur in principle?

(c) Consider a multi-level system and give an argument as to whether the diagrams of part (a) correspond to physically independent processes or not.

(d) In a single diagram of your choice from part (a), label all states and frequencies and write down the perturbation expression for its contribution to the second-order density matrix element $\rho^{(2)}$.

5.14. In general, in a closed system of two-level atoms subjected to standing-wave excitation at frequency ω, two holes are burned in the velocity profile at values of v where light is shifted into exact resonance with the moving atoms by the Doppler effect. If detuning is reduced toward zero, the holes converge to the center of the distribution, eventually overlapping (when $\Delta = 0$ and $\mathrm{v} = 0$) to yield a pronounced (saturation) dip in the absorption. Show that for $\Delta = 0$

(a) at low intensities ($\Omega \ll \Gamma$) the spatially averaged absorption is indeed lower at $\mathrm{v} = 0$ than at $\mathrm{v} = \Gamma/20k$ near the center of the distribution. To do this calculate R, the ratio of absorption at $\mathrm{v} = 0$ to that at $\mathrm{v} = \Gamma/20k$, and show that $R < 1$,

(b) at intermediate intensities ($\Omega' = \Gamma/2$) exactly the opposite is true and $R > 1$,

(c) at very high intensities ($\Omega' \gg \Gamma$) the ratio of resonant absorption at $\mathrm{v} = 0$ to that at $\mathrm{v} = \Omega'/k$ is roughly proportional to intensity.

(d) Briefly explain why the ratio of saturated absorption at $\mathrm{v} = 0$ and $\mathrm{v} \neq 0$ reverses between low and intermediate intensities in parts (a) and (b).

5.15. Consider a system of three-level atoms undergoing two-photon absorption that is only slightly detuned from intermediate state $|2\rangle$, as shown in the accompanying figure. The light field is a standing wave composed of two counter-propagating, traveling waves of equal amplitude and frequency $\omega \cong \omega_{13}/2$. Atoms in the gas may interact twice with a given traveling wave or they may absorb one photon from the forward wave and one from the backward wave to complete the transition from state $|1\rangle$ to state $|3\rangle$.

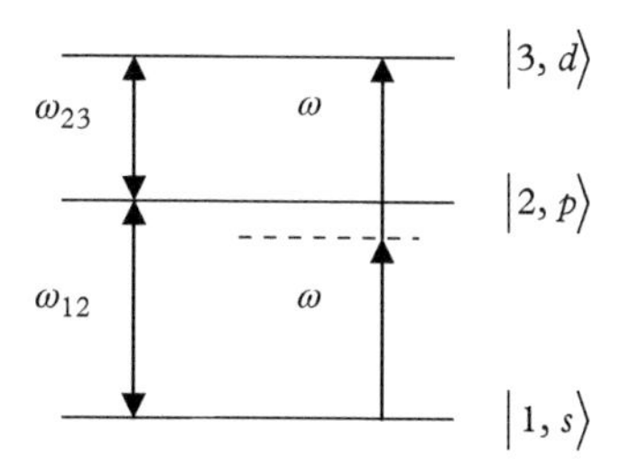

(a) By considering how atoms of velocity v interact with the light field above, justify the following forms for first-order and second-order coherences in the system:

$$\rho_{12}^{(1)} = \tilde{\rho}_+ e^{i(\omega t - kz)} + \tilde{\rho}_- e^{i(\omega t + kz)}; \rho_{13}^{(2)} = \tilde{\rho}_{++} e^{2i(\omega t - kz)} + \tilde{\rho}_{--} e^{2i(\omega t + kz)} + \tilde{\rho}_{+-} e^{2i(\omega t)}.$$

(b) Using the coherences of part (a) calculate the amplitudes $\tilde{\rho}_{+-}$, $\tilde{\rho}_{+-}$, $\tilde{\rho}_{+-}$ for all three processes or "channels" that can mediate a transition to state $|3\rangle$, using second-order perturbation theory.

(c) Give a physical explanation as to why the resonant (two-photon) denominator of only one process is independent of v, giving rise to an absorption spectrum with a sub-Doppler linewidth.

5.16. Equation (5.2.39) gives an expression for the average population difference of states in a system of moving two-level atoms subjected to standing wave excitation. Using the parameters $\bar{N} = \exp[-k^2 v^2/\Gamma^2]$, $\Gamma = \gamma_1 = \gamma_2 \equiv \gamma = 1$, $k = 1$, and $\Delta = 0$, plot the population difference $\bar{\rho}_{22} - \bar{\rho}_{11}$ versus atomic velocity kv from v = −200 to v = +200 for (a) $\Omega = \gamma$, and (b) $\Omega = 2\gamma$. (c) Compare results from (a) and (b) and explain any differences.

5.17. A three-level system is continuously subjected to two fields that connect levels 1 and 3 via a two-photon resonance, as depicted in the figure. The single-photon transitions $1 \to 2$ and $2 \to 3$ have non-zero, dipole moment matrix elements, whereas the $1 \to 3$ transition is forbidden. The resonant Rabi frequencies for the fields on the $1 \to 2$ and $2 \to 3$ transitions are $\Omega_1(t) \equiv \mu_{12}\tilde{E}_1(t)/\hbar$ and $\Omega_2(t) \equiv \mu_{23}\tilde{E}_2(t)/\hbar$, respectively. In the interaction picture, the effective Hamiltonian of the system is found to be expressible in the form

$$\hat{H}_{eff} = -\frac{1}{2}\hbar \begin{bmatrix} 0 & \Omega_1(t) & 0 \\ \Omega_1(t) & 2\Delta(t) & \Omega_2(t) \\ 0 & \Omega_2(t) & 0 \end{bmatrix},$$

where $\Delta(t) = \Delta_{12}(t) = -\Delta_{23}(t)$ is a detuning condition that assures two-photon resonance, as shown in the diagram.

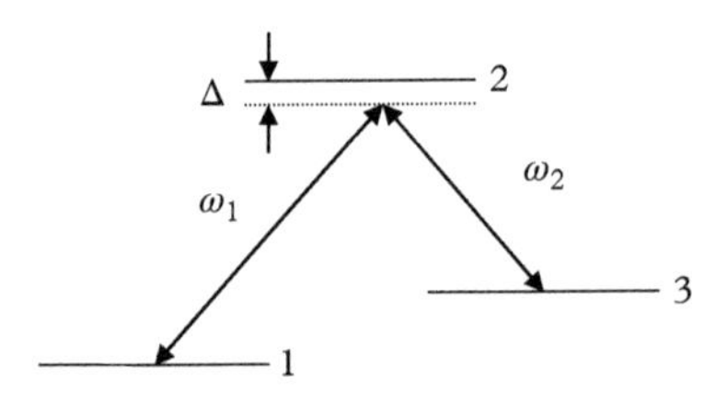

(a) Solve for the *eigenvalues* of the system. (Do not solve for any eigenvectors as yet!)

(b) Show that the eigenvector $\bar{V}_1 = a_1|1\rangle + b_1|2\rangle + c_1|3\rangle$ for the eigenvalue of *smallest* magnitude can be expressed in the form $|u_1(t)\rangle = \cos\theta(t)|1\rangle - \sin\theta(t)|3\rangle$ and determine the corresponding mixing angle θ(t).

(c) Bearing in mind the result of part (b), solve the Schrödinger equation explicitly in the interaction picture for $|\Psi(t)\rangle$, the general wavefunction of the system at time t, if it is prepared at $t = 0$ in state $|u_1(0)\rangle = 0.7071|1\rangle - 0.7071|3\rangle$.

(d) Calculate the probability of excitation from state $|u_1(t)\rangle$ to state $|2\rangle$ at time t, presuming the time variation of the fields $\Omega_1(t)$ and $\Omega_2(t)$ is known.

(e) Find the remaining two eigenvectors $\bar{V}_i = a_i |1\rangle + b_i |2\rangle + c_i |3\rangle$ $(i = 2, 3)$.

5.18. Show that the coherence $\rho_{1j}^{(n)}$ on a multi-photon transition of any order n that originates from the ground state equals the occupation probability ρ_{jj} of the terminal state j if the process is parametric (i.e., if there is negligible ground state depletion).

5.19. An optical interaction of the form $V = -\frac{1}{2}\mu_{12}E_0 \exp(i\omega t) + c.c.$ couples levels 1 and 2 of the three-level system in the figure, in which all excited states undergo spontaneous decay to lower levels.

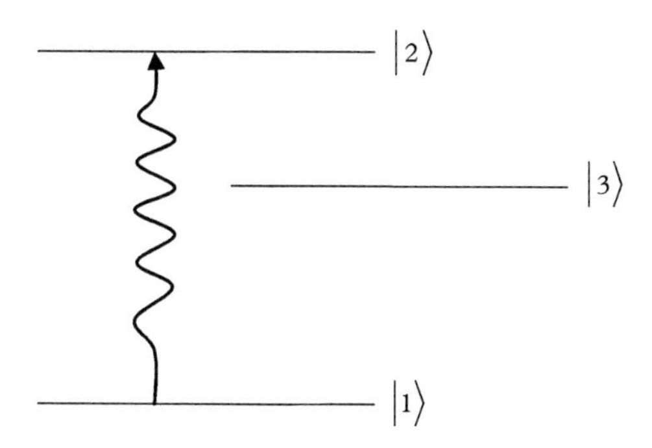

(a) Write down and solve the equations of motion for elements of the density matrix and determine the ratio of populations in states 3 and 2 under steady-state conditions.

(b) In the limit that the decay rate of state 3 approaches zero and steady excitation is maintained, what are the populations of all three states?

5.20. Find the condition for complete transparency in the vicinity of an ultranarrow Fano resonance ($\Gamma_{12} \cong 0$).

6

Quantized Fields and Coherent States

When optical fields are weak, spontaneous processes and statistical fluctuations which we have not yet considered assume greater importance. Additionally, interactions of light with atoms exactly on resonance alter the static part of the Hamiltonian significantly in a dynamic way, even while transitions between the "renormalized" and shifted energy levels are taking place. In these limits new phenomena emerge that require careful attention to the "quantized" or particle-like nature of the electromagnetic field as well as strong coupling of light and matter. Hence this chapter begins with the formal quantization of the electromagnetic field interaction—introduction of a particle-like description of light—and then uses it to cover subjects where quantum optics provides insights and sometimes unique agreement with experimentally measured dynamics.

6.1 Quantization of the Electromagnetic Field

The classical description of fields is best formulated in terms of a vector potential $\bar{A}$ and a scalar potential φ that simplify the relationship between fields and sources. By imposing constraints on the potentials, one can guarantee that the fields $\bar{E}$ and $\bar{B}$ determined by them always satisfy Maxwell's equations (see Appendix C). Since Maxwell's equations represent the distillation of all that was known about electromagnetic phenomena before the advent of quantum mechanics, the quantum mechanical version of electrodynamics should also satisfy Maxwell's equations.

In developing quantum electrodynamics, we therefore begin with Eqs. (C.3) and (C.7), by writing

$$\bar{B} = \bar{\nabla} \times \bar{A}, \tag{6.1.1}$$

$$\bar{E} = -\frac{\partial \bar{A}}{\partial t} - \bar{\nabla}\varphi. \tag{6.1.2}$$

In vacuum $\bar{B} = \mu_0 \bar{H}$ and $\bar{D} = \varepsilon_0 \bar{E}$. Use of these expressions in the cross-substituted curl equations of classical electrodynamics (Chapter 1) then yields a wave equation for the vector potential:

$$\nabla^2 \bar{A} - \mu_0 \varepsilon_0 \frac{\partial^2 \bar{A}}{\partial t^2} - \bar{\nabla}\left(\bar{\nabla} \cdot \bar{A}\right) = -\mu_0 \bar{J} + \mu_0 \varepsilon_0 \bar{\nabla} \frac{\partial \varphi}{\partial t}. \tag{6.1.3}$$

Lectures on Light. Second Edition. Stephen C. Rand.
© Stephen C. Rand 2016. Published in 2016 by Oxford University Press.

Some arbitrariness is associated with the definition of the potentials. This is evident from the fact that the fields are not altered if one makes the gauge transformations

$$\bar{A}' = \bar{A} + \bar{\nabla}\Phi, \tag{6.1.4}$$

$$\varphi' = \varphi - \frac{\partial \Phi}{\partial t}, \tag{6.1.5}$$

where Φ is an arbitrary scalar function. Hence the selection of gauge is arbitrary, and for convenience we can choose the so-called Coulomb gauge in which $\bar{\nabla} \cdot \bar{A} = 0$. In this gauge, the vector potential becomes solenoidal or "transverse" like the magnetic field $(\bar{\nabla} \cdot \bar{B} = 0)$, and the static scalar potential φ is directly related to any charge distribution in the region of interest $(\nabla^2\varphi = -\bar{\nabla} \cdot \bar{E} = -\rho_V/\varepsilon_0)$. This shows that the volumetric charge density ρ_V determines both the scalar potential and the electric field $\bar{E}$. In free space Eq. (6.1.3) reduces to

$$\left(\nabla^2 - \frac{1}{c^2}\frac{\partial^2}{\partial t^2}\right)\bar{A} = -\mu_0 \bar{J}. \tag{6.1.6}$$

This equation shows that the current $\bar{J}$ determines both the magnitude and direction of the vector potential and the magnetic flux density $\bar{B} = \bar{\nabla} \times \bar{A}$ (Figure 6.1).

The scalar and vector potentials are related to charge and current distributions at the (primed) source coordinates:

$$\varphi(\bar{r}) = (4\pi\varepsilon_0)^{-1} \int \frac{\rho_V(\bar{r}')dV'}{|\bar{r} - \bar{r}'|}, \tag{6.1.7}$$

$$\bar{A}(\bar{r}, t) = (\mu_0/4\pi) \int \frac{\bar{J}(\bar{r}')dV'}{|\bar{r} - \bar{r}'|}. \tag{6.1.8}$$

Due to its transverse character, the electric field may be divided into two parts (transverse and longitudinal),

$$\bar{E} = \bar{E}_\perp + \bar{E}_\parallel, \tag{6.1.9}$$

with the properties

$$\bar{\nabla} \cdot \bar{E}_\perp = 0, \tag{6.1.10}$$

$$\bar{\nabla} \times \bar{E}_\parallel = 0. \tag{6.1.11}$$

Then from Eq. (6.1.2) we obtain the relations between these field components and the potentials:

$$\bar{E}_\perp = -\frac{\partial \bar{A}}{\partial t}, \tag{6.1.12}$$

$$\bar{E}_\parallel = -\bar{\nabla}\varphi. \tag{6.1.13}$$

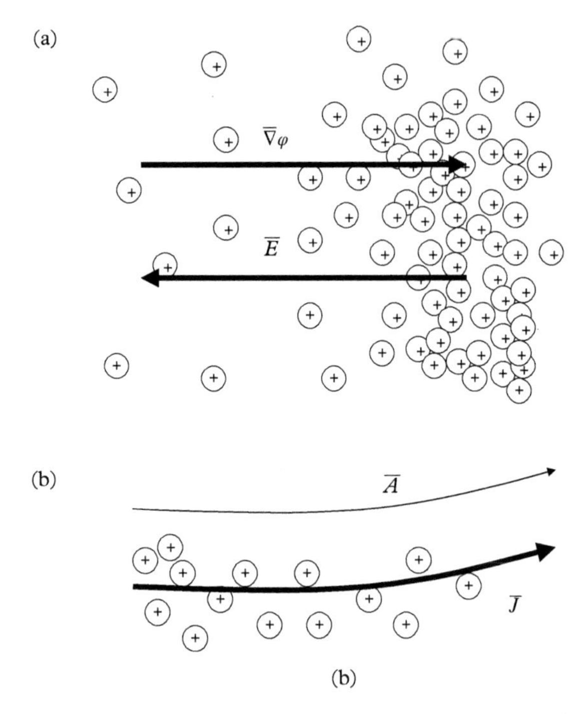

Figure 6.1 *Illustrations of the vectorial relationships between (a) a static charge distribution that generates a potential gradient and consequently an electric field $\bar{E} = -\bar{\nabla}\varphi$, and (b) a flow of charges that produces a vector potential and an associated magnetic flux density $\bar{B} = \bar{\nabla} \times \bar{A}$.*

$\bar{E}_{\parallel}$ is a longitudinal electric field which reflects Coulomb interactions between charges. This term is of no interest in far-field radiation problems. Therefore, we turn to the problem of quantizing the energy of the transverse field by focusing on $\bar{A}$, $\bar{E}_{\perp}$, and $\bar{B}$.

A charged particle in the electromagnetic field has a Hamiltonian given by the sum of its kinetic and potential energies, with the kinetic term expressed in terms of the canonical momentum:

$$H = \frac{1}{2m}\left(\bar{p} - q\bar{A}\right)^2 + q\varphi. \tag{6.1.14}$$

In Hamilton's formulation, the equations of motion are

$$\dot{\bar{r}} = \frac{\partial H}{\partial \bar{p}} = \frac{\bar{p} - q\bar{A}}{m} = \bar{v}, \tag{6.1.15}$$

$$\dot{\bar{p}} = -\bar{\nabla} H. \tag{6.1.16}$$

Together with Eq. (6.1.14), Eq. (6.1.16) yields the Lorentz force law

$$\dot{\bar{v}} = \frac{1}{m}\left[\dot{\bar{p}} - q\frac{\partial}{\partial t}\bar{A} - q\left(\bar{v}\cdot\bar{\nabla}\right)\bar{A}\right]$$

$$= -\frac{q}{m}\left[\bar{\nabla}\phi + \frac{\partial}{\partial t}\bar{A} - \bar{\nabla}\left(\bar{v}\cdot\bar{A}\right) + \left(\bar{v}+\bar{\nabla}\right)\bar{A}\right]$$

$$= \frac{q}{m}\left[\bar{E} + \bar{v}\times\bar{B}\right]. \tag{6.1.17}$$

The process of quantization of conjugate pairs of variables like $(\bar{r}, \bar{p})$ proceeds with the introduction of a commutator consistent with their equation of motion. Consequently, the electromagnetic fields $\bar{E}$ and $\bar{B}$ that exert forces on charges will be quantized in a way which satisfies Eq. (6.1.17). Instead of working directly with Eq. (6.1.17) however, we shall make use of an expression for the energy density of electromagnetic fields (derivable from Maxwell's equations) together with the force law.

To do this, we first expand $\bar{A}$ in terms of the modes of a cavity of volume V. Instead of using standing waves for the expansion, we introduce running waves that are subject to periodic boundary conditions:

$$\bar{A}\left(\bar{r}, t\right) = \frac{-i}{\sqrt{\varepsilon_0 V}}\sum_{k,\lambda}\bar{c}_{k\lambda}(t)\exp(i\bar{k}\cdot\bar{r}) + c.c. \tag{6.1.18}$$

Here $\bar{k}$ is the wavevector and $\lambda = 1, 2$ is an index for polarization of the field. The initial phase factor of $-i$ in Eq. (6.1.18) is arbitrary, but ensures convenient forms of the electric and magnetic fields. For the running waves in Eq. (6.1.18) to be solutions of the wave equation in free space, they must be transverse in character. Hence the condition may be imposed that $\bar{\nabla}\cdot\bar{A} = 0$, or

$$\bar{k}\cdot\bar{c}_{k\lambda} = 0. \tag{6.1.19}$$

The wave amplitude $\bar{c}_{k\lambda}$ must be perpendicular to $\bar{k}$, so there can be only two independent polarizations $\hat{e}_{k\lambda}$ for each $\bar{k}$. It follows from Eq. (6.1.6) that the amplitudes of right-going wave solutions must have the form

$$\bar{c}_{k\lambda}(t) = \hat{e}_{k\lambda}c_{k\lambda}(0)\exp(-i\omega_k t). \tag{6.1.20}$$

The dispersion relation $\omega_k = ck$ of Section 1.2 has also been assumed.

On the basis of Eqs. (6.1.12) and (6.1.20), the transverse (radiation zone) electric field is

$$\bar{E}_\perp = \frac{-1}{\sqrt{\varepsilon_0 V}}\sum_{k,\lambda}\left[\omega_k\bar{c}_{k\lambda}(t)\exp(i\bar{k}\cdot\bar{r}) + c.c.\right]. \tag{6.1.21}$$

The magnetic flux density may similarly be found from Eq. (6.1.1):

$$\bar{B} = \bar{\nabla}\times\bar{A}\left(\bar{r}, t\right) = \frac{-1}{\sqrt{\varepsilon_0 V}}\times\sum_{k,\lambda}\left[\bar{k}\times\bar{c}_{k\lambda}(t)\exp(i\bar{k}\cdot\bar{r}) + c.c.\right]. \tag{6.1.22}$$

Since the energy density of the radiation field is

$$U_{rad} = \left(\varepsilon_0 E_\perp^2 + B^2/\mu_0\right)/2, \tag{6.1.23}$$

the next step of the quantization procedure is to substitute Eqs. (6.1.21) and (6.1.22) into the energy density of Eq. (6.1.23) and to find the total energy for the cavity volume V. Upon substitution, the Dirac delta function,

$$\frac{1}{V}\int dV \exp(i[\bar{k}-\bar{k}']\cdot\bar{r}) = \delta\left(\bar{k}-\bar{k}'\right), \tag{6.1.24}$$

may be recognized and used to simplify the vector product,

$$\sum_{k',\lambda'}\left(\bar{k}\times\bar{c}_{k\lambda}\right)\cdot\left(\bar{k}'\times\bar{c}^*_{k'\lambda'}\right)\delta(\bar{k}-\bar{k}') = k^2\left|c_{k\lambda}\right|^2, \tag{6.1.25}$$

that appears in the volume integral of the energy density of the radiation field. One finds

$$U_{field} = \frac{1}{2}\int dV\left(\varepsilon_0 E_\perp^2 + B^2/\mu_0\right) = 2\sum_{k,\lambda}\omega_k^2\left|c_{k\lambda}\right|^2. \tag{6.1.26}$$

Now, to quantize the energy in Eq. (6.1.26), scalar variables $Q_{k\lambda}$ and $P_{k\lambda}$ are introduced using the definition

$$\bar{c}_{k\lambda} = \frac{1}{2}\left(Q_{k\lambda} + \frac{iP_{k\lambda}}{\omega_k}\right)\hat{\varepsilon}_{k\lambda}, \tag{6.1.27}$$

where $\hat{\varepsilon}_{k\lambda}$ represent the circular polarization basis vectors of Eq. (1.2.24) with the property

$$\hat{\varepsilon}_{k\lambda}\cdot\hat{\varepsilon}^*_{k\lambda'} = \delta_{\lambda\lambda'}. \tag{6.1.28}$$

The change of variables in Eq. (6.1.27) is useful because U_{field} can then be written in a form recognizable as a sum of simple harmonic oscillator Hamiltonians (compare Eq. (2.2.19)):

$$U_{field} = \sum_{k,\lambda}\frac{1}{2}\left[P_{k\lambda}^2 + \omega_k^2 Q_{k\lambda}^2\right] \equiv \sum_{k,\lambda} H_{k\lambda}. \tag{6.1.29}$$

Using Eq. (6.1.20), it can readily be shown that $Q_{k\lambda}$ and $P_{k\lambda}$ also obey Hamilton's equations of motion:

$$\dot{Q}_{k\lambda} = P_{k\lambda} = \frac{\partial H_{k\lambda}}{\partial P_{k\lambda}}, \tag{6.1.30}$$

$$\dot{P}_{k\lambda} = -\omega_k^2 Q_{k\lambda} = -\frac{\partial H_{k\lambda}}{\partial Q_{k\lambda}}. \tag{6.1.31}$$

This demonstrates that not only can U_{field} be reduced to a sum of Hamiltonians of harmonic oscillator form, but $P_{k\lambda}$ and $Q_{k\lambda}$ are the canonical variables of the "oscillators." Hence, $P_{k\lambda}$ and $Q_{k\lambda}$ are the natural variables to use in quantizing our field theory and their indices tell us that the oscillators in question correspond to the modes of free space. We next postulate the commutator for the corresponding operators to be

$$\left[\hat{Q}_{k\lambda}, \hat{P}_{k'\lambda'}\right] = i\hbar\delta_{kk'}\delta_{\lambda\lambda'}, \tag{6.1.32}$$

and introduce some operators that are convenient for exploring the energy structure of the electromagnetic field, namely

$$\hat{a}^-_{k\lambda} = \sqrt{\frac{\omega_k}{2\hbar}}\left(Q_{k\lambda} + i\frac{P_{k\lambda}}{\omega_k}\right), \tag{6.1.33}$$

$$\hat{a}^+_{k\lambda} = \sqrt{\frac{\omega_k}{2\hbar}}\left(\hat{Q}_{k\lambda} - i\frac{\hat{P}_{k\lambda}}{\omega_k}\right). \tag{6.1.34}$$

Exercise: Verify using Eq. (6.1.32) that the operators $\hat{a}^-, \hat{a}^+$ satisfy the commutation relation

$$\left[\hat{a}^-_{k\lambda}, \hat{a}^+_{k\lambda'}\right] = \delta_{kk'}\delta_{\lambda\lambda'}, \tag{6.1.35}$$

and confirm the following quantum mechanical expressions for electromagnetic field quantities:

$$A(\bar{r}, t) = -i\sum_{k,\lambda}\sqrt{\frac{\hbar}{2\varepsilon_0\omega_k V}}\left[\hat{a}^-_{k\lambda}(t)\hat{\varepsilon}^-_{k\lambda}\exp(i\bar{k}\cdot\bar{r}) - \hat{a}^+_{k\lambda}(t)\hat{\varepsilon}^+_{k\lambda}\exp(-i\bar{k}\cdot\bar{r})\right], \tag{6.1.36}$$

$$E_\perp(\bar{r}, t) = \sum_{k,\lambda}\sqrt{\frac{\hbar\omega_k}{2\varepsilon_0 V}}\left[\hat{a}^-_{k\lambda}(t)\hat{\varepsilon}^-_{k\lambda}\exp(i\bar{k}\cdot\bar{r}) + \hat{a}^+_{k\lambda}(t)\hat{\varepsilon}^+_{k\lambda}\exp(-i\bar{k}\cdot\bar{r})\right], \tag{6.1.37}$$

$$B(\bar{r}, t) = \sum_{k,\lambda}\sqrt{\frac{\hbar}{2\varepsilon_0\omega_k V}}\left[\hat{a}^-_{k\lambda}(t)\left(\bar{k}\times\hat{\varepsilon}^-_{k\lambda}\right)\exp(i\bar{k}\cdot\bar{r}) + \hat{a}^+_{k\lambda}(t)\left(\bar{k}\times\hat{\varepsilon}^+_{k\lambda}\right)\exp(-i\bar{k}\cdot\bar{r})\right], \tag{6.1.38}$$

$$U_{field} = \sum_{k,\lambda}\hat{H}_{k\lambda} = \sum_{k,\lambda}\hbar\omega_k\left(\hat{a}^+_{k\lambda}\hat{a}^-_{k\lambda} + \frac{1}{2}\right). \tag{6.1.39}$$

Exercise: Show that in free space Eq. (6.1.38) is equivalent to

$$B(\bar{r}, t) = i\sum_{k,\lambda}\sqrt{\frac{\mu_0\hbar\omega_k}{2V}}\left[\hat{a}^-_{k\lambda}(t)\hat{\varepsilon}^-_{k\lambda}\exp(i\bar{k}\cdot\bar{r}) - \hat{a}^+_{k\lambda}(t)\hat{\varepsilon}^+_{k\lambda}\exp(-i\bar{k}\cdot\bar{r})\right]. \tag{6.1.40}$$

Using the operators of Eqs. (6.1.33) and (6.1.34), the Hamiltonian operator for the entire field assumes a particularly compact form:

$$\hat{H}_{field} = \sum_{k,\lambda} \hbar\omega_k \left(\hat{a}^+_{k\lambda} \hat{a}^-_{k\lambda} + \frac{1}{2} \right). \tag{6.1.41}$$

Using Eqs. (6.1.35) and (6.1.41), we can find the additional commutation relations,

$$\left[\hat{H}, \hat{a}^-\right] = -\hbar\omega\hat{a}^-, \tag{6.1.42}$$

$$\left[\hat{H}, \hat{a}^+\right] = \hbar\omega\hat{a}^+, \tag{6.1.43}$$

applicable to a particular single mode of the field, and the eigenvalues and eigenstates of the quantized electromagnetic field can be determined.

Let the eigenstates be $|n\rangle$ and the corresponding eigenvalues be E_n, so that

$$\hat{H}|n\rangle = E_n|n\rangle\,. \tag{6.1.44}$$

The energy of the field state $\hat{a}^- |n\rangle$ is different from E_n, and may be found by substituting Eq. (6.1.42) into Eq. (6.1.44):

$$\begin{aligned} \hat{H}\hat{a}^- |n\rangle &= \left(\hat{a}^- \hat{H} - \hbar\omega\hat{a}^-\right) |n\rangle \\ &= (E_n - \hbar\omega)\hat{a}^- |n\rangle. \end{aligned} \tag{6.1.45}$$

The action of $\hat{H}$ on state $\hat{a}^- |n\rangle$ reveals that its eigenvalue is $E_n - \hbar\omega$. Evidently, the operator $\hat{a}^-$ reduces the energy eigenvalue of the field by an amount $\hbar\omega$, making it an energy "lowering" or "annihilation" operator. Repeated application of this operator therefore produces eigenstates with progressively smaller eigenvalues until the state of lowest energy is reached.

The lowest eigenvalue is positive. This is readily confirmed by examining the expectation value of the Hamiltonian for an arbitrary state $|\psi\rangle$:

$$\langle\psi|\, \hbar\omega \left(\hat{a}^+\hat{a}^- + \frac{1}{2} \right) |\psi\rangle = \hbar\omega\, \langle\psi|\, \hat{a}^+\hat{a}^- \,|\psi\rangle + \frac{1}{2}\hbar\omega. \tag{6.1.46}$$

The expectation value $\langle\psi|\, \hat{a}^+\hat{a}^- \,|\psi\rangle$ can be interpreted as the probability $\langle\psi'|\, \psi'\rangle$ of being in state $|\psi'\rangle = \hat{a}^- |\psi\rangle$. All state occupation probabilities are positive definite. Since ω is also positive, the entire right side of Eq. (6.1.46) must be positive. The eigenvalue E_n is therefore positive.

Now consider the lowest energy eigenstate of the field $|\psi_0\rangle = |0\rangle$, where the notation simply labels the state by its quantum number n. The lowest energy state has $n = 0$.

Action of $\hat{a}^-$ on $|0\rangle$ must yield zero, since otherwise $|0\rangle$ would not be the state of lowest energy:

$$\hat{a}^- |0\rangle = 0. \tag{6.1.47}$$

The lowest possible energy in each mode of the field can now be calculated, with the result

$$\langle 0|\hbar\omega\left(\hat{a}^+\hat{a}^- + \frac{1}{2}\right)|0\rangle = \hbar\omega\langle 0|\hat{a}^+\hat{a}^-|0\rangle + \frac{1}{2}\hbar\omega = \frac{1}{2}\hbar\omega. \tag{6.1.48}$$

Hence the energy of the zero state is

$$E_0 \equiv \langle 0|\hat{H}_{field}|0\rangle = \frac{1}{2}\hbar\omega. \tag{6.1.49}$$

Application of the commutation relation (Eq. (6.1.43)) can now be used to generate all states of higher energy, together with their eigenvalues. For example, the $n = 1$ state is

$$\begin{aligned}\hat{H}\,|1\rangle &= \hat{H}\hat{a}^+\,|0\rangle = \left(\hat{a}^+\hat{H} + \hbar\omega\hat{a}^+\right)|0\rangle \\ &= \left(\frac{1}{2}\hbar\omega\hat{a}^+ + \hbar\omega\hat{a}^+\right)|0\rangle \\ &= \frac{3}{2}\hbar\omega\hat{a}^+\,|0\rangle\,. \end{aligned}\tag{6.1.50}$$

Repeating this procedure n times yields

$$\hat{H}|n\rangle = \hat{H}\left(\hat{a}^+\right)^n|0\rangle = \left(n + \frac{1}{2}\right)\hbar\omega\left(\hat{a}^+\right)^n|0\rangle. \tag{6.1.51}$$

The eigenvalues of the field are therefore

$$E_n = \left(n + \frac{1}{2}\right)\hbar\omega. \tag{6.1.52}$$

According to Eq. (6.1.45), the state $\hat{a}^-|n\rangle$ is an eigenstate of the Hamiltonian. If such a state is to be consistently normalized, we need to find a scalar s_n such that

$$\hat{a}^-\,|n\rangle = s_n\,|n-1\rangle\,. \tag{6.1.53}$$

From Eq. (6.1.51) we see that

$$\left(\hat{a}^+\hat{a}^- + \frac{1}{2}\right)\hbar\omega\,|n\rangle = \left(n + \frac{1}{2}\right)\hbar\omega\,|n\rangle\,,$$

or

$$\langle n|\hat{a}^{+}\hat{a}^{-}|n\rangle = n\langle n|n\rangle = n. \tag{6.1.54}$$

Multiplying Eq. (6.1.53) by its adjoint we find

$$s_n^2\langle n-1|n-1\rangle = \langle n|\hat{a}^{+}\hat{a}^{-}|n\rangle = n, \tag{6.1.55}$$

or

$$s_n = \sqrt{n}. \tag{6.1.56}$$

The result is

$$\hat{a}^{-}|n\rangle = \sqrt{n}\,|n-1\rangle. \tag{6.1.57}$$

By analyzing the effect of the adjoint operator $\hat{a}^{+}|n\rangle = s_{n+1}|n+1\rangle$, one finds

$$s_{n+1}^2\langle n+1|n+1\rangle = \langle n|\hat{a}^{-}\hat{a}^{+}|n\rangle = \langle n|(\hat{a}^{+}\hat{a}^{-}+1)|n\rangle = (n+1). \tag{6.1.58}$$

Therefore we also find

$$s_{n+1} = \sqrt{n+1}, \tag{6.1.59}$$

$$\hat{a}^{+}|n\rangle = \sqrt{n+1}\,|n+1\rangle. \tag{6.1.60}$$

When written in terms of the "creation" or "raising" operator $\hat{a}^{+}$, the normalized eigenstates are therefore

$$|n\rangle = \frac{1}{\sqrt{n!}}\left(\hat{a}^{+}\right)^{n}|0\rangle, \tag{6.1.61}$$

and may be thought of as states with n quanta of energy $\hbar\omega$ in the mode. Such states are called photon number or Fock states. Although the allowed energy eigenvalues for each mode are now discrete, as shown in Figure 6.2, the energy expectation values for multimode fields can take on essentially any value if they consist of superpositions of a great many photon number eigenstates. Also the lowest energy state does not correspond to occupation of the mode by one quantum, since its energy is only half the requisite energy per photon. This is the zero-point energy of the field.

In this treatment, the Pauli exclusion principle has been ignored. This is because the field states are equally spaced, making them bosonic in character. More than one quantum of excitation may exist per mode. In half-integer spin systems the statistics are different. Only one excitation per available state is permitted. Hence in fermionic

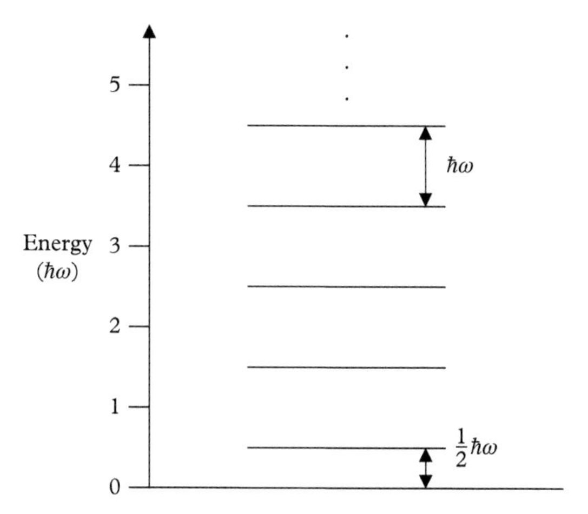

Figure 6.2 *Energy structure of the quantized electromagnetic field for a single mode, showing equally spaced levels above the zero-point energy of* $\frac{1}{2}\hbar\omega$*, the energy of the vacuum field.*

systems great care must be exercised in the construction of appropriate energy states (see Appendix E).

Exercise:

(a) Show that the expectation value of the electric field in a Fock state is zero. That is, show that

$$\langle n| \hat{E} |n\rangle = 0. \tag{6.1.62}$$

(b) Show that the expectation of the square of the electric field in a number state is not zero. This result indicates that the energy density is not zero and there are field fluctuations even in the lowest energy state of the field, although it is unoccupied. These are known as vacuum fluctuations.

$$\langle n| \hat{E}^2 |n\rangle \neq 0. \tag{6.1.63}$$

The creation and annihilation operators introduced in this section were based on analysis of the energy density of the electromagnetic field. The associated concept of photons consequently refers to a discrete measure of the energy density of the mode volume at a specific wavelength. Notice that there was no need to define a position operator for this particle-like quantum of energy. This is fortunate because the photon has wavelike properties, making it difficult to answer the question "Where is the photon?" in configuration space. Nevertheless, some perspective on the spatial extent of photons can be obtained in the form of an uncertainty relation, as discussed in Ref. [6.1].

6.2 Spontaneous Emission

In this section a single mode calculation of light emission induced by the quantized electromagnetic field is presented. Even when the radiation field contains no photons, emission is found to be induced by vacuum field fluctuations. This explains the origin of spontaneous emission but fails to give a proper account of the decay behavior of real atoms, because a single mode treatment also introduces spontaneous absorption, which is not observed in nature. This shortcoming is remedied (in Section 6.3) by taking into account the many modes of space available for spontaneous emission that are not available in the absorption process.

With both quantized atomic states $|\psi_{atom}\rangle = |\psi_m\rangle$ and quantized field states $|\psi_{field}\rangle = \sum_n c_n |n\rangle$ in hand, the construction of completely quantized state vectors for the system consisting of atom plus field is straightforward. We begin by writing down an expansion in terms of atom–field product states,

$$|\psi>= \sum_{m,n} c_{mn} |\psi_m\rangle |n\rangle , \tag{6.2.1}$$

and consider an atom with two states labeled 1 and 2. The system Hamiltonian is

$$\hat{H} = \hat{H}_{atom} + \hat{H}_{field} + \hat{H}_{\text{int}}. \tag{6.2.2}$$

For interaction of the atom with a single mode field, the first two components of $\hat{H}$ are simply

$$\hat{H}_{atom} = \hbar \begin{pmatrix} \omega_2 & 0 \\ 0 & \omega_1 \end{pmatrix}, \tag{6.2.3}$$

$$\hat{H}_{field} = \hbar\omega \left(\hat{a}^+\hat{a}^- + \frac{1}{2} \right). \tag{6.2.4}$$

As in Eq. (2.3.8), the dipole moment operator is $\hat{\mu} = \mu(\hat{\sigma}^+ + \hat{\sigma}^-)\hat{E}_\mu$. Since the interaction of light with the atom takes place at the origin in the point dipole approximation, the electric field operator (Eq. (6.1.35)) reduces to $\hat{E} = \sum_{k,\lambda} \sqrt{\hbar\omega_k/2\varepsilon_0 V}[\hat{a}^-_{k\lambda}(t) + \hat{a}^+_{k\lambda}(t)]\hat{e}_{k\lambda}$ for linear (real) polarization in the direction $\hat{e}_{k\lambda}$, so for this case the last component of the total Hamiltonian can simply be written

$$\hat{H}_{\text{int}}(t) = -\mu\xi \left(\hat{\sigma}^- + \hat{\sigma}^+\right) \left(\hat{a}^-(t) + \hat{a}^+(t)\right) \hat{e}_\mu \cdot \hat{e}_{k\lambda}, \tag{6.2.5}$$

where $\xi \equiv \sqrt{\hbar\omega/2\varepsilon_0 V}$ is the electric field per photon and μ is the electric dipole matrix element. In many materials the induced dipole is parallel to the field direction. Under these circumstances one has the additional simplification that $\hat{e}_\mu \cdot \hat{e}_{k\lambda} = 1$. However, in crystals the induced dipole may not be parallel to the field, in which case this simplification is unacceptable.

Equation (6.2.5) is the coupling Hamiltonian for an atom interacting with a single-mode field polarized along the direction $\hat{e}$. In it appear the combinations of Pauli spin matrices $\hat{\sigma}^+$ and $\hat{\sigma}^-$ encountered in Chapter 2:

$$\hat{\sigma}^- = \frac{1}{2}(\hat{\sigma}_x - i\hat{\sigma}_y) = \begin{pmatrix} 0 & 0 \\ 1 & 0 \end{pmatrix}, \tag{6.2.6}$$

$$\hat{\sigma}^+ = \frac{1}{2}(\hat{\sigma}_x + i\hat{\sigma}_y) = \begin{pmatrix} 0 & 1 \\ 0 & 0 \end{pmatrix}. \tag{6.2.7}$$

$\hat{\sigma}^+$ and $\hat{\sigma}^-$ change the occupation of states of the atom in the same way that $\hat{a}^+, \hat{a}^-$ change the occupation number of states of the electromagnetic field. This was confirmed in Eqs. (2.3.6) and (2.3.7):

$$\hat{\sigma}^- \begin{pmatrix} 1 \\ 0 \end{pmatrix} = \begin{pmatrix} 0 \\ 1 \end{pmatrix}; \quad \hat{\sigma}^+ \begin{pmatrix} 0 \\ 1 \end{pmatrix} = \begin{pmatrix} 1 \\ 0 \end{pmatrix}. \tag{6.2.8}$$

Note however that atomic operators only act on atomic states. They do not have any effect on states of the field. Similarly, field operators do not change the state of the atom, only the state of the field.

In the interaction representation, the interaction Hamiltonian becomes

$$\hat{H}^I_{\text{int}} = \hbar g \left\{ \hat{a}^- \hat{\sigma}^+ \exp[-i(\omega - \omega_L)t] + h.c. + \hat{a}^- \hat{\sigma}^- \exp[-i(\omega + \omega_L)t] + h.c. \right\}, \tag{6.2.9}$$

where $g \equiv -(\mu\xi/\hbar)$. Utilizing the rotating wave approximation (RWA), this simplifies to

$$\hat{H}^I_{\text{int}} = \hbar g \left\{ \hat{a}^- \hat{\sigma}^+ \exp[-i(\omega - \omega_L)t] + h.c. \right\}. \tag{6.2.10}$$

The omitted terms are called non-secular because they do not conserve energy. The final Hamiltonian in Eq. (6.2.10) only causes transitions between the states $|2\rangle\,|n\rangle$ and $|1\rangle\,|n+1\rangle$ where a photon is added to the field as the atom makes a downward transition, thus conserving energy. States $|2\rangle\,|n+1\rangle$ and $|1\rangle\,|n\rangle$ can therefore be ignored and Eq. (6.2.1) becomes

$$|\psi^I\rangle = c_{2,n}\,|2n\rangle + c_{1,n+1}\,|1, n+1\rangle\,. \tag{6.2.11}$$

In the interaction picture the Schrödinger equation is

$$\frac{d}{dt}\,|\psi^I\rangle = -\frac{i}{\hbar}\hat{H}^I_{\text{int}}\,|\psi^I\rangle\,. \tag{6.2.12}$$

Substitution of Eqs. (6.2.10) and (6.2.11) into Eq. (6.2.12) yields

$$\frac{d}{dt}\left|\psi^I\right\rangle = -ig\left[\hat{a}^-\hat{\sigma}^+\exp[-i(\omega-\omega_L)t] + h.c.\right]\left\{c_{2,n}\left|2n\right\rangle + c_{1,n+1}\left|1,n+1\right\rangle\right\}. \quad (6.2.13)$$

From Eq. (6.2.11) we can also write

$$\frac{d}{dt}\left|\psi^I\right\rangle = \dot{c}_{2,n}\left|2n\right\rangle + \dot{c}_{1,n+1}\left|1,n+1\right\rangle. \quad (6.2.14)$$

Projecting both Eq. (6.2.13) and Eq. (6.2.14) onto $\langle 2n|$ now gives the coefficient

$$\dot{c}_{2,n} = -igc_{1,n+1}\sqrt{n+1}\exp[-i(\omega-\omega_L)t]. \quad (6.2.15)$$

Similarly, projection onto $\langle 1,n+1|$ yields the coefficient

$$\dot{c}_{1,n+1} = -igc_{2,n}\sqrt{n+1}\exp[i(\omega-\omega_L)t]. \quad (6.2.16)$$

First-order perturbation solutions of Eqs. (6.2.15) and (6.2.16) for an atom initially in the excited state give the emission probability. At short times this is the same as the probability of finding the atom in the lower state. Hence it is equal to

$$\left|c_{1,n+1}\right|^2 \doteq g^2(n+1)t^2\left[\frac{\sin^2(\omega-\omega_L)t/2}{[(\omega-\omega_L)t/2]^2}\right]. \quad (6.2.17)$$

Notice that even with no photons present ($n = 0$), there is some probability of finding that the atom has decayed. Since there is no external field present, this accounts for the well-known phenomenon of spontaneous emission. Unfortunately the theory is flawed, because the absorption probability for an atom initially in the ground state is also non-zero. This would cause spontaneous absorption (Figure 6.3), which is not observed in nature. That is, we would calculate

$$\left|c_{2,n}\right|^2 \doteq g^2(n+1)t^2\left[\frac{\sin^2(\omega-\omega_L)t/2}{[(\omega-\omega_L)t/2]^2}\right]. \quad (6.2.18)$$

This flaw in the theory is not due to the perturbation approach we chose. The exact Rabi solution predicts the same (unphysical) behavior. In emission, the Rabi results are

$$c_{2,n} = \cos\left(g\sqrt{(n+1)}t\right), \quad (6.2.19)$$

$$c_{1,n+1} = -i\sin\left(g\sqrt{(n+1)}t\right), \quad (6.2.20)$$

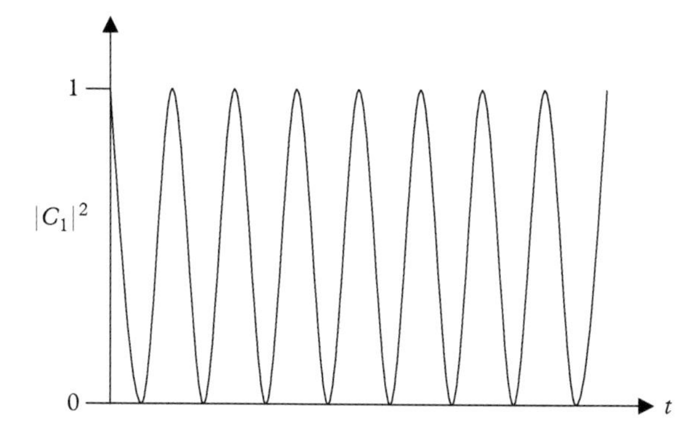

Figure 6.3 *Qualitative behavior of spontaneous absorption predicted by an exact Rabi treatment of a two-level atom interacting with a single mode of the electromagnetic field.*

and in absorption one finds

$$c_{2,n} = -i \sin \left(g\sqrt{(n+1)}t \right), \tag{6.2.21}$$

$$c_{1,n+1} = \cos \left(g\sqrt{(n+1)}t \right). \tag{6.2.22}$$

The theory obviously needs refinement. An atom cannot absorb a quantum spontaneously from the vacuum state without producing a state of the field with one less photon. Such a state would necessarily be of lower energy than the vacuum state, but $|0\rangle$ is already defined to be the lowest energy eigenstate of the field. Hence this violates a key premise, and we need to pursue a remedy of this problem in Section 6.3 by investigating a multi-mode version of this analysis.

6.3 Weisskopf–Wigner Theory

So far, our theory of transition rates and decay involves only a single mode of the field, and emission is assumed to be perfectly monochromatic. However, to be in accord with the Heisenberg uncertainty principle, there must be a spread in emitted frequencies for any process that takes place over a finite time interval. That is, the emission spectrum must have a finite bandwidth, something that can only be included through a multi-mode treatment.

To extend the treatment to more than one mode, the Hamiltonian must be written as a summation over all possible modes s. The wavefunction must also be written as a summation over all possible states of the atom m and mode occupation numbers n_s:

$$\hat{H}^I_{\text{int}} = \sum_s \hbar g_s \left[\hat{\sigma}^+ \hat{a}_s \exp[-i(\omega_s - \omega)t] + h.c. \right], \tag{6.3.1}$$

$$|\psi^I\rangle = \sum_\alpha \sum_{n_1} \sum_{n_2} \cdots \sum_{n_s} \cdots C_{\alpha;n_1,n_2\ldots n_s\ldots} \, |\alpha; n_1, n_2, \ldots n_s, \ldots\rangle \,. \tag{6.3.2}$$

In Eq. (6.3.2), the sum over energy states includes $\alpha = 1, 2$ and the sum over photon occupation runs over the values $n_s = 0, 1, 2, 3 \ldots \infty$. The field operators $\hat{a}_s^-, \hat{a}_s^+$ for mode s affect only the occupation of the individual mode s. So, for example,

$$\begin{aligned}\hat{a}_s^- |\alpha; n_1 n_2 \ldots n_s \ldots\rangle &= \hat{a}_s^- |\alpha\rangle |n_1\rangle |n_2\rangle \ldots |n_s\rangle \ldots \\ &= |\alpha\rangle |n_1\rangle |n_2\rangle \ldots \hat{a}_s^- |n_s\rangle \ldots \\ &= \sqrt{n_s} |\alpha; n_1, n_2 \ldots (n_s - 1) \ldots\rangle .\end{aligned} \tag{6.3.3}$$

As before, transition probability amplitudes are calculated according to

$$\begin{aligned}&\dot{C}_{\alpha;n_1 n_2 \ldots n_s \ldots} \\ &= -\frac{i}{\hbar} \sum_\beta \sum_{m_1, m_2, \ldots m_s \ldots} \langle \alpha; n_1, n_2, \ldots n_s \ldots | \sum_s (\hat{H}_{\text{int}}^I)_s | \beta; m_1, m_2 \ldots m_s \ldots \rangle C_{\beta;m_1,m_2 \ldots m_s \ldots}.\end{aligned} \tag{6.3.4}$$

The interaction matrix element can be simplified:

$$\begin{aligned}&\langle \alpha; n_1, n_2 \ldots n_s \ldots | \sum_r (\hat{H}_{\text{int}}^I) | \beta; m_1, m_2 \ldots m_r \ldots \rangle \\ &\quad = \sum_r \langle n_r | \langle \alpha | (\hat{H}_{\text{int}}^I)_r | \beta \rangle | m_r \rangle \langle n_1, n_2 \cdots n_{r-1}, n_{r+1} \cdots | m_1, m_2 \cdots m_{r-1}, m_{r+1} \cdots \rangle \\ &\quad = \sum_r \langle \alpha; n_r | (\hat{H}_{\text{int}}^I)_r | \beta; m_r \rangle \delta_{n_1 m_1} \delta_{n_2 m_2} \ldots \delta_{n_{r-1} m_{r-1}} \delta_{n_{r+1} m_{r+1}} \cdots\end{aligned} \tag{6.3.5}$$

so that

$$\dot{C}_{\alpha;n_1,n_2 \ldots n_s \ldots} = -\frac{i}{\hbar} \sum_\beta \sum_r \sum_{m_r} \langle \alpha; n_r | \, (H_{\text{int}}^I)_r \, | \beta; m_r \rangle \, C_{\beta;n_1,n_2 \ldots m_r \ldots}. \tag{6.3.6}$$

Note that the interaction matrix element is zero if H_{int} has the form in Eq. (6.3.1) and the atomic state remains unchanged ($\alpha = \beta$). So the final equations of motion (for $\alpha \neq \beta$) are

$$\dot{C}_{2;n_1,n_2 \ldots n_s \ldots} = -i \sum_r g_r \sqrt{n_r + 1} \exp[-i(\omega_r - \omega)t] C_{1;n_1,n_2 \ldots (n_r+1) \ldots} \tag{6.3.7}$$

$$\dot{C}_{1,n_1,n_2 \ldots n_s ..} = -i \sum_r g_r \sqrt{n_r} \exp[i(\omega_r - \omega)t] C_{2,n_1 \ldots (n_r-1) \ldots} \tag{6.3.8}$$

Now, returning to the emission problem, assume that initially there are no photons in any mode, and that the system is in the upper state $|2\rangle$. Then the initial condition and equations of motion are

$$C_{2,\{0\}}(0) = 1, \tag{6.3.9}$$

$$\dot{C}_{2,\{0\}} = -i\sum_r g_r \exp[-i(\omega_r - \omega_0)t]C_{1,\{1_r\}}, \tag{6.3.10}$$

$$\dot{C}_{1,\{1_s\}} = -ig_s \exp[i(\omega_s - \omega_0)t]C_{2,\{0\}}, \tag{6.3.11}$$

where $\omega_0 = \omega_2 - \omega_1$. The formal solution of Eq. (6.3.11) , namely

$$C_{1,\{1_r\}}(t) = -ig_r \int_0^t \exp[i(\omega_r - \omega_0)t']C_{2,\{0\}}\left(t'\right) dt', \tag{6.3.12}$$

may be used in Eq. (6.3.10) to find

$$\dot{C}_{2,\{0\}} = -\sum_r g_r^2 \int_0^t \exp[-i(\omega_r - \omega_0)(t - t')]C_{2,\{0\}}\left(t'\right) dt'. \tag{6.3.13}$$

The summation over r accounts for all the modes into which emission may occur. We assume the modes are closely spaced in frequency, and replace the summation over r by integration over frequency ω'. That is, we make the replacement

$$\sum_r \to \int D(\omega')d\omega', \tag{6.3.14}$$

where $D(\omega)$, the number of final radiation states per unit frequency interval in a volume V, is called the density of states or the density of modes:

$$D(\omega) = \frac{V\omega^2}{\pi^2 c^3}. \tag{6.3.15}$$

With these changes, Eq. (6.3.13) becomes

$$\dot{C}_{2,\{0\}}(t) = -\int d\omega' g^2\left(\omega'\right) D(\omega') \int_0^t \exp[-i(\omega' - \omega_0)(t - t')]C_{2,\{0\}}\left(t'\right)dt. \tag{6.3.16}$$

We now assume that $g^2(\omega')$ and $D(\omega')$ vary slowly with frequency. These quantities can then be removed from inside the integral in Eq. (6.3.16):

$$\dot{C}_{2,\{0\}} = -g^2(\omega)D(\omega) \int d\omega' \int_0^t \exp[-i(\omega' - \omega)(t - t')]C_{2,\{0\}}(t')dt'. \tag{6.3.17}$$

For a given mode frequency ω', the final integral is dominated by contributions at $t = t'$. Hence

$$\int d\omega' \int_0^t \exp[-i(\omega' - \omega)(t - t')] C_{2,\{0\}}(t') dt' = C_{2,\{0\}}(t) \left\{ \pi\delta\left(\omega' - \omega_0\right) - iP\left(\frac{1}{\omega' - \omega_0}\right) \right\}. \tag{6.3.18}$$

In Eq. (6.3.18), P denotes the principal part of the integral. This equation reveals that the relaxation process has real and imaginary parts given by

$$\mathrm{Re}\left\{\dot{C}_{2,\{0\}}(t)\right\} = -\pi g^2(\omega) D(\omega) C_{2,\{0\}}(t), \tag{6.3.19}$$

$$\mathrm{Im}\left\{\dot{C}_{2,\{0\}}(t)\right\} = g^2(\omega) D(\omega) P\left(\frac{1}{\omega' - \omega}\right) C_{2,\{0\}}(t). \tag{6.3.20}$$

The real part of $\dot{C}_{2,\{0\}}$ describes spontaneous relaxation of the system. Equation (6.3.19) has the form

$$\dot{C}_{2,\{0\}}(t) = -\frac{1}{2}\gamma C_{2,\{0\}}(t), \tag{6.3.21}$$

where the decay rate is given by

$$\gamma \equiv 2\pi g^2(\omega) D(\omega) = \frac{\mu^2\omega^3}{\varepsilon_0 \pi \hbar c^3}. \tag{6.3.22}$$

The interaction constant g^2 in Eq. (6.3.22) depends on the square of the scalar product between the field and induced dipole direction, as indicated in Eq. (6.2.5). This contributes a factor of $\cos^2\theta$ to the decay rate which must be averaged over θ. Since the vacuum field is unpolarized the result is $< \cos^2\theta > = 1/3$, so the spontaneous emission rate is $\gamma_{sp} = \gamma/3$. Solution of Eq. (6.3.21) then yields exponential decay of the occupation probability itself:

$$\left|C_{2,\{0\}}(t)\right|^2 = \exp(-\gamma_{sp} t). \tag{6.3.23}$$

This justifies our earlier assumption that upper state probability decays with a lifetime $\tau = 1/\gamma_{sp}$. Additional details may be found in Ref. [6.2]. The imaginary part of $\dot{C}_{2,\{0\}}$ in Eq. (6.3.20) yields a shift of the energy level. For more discussion of driven and undriven dissipation processes and associated frequency shifts (the AC Stark shift and Lamb shift, respectively) see Appendix F and Refs. [6.3, 6.4].

Exercise: Show that spontaneous absorption no longer occurs with a multimode treatment, by calculating $\dot{C}_2(0)$ for the initial condition $C_1(0) = 1$.

6.4 Coherent States

In previous sections we found that annihilation and creation operators $\hat{a}^-$, $\hat{a}^+$ changed the occupation of number states used to describe the field. Now we construct states of the field which "cohere" when acted upon by the non-Hermitian operator $\hat{a}^-$. These new states are normalized eigenstates of the operator $\hat{a}^-$ that remain essentially unchanged as they propagate in space and time. Because they retain their basic form, they are said to be coherent and closely describe the properties of laser beams, for example.

A coherent state $|\alpha\rangle$ is defined to be an eigenstate of the annihilation operator. It maintains its essential character despite changes in the number of photons of which it is composed. Thus

$$\hat{a}^- |\alpha\rangle = \alpha |\alpha\rangle . \tag{6.4.1}$$

Eigenstates of field operators must be based on the operator $\hat{a}^-$ because $\hat{a}^+$ does not have eigenstates. A discussion of this point may be found in Ref. [6.5] and it means the action of the creation operator on a coherent state can only be defined when it acts to the left, since this is the conjugate of Eq. (6.4.1):

$$\langle\alpha| a^+ = \langle\alpha| \alpha^*. \tag{6.4.2}$$

We shall insist that $|\alpha\rangle$ be normalized:

$$\langle\alpha|\alpha\rangle = 1. \tag{6.4.3}$$

However, because they are eigenstates of a non-Hermitian operator, different coherent states $|\alpha\rangle$ and $|\alpha'\rangle$ are not orthogonal, although they do form a complete set. Here we merely seek the relationship between $|\alpha\rangle$ and our earlier number states $|n\rangle$:

$$|\alpha\rangle = \sum_{n=0}^{\infty} |n\rangle \langle n|\alpha\rangle, \tag{6.4.4}$$

$$\hat{a}^- |\alpha\rangle = \sum_{n=0}^{\infty} \hat{a}^- |n\rangle \langle n|\alpha\rangle. \tag{6.4.5}$$

Insertion of the earlier result from Eq. (6.1.54) yields

$$\hat{a}^- |\alpha\rangle = \sum_{n=1}^{\infty} \langle n|\alpha\rangle \sqrt{n}\, |n-1\rangle . \tag{6.4.6}$$

By replacing n with $n+1$, the sum can be rewritten as

$$\hat{a}^- |\alpha\rangle = \sum_{n=0}^{\infty} \langle n+1|\alpha\rangle \sqrt{n+1}\, |n\rangle. \tag{6.4.7}$$

Combining Eqs. (6.4.5) and (6.4.7) we obtain

$$\sum_{n=0} \sqrt{n+1}\,|n\rangle\,\langle n+1|\alpha\rangle = \sum_{n=0}^{\infty} \hat{a}^{-}\,|n\rangle\,\langle n|\alpha\rangle = \alpha|\alpha\rangle. \tag{6.4.8}$$

Multiplying through by $\langle m|$ and using the orthogonality of the number states ($\langle m|\, n\rangle = \delta_{mn}$), one finds

$$\langle n+1|\alpha\rangle = \frac{\langle n|\alpha\rangle}{\sqrt{n+1}}\alpha. \tag{6.4.9}$$

All the coefficients $\langle n|\alpha\rangle$ for the expansion in Eq. (6.4.4) can be found by using Eq. (6.4.9) as a recursion relation to build up subsequent coefficients from the first one, which is $\langle 0|\alpha\rangle$:

$$\langle n\,|\alpha\rangle = \frac{\alpha^n}{\sqrt{n!}}\,\langle 0\,|\alpha\rangle\,. \tag{6.4.10}$$

Substituting Eq. (6.4.10) into Eq. (6.4.4) we find

$$|\alpha\rangle = c_0 \sum_{n=0}^{\infty} \frac{\alpha^n}{\sqrt{n!}}\,|n\rangle\,, \tag{6.4.11}$$

where $c_0 \equiv <0\,|\,\alpha>$. But what is c_0? To find this coefficient explicitly, we make use of the normalization condition $\langle\alpha|\alpha\rangle = 1$:

$$\langle\alpha|\alpha\rangle = |c_0|^2 \sum_{m,n} \frac{(\alpha^*)^m\,\alpha^n}{\sqrt{m!n!}} <m\,|\,n>$$

$$= |c_0|^2 \sum_{n} \frac{|\alpha|^{2n}}{n!} = |c_0|^2 \exp(|\alpha|^2). \tag{6.4.12}$$

Setting Eq. (6.4.12) equal to one, the coefficient is found to be

$$c_0 = \exp\left(-\frac{1}{2}\,|\alpha|^2\right). \tag{6.4.13}$$

Consequently, the normalized form of the coherent state is

$$|\alpha\rangle = \exp\left(-\frac{1}{2}\,|\alpha|^2\right) \sum_{n} \frac{\alpha^n}{\sqrt{n!}}\,|n\rangle = \exp\left(-\frac{1}{2}\,|\alpha|^2\right) \sum_{n} \frac{(\alpha a^+)^n}{\sqrt{n!}}\,|0\rangle\,, \tag{6.4.14}$$

or

$$|\alpha\rangle = \hat{D}(\alpha)|0\rangle = e^{-\frac{1}{2}|\alpha|^2} e^{\alpha a^+}\,|0\rangle\,. \tag{6.4.15}$$

Exercise: Using the Campbell–Baker–Haussdorf or Weyl formula (see Problem 6.1) in the form

$$\exp\left(\hat{c} + \hat{d} + \frac{1}{2}\left[\hat{c}, \hat{d}\right]\right) = \exp(\hat{c})\exp(\hat{d})$$

or

$$\exp(\hat{c} + \hat{d}) = \exp(-[\hat{c}, \hat{d}]/2)\exp(\hat{c})\exp(\hat{d}) \tag{6.4.16}$$

for any operators $\hat{c}$ and $\hat{d}$ that obey $[[\hat{c}, \hat{d}\,], \hat{c}] = [[\hat{c}, \hat{d}\,], \hat{d}\,] = 0$, together with commutation relations of the raising and lowering operators, show that Eq. (6.4.15) can also be expressed in the form

$$|\alpha\rangle = e^{\alpha a^{+} - \alpha * a^{-}}|0\rangle\,. \tag{6.4.17}$$

Now consider some of the properties of coherent states, as defined by Eq. (6.4.14). The proportion of a coherent state that consists of some particular number state is given by the projection (or inner product) of the one on the other:

$$\langle n|\alpha\rangle = \exp\left(-|\alpha|^2/2\right)\frac{\alpha^n}{\sqrt{n!}}. \tag{6.4.18}$$

Hence

$$|\langle n|\alpha\rangle|^2 = \exp\left(-|\alpha|^2\right)\frac{|\alpha|^{2n}}{n!}. \tag{6.4.19}$$

The quantity $|\langle n|\alpha\rangle|^2$ is also the probability of finding n photons in the state $|\alpha\rangle$. The probability distribution of a coherent state, given by Eq. (6.4.19), is Poissonian rather than Gaussian in form.

Let us examine more closely how the state $|\alpha\rangle$ evolves in time. By making use of what we have learned about the way the ladder operators $\hat{a}$ and $\hat{a}^{+}$ act on number states, we find

$$\begin{aligned}
|\alpha(t)\rangle &= \exp(-iHt/\hbar)\,|\alpha(0)\rangle \\
&= \exp[-i\hbar\omega(\hat{a}^{+}\hat{a}^{-})t/\hbar]\left\{\exp\left(-\frac{1}{2}|\alpha|^2\right)\sum_n \frac{\alpha^n}{\sqrt{n!}}|n\rangle\right\} \\
&= \sum_n \frac{[\alpha\exp(-i\omega t)]^n}{\sqrt{n!}}\exp\left(-\frac{1}{2}|\alpha|^2\right)|n\rangle \\
&= |\alpha\exp(-i\omega t)\rangle\,.
\end{aligned} \tag{6.4.20}$$

As time progresses, the eigenvalue of state $|\alpha\rangle$ merely acquires a phase factor. That is, it evolves coherently. A more complete proof that $|\alpha\rangle$ evolves coherently in time, in which H is the general Hamiltonian of a forced harmonic oscillator, may be found in Ref. [6.6].

An interesting point is that expectation values for the electric field and the field fluctuations in a coherent state both differ from those of an electric field that is a number state:

$$\langle\alpha|\,\hat{E}\,|\alpha\rangle = \xi\hat{\varepsilon}\left[\langle\alpha|\,\hat{a}^{-}\,|\alpha\rangle + \langle\alpha|\,\hat{a}^{+}\,|\alpha\rangle\right] = \xi\hat{\varepsilon}(\alpha+\alpha^{*}) \neq 0. \tag{6.4.21}$$

The amplitude of the field in Eq. (6.4.21) depends on the complex quantity α. The phase of the field is the phase of α and the magnitude depends on $|\alpha|$, which is proportional to the square root of the number of photons.

Exercise: Show that,

$$\langle\alpha|\,E^2\,|\alpha\rangle = \frac{\hbar\omega}{2\varepsilon_0 V}\left(1-\left[\alpha-\alpha^{*}\right]^2\right). \tag{6.4.22}$$

Use this result to show that the mean square fluctuation of the field is

$$\left\langle\Delta E^2\right\rangle \equiv \langle\alpha|\,E^2\,|\alpha\rangle - (\langle\alpha|\,E\,|\alpha\rangle)^2 = \frac{\hbar\omega}{2\varepsilon_0 V}. \tag{6.4.23}$$

Equation (6.4.23) is an important result which differs from our earlier one for photon number states. There the mean square fluctuation tended to infinity as the photon number n increased. Here we find that quantum fluctuations in a coherent state are just zero-point fluctuations of the vacuum which exhibit no dependence on E or α.

Notice that the expectation value of the field in a coherent state, given by Eq. (6.4.21) for a single mode, returns the form of a classical plane wave as time evolves (see Eq. (6.4.20)). Hence, there must be a very close correspondence between a single mode coherent field and a classical plane wave. Multimode coherent states correspond to superpositions of plane waves. Because the mean square fluctuation of the field is independent of E and α according to Eq. (6.4.23), the noise of the field diminishes relative to the mean number of photons in the coherent state increases, as indicated in Figure 6.4. In view of the properties of coherent states described earlier, they can be said to resemble well-defined classical fields more and more as their intensity (proportional to the number of photons) increases.

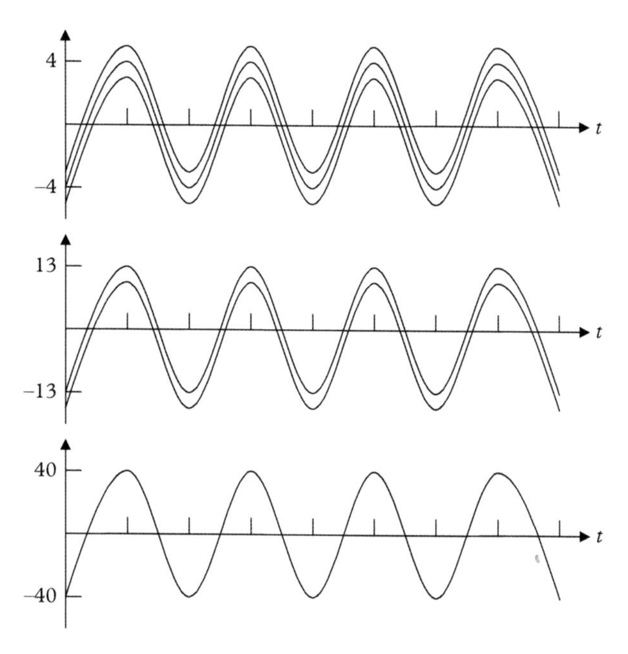

Figure 6.4 *A schematic illustration of the relative reduction of noise in a coherent state that accompanies an increase of the strength of the field. From top to bottom, the mean photon number $|\alpha|^2$ is 4, 40, and 400, respectively. The vertical width of the traces reflects the uncertainty in the field value, and includes both the amplitude uncertainty due to Δn and the phase uncertainty. (After Ref. [6.7].)*

Now let us examine uncertainties of the conjugate variables Q, P defined by Eqs. (6.1.30) and (6.1.31) which characterize the electromagnetic field in photon number and coherent states. The inverses of Eqs. (6.1.33) and (6.1.34) yield

$$\hat{Q} = \sqrt{\frac{\hbar}{2\omega}}\,(\hat{a}^- + \hat{a}^+), \tag{6.4.24}$$

$$\hat{P} = -i\sqrt{\frac{\hbar\omega}{2}}\,(\hat{a}^- - \hat{a}^+). \tag{6.4.25}$$

Exercise: Show that for a field in a number state $|n\rangle$ the fluctuations of $<\hat{P}>$ and $<\hat{Q}>$ have the following properties:

$$(\Delta Q)_n^2 \equiv \langle n|\hat{Q}^2|n\rangle - \langle n|\hat{Q}|n\rangle^2 = \frac{\hbar}{2\omega}(2n+1), \tag{6.4.26}$$

$$(\Delta P)_n^2 \equiv \langle n|\hat{P}^2|n\rangle - \langle n|\hat{P}|n\rangle^2 = \frac{\hbar\omega}{2}(2n+1), \tag{6.4.27}$$

$$(\Delta Q \Delta P)_n = \frac{\hbar}{2}(2n+1). \tag{6.4.28}$$

The product of uncertainties ΔP and ΔQ obeys an uncertainty principle, since for all values of n we find $(\Delta Q \Delta P) \geq \hbar/2$. However, only for the vacuum state do we find the minimum uncertainty condition $(\Delta Q \Delta P) = \hbar/2$ via Eq. (6.4.28).

With Eqs. (6.4.24) and (6.4.25) it can be shown that for a coherent state we have slightly different relations:

$$\langle \alpha | \hat{Q} | \alpha \rangle = \left(\frac{\hbar}{2\omega} \right)^{1/2} (\alpha^* + \alpha), \tag{6.4.29}$$

$$\langle \alpha | \hat{Q}^2 | \alpha \rangle = \frac{\hbar}{2\omega} \left([\alpha + \alpha *]^2 + 1 \right), \tag{6.4.30}$$

$$\langle \alpha | \hat{P} | \alpha \rangle = i \left(\frac{\hbar}{2\omega} \right)^{1/2} (\alpha^* + \alpha), \tag{6.4.31}$$

$$\langle \alpha | \hat{P}^2 | \alpha \rangle = \frac{\hbar\omega}{2} \left([\alpha - \alpha *]^2 - 1 \right), \tag{6.4.32}$$

$$(\Delta Q)_c^2 = \langle \alpha | \hat{Q}^2 | \alpha \rangle - \langle \alpha | \hat{Q} | \alpha \rangle^2 = \frac{\hbar}{2\omega}, \tag{6.4.33}$$

$$(\Delta P)_c^2 = <\alpha \left| \hat{P}^2 \right| \alpha> - <\alpha \left| \hat{P} \right| \alpha>^2 = \frac{\hbar\omega}{2}, \tag{6.4.34}$$

$$(\Delta Q \Delta P)_c = \frac{\hbar}{2}. \tag{6.4.35}$$

Notice that according to Eq. (6.4.35) a coherent state is always a minimum uncertainty state, whereas among the number states only the vacuum state has minimum uncertainty.

Since each mode of the radiation field has a complex amplitude, there must be two "quadrature" components associated with real and imaginary parts of the field operators $\hat{a}^-, \hat{a}^+$. This realization leads to an interesting discovery. So let us now write

$$\hat{a}^- = \frac{1}{\sqrt{2}} (\hat{a}_1 + i\hat{a}_2); \; \hat{a}^+ = \frac{1}{\sqrt{2}} (\hat{a}_1 - i\hat{a}_2); \tag{6.4.36}$$

$$\hat{a}_1 = \frac{1}{\sqrt{2}} (\hat{a}^+ + \hat{a}^-); \; \hat{a}_2 = \frac{i}{\sqrt{2}} (\hat{a}^+ - \hat{a}^-), \tag{6.4.37}$$

where subscripts refer to the two quadratures. With these definitions we find

$$\hat{Q} = \sqrt{\frac{\hbar}{2\omega}} (\hat{a}^+ + \hat{a}^-) = \sqrt{\frac{\hbar}{\omega}} \hat{a}_1, \tag{6.4.38}$$

$$\hat{P} = i \sqrt{\frac{\hbar\omega}{2}} (\hat{a}^+ - \hat{a}^-) = \sqrt{\hbar\omega} \hat{a}_2. \tag{6.4.39}$$

We can show explicitly that $\hat{a}_1$ and $\hat{a}_2$ correspond to orthogonal degrees of freedom in complex space by writing down the electric field operator in terms of these quadrature components:

$$\hat{E}(\bar{r}, t) = \sqrt{\frac{\hbar\omega}{2\varepsilon_0 V}}\hat{\varepsilon}\left[\hat{a}^- \exp[-i(\omega t - \bar{k}\cdot\bar{r})] + \hat{a}^+ \exp[i(\omega t - \bar{k}\cdot\bar{r})]\right]$$

$$= \sqrt{\frac{\hbar\omega}{\varepsilon_0 V}}\hat{\varepsilon}[\hat{a}_1 \cos\phi + \hat{a}_2 \sin\phi], \quad (6.4.40)$$

where $\phi = \omega t - \bar{k}\cdot\bar{r}$.

It is interesting to compare the variances exhibited in the two quadratures of photon number representations of coherent states (Figure 6.5). Since we already know that $[\hat{a}^-, \hat{a}^+] = 1$, it is straightforward to show that

$$\left[\hat{a}_1, \hat{a}_2\right] = i. \quad (6.4.41)$$

Exercise: Show that if two variables A, B obey a commutator relation of the form $[A, B] = ic$, where c is a constant, the product of their uncertainties is given by

$$\Delta A \Delta B \geq c/2. \quad (6.4.42)$$

On the basis of Eqs. (6.4.41) and (6.4.42), a photon number state $|n\rangle$ has fluctuations of the quadrature components that obey the uncertainty relation

$$(\Delta a_1 \Delta a_2)_n \geq 1/2. \quad (6.4.43)$$

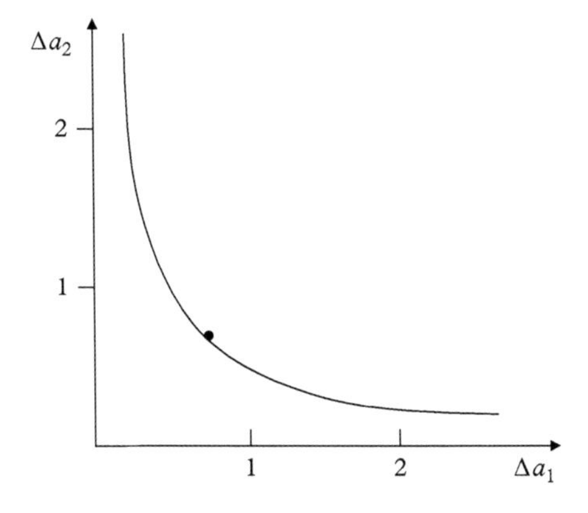

Figure 6.5 *The relationship between variances of the two quadratures of generalized coherent states. The dot corresponds to a coherent state with equal variances. All other points on the curve correspond to squeezed coherent states.*

For a coherent state, the uncertainty product assumes the minimum value. This means

$$(\Delta a_1 \Delta a_2)_c = 1/2. \tag{6.4.44}$$

Ordinarily the variances in a_1 and a_2 are equal. However, the implication of Eq. (6.4.44) is that in a broad class of coherent states called squeezed states, which may be generated by creating correlations between the quadratures, the fluctuations in one quadrature may be reduced below the noise level of an ordinary (minimum uncertainty) coherent state. The noise in the second quadrature correspondingly must increase above the minimum level to maintain agreement with Eq. (6.4.44), as illustrated in Figure 6.5. This provides a way of "beating" the uncertainty principle in at least one quadrature.

The quadrature operators $\hat{a}_1$ and $\hat{a}_2$ that exhibit correlated noise properties can refer not only to the amplitude of the electromagnetic field, but also to its phase in principle. Hence a generalized coherent state of the field may have phase noise or amplitude noise that descends below the shot-noise limit in one quadrature, as illustrated in Figure 6.6.

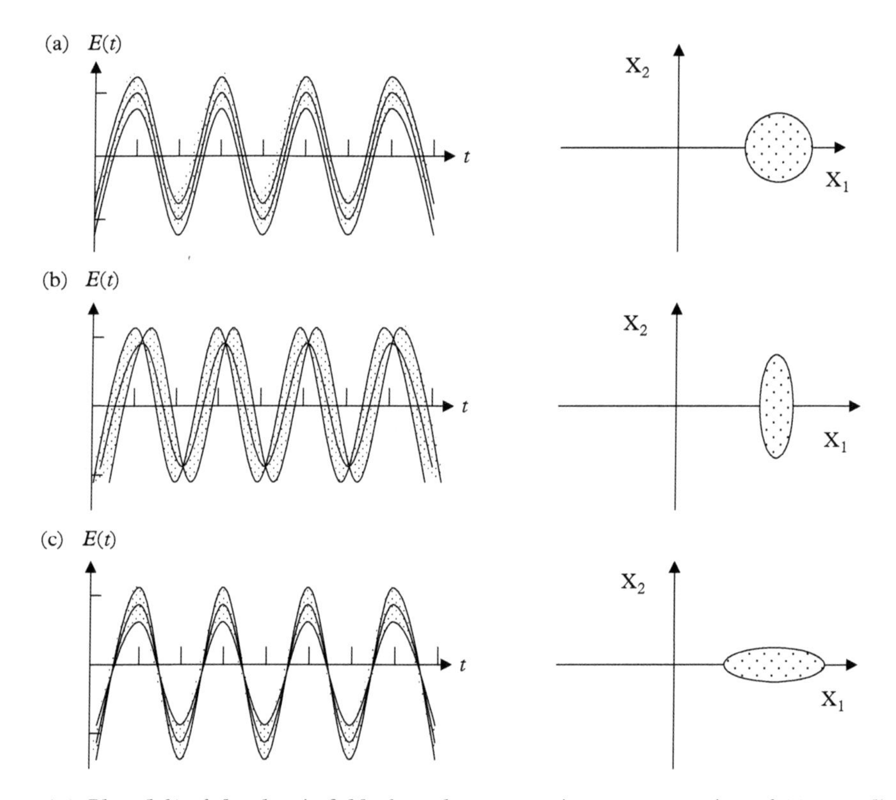

Figure 6.6 *Plots (left) of the electric field of an electromagnetic wave versus time of (a) an ordinary coherent state, (b) a coherent state with squeezed amplitude fluctuations, and (c) a coherent state with squeezed phase fluctuations (after Ref. [6.7]). Loci (right) of the tip of the electric field vector at a single point in time, corresponding to the peak of $E(t)$ for each case.*

Methods for controlling amplitude or phase fluctuations of "squeezed" light are discussed further in Chapter 7. These methods are based on nonlinear optical interactions that go beyond the intended scope of this book, so the interested reader should consult the references at the end of Chapter 7. Experimental realization of squeezed states of light have been the basis for efforts to send information from one point to another below the standard noise limits of communication systems. By reducing the noise in isolated quadratures below the shot noise limit, squeezing also offers potential improvements in the precision of optical measurements based on interferometry [6.8], where the objective is invariably to measure the positions or shifts of optical fringes with the best possible signal-to-noise ratio.

6.5 Statistics

6.5.1 Classical Statistics of Light

Statistical properties of light are determined by correlations between electric field amplitudes at different locations or times. They vary greatly from one light source to another and can be compared by calculating the first-order correlation function of the field,

$$<E^*(t)E(t+\tau)> = \frac{1}{T}\int_0^T E^*(t)E(t+\tau)dt, \tag{6.5.1}$$

or its normalized counterpart called the degree of first-order temporal coherence:

$$<g^{(1)}(\tau)> = \frac{<E^*(t)\,E(t+\tau)>}{<E^*(t)\,E(t)>}. \tag{6.5.2}$$

$g^{(1)}(\tau)$ should not be confused with the coupling term in earlier expressions of the interaction Hamiltonian (Eqs. (6.2.9) or (6.3.1)). The quantity $g^{(1)}(\tau)$ is intimately related (by the Wiener–Khintchine theorem) to the power spectrum of the light $F(\omega)$:

$$\begin{aligned} F(\omega) &= \frac{|E(\omega)|^2}{\int_{-\infty}^{+\infty} |E(\omega)|^2\,d\omega} = \frac{\frac{1}{4\pi^2}\int_0^T\int_{-\infty}^{+\infty} E^*(t)E(t+\tau)\exp(i\omega\tau)dtd\tau}{(T/2\pi)<E^*(t)E(t)>} \\ &= \frac{(T/4\pi^2)\int_{-\infty}^{+\infty} <E^*(t)E(t+\tau)>\exp(i\omega\tau)d\tau}{(T/2\pi)<E^*(t)E(t)>} \\ &= \frac{1}{2\pi}\int_{-\infty}^{+\infty} g^{(1)}(\tau)\exp(i\omega\tau)d\tau. \end{aligned} \tag{6.5.3}$$

Here the period T over which timc averaging is performed is assumed to be longer than both the optical period and the coherence or correlation time τ_c, which is a characteristic time of decay of field correlations due to random phase disruptions or fluctuations at the source.

Consider some of the basic characteristics of a classical light source. The source could be a star with an emission spectrum, centered at frequency ω_0, which is broadened by the effects of inter-atomic collisions. From moment to moment, the phase of the electric field experiences random jumps, due to the chaotic conditions within the source. Hence the first-order correlation of the total field emitted by N identical atoms is

$$
\begin{aligned}
<E^*(t)E(t+\tau)> &= E_0^2 \exp(i\omega_0\tau)<\left\{e^{-i\phi_1(t)} + \cdots + e^{-i\phi_N(t)}\right\}\left\{e^{i\phi_1(t+\tau)} + \cdots + e^{i\phi_N(t+\tau)}\right\}> \\
&= E_0^2 \exp(i\omega_0\tau) \sum_{i=1}^{N} < \exp\left[i\left\{\phi_i(t+\tau) - \phi_i(t)\right\}\right]>. \qquad (6.5.4)
\end{aligned}
$$

Once any atom jumps to a new phase of its oscillation, it ceases to contribute to the phase correlation. Hence for delays exceeding the period over which atom i evolves harmonically, that is, for $\tau \gg \tau_i$, one finds $< \exp[i\{\phi_i(t+\tau) - \phi_i(t)\}] > = 0$. In the opposite limit, when $\tau \ll \tau_i$, atom i maintains the same phase as τ is varied, and the result is $< \exp[i\{\phi_i(t+\tau) - \phi_i(t)\}] > = 1$. In reality there is a distribution $p(\tau_i)$ of the constant phase periods for each atom, so the phase correlation function in Eq. (6.5.4) is proportional to the probability that the atom does not undergo a phase jump.

We shall assume the phase jump process is Markoffian. Then, the probability that the atom maintains constant phase for a period of time τ_i decreases exponentially with time according to

$$
p(\tau_i) = \exp(-\tau_i/\tau_c). \qquad (6.5.5)
$$

In this expression, τ_c is the average interval over which atoms emit with constant phase, or the optical coherence time. The phase correlation function for atom i may then be written as

$$
< \exp\left[i\left\{\phi_i(t+\tau) - \phi_i(t)\right\}\right] > = \frac{1}{\tau_c}\int_{\tau}^{\infty} p(\tau_i)d\tau_i. \qquad (6.5.6)
$$

For N atoms Eq. (6.5.4) then yields

$$
<E^*(t)E(t+\tau)> = NE_0^2 \exp\left\{i\omega_0\tau - (\tau/\tau_c)\right\}, \qquad (6.5.7)
$$

or

$$
g^{(1)}(\tau) = \exp\left\{i\omega_0\tau - (\tau/\tau_c)\right\}. \qquad (6.5.8)
$$

This is the first-order correlation of light from a source in which the probability that phase remains constant decreases exponentially constant phase interval. The magnitude of this function is plotted in Figure 6.7.

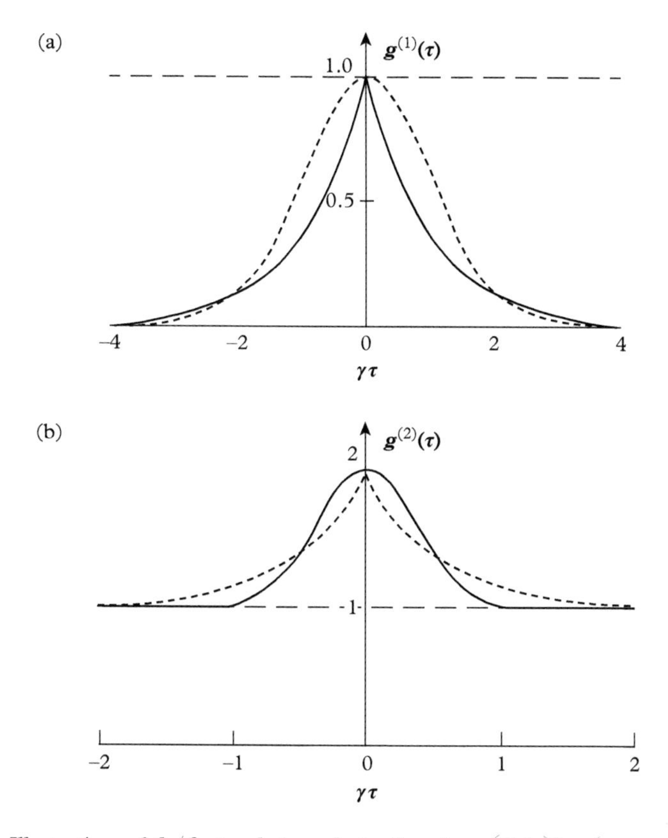

Figure 6.7 *(a) Illustrations of the first-order correlation function of light from sources with Lorentzian (solid curve) and Gaussian (dashed) phase distributions. (b) The second-order correlation functions of Lorentzian (dashed curve) and Gaussian (solid) light sources. The scaling factor on the horizontal axis is the dephasing rate* $\gamma \equiv 1/\tau_c$. *(Adapted from Ref. [6.9].)*

Statistical properties of light can be investigated through field measurements at different locations in either time or space. It is therefore useful to generalize the concept of first-order degree of coherence to include correlations between fields measured at different space–time points (r_1, t_1) and (r_2, t_2):

$$g^{(1)}\left(r_1, t_1; r_2, t_2\right) = \frac{< E^*\left(r_1, t_1\right) E\left(r_2, t_2\right) >}{< \left|E\left(r_1, t_1\right)\right|^2 \left|E\left(r_2, t_2\right)\right|^2 >^{1/2}}. \tag{6.5.9}$$

Now if a light field consists of contributions from two sources,

$$E(r, t) = u_1 E\left(r_1, t_1\right) + u_2 E\left(r_2, t_2\right), \tag{6.5.10}$$

as in a Young's interference experiment, the intensity of light at an arbitrary field point (r, t) averaged over a single period would be

$$I(r,t) = \frac{1}{2}\varepsilon_0 c\,|E(r,t)|^2$$
$$= \frac{1}{2}\varepsilon_0 c\left\{|u_1|^2\,|E(r_1,t_1)|^2 + |u_2|^2\,|E(r_2,t_2)|^2 + 2Re u_1^* u_2\left[E^*(r_1,t_1)E(r_2,t_2)\right]\right\}. \tag{6.5.11}$$

Averaged over times long compared to the coherence time τ_c, the intensity is

$$\langle I(r_1,t_1)\rangle = \frac{1}{2}\varepsilon_0 c\left\{\left\langle |u_1|^2\,|E(r_1,t_1)|^2\right\rangle + \left\langle |u_2|^2\,|E(r_2,t_2)|^2\right\rangle + 2Re u_1^* u_2\left\langle E^*(r_1,t_1)E(r_2,t_2)\right\rangle\right\}. \tag{6.5.12}$$

Notice that first-order coherence contributes explicitly to the last term in the ensemble average and governs the formation of fringes in classical interference experiments. Numerically, $g^{(1)}(r_1,t_1;r_2,t_2)$ is in fact equal to the conventional visibility of fringes $\left(V \equiv \frac{(I_{\max}-I_{\min})}{(I_{\max}+I_{\min})}\right)$. Conversely, Young's interference patterns determine the degree of first-order field coherence.

Intensity correlations yield what is known as the second-order coherence function:

$$g^{(2)}(r_1,t_1;r_2,t_2;r_2,t_2;r_1,t_1) = \frac{<E^*(r_1,t_1)\,E^*(r_2,t_2)\,E(r_2,t_2)\,E(r_1,t_1)>}{<\left|E(r_1,t_1)\right|^2><\left|E(r_2,t_2)^2\right|>} \tag{6.5.13}$$

or

$$g^{(2)} = \frac{<\bar{I}(r_1,t_1)\,\bar{I}(r_2,t_2)>}{<\bar{I}(r_1,t_1)><\bar{I}(r_2,t_2)>}. \tag{6.5.14}$$

Historically, the measurement of $g^{(2)}$ by the intensity correlation technique of Hanbury-Brown and Twiss [6.10] made it possible to obtain statistical information on field correlations without the need for interferometric stability. However, $g^{(2)}$ only contains information about energy dynamics of the field (photon arrival times) and unlike $g^{(1)}$ yields no information about phase fluctuations.

Before turning to statistics of quantized fields, let us consider the limiting values for $g^{(1)}$ and $g^{(2)}$ for classical (unquantized) light sources. First we reiterate the earlier point that fields that maintain their phase for an average time of τ_c are completely correlated at short delay times ($\tau \ll \tau_c$) and are completely uncorrelated at long delay times ($\tau \gg \tau_c$). Hence $g^{(1)}$ takes on the limiting values

$$g^{(1)}(\tau) \to 0, \quad \text{when } \tau \gg \tau_c, \tag{6.5.15}$$

$$g^{(1)}(\tau) \to 1, \quad \text{when } \tau \ll \tau_c, \tag{6.5.16}$$

and its range is

$$0 \leq g^{(1)} \leq 1. \tag{6.5.17}$$

To establish the allowed range for $g^{(2)}$, we note that by applying the Cauchy–Schwartz inequality to intensity measurements at times t_1 and t_2, we find the inequality

$$\bar{I}^2(t_1) + \bar{I}^2(t_2) \geq 2\bar{I}(t_1)\bar{I}(t_2). \tag{6.5.18}$$

For N pairs of intensity measurements, the statistical averages of the left and right sides of Eq. (6.5.18) yield the inequality

$$\langle \bar{I}^2(t) \rangle \geq \langle \bar{I}(t) \rangle^2. \tag{6.5.19}$$

Thus we are assured that the numerator in Eq. (6.5.14) always equals or exceeds the denominator. Consequently, the range of $g^{(2)}$ is

$$g^{(2)}(0) \geq 1. \tag{6.5.20}$$

Because intensity is positive definite and there is no upper limit for $g^{(2)}$, its range is

$$1 \leq g^{(2)}(\tau) \leq \infty. \tag{6.5.21}$$

Two examples of the second-order correlation function of classical sources are shown in Figure 6.7(b). We shall see that quite different shapes and limits are predicted by quantum theory for the second-order coherence of quantum sources in Section 6.5.2.

6.5.2 Quantum Statistics of Light

Since the electric field is real, the electric field operator is a sum of Hermitian conjugates.

$$\hat{E}(r,t) = \hat{E}^+(r,t) + \hat{E}^-(r,t), \tag{6.5.22}$$

where

$$\hat{E}^+(r,t) = \sum_k (\hbar\omega_k/2\varepsilon_0 V)^{1/2}\, \hat{\varepsilon}_k^- \hat{a}_k^- \exp\left(-i\omega_k t + i\bar{k}\cdot\bar{r}\right) \tag{6.5.23}$$

and

$$\hat{E}^-(r,t) = \sum_k (\hbar\omega_k/2\varepsilon_0 V)^{1/2}\, \hat{\varepsilon}_k^+ \hat{a}_k^+ \exp\left(i\omega_k t + i\bar{k}\cdot\bar{r}\right). \tag{6.5.24}$$

To analyze the quantum degree of coherence in Young's interference experiment, we introduce the field operators

$$\hat{E}^{\pm}(r,t) = u_1\hat{E}^{\pm}(r_1,t_1) + u_2\hat{E}^{\pm}(r_2,t_2) \tag{6.5.25}$$

in place of Eq. (6.5.10) and calculate the expectation value of the intensity operator quantum mechanically:

$$<\hat{I}(r,t)> \equiv <\hat{E}^{-}\hat{E}^{+}> = 2\varepsilon_0 c\{|u_1|^2 <\hat{E}^{-}(r_1,t_1)\hat{E}^{+}(r_1,t_1)> + |u_2|^2 <\hat{E}^{-}(r_2,t_2)\hat{E}^{+}(r_2,t_2)>$$
$$+ u_1^* u_2 <\hat{E}^{-}(r_1,t_1)\hat{E}^{+}(r_2,t_2)> + u_1 u_2^* <\hat{E}^{-}(r_2,t_2)\hat{E}^{+}(r_1,t_1)>\}. \tag{6.5.26}$$

To find the correlation in terms of field operators, we turn to the density matrix and follow Eq. (3.6.17) by calculating $< \hat{E}^{-}\hat{E}^{+} >= Tr\left[\hat{\rho}\hat{I}\right]$, where $\hat{I}$ is the intensity operator.

In the case of the Young's interference experiment, wave amplitudes emanating from two closely spaced slits are assumed to have the same magnitude but different phase. Hence, the last two terms in $<\hat{I}>$ are complex conjugates, and we can write

$$<\hat{I}(r,t)> = 2\varepsilon_0 c\{|u_1|^2 <\hat{E}^{-}(r_1,t_1)\hat{E}^{+}(r_1,t_1)> + |u_2|^2 <\hat{E}^{-}(r_2,t_2)\hat{E}^{+}(r_2,t_2)>$$
$$+ 2\mathrm{Re}u_1^* u_2 <\hat{E}^{-}(r_1,t_1)\hat{E}^{+}(r_2,t_2)>\}. \tag{6.5.27}$$

The two pinholes create or annihilate photons in the cavity formed by the lens and the aperture plane of the double slit experiment. For a given detector position, two modes contribute to the signal:

$$\hat{a}^{+} = \left(\hat{a}_1^{+} + \hat{a}_2^{+}\right)/\sqrt{2}, \tag{6.5.28}$$

$$\hat{a}^{-} = \left(\hat{a}_1^{-} + \hat{a}_2^{-}\right)/\sqrt{2}. \tag{6.5.29}$$

Hence the field operator contains two contributions:

$$\hat{E}^{+}(r,t) \propto u_1\hat{a}_1^{-}\exp(iks_1) + u_2\hat{a}_2^{-}\exp(iks_2), \tag{6.5.30}$$

where s_1 and s_2 are the distances of slit 1 and 2 from the detector. Correspondingly, the propagation time from slits 1 and 2 to the detector are

$$t_1 = t - (s_1/c), \tag{6.5.31}$$

$$t_2 = t - (s_2/c). \tag{6.5.32}$$

Thus for an incident beam with n photons, the intensity of the interference pattern is

$$
\begin{aligned}
<\hat{I}(r,t)> &\propto |u_1|^2 \langle n|\hat{a}_1^+\hat{a}_1^-|n\rangle + |u_2|^2 \langle n|\hat{a}_2^+\hat{a}_2^-|n\rangle \\
&\quad + 2u_1^* u_2 \cos\{k(s_1 - s_2)\} \langle n|\hat{a}_1^+\hat{a}_2^-|n\rangle \\
&\propto n\left\{\frac{1}{2}\left(|u_1|^2 + |u_2|^2\right) + \text{Re}\, u_1^* u_2 \cos k(s_1 - s_2)\right\}. \qquad (6.5.33)
\end{aligned}
$$

An interesting aspect of the result in Eq. (6.5.33) is that the total intensity including the interference term is proportional to n. Consequently even if $n = 1$ the interference effect and its visibility remains unchanged. This indicates that even if there is only a single photon in the apparatus, interference may still be observed.

The quantum degree of first-order coherence is

$$
g^{(1)}(r_1, t_1; r_2, t_2) = \frac{<\hat{E}^-(r_1,t_1)\,\hat{E}^+(r_2,t_2)>}{\left[<\hat{E}^-(r_1,t_1)\,\hat{E}^+(r_1,t_1)><\hat{E}^-(r_2,t_2)\,\hat{E}^+(r_2,t_2)>\right]^{1/2}}, \qquad (6.5.34)
$$

and its range is the same as the classical $g^{(1)}$. The quantum degree of second-order coherence is

$$
g^{(2)}(r_1, t_1; r_2, t_2; r_2, t_2; r_1, t_1) = \frac{<\hat{E}^-(r_1,t_1)\,\hat{E}^-(r_2,t_2)\,\hat{E}^+(r_2,t_2)\,\hat{E}^+(r_1,t_1)>}{<\hat{E}^-(r_1,t_1)\,\hat{E}^+(r_1,t_1)><\hat{E}^-(r_2,t_2)\,\hat{E}^+(r_2,t_2)>}. \qquad (6.5.35)
$$

Notice that the second-order coherence $g^{(2)}$ contains only products of field operators and their Hermitian conjugates. As a result, $g^{(2)}$ must be positive definite. That is,

$$
0 \le g^{(2)}(\tau) \le \infty. \qquad (6.5.36)
$$

Even for zero time delay ($\tau = 0$) it is not possible to prove $g^{(2)}(\tau) \ge 1$ as in the classical case. Consequently, light which exhibits second-order coherence in the range

$$
0 \le g^{(2)}(\tau) \le 1 \qquad (6.5.37)
$$

is deemed "non-classical" light. Figure 6.8 shows some examples of non-classical correlations.

What light sources actually exhibit values in the range $0 \le g^{(2)}(\tau) \le 1$? Do sources of "non-classical" light exist? Consider a single mode Fock state in which the number of photons is determined by $<\hat{n}> = <\hat{a}^+\hat{a}^->$. Its second-order coherence function is

$$
g^{(2)}(\tau) = \frac{<\hat{a}^+\hat{a}^+\hat{a}^-\hat{a}^->}{<\hat{a}^+\hat{a}^->^2}. \qquad (6.5.38)
$$

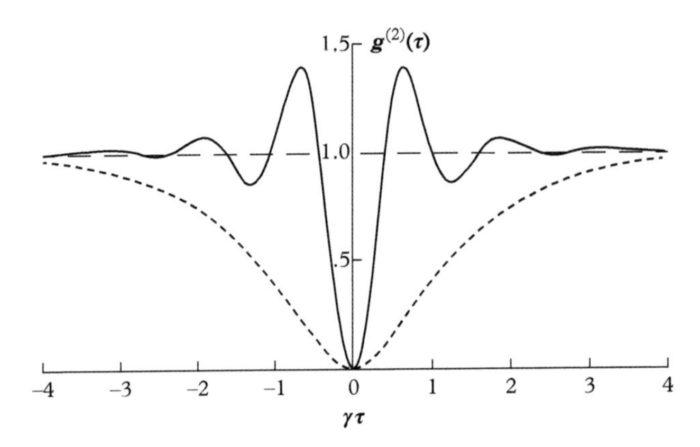

Figure 6.8 *Quantum second-order correlation functions for non-classical light ($g^{(2)} < 1$) such as that from an atom that is weakly excited (dashed curve; $\Omega \ll \gamma$) or strongly driven (solid curve; $\Omega = 5\gamma$). There is no collisional dephasing, so $\gamma = 1/\tau_{sp}$ is determined solely by the radiative decay rate. (Adapted from figures in Ref. [6.9].)*

By calculating Eq. (6.5.35) explicitly, it is easy to show that the Fock state $|n\rangle$ has a second-order correlation function given by

$$g^{(2)}(\tau) = \begin{cases} (n-1)/n, & n \geq 2, \\ 0, & n = 0, 1. \end{cases} \tag{6.5.39}$$

This value for $g^{(2)}(\tau)$ is always less than unity, showing that a photon number state exemplifies one characteristic of non-classical light. Even at zero time delay, correlations lie in the range $0 \leq g^{(2)}(\tau) \leq 1$, which is less than the uncorrelated light value $g^{(2)}(0) = 1$. This is because the arrival times of photons from non-classical light sources are not clustered together. They are said to be "anti-bunched." That is, a second photon cannot emerge from such a source immediately after emission of a first photon.

A concrete example of a non-classical source of light is an atom. Atoms cannot emit a second photon immediately after a first because each radiative event requires decay to a lower energy state. Decay takes time and further emission cannot take place until the atom returns to the ground state and is re-excited. The quantum degree of second-order correlation corresponding to this behavior is illustrated in Figure 6.8. Only individual luminescent centers that obey anti-bunching photon statistics exhibit a correlation function that is zero at zero delay, like that in Figure 6.8. The fluorescence of an ensemble of atoms is the incoherent sum of emission events that are not correlated at all, so their correlation function never drops to zero.

6.6 Quantized Reservoir Theory

6.6.1 The Reduced Density Matrix

It often happens that we regard one component of a system as more "interesting" than other parts, or one part of a problem more important than another. We may wish to follow the behavior of one part while ignoring the rest of the system—possibly a large reservoir—or we may wish to examine the interaction of the system with a reservoir without solving for any reservoir dynamics. Fortunately this is always possible using the "reduced" density matrix. In this section, we introduce reduced density matrices and then utilize them in examining the very important problem of a two-level atom interacting with a simple, quantized reservoir. This problem serves as a model for others.

Consider a system AB with two components A and B which do not interact (Figure 6.9). The components A and B occupy two independent mathematical spaces, and the state of the combined system $|\varphi_{ab}\rangle$ is just the uncorrelated product of component wavefunctions,

$$|\varphi_{ab}\rangle = |\varphi_a\rangle|\varphi_b\rangle \,. \tag{6.6.1}$$

Hence the systems A , B, and AB are pure states:

$$\rho(A) = |\varphi_a\rangle\langle\varphi_a| \,, \tag{6.6.2}$$

$$\rho(B) = |\varphi_b\rangle\langle\varphi_b| \,, \tag{6.6.3}$$

$$\rho(AB) = |\varphi_{ab}\rangle\langle\varphi_{ab}| \,. \tag{6.6.4}$$

We now show that the expectation of any operator $O(A)$ which acts only on wavefunctions in subsystem A can be calculated from a reduced density matrix for A which does not depend on reservoir variables:

$$\begin{aligned}
\langle 0(A)\rangle &= \mathrm{Tr}\,\{O(A)\rho(AB)\} \\
&= \sum_{ab} \langle\varphi_{ab}|O(A)\rho(AB)|\,\varphi_{ab}\rangle \\
&= \sum_{aba'b'} \langle\varphi_{ab}|O(A)|\,\varphi_{a'b'}\rangle\langle\varphi_{a'b'}\,|\rho(AB)|\,\varphi_{ab}\rangle \\
&= \sum_{aba'} \langle\varphi_{ab}|O(A)|\varphi_{a'}\rangle \sum_{b'} \delta_{bb'}\,\langle\varphi_{a'}|\langle\varphi_{b'}\,|\rho(AB)|\,\varphi_b\rangle|\varphi_a\rangle \\
&= \sum_{aa'} \langle\varphi_a|O(A)|\,\varphi_{a'}\rangle \sum_{b} \langle\varphi_{a'}|\langle\varphi_b|\,(AB)|\varphi_b\rangle|\varphi_a\rangle \\
&= \sum_{a} \langle\varphi_a|O(A)\,Tr_B\rho(AB)|\varphi_a\rangle,
\end{aligned} \tag{6.6.5}$$

where we have used $Tr_B \rho(AB) \equiv \sum_b \langle \varphi_b | \rho(AB) | \varphi_b \rangle$. Notice that this result has the form

$$\langle O(A) \rangle \equiv \sum_a \langle \varphi_a | O(A) \, \rho(A) | \varphi_a \rangle , \tag{6.6.6}$$

if we define a reduced density matrix $\rho(A)$for subsystem A by the relation

$$\rho(A) \equiv Tr_B \rho(AB). \tag{6.6.7}$$

Similarly, it is easy to show that

$$\langle O(B) \rangle \equiv \sum_b \langle \varphi_b | O(B) \rho(B) | \varphi_b \rangle, \tag{6.6.8}$$

where

$$\rho(B) \equiv Tr_A \rho(AB). \tag{6.6.9}$$

The derivation of results (6.6.7) and (6.6.9) for systems in which the subsystems A and B interact is more cumbersome, but the results are the same [6.11]. Reduced density matrices are still found to be adequate for calculating variables of the subsystems alone (such as $<O(A)>$, $<O(B)>$). The importance of this is that the time development of reduced density operators can be obtained without calculating reservoir variables. Thus we can choose theoretically to include or ignore dynamic processes of a host medium B in which an optical center A is located in the same way that the strength of the optical interaction between light and center A can be adjusted experimentally by changing the detunings with respect to resonances in subsystems A and B. This idea is depicted schematically in Figure 6.9.

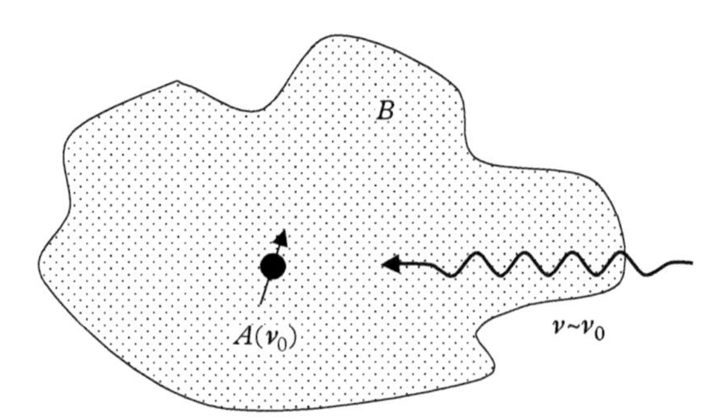

Figure 6.9 *A schematic drawing of a compound system comprised of an optical center A with a resonance at frequency ν_0 in a host medium B, called the reservoir. Experimentally, the optical interaction can be targeted to A by tuning the light frequency near ν_0. Theoretically, the operators describing the dynamics of A can be determined while ignoring dynamics of B by tracing over the reservoir states.*

Take the compound system Hamiltonian to be

$$H_{AB} = H_A + H_B + V, \tag{6.6.10}$$

where V represents an interaction between systems A and B. The equation of motion is

$$i\hbar\dot{\rho}_{AB}(t) = [H_{AB}, \rho_{AB}(t)]. \tag{6.6.11}$$

However, for the subsystem A we can write

$$\begin{aligned} i\hbar\dot{\rho}_A(t) &= i\hbar Tr_B \dot{\rho}_{AB}(t) \\ &= Tr_B\,[(H_A + H_B + V), \rho_{AB}(t)]. \end{aligned} \tag{6.6.12}$$

Making use of the results

$$Tr_B\,[H_A, \rho_{AB}(t)] = [H_A, Tr_B \rho_{AB}(t)] = [H_A, \rho_A(t)]\,, \tag{6.6.13}$$

$$\begin{aligned} Tr_B\,[H_B, \rho_{AB}(t)] &= \sum_B [<\varphi_B\,|H_B\rho_{AB}(t)|\,\varphi_B> - <\varphi_B\,|\rho_{AB}(t)H_B|\,\varphi_B>] \\ &= \sum_B [E_B<\varphi_B\,|\rho_{AB}(t)|\,\varphi_B> - <\varphi_B\,|\rho_{AB}(t)|\,\varphi_B>E_B] = 0, \end{aligned} \tag{6.6.14}$$

one obtains dynamic equations for the subsystems. In the Schrödinger representation,

$$i\hbar\dot{\rho}_A(t) = [H_A, \rho_A] + Tr_B\,[V, \rho_{AB}(t)]\,, \tag{6.6.15}$$

$$i\hbar\dot{\rho}_B(t) = [H_B, \rho_B] + Tr_A\,[V, \rho_{AB}(t)]\,. \tag{6.6.16}$$

In the interaction picture,

$$i\hbar\dot{\rho}_A(t) = Tr_B\,[V(t), \rho_{AB}(t)], \tag{6.6.17}$$

$$i\hbar\dot{\rho}_B(t) = Tr_A\,[V(t), \rho_{AB}(t)]. \tag{6.6.18}$$

To solve for $\rho_A(t)$ and $\rho_B(t)$ when $V \neq 0$, that is when the subsystems are correlated by virtue of an interaction V, we can imagine a perturbative approach which assumes the coupling makes only a small addition to the density matrix of the uncoupled subsystems:

$$\rho_{AB}(t) = \rho_A(t)\rho_B(t) + \rho_C(t). \tag{6.6.19}$$

The idea behind Eq. (6.6.19) is that $\rho_C(t)$ represents a weak correlation which develops in time between subsystems A and B because of their interaction:

$$\begin{aligned} Tr_B\left[V(t),\rho_{AB}(t)\right] &= Tr_B\left[V(t),\rho_A(t)\rho_B(t)\right] + Tr_B\left[V(t),\rho_C(t)\right] \\ &= Tr_B\left[V(t),\rho_B(t)\right]\rho_A(t) + Tr_B\left[\rho_B(t)V(t),\rho_A(t)\right] + Tr_B\left[V(t),\rho_C(t)\right] \\ &= Tr_B\left[V(t)\rho_B(t),\rho_A(t)\right] + Tr_B\left[V(t),\rho_C(t)\right] \\ &= \left[V_A(t),\rho_A(t)\right] + Tr_B\left[V(t),\rho_C(t)\right]. \end{aligned} \tag{6.6.20}$$

In obtaining Eq. (6.6.20) we have used the fact that operators commute under a trace, and have also defined

$$V_A(t) \equiv Tr_B\left[V(t),\rho_B(t)\right]. \tag{6.6.21}$$

$V_A(t)$ is a self-consistent correction to the energy of subsystem A. We finally obtain

$$i\hbar\dot{\rho}_A(t) = \left[V_A(t),\rho_A(t)\right] + Tr_B\left[V(t),\rho_C(t)\right], \tag{6.6.22}$$

$$i\hbar\dot{\rho}_B(t) = \left[V_B(t),\rho_B(t)\right] + Tr_A\left[V(t),\rho_C(t)\right]. \tag{6.6.23}$$

To solve these equations we need an equation for the correlation operator ρ_C. This is found by differentiating its defining relation:

$$\begin{aligned} i\hbar\dot{\rho}_C(t) &= i\hbar\left[\dot{\rho}_{AB}(t) - \dot{\rho}_A\rho_B - \rho_A\dot{\rho}_B\right] \\ &= \left[V,\rho_A\rho_B+\rho_C\right] - \left[V_A,\rho_A\right]\rho_B - Tr_B\left[V,\rho_C\right]\rho_B - \rho_A\left[V_B,\rho_B\right] - \rho_A Tr_A\left[V,\rho_C\right]. \end{aligned} \tag{6.6.24}$$

By restricting ourselves to two-level atoms or simple harmonic oscillators (SHOs) interacting with a reservoir of SHOs we can set

$$V_A = V_B = 0. \tag{6.6.25}$$

We shall neglect higher order terms involving the commutator $[V,\rho_C]$ and may assume that B is a reservoir which is too large to undergo significant change upon energy input from A, so that

$$\dot{\rho}_B(t) \approx 0. \tag{6.6.26}$$

Hence Eq. (6.6.24) simplifies to

$$i\hbar\dot{\rho}_C(t) = \left[V(t),\rho_A(t)\rho_B(t)\right]. \tag{6.6.27}$$

If systems A and B are initially uncorrelated ($\rho_C(t_0) = 0$), the formal solution of Eq. (6.6.27) is

$$\rho_C(t) = -\frac{i}{\hbar}\int_{t_0}^{t} dt' \left[V\left(t'\right), \rho_A\left(t'\right)\rho_B\left(t'\right)\right]. \tag{6.6.28}$$

This solution provides a starting point for iterative solutions of ρ_C in terms of ρ_A, ρ_B when A and B are on an equal footing, or for simplified solutions when one subsystem is a large reservoir:

$$\rho_C(t) \doteq -\frac{i}{\hbar}\int_{t_0}^{t} dt' \left[V\left(t'\right), \rho_A\left(t'\right)\rho_B\left(t_0\right)\right]. \tag{6.6.29}$$

Equation (6.6.26) is called the reservoir assumption. Notice that it has been used to write Eq. (6.6.29). Together with Eq. (6.6.28) it allows us to write the decay law for subsystem A simply as

$$\dot{\rho}_A(t) = -\frac{i}{\hbar}\int_{t_0}^{t} dt'\, Tr_B\left[V(t), \left[V(t'), \rho_A(t')\rho_B(t_0)\right]\right]. \tag{6.6.30}$$

Equation (6.6.30) is a key result of this section that can be used to evaluate decay of a radiation field or atom interacting with an ensemble of oscillators. The ensemble may be a reservoir of phonons, photons or cavity modes. As an example of its use we now treat the problem of a two-level atom coupled to modes of an unstructured vacuum, re-deriving the Weisskopf–Wigner result for spontaneous emission.

6.6.2 Application of the Reduced Density Matrix

Quantum mechanically, an atom makes transitions described by creation and annihilation operators $\hat{\sigma}^+$ and $\hat{\sigma}^-$ when its excitation frequency is close to the atomic resonant frequency ω_0. The simple harmonic oscillators of the reservoir in which it is embedded are similarly described by creation and annihilation operators $\hat{b}_k^+$ and $\hat{b}_k^-$, where k is the mode index, and they have corresponding excitation frequencies designated by ω_k. Multi-component systems like this must conserve energy as they undergo internal changes. Hence the excitation of an atom must be accompanied by an energy-conserving de-excitation of an oscillator, and vice versa. To simplify notation, we dispense with carets over operators.

In the RWA, the coupling Hamiltonian in the interaction picture is

$$V = \sum_k \hbar g_k \left[\sigma^- b_k^+ \exp[-i(\omega_0 - \omega_k)t] + h.c.\right]. \tag{6.6.31}$$

We assume that the reservoir consists of a thermal equilibrium distribution of vacuum modes. The density operator for a single mode k of thermal radiation is

$$\rho_{kk} = \frac{\exp\left(-\hbar\omega_k b_k^+ b_k/k_B T\right)}{Tr\left\{\exp\left(-\hbar\omega_k b_k^+ b_k/k_B T\right)\right\}}$$
$$= \exp\left(-\hbar\omega_k b_k^+ b_k/k_B T\right)[1 - \exp(-\hbar\omega_k/k_B T)], \tag{6.6.32}$$

where the second line makes use of the binomial expansion. Hence for all modes

$$\rho_B = \prod_k \exp\left(-\hbar\omega_k b_k^+ b_k/k_B T\right)[1 - \exp(-\hbar\omega_k/k_B T)]. \tag{6.6.33}$$

To work out, say, the first term for $\dot{\rho}_A(t)$ in Eq. (6.6.30) we need to evaluate the following trace:

$$Tr_B\{V(t)V(t')\rho_A(t')\rho_B(t_0)\}$$

$$= \hbar^2 \sum_B \langle\psi_B| \left\{ \sum_{j\ell} g_\ell g_j \sigma b_\ell^+ \sigma b_j^+ \exp(-2i\omega_0 t + i\omega_l t + i\omega_l t')\rho_A(t')\rho_B \right.$$

$$+ \sum_{j\ell} g_\ell g_j [\sigma b_\ell^+ \sigma^+ b_j \exp[-i(\omega_0 - \omega_l)t + i(\omega_0 - \omega_j)t']\rho_A(t')\rho_B + h.c.]$$

$$\left. + \sum_{j\ell} g_\ell g_j \sigma^+ b_\ell \sigma^+ b_j \exp(2i\omega_0 t - i\omega_l t - i\omega_l t')\rho_A(t')\rho_B \right\} |\psi_B\rangle$$

$$= \hbar^2 \sum_B \langle\psi_B| \sum_{j\ell} g_\ell g_j [\sigma b_\ell^+ \sigma^+ b_j \exp[-i(\omega_0 - \omega_l)t + i(\omega_0 - \omega_j)t']\rho_A(t')\rho_B + h.c.]|\psi_B\rangle$$

$$= \hbar^2 \sum_{j\ell} g_\ell g_j \sigma\sigma^+ Tr_B(b_\ell^+ b_j \rho_B)\rho_A(t') \exp[-i(\omega_0 - \omega_l)t + i(\omega_0 - \omega_j)t'] + h.c. \tag{6.6.34}$$

In obtaining Eq. (6.6.34), the high-frequency contributions have been dropped, consistent with the RWA.

Exercise: Show the following results, where $\bar{n}_k$ is the average occupation of reservoir mode k:

$$Tr_B[b_j^+ b_k \rho_B] = \bar{n}_k \delta_{jk}, \tag{6.6.35}$$

$$Tr_B[b_j b_k^+ \rho_B] = (\bar{n}_k + 1)\delta_{jk}, \tag{6.6.36}$$

$$Tr_B[b_j b_k \rho_B] = <b_j b_k>_B = 0, \tag{6.6.37}$$

$$Tr_B[b_j^+ b_k^+] = <b_j^+ b_k^+>_B = 0. \tag{6.6.38}$$

Making use of the results of the previous exercise, Eq. (6.6.34) can be simplified further:

$$Tr_B\left[V(t)V\left(t'\right)\rho_A\left(t'\right)\rho_B\left(t_0\right)\right]=\hbar^2\sum_k g_k^2\Big\{\left[\sigma^+\sigma\rho_A(t')\,(\bar{n}_k+1)\exp[i(\omega_0-\omega_k)(t-t')]+\right.$$
$$\left.+\,\sigma\sigma^+\rho_A(t')\bar{n}_k\exp[-i(\omega_0-\omega_k)(t-t')]\right]+h.c.\Big\}.\qquad(6.6.39)$$

Substituting this expression and similar terms for the remaining commutator contributions into the equation for $\rho_A(t)$, we find

$$\dot{\rho}_A(t)=-\int_{t_0}^{t}dt'\sum_k g_k^2\Big\{\sigma^+\sigma\rho_A(t')\,(\bar{n}_k+1)\exp[i(\omega_0-\omega_k)(t-t')]$$
$$+\,\sigma\sigma^+\rho_A(t')\bar{n}_k\exp[-i(\omega_0-\omega_k)(t-t')]-$$
$$-\,\sigma^+\rho_A(t')\sigma\,(\bar{n}_k+1)\exp[i(\omega_0-\omega_k)(t-t')]+$$
$$+\,\sigma\rho_A(t')\sigma^+\bar{n}_k\exp[-i(\omega_0-\omega_k)(t-t')]\Big\}+h.c.\qquad(6.6.40)$$

We proceed by noting that the summation over boson states can be converted in the usual way to an integration over frequency by introducing the density of reservoir oscillator states $D(\omega)$:

$$\sum_k\rightarrow\int d\omega D(\omega).\qquad(6.6.41)$$

We assume that the density of states $D(\omega)$, the coupling strength $g(\omega)$, and the mode occupation probability $\bar{n}(\omega)$ all change little over the frequency range of the atomic resonance. These factors may then be treated as having the constant value

$$D(\omega)\bar{n}(\omega)g^2(\omega)=D(\omega_0)\bar{n}(\omega_0)g^2(\omega_0),\qquad(6.6.42)$$

and may be factored from the integral. Additionally, the integrand is dominated by contributions to the exponential factors at times $t'\approx t$. Consequently we may set $\rho_A(t')=\rho_A(t)$, which is equivalent to the Markoff approximation made in our earlier analysis of dephasing (Section 3.7). Finally we note

$$\int_{t_0}^{t}dt'\exp[\pm i(\omega_0-\omega)(t-t')]=\pi\delta\left(\omega_0-\omega\right).\qquad(6.6.43)$$

Therefore,

$$\dot{\rho}_A(t) = -\pi D(\omega_0) g^2(\omega_0) \left\{ \bar{n} \left[\sigma\sigma^+ \rho_A(t) - \sigma^+ \rho_A(t)\sigma \right] + \right.$$
$$\left. + (\bar{n}+1) \left[\sigma^+\sigma\rho_A(t) - \sigma\rho_A(t)\sigma^+ \right] \right\} + h.c. \tag{6.6.44}$$

By identifying the decay constant for the transition at frequency ω_0 as

$$\gamma \equiv 2\pi D(\omega_0) g^2(\omega_0), \tag{6.6.45}$$

we obtain a key result of this section:

$$\dot{\rho}_A(t) = -\frac{1}{2}\gamma \left\{ \bar{n} \left[\sigma\sigma^+ \rho_A(t) - \sigma^+ \rho_A(t)\sigma \right] + \left\{ (\bar{n}+1) \left[\sigma^+\sigma\rho_A(t) - \sigma\rho_A(t)\sigma^+ \right] + h.c. \right\} \right. . \tag{6.6.46}$$

Notice that for a reservoir with no excitations ($\bar{n} = 0$), an initially excited atom will decay from its upper state 2 according to the law

$$\dot{\rho}^A_{22}(t) = -\frac{1}{2}\gamma \langle 2 | \sigma^+\sigma\rho_A(t) - \sigma\rho_A(t)\sigma^+ | 2 \rangle + h.c.$$
$$= -\frac{1}{2}\gamma \langle 2 | \sigma^+\sigma\rho_A(t) | 2 \rangle - \frac{1}{2}\gamma \langle 2 | \rho^+_A(t)\sigma^+\sigma | 2 \rangle . \tag{6.6.47}$$

Recall that $\rho = \rho^+$ for diagonal matrix elements like those in Eq. (6.6.47). So Eq. (6.6.47) yields

$$\dot{\rho}^A_{22}(t) = -\gamma \rho^A_{22}(t). \tag{6.6.48}$$

Equation (6.6.48) shows that quantized reservoir theory based on the reduced density matrix reproduces the Wigner–Weisskopf spontaneous emission decay law derived earlier. As illustrated in Figure 6.10, it demonstrates that the decay of an atom (A) takes place via coupling with available modes of the reservoir (modes of space B) whose dynamics can be ignored. If all modes of space are available, the decay proceeds exponentially at the natural decay rate γ. However, by placing the atom near one or more conducting surfaces, the decay rate can be altered significantly. As suggested by the diagram in Figure 6.10(b), the presence of a cavity diminishes the density of states $D(\omega)$ available for the emission process if the separation of the plates is less than half the emission wavelength [6.12]. Thus spontaneous emission is inhibited. On the other hand, for large spacings, resonant cavity modes can enhance the decay rate, yielding a decay rate proportional to the cavity Q factor [6.13]. Inhibition and enhancement of spontaneous emission by manipulation of the density of states has been discussed and demonstrated experimentally by many groups [6.9, 6.14, 6.15].

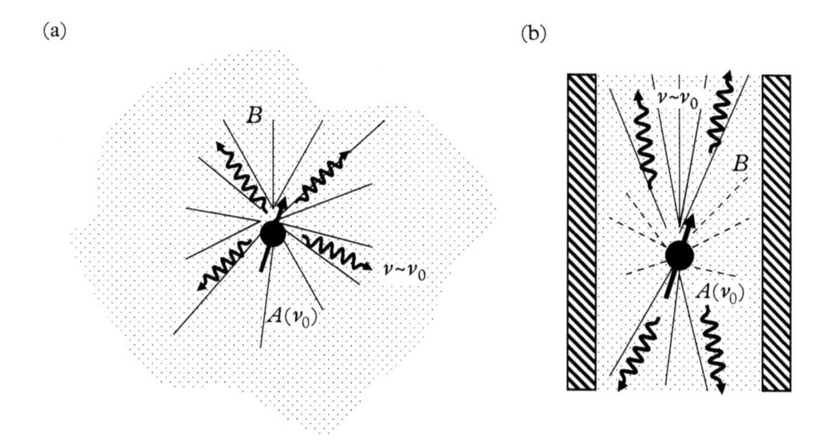

Figure 6.10 *(a) Radiative coupling of an excited atom to all modes of space. (b) Radiative coupling of an atom to modes of space restricted by the presence of conducting surfaces.*

Using the formalism of quantum relaxation theory one can account for the prevalence of exponential decay processes in nature. The formulation in the previous paragraph is applicable to any situation in which two-level systems are coupled to a reservoir of simple harmonic oscillators, or vice versa. Although the development of this section has focused on the implications of atom–reservoir coupling for decay, a broader discussion brings to light four fundamental phenomena underlying such coupling. Appendix F presents the close links between spontaneous decay, stimulated emission, the Lamb shift, and the AC Stark shift.

6.7 Resonance Fluorescence

When an atom is excited exactly on resonance, the conditions needed to apply perturbation theory are not met, and its predictions fail. The need for a different approach is illustrated in this section, where the resonance fluorescence spectrum is found to differ significantly from the predictions of perturbation theory, because resonant light dynamically alters the eigenfunctions and eigenenergies of the atom. The original work was done by Mollow [6.16] but the development presented here follows Ref. [6.17] and shows that strong-field fluorescence from a two-level system consists of three lines (the Mollow triplet), as indicated in Figure 6.11. It provides useful perspective on what went right and what went wrong in our preliminary discussion of strong-field effects in Section 3.4, based on perturbation theory. The experiments [6.18–6.20] in which the resonance fluorescence spectrum was first observed also illustrate the fact that multi-level atoms like sodium can indeed be converted into true two-level systems by carefully restricting the accessible optical transitions to just one, as described in Section 3.11. For sodium, the transition $3^2S_{1/2}$ $(F = 2) \leftrightarrow 3^2P_{3/2}$ $(F = 3)$ was selected, since excitation to $F = 3$ can only be followed by emission back to the initial state $F = 2$ according to the selection rule for one-photon transitions among hyperfine states ($\Delta F = 0, \pm 1$).

6.7.1 Fluorescence of Strongly Driven Atoms

Fluorescence spectra are frequency-domain versions (Fourier transforms) of the power delivered to or taken away from the optical field by various modes in real time. Hence we begin with an expression for the rate of change of energy in each mode and polarization state, labeled by k and λ respectively, versus time. According to Eq. (6.1.38) the total emitted power can be written as

$$P(t) = \frac{\partial}{\partial t} \sum_{k,\lambda} \hbar\omega_k \left\langle \hat{a}^+_{k\lambda}(t)\hat{a}^-_{k\lambda}(t)\right\rangle. \tag{6.7.1}$$

To evaluate $P(t)$, we first need to find the equation of motion for $\left\langle \hat{a}^+(t)\hat{a}^-(t)\right\rangle$. For this we require the system Hamiltonian comprised of the field, atom, and interaction energies. We drop the zero-point energy of the field since it does not contribute to the equations of motion, and choose the zero of energy for the atom midway between its two states. In general the interaction may involve more than one mode, hence the modes are specified by wavenumber k and polarization index λ. The total Hamiltonian is

$$H = \sum_{k,\lambda} \hbar\omega_k \hat{a}^+_{k\lambda}\hat{a}^-_{k\lambda} + \frac{1}{2}\hbar\omega_0\hat{\sigma}_3 + \hbar\sum_{k\lambda} g_{k\lambda}\left(\hat{\sigma}^+\hat{a}^-_{k\lambda} + \hat{\sigma}^-\hat{a}^+_{k\lambda}\right). \tag{6.7.2}$$

The Heisenberg picture has been adopted and non-secular (time-varying, energy-non-conserving) terms in the atom–field interaction, $-\bar{\mu}\cdot\bar{E}(t) = \hbar g_{k\lambda}(\hat{\sigma}^+e^{i\omega t}+\hat{\sigma}^-e^{-i\omega t})(\hat{a}^-e^{-i\omega t}+\hat{a}^+e^{i\omega t})$, have been dropped. The explicit expression for the atom–field coupling strength is

$$g_{k\lambda} = g_k(\hat{\mu}\cdot\hat{\varepsilon})_{k\lambda} = -\mu_0\left(\frac{\omega_k}{2\hbar\varepsilon_0 V}\right)^{1/2}\hat{\mu}\cdot\hat{\varepsilon}_{k\lambda}. \tag{6.7.3}$$

Omission of the non-secular terms is equivalent to the RWA in Eq. (6.7.2).

Exercise: Using Eq. (2.4.24) and commutation properties of operators that do not commute, verify that the equations of motion in the Heisenberg picture describing the interplay of field, atomic polarization, and population in resonance fluorescence are given by

$$\dot{\hat{a}}^-_{k\lambda} = -i\omega_k\hat{a}^-_{k\lambda} - ig_{k\lambda}\hat{\sigma}^-, \tag{6.7.4}$$

$$\dot{\hat{\sigma}}^- = -i\omega_0\hat{\sigma}^- + i\sum_{k,\lambda} g_{k\lambda}\hat{\sigma}_3\hat{a}^-_{k\lambda}, \tag{6.7.5}$$

$$\dot{\hat{\sigma}}_3 = 2i\sum_{k,\lambda} g_{k\lambda}\left(\hat{\sigma}^-\hat{a}^+_{k\lambda} - \hat{\sigma}^+\hat{a}^-_{k\lambda}\right). \tag{6.7.6}$$

The conjugate equations for the field and atom lowering operators are

$$\dot{\hat{a}}^+_{k\lambda} = i\omega_k \hat{a}^+_{k\lambda} + ig_{k\lambda}\hat{\sigma}^+, \tag{6.7.7}$$

$$\dot{\hat{\sigma}}^+ = i\omega_0\hat{\sigma}^+ - i\sum_{k,\lambda} g_{k\lambda}\hat{\sigma}_3\hat{a}^+_{k\lambda}. \tag{6.7.8}$$

Note that Eqs. (6.7.4)–(6.7.6) consist of three coupled equations in the operators $\hat{a}^-$, $\hat{\sigma}^-$, and $\hat{\sigma}_3$ that represent the three physical parts of the system. Not surprisingly, as we proceed to solve these equations, we shall encounter three coupled equations involving three related correlation functions.

The next step is to proceed with formal integration of Eq. (6.7.4). To simplify notation we dispense with operator carets and introduce the slowly varying quantities $\tilde{a}^-_{k\lambda}(t) = a^-_{k\lambda}\exp[i\omega_k t]$, $\tilde{a}^+_{k\lambda}(t) = a^+_{k\lambda}\exp[-i\omega_k t]$, $\tilde{\sigma}^-(t) = \sigma^-\exp[i\omega_k t]$, and $\tilde{\sigma}^+(t) = \sigma^+(t)\exp[-i\omega_k t]$. Steady-state conditions are not assumed as yet, so that Eq. (6.7.4) becomes

$$\dot{\tilde{a}}^-_{k\lambda}(t') = -ig_{k\lambda}\tilde{\sigma}^-(t'). \tag{6.7.9}$$

Upon integration between the limits 0 and t, this yields

$$\tilde{a}^-_{k\lambda}(t) = \tilde{a}^-_{k\lambda}(0) - ig_{k\lambda}\int_0^t dt'\tilde{\sigma}^-(t'). \tag{6.7.10}$$

Multiplying through by $\exp(-i\omega_k t)$, one finds

$$\begin{aligned} a^-_{k\lambda}(t) &= a^-_{k\lambda}(0)\exp(-i\omega_k t) - ig_{k\lambda}\int_0^t dt'\tilde{\sigma}^-(t')\exp(-i\omega_k t) \\ &= a^-_{k\lambda}(0)\exp(-i\omega_k t) - ig_{k\lambda}\int_0^t dt'\sigma^-(t')\exp[-i\omega_k(t-t')]. \end{aligned} \tag{6.7.11}$$

The Hermitian conjugate operator is

$$a^+_{k\lambda}(t) = a^+_{k\lambda}(0)\exp(i\omega_k t) + ig_{k\lambda}\int_0^t dt'\sigma^+(t')\exp[i\omega_k(t-t')]. \tag{6.7.12}$$

Next, we describe the evolution of the atomic operators in terms of slowly varying amplitudes referenced to the resonance frequency of the atom ω_0. This is a matter of convenience that is intended to make it easier to interpret the final results:

$$s^-(t) = \tilde{\sigma}^-(t)\exp[i\Delta' t] = \sigma^-\exp[i\omega_k t]\exp[i\Delta' t], \tag{6.7.13}$$

$$s^+(t') = \tilde{\sigma}^+(t)\exp[-i\Delta' t'] = \sigma^+(t')\exp[-i\omega_k t']\exp[-i\Delta' t']. \tag{6.7.14}$$

Here $\Delta' = \omega - \omega_k$. In terms of the slowly varying amplitudes s^- and s^+, the radiated power obtained by substituting Eqs. (6.7.11)–(6.7.12) and (6.7.4)–(6.7.5) into Eq. (6.7.1) is

$$\begin{aligned} P(t) &= \sum_{k,\lambda} \hbar\omega_k \left\langle \dot{a}^+_{k\lambda}(t) a^-_{k\lambda}(t) + a^+_{k\lambda}(t) \dot{a}^-_{k\lambda}(t) \right\rangle \\ &= -\sum_{k\lambda} i g_{k\lambda} \hbar\omega_k \left[\left\langle a^+_{k\lambda}(0)\sigma^-(t) \right\rangle e^{i\omega_k t} - \left\langle \sigma^+(t) a^-_{k\lambda}(0) \right\rangle e^{-i\omega_k t} \right] + \\ &\quad + 2\mathrm{Re} \sum_{k\lambda} g^2_{k\lambda} \hbar\omega_k \int_0^t dt' \left\langle s^+(t') s^-(t) \right\rangle e^{i\Delta'(t-t')}. \end{aligned} \tag{6.7.15}$$

The first two terms on the right side of Eq. (6.7.15) give the rate of change of field energy due to stimulated emission and absorption. The last term involves only atomic operators and represents power scattered quasi-elastically out of the incident beam as resonance fluorescence. It is this term that is of most interest here. Making use of the density of modes to replace the sum over k, λ by an integral, one finds that the fluorescence is determined entirely by a polarization correlation function:

$$\begin{aligned} P_s(t) &= 2\mathrm{Re} \sum_{k\lambda} g^2_{k\lambda} \hbar\omega_k \int_0^t dt' \left\langle s^+(t') s^-(t) \right\rangle e^{i\Delta'(t-t')} \\ &= \frac{1}{4\pi\varepsilon_0} \frac{4\mu^2}{3c^3} \mathrm{Re} \int_0^\infty d\omega'' \left(\omega''\right)^4 \int_0^t dt' \left\langle s^+(t') s^-(t) \right\rangle e^{i\Delta''(t-t')}. \end{aligned} \tag{6.7.16}$$

Here ω'' is a representative mode frequency near that of the driving field. The detuning has been redefined accordingly as $\Delta'' \equiv \omega - \omega''$. Since we expect the spectrum to be dominated by quasi-elastic frequencies $\omega \cong \omega''$, the frequency factor $(\omega'')^4$ in Eq. (6.7.16) is well approximated by ω_0^4 and may be removed from the integral, with the result

$$P_s(t) \cong \frac{\mu_{12}^2 \omega_0^4}{3\pi\varepsilon_0 c^3} \mathrm{Re} \int_0^\infty d\omega'' \int_0^t dt' \left\langle s^+(t') s^-(t) \right\rangle \exp[i(\omega - \omega'')(t - t')]. \tag{6.7.17}$$

In the stationary regime ($t \to \infty$), the two-time correlation function, $\left\langle s^+(t') s^-(t) \right\rangle$, only depends on the difference between t and t'. Hence we define $\tau = t - t'$ and re-express the steady-state power as

$$P_s(\infty) \cong \frac{\mu_{12}^2 \omega_0^4}{3\pi\varepsilon_0 c^3} \mathrm{Re} \int_0^\infty d\omega \int_0^\infty d\tau \left\langle s^+(t_0) s^-(t_0 + \tau) \right\rangle \exp[i(\omega - \omega_0)\tau], \tag{6.7.18}$$

where t_0 is any time in the stationary regime. In this limit, the power spectrum is

$$P(\omega) \propto 2\mathrm{Re} \int_0^\infty d\tau \left\langle s^+(t_0) s^-(t_0 + \tau) \right\rangle \exp[i(\omega - \omega'')\tau], \tag{6.7.19}$$

and is entirely determined by the Fourier transform of the first-order correlation function $g(\tau)$ of the atomic operator. This result reflects a fundamental relationship between time-domain fluctuations and frequency-domain power spectra, known as the Wiener–Khintchine theorem [6.23]. To calculate the resonance fluorescence spectrum we therefore need to determine

$$g(\tau) \equiv \langle s^+(t_0)s^-(t_0+\tau)\rangle . \tag{6.7.20}$$

This is the heart of the resonance fluorescence problem and requires solution of Eqs. (6.7.4)–(6.7.6) as an interdependent set of equations, with all modes participating.

Rewriting Eqs. (6.7.5) and (6.7.6) using the earlier solution for $a_{k\lambda}^-$ from Eq. (6.7.11) one finds

$$\dot{\sigma}^-(t) = -i\omega_0\sigma^-(t) - \frac{i}{\hbar}\mu \cdot \sigma_3(t)E^+(t), \tag{6.7.21}$$

$$\dot{\sigma}_3(t) = (2i/\hbar)\mu \cdot \left[E^+(t)\sigma^+(t) - E^-(t)\sigma^-(t)\right], \tag{6.7.22}$$

where

$$E^+(t_0) = \sum_{k\lambda}\left(\frac{\hbar\omega_k}{2\varepsilon_0 V}\right)^{1/2} a_{k\lambda}^-(t)\hat{\varepsilon}_{k\lambda} \tag{6.7.23}$$

and

$$E^-(t) = \left[E^+(t)\right]^+ . \tag{6.7.24}$$

Now notice that the electric field operator $E^+(t)$ in Eq. (6.7.23) consists of two parts, since $a_{k\lambda}^-(t)$ consists of the two parts given in Eq. (6.7.11). Consequently,

$$\begin{aligned} E^+(t) &= \sum_{k\lambda}\left(\frac{\hbar\omega_k}{2\varepsilon_0 V}\right)^{1/2} a_{k\lambda}^-(0)\exp(-i\omega_k t)\hat{\varepsilon}_{k\lambda} \\ &\quad - i\sum_{k\lambda}\left(\frac{\hbar\omega_k}{2\varepsilon_0 V}\right)^{1/2} \hat{\varepsilon}_{k\lambda} g_{k\lambda}\int_0^t dt'\sigma^-(t')\exp[-i\omega_k(t-t')] \\ &\equiv E_0^+(t) + E_{RR}^+(t). \end{aligned} \tag{6.7.25}$$

The first term in this result, $E_0^+(t)$, is the solution of the homogeneous Maxwell equation. That is, it is the field without any modification due to the atom. The second term is the modification due to the presence of the atom, called the radiation reaction field E_{RR}^+. This is clear from the absence of field operators in the second integral.

In classical electrodynamics the reaction of an atom to its own radiation can be estimated from Newton's third law stating that for every action there is an equal and opposite reaction. Even though we have undertaken to calculate the emission spectrum in a fully quantum mechanical manner, we are guided by correspondence with classical concepts. Hence in the point dipole approximation, we assume the radiation reaction field is given by the operator for the corresponding classical field expression [6.20]:

$$E_{RR}(t) = E^{+}_{RR}(t) + E^{-}_{RR}(t) = \frac{1}{4\pi\varepsilon_0}\left[\frac{2}{3c^3}\dddot{\sigma}_x - \frac{4K_c}{3\pi c^2}\ddot{\sigma}_x\right]\mu \tag{6.7.26}$$

K_c is a high-frequency cutoff from the non-relativistic theory of point dipoles [6.24]. By assuming harmonic oscillation of $\sigma^-(t)$ and $\sigma^+(t)$ at frequency ω_0 and rewriting the Pauli matrix as $\sigma_x\mu = \left[\sigma^- + \sigma^+\right]\mu$, Eq. (6.7.26) can be re-expressed as

$$E_{RR}(t) \doteq \frac{i\omega_0^3}{6\pi\varepsilon_0 c^3}\left[\sigma^-(t) - \sigma^+(t)\right]\mu + \frac{K_c\omega_0^2}{3\pi^2\varepsilon_0 c^2}\left[\sigma^-(t) + \sigma^+(t)\right]\mu. \tag{6.7.27}$$

With this result in hand, we can write

$$\mu \cdot E^{+}_{RR}(t) \cong (i\beta + \gamma)\,\hbar\sigma^-(t), \tag{6.7.28}$$

where $\beta \equiv \mu^2\omega_0^3/6\pi\varepsilon_0\hbar c^3 = \gamma_{sp}/2$ and $\gamma \equiv K_c\mu^2\omega_0^2/3\pi^2\varepsilon_0\hbar c^2$. Use of Eq. (6.7.28) in Eq. (6.7.21) now gives an equation for the transition operator of the atom that includes radiation reaction:

$$\dot{\sigma}^-(t) = -i\omega_0\sigma^-(t) + i\,(i\beta + \gamma)\,\sigma^-(t) - \frac{i}{\hbar}\mu \cdot \sigma_3(t)E_0^+(t). \tag{6.7.29}$$

The result $\sigma_3\sigma^- = -\sigma^-$ has been used to simplify Eq. (6.7.29). Equations (6.7.21) and (6.7.22) can now be written in a simple form:

$$\dot{\sigma}^-(t) = -i\,(\omega_0 - \gamma - i\beta)\,\sigma^-(t) - \frac{i}{\hbar}\mu \cdot \sigma_3(t)E_0^+(t), \tag{6.7.30}$$

$$\dot{\sigma}_3(t) = -2\beta\,(1 + \sigma_3(t)) - \frac{2i}{\hbar}\mu \cdot \left[E^-(t)\sigma^-(t) - \sigma^+(t)E^+(t)\right]. \tag{6.7.31}$$

Let us return to the problem of calculating $g(\tau)$, for which we need to work with expressions written in terms of slowly varying operators $s^{\pm} = \sigma^{\pm}\exp(\mp i\omega t)$. Equations (6.7.30) and (6.7.31) become

$$\dot{s}^-(t) = -i\,(\omega_0 - \omega - i\beta)\,s^-(t) - \frac{i}{\hbar}\mu \cdot \sigma_3(t)E^+(t)\exp(i\omega t), \tag{6.7.32}$$

$$\dot{\sigma}_3(t) = -2\beta\,(1 + \langle\sigma_3(t)\rangle) + \frac{2i}{\hbar}\mu \cdot \left[s^+(t)E_0^+(t)\exp(i\omega t) - E_0^-(t)s^-\exp(-i\omega t)\right]. \tag{6.7.33}$$

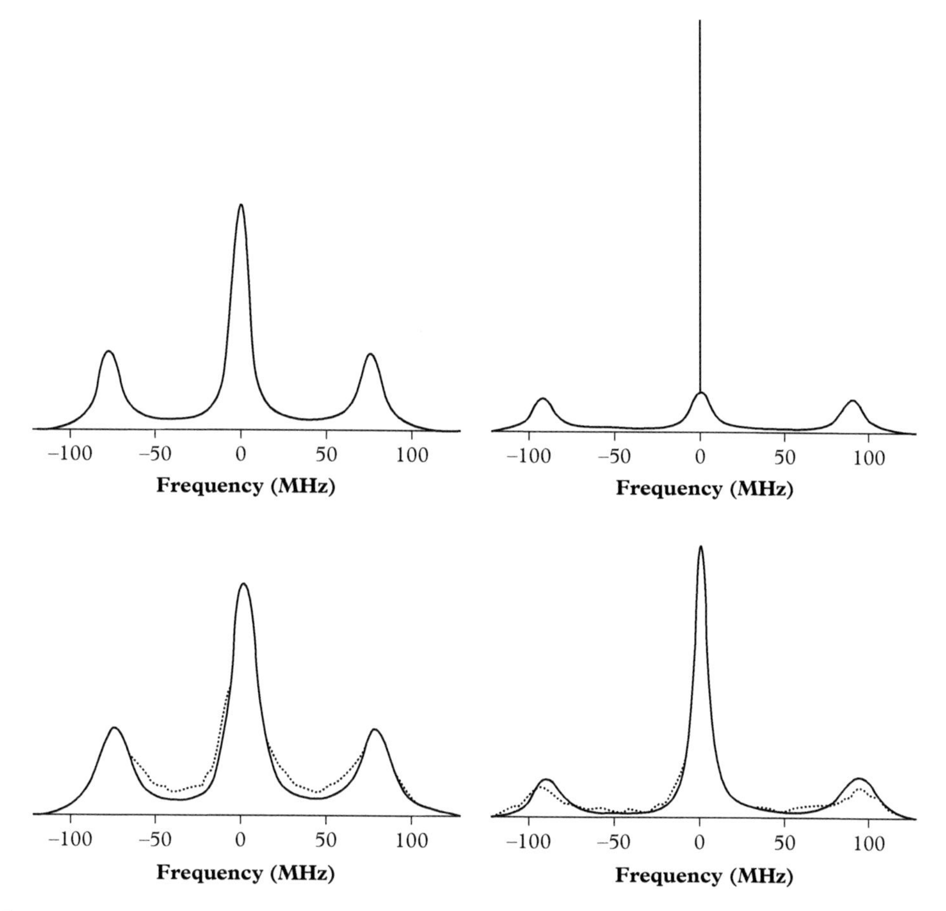

Figure 6.11 *Calculated (top) and observed (bottom) resonance fluorescence spectra of sodium atoms. The solid curves in the experimental traces are best fits to data taken exactly on resonance ($\Omega_R = \Omega$ at left) and at a finite detuning off resonance ($\Omega_R = 7.8\Omega$ at right). (After Refs. [6.16, 6.21, 6.22] (theory), [6.18–6.20] (experiment).)*

In order for the expectation value of $E_0^+(t)$ in Eq. (6.7.33) to correctly yield the amplitude of a coherent state of the form $<E_0(t)> = E_0 \sin \omega t$, we assume that

$$E_0^+(t)\,|\psi\rangle = \frac{i}{2}E_0 \exp(-i\omega t)\,|\psi\rangle\,. \tag{6.7.34}$$

Then the equation of motion for $g(\tau)$ becomes

$$\begin{aligned}\frac{\partial}{\partial\tau}g(\tau) &= -i(\omega_0-\omega-i\beta)g(\tau) + \frac{\mu E_0}{2\hbar}\left\langle s^+(t_0)\sigma_3(t+\tau)\right\rangle \\ &= -i(\omega_0-\omega-i\beta)g(\tau) + \frac{\Omega_R}{2}h(\tau),\end{aligned} \tag{6.7.35}$$

where we have defined

$$h(\tau) \equiv \langle s^+(t_0)\sigma_3(t_0+\tau)\rangle. \tag{6.7.36}$$

From (6.7.31) we can also work out an equation of motion for $h(\tau)$, the second correlation function to appear in our expressions:

$$\begin{aligned}\frac{\partial}{\partial\tau}h(\tau) &= -2\beta\,\langle s^-(t_0)\rangle - \Omega_R\langle s^+(t_0)s^+(t_0+\tau)\rangle + \\ &\quad -(2i\mu/\hbar)\,\langle s^+(t_0)E_0^-(t_0+\tau)s^-(t_0+\tau)\rangle \exp[-i\omega(t_0+\tau)].\end{aligned} \tag{6.7.37}$$

It can be shown [6.9] that the commutator of atom and field operators in Eq. (6.7.37) is zero for delay times $\tau \gg \omega_0^{-1}$. That is, $[s^+(t_0), E_0^-(t_0+\tau)] \approx 0$. So the last term in Eq. (6.7.37) simplifies to

$$-\Omega_R\langle s^+(t_0)s^-(t_0+\tau)\rangle = -\Omega_R g(\tau)\,. \tag{6.7.38}$$

Hence the equation for the second of our correlation functions, $h(\tau)$, becomes

$$\begin{aligned}\frac{\partial}{\partial\tau}h(\tau) &= -2\beta h(\tau) - 2\beta\langle s^+(t_0)\rangle - \Omega_R\langle s^+(t_0)s^+(t+\tau)\rangle - \Omega_R g(\tau) \\ &= -2\beta h(\tau) - 2\beta\langle s^+(t_0)\rangle - \Omega_R f(\tau) - \Omega_R g(\tau).\end{aligned} \tag{6.7.39}$$

In Eq. (6.7.39) we encounter a third and final correlation function, defined by

$$f(\tau) \equiv \langle s^+(t_0)s^+(t_0+\tau)\rangle\,. \tag{6.7.40}$$

The equation of motion for $f(\tau)$, the last needed to solve the resonance fluorescence problem, is obtained from Eq. (6.7.32):

$$\begin{aligned}\frac{\partial}{\partial\tau}f(\tau) &= i(\omega_0-\omega+i\beta)f(\tau) + i(\mu/\hbar)\,\langle s^+(t_0)E_0^-(t_0+\tau)\rangle \exp[-i\omega(t_0+\tau)] \\ &= i(\omega_0-\omega+i\beta)f(\tau) + (\Omega_R/2)h(\tau).\end{aligned} \tag{6.7.41}$$

Finally, with Eqs. (6.7.35), (6.7.39), and (6.7.41), we have a closed set of equations for the correlation functions $g(\tau)$, $h(\tau)$, and $f(\tau)$ which collectively determine the emission spectrum. Gathering them together, we have

$$\left[\frac{\partial}{\partial\tau} + i\,(\Delta - i\beta)\right]g(\tau) = \frac{\Omega_R}{2}h(\tau), \tag{6.7.42}$$

$$\left[\frac{\partial}{\partial\tau} + 2\beta\right]h(\tau) = 2\beta\langle s^+(t_0)\rangle - \Omega_R g(\tau) - \Omega_R f(\tau), \tag{6.7.43}$$

$$\left[\frac{\partial}{\partial\tau} - i\,(\Delta + i\beta)\right]f(\tau) = \frac{\Omega_R}{2}h(\tau), \tag{6.7.44}$$

where $\Delta \equiv \omega_0 - \omega$. We must solve these equations subject to the initial conditions

$$g(0) = \left\langle s^+(t_0)s^-(t_0)\right\rangle, \tag{6.7.45}$$

$$h(0) = \left\langle s^+(t_0)\sigma_3(t_0)\right\rangle, \tag{6.7.46}$$

$$f(0) = \left\langle s^+(t_0)s^+(t_0)\right\rangle = 0. \tag{6.7.47}$$

Values of the initial correlations $g(0) = \left\langle s^+s^-\right\rangle = \frac{1}{2}\left(1 + \langle\sigma_3(t_0)\rangle\right)$ and $h(0) = \left\langle s^+\sigma_3\right\rangle = \left\langle s^+(t_0)\right\rangle$ can be found using the algebra of the Pauli spin operators. It is also easy to work out $\left\langle s^+(t_0)\right\rangle$ and $\langle\sigma_3(t_0)\rangle$ from Eqs. (6.7.32) and (6.7.33), by taking expectation values of each equation and setting the time derivatives equal to zero for steady-state solutions. This yields the results

$$g(0) = \left(\Omega_R^2/4\left[\Delta^2 + \frac{1}{2}\beta^2\Omega_R^2\right]\right), \tag{6.7.48}$$

$$h(0) = \Omega_R(\beta + i\Delta)/2[\Delta^2 + \beta^2 + (\Omega_R^2/2)]. \tag{6.7.49}$$

The results of these calculations are summarized in Figures 6.11, 6.12, and the following two subsections.

6.7.1.1 *On Resonance*

For $\Delta = 0$ at high intensities ($\Omega_R \gg \beta$), the solution of Eq. (6.7.42) that takes into account the condition (6.7.45) is

$$g(\tau) \cong \frac{1}{2}\left[\exp(-\beta\tau) + \exp(-3\beta\tau/2)\cos\Omega_R\tau\right] + (\beta/\Omega_R)^2. \tag{6.7.50}$$

The power spectrum of resonance fluorescence is obtained from the Fourier transform of the first-order correlation function as expressed by Eq. (6.7.19):

$$P(\omega) = \frac{3\beta/8}{(\omega-\omega_L-\Omega_R)^2 + 9\beta^2/4} + \frac{\beta/2}{(\omega-\omega_L)^2+\beta^2} + \frac{3\beta/8}{(\omega-\omega_L+\Omega_R)^2+9\beta^2/4} + 2\pi\left(\beta/\Omega_R\right)^2\delta\left(\omega-\omega_L\right). \tag{6.7.51}$$

This is the exact expression for the spectrum of resonance fluorescence. In it, the frequency of the laser used to excite the atom has been relabeled $\omega_L = \omega''$. At high fields ($\Omega \gg \beta$) the spectrum consists of a delta function plus a three-component spectrum shown in the plots of Figure 6.11. At low fields ($\Omega \ll \beta$) it consists of the delta function alone, as discussed by Heitler [6.25]. While the three-line, resonance fluorescence spectrum was first observed over thirty years ago [6.18–6.20], it is only recently that these features have been reported in the resonance fluorescence of individual molecules [6.26].

6.7.1.2 *Off Resonance*

If incident light is detuned from resonance, the scattered light spectrum has a significant dependence on bandwidth. When the resonance is displaced beyond the weak tail of the incident light spectral distribution, scattering only takes place at the incident frequency. In Figure 6.12, this results in an intense narrow peak at the laser frequency ω_L. In the foreground of the figure, the laser is assumed to have negligible bandwidth compared to the atomic linewidth and it is detuned by five times the transition linewidth. In traces that move progressively toward the back of the figure, the bandwidth of the incident light increases. Because the excitation begins to overlap the resonance frequency more and more as its bandwidth grows, the fluorescent emission that follows absorption by the atom grows in intensity. The atomic emission is broadened by the radiative lifetime and centered on the resonant frequency ω_0.

6.7.2 Coherence of Strongly Driven Two-Level Atoms

The general solution for the first-order correlation function $g^{(1)}(\tau)$ in Eq. (6.7.50) provides the quantum degree of first-order coherence of two-level atoms driven strongly

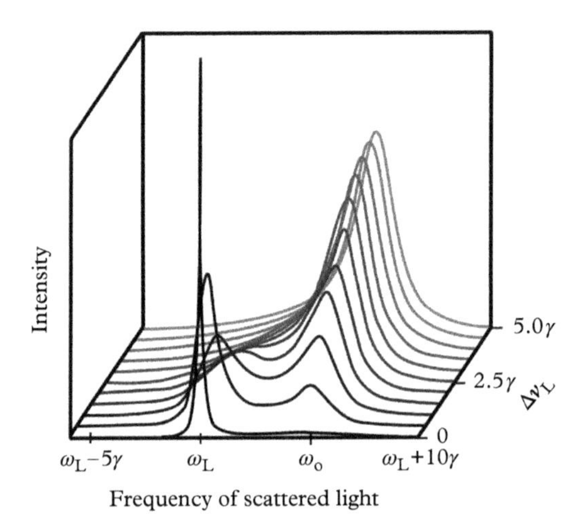

Figure 6.12 *Near-resonance fluorescence spectrum at low intensity as a function of the bandwidth $\Delta\nu_L$ of incident light. The frequency of the incident light is detuned from resonance by five natural linewidths ($\Delta = \omega_0 - \omega = 5\gamma$). The triplet spectrum is absent. (Illustration adapted from [6.27].)*

on resonance. This function was plotted earlier as the solid curve in Figure 6.7 and is the correlation function of the electric fields. The calculation of the second-order correlation function $g^{(2)}(\tau)$ of driven atoms, which is the correlation of intensities, requires additional considerations.

Resonance fluorescence is radiation from electric dipole oscillations of driven atoms. Hence it is determined by correlation properties of the atomic transition operators $\hat{\sigma}^{\pm}$, not the field operators $\hat{E}^{\pm}$. These operators determine both the first-order coherence as specified by Eq. (6.7.18) and the second-order coherence as follows. Since the atomic operators are already normalized, the quantum degree of second-order coherence is simply

$$\begin{aligned} g^{(2)}(\tau) &= \langle \hat{\sigma}^+(t_0)\hat{\sigma}^+(t_0+\tau)\hat{\sigma}^-(t_0+\tau)\hat{\sigma}^-(t_0)\rangle \\ &= \langle s^+(t_0)s^+(t_0+\tau)s^-(t_0+\tau)s^-(t_0)\rangle . \end{aligned} \tag{6.7.52}$$

This expression may be reduced, using properties of the Pauli matrices:

$$\begin{aligned} s^+(t_0+\tau)s^-(t_0+\tau) &= \sigma^+(t_0+\tau)\sigma^-(t_0+\tau) \\ &= \frac{1}{2} + \sigma_z(t_0+\tau). \end{aligned} \tag{6.7.53}$$

Substitution of Eq. (6.7.53) into Eq. (6.7.52) then results in the expression

$$g^{(2)}(\tau) = \frac{1}{2}g^{(1)}(0) + \frac{1}{2}G(\tau), \tag{6.7.54}$$

where $g^{(1)}(0)$ is given by Eq. (6.7.50) and the remaining correlation function,

$$G(\tau) \equiv \langle s^+(t_0)\sigma_z(t_0+\tau)s^-(t_0)\rangle , \tag{6.7.55}$$

is still to be determined.

Just as in Section 6.7.1, where a set of differential equations for three related correlation functions was solved to obtain $g(\tau)$, the solution for $g^{(2)}(\tau)$ can be found by solving a set of equations for the three inter-related correlation functions:

$$F(\tau) \equiv \langle s^+(t_0)s^+(t_0+\tau)s^-(t_0)\rangle , \tag{6.7.56}$$

$$G(\tau) \equiv \langle s^+(t_0)\sigma_z(t_0+\tau)s^-(t_0)\rangle , \tag{6.7.57}$$

$$H(\tau) \equiv \langle s^+(t_0)s^-(t_0+\tau)s^-(t_0)\rangle , \tag{6.7.58}$$

which are determined by the operators s^+, s^-, and σ_z. From the Heisenberg equations of motion for the various atomic operators, it can be shown that

$$\frac{\partial F(\tau)}{\partial \tau} = i(\Delta + i\beta)F(\tau) + (\Omega_R/2)G(\tau), \tag{6.7.59}$$

$$\frac{\partial G(\tau)}{\partial \tau} = -2\beta g(0) - 2\beta G(\tau) - \Omega_R(F(\tau) + H(\tau)), \tag{6.7.60}$$

$$\frac{\partial H(\tau)}{\partial \tau} = -i(\Delta - i\beta)H(\tau) + (\Omega_R/2)G(\tau). \tag{6.7.61}$$

These equations may be solved [6.28], using the initial conditions

$$F(0) = \langle s^+(t_0)s^+(t_0)s^-(t_0)\rangle = 0, \tag{6.7.62}$$

$$G(0) = \langle s^+(t_0)\sigma_z(t_0)s^-(t_0)\rangle = -\langle s^+(t_0)s^-(t_0)\rangle = -g(0), \tag{6.7.63}$$

$$H(0) = \langle s^+(t_0)s^-(t_0)s^-(t_0)\rangle = 0, \tag{6.7.64}$$

to find an exact solution for $g^{(2)}(\tau)$. At resonance ($\Delta = 0$), the result is

$$g^{(2)}(\tau) = g(0)^2\left[1 - \exp\left(\frac{-3\beta\tau}{2}\right)\left(\cos(\Omega''\tau) + \frac{3\beta}{2\Omega''}\sin(\Omega''\tau)\right)\right] \tag{6.7.65}$$

where

$$\Omega'' \equiv \sqrt{\Omega_R^2 - (\beta^2/4)}. \tag{6.7.66}$$

For $\tau = 0$, the quantum degree of second-order coherence clearly takes on non-classical values, since $g^{(2)}(0) = 0$. This quantum effect was evident in the earlier plot of this function for a single atom presented in Figure 6.8(a). There is an "anti-bunching" effect in the emission rate of fluorescence photons from single atoms. The photon statistics of driven atoms reflects their internal quantum structure and their inability to emit a second photon immediately after a first. A finite time is required for re-excitation of an atom before it can emit again, automatically making it a "non-classical" source of light.

6.8 Dressed Atom Theory

6.8.1 Strong Coupling of Atoms to the Electromagnetic Field

An elegant approach to the analysis of resonance fluorescence that is quite a bit simpler than that presented in Section 6.7 is possible by diagonalizing the complete Hamiltonian, including the light field [6.29]. This method is referred to as dressed atom theory.

Consider a quantum system consisting of a two-level atom in a near-resonant radiation field. Initially we neglect the interaction between the two parts of the system, and identify the "bare" states of the uncoupled system. Written as the product of atomic and number states, these are

$$|1,n\rangle = |1\rangle|n\rangle \tag{6.8.1}$$

and

$$|2,n-1\rangle = |2\rangle|n-1\rangle . \tag{6.8.2}$$

These states are eigenstates of the atom-plus-field Hamiltonian $H_{atom} + H_{field}$ with eigenenergies

$$E_{1,n} = -\frac{\hbar\omega_0}{2} + n\hbar\omega, \tag{6.8.3}$$

$$E_{2,n-1} = \frac{\hbar\omega_0}{2} + (n-1)\,\hbar\omega, \tag{6.8.4}$$

and a transition energy of

$$E_{2,n-1} - E_{1,n} = \hbar\,(\omega_0 - \omega) = \hbar\Delta. \tag{6.8.5}$$

The two uncoupled states have the same energy (are degenerate) when $\Delta = 0$.

Next we introduce the interaction between the field and atom, and attempt to find an exact description of the system in terms of coupled states, one which takes the atom–field interaction into account. These states must be constructed to be eigenstates of the entire Hamiltonian:

$$H = H_{atom} + H_{field} + H_{\text{int}}. \tag{6.8.6}$$

The photons may be pictured as "dressing" the atom, and near resonance or at very high intensities it is impossible to distinguish which photons interact with the atom, or how many interactions take place. The available states are different from those of the bare atom by virtue of the strong interaction H_{int}. The task is therefore to find new coupled or dressed states $|D\rangle$ of the system,

$$|D\rangle = c_1|1,n\rangle + c_2|2,n-1\rangle , \tag{6.8.7}$$

which satisfy the energy equation

$$H\,|D\rangle = \left(H_{atom} + H_{field} + H_{\text{int}}\right)|D\rangle = E_D|D\rangle . \tag{6.8.8}$$

Now $|1,n\rangle$ and $|2,n-1\rangle$ are orthonormal, so upon substitution of Eq. (6.8.7) into Eq. (6.8.8) we immediately obtain

$$c_1\langle 1,n\,|H|\,1,n\rangle + c_2\langle 1,n\,|H|\,2,n-1\rangle = E_D c_1, \tag{6.8.9}$$

$$c_1\langle 2,n-1\,|H|\,1,n\rangle + c_2\langle 2,n-1\,|H|\,2,n-1\rangle = E_D c_2. \tag{6.8.10}$$

By explicit evaluation of the matrix elcment using the Hamiltonian of Eq. (6.2.10), we find

$$\langle 2, n-1 \,|H|\, 1, n\rangle = \hbar \left\langle 2, n-1 \left| g\hat{a}\hat{\sigma}^{+} + g^{*}\hat{a}^{+}\hat{\sigma}^{-} \right| 1, n \right\rangle = \hbar g\sqrt{n}, \tag{6.8.11}$$

and the two equations (6.8.9) and (6.8.10) simplify to

$$c_1\,(E_1 - E_D) + c_2\hbar g^{*}\sqrt{n} = 0, \tag{6.8.12}$$

$$c_1\hbar g\sqrt{n} + c_2\,(E_2 - E_D) = 0. \tag{6.8.13}$$

The solution of the secular equation for coupled Eqs. (6.8.12) and (6.8.13) then yields

$$\begin{aligned} E_D &= \frac{1}{2}\,(E_1 + E_2) \pm \frac{1}{2}\sqrt{(E_1 + E_2)^2 - 4\left(E_1E_2 - \hbar^2\,|g|^2\,n\right)} \\ &= \left(n - \frac{1}{2}\right)\hbar\omega \pm \frac{1}{2}\hbar\sqrt{\Delta^2 + 4\,|g|^2\,n} \\ &= \left(n - \frac{1}{2}\right)\hbar\omega \pm \frac{1}{2}\hbar\Omega_R, \end{aligned} \tag{6.8.14}$$

where $\Omega_R \equiv \sqrt{\Delta^2 + 4\,|g|^2\,n}$ is the generalized Rabi frequency. The quantities E_D are the eigenvalues of the dressed states depicted in Figure 6.13. To find the corresponding eigenstates$|D\rangle$, it is convenient to define

$$\sin 2\theta = \frac{2\,|g|\,\sqrt{n}}{\Omega_R} \equiv \frac{\Omega}{\Omega_R}, \tag{6.8.15}$$

$$\cos 2\theta \equiv \frac{\Delta}{\Omega_R}. \tag{6.8.16}$$

With $E_D = E_{+}(n)$, we proceed to solve for the coefficients c_1 and c_2, obtaining

$$\frac{1}{2}c_1\,(\Delta + \Omega_R) = c_2 g^{*}\sqrt{n}, \tag{6.8.17}$$

$$\frac{1}{4}\,|c_1|^2\,(\Delta + \Omega_R)^2 = |c_2|^2\,|g|^2\,n. \tag{6.8.18}$$

Exercise: Show that by imposing the normalization condition $|c_1|^2 + |c_2|^2 = 1$, Eq. (6.8.18) can be solved to yield the following expressions for the coefficients:

$$|c_1|^2 = \sin^2\theta, \tag{6.8.19}$$

$$|c_2|^2 = \cos^2\theta, \tag{6.8.20}$$

and that these expressions are consistent with Eqs. (6.8.15) and (6.8.16).

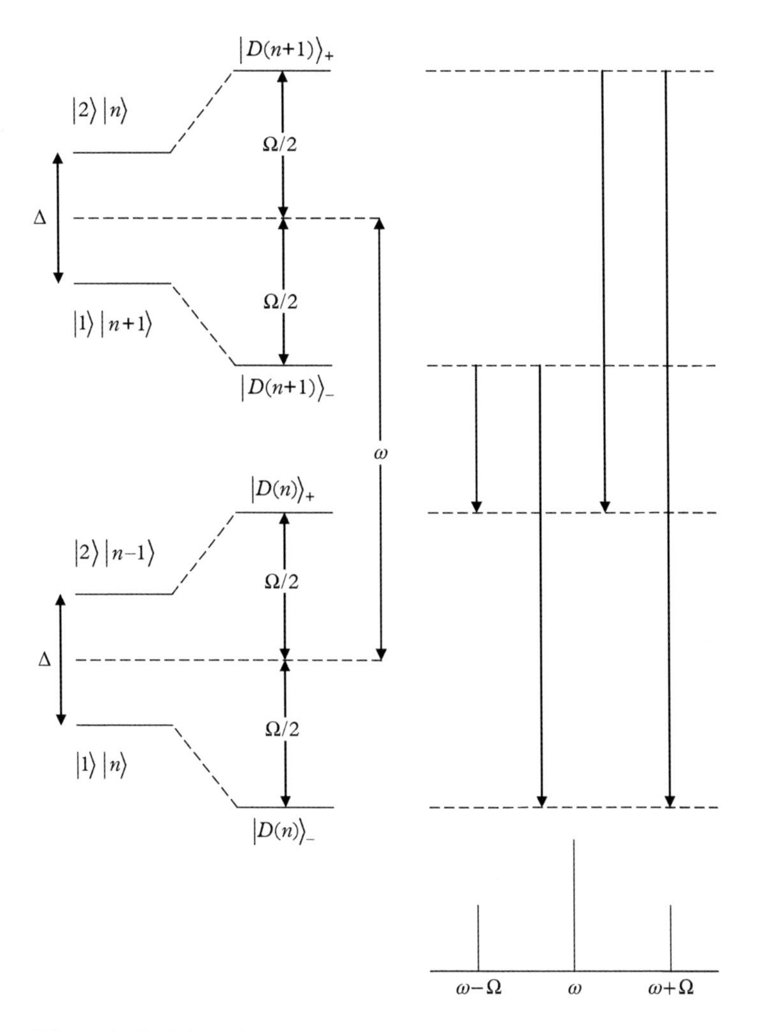

Figure 6.13 *Schematic depiction of dressed atom energy levels and the transitions between them. The resonance fluorescence spectrum predicted by the model is shown in the lower right, directly beneath the transitions giving rise to it. (After Ref. [6.11].)*

Given the results in Eqs. (6.8.19) and (6.8.20), the first dressed state can be written as:

$$|D(n)\rangle_+ = \sin\theta\,|1,n\rangle + \cos\theta\,|2,n-1\rangle\,. \tag{6.8.21}$$

With $E_D = E_-(n)$, we find a second solution:

$$|c_1|^2 = \cos^2\theta, \tag{6.8.22}$$

$$|c_2|^2 = \sin^2\theta. \tag{6.8.23}$$

In order for $|D(n)\rangle_-$ to be orthogonal to $|D(n)\rangle_+$, the signs of the coefficients must be chosen such that

$$|D(n)\rangle_- = \cos\theta\,|1,n\rangle - \sin\theta\,|2,n-1\rangle\,. \tag{6.8.24}$$

Notice that the dressed states $|D\rangle_+$, $|D\rangle_-$ are not degenerate on resonance, the way the uncoupled states were. For $\Delta = 0$, we now find $\Omega_R = \Omega$ $(2\theta = \pi/2)$ and

$$E_+(n) - E_-(n) = \hbar\Omega, \tag{6.8.25}$$

$$|D(n)\rangle_\pm = \frac{1}{\sqrt{2}}\left[|1,n\rangle \pm |2,n-1\rangle\right]. \tag{6.8.26}$$

Based on Eq. (6.8.26), a complex picture of the states and energies of the atom+field system now emerges. To begin with, there is an infinite ladder of dressed states. To see this, consider two bare states that are slightly different from the initial basis states taken previously. If we had started with states

$$|1,n+1\rangle = |1\rangle\,|n+1\rangle\,, \tag{6.8.27}$$

$$|2,n\rangle = |2\rangle\,|n\rangle\,, \tag{6.8.28}$$

differing from our original bare states by the presence of an added photon in the particular mode of interest of the radiation field, we would have found dressed states $|D(n+1)\rangle_+$, $|D(n+1)\rangle_-$ with energies

$$E_\pm(n+1) = \left[(n+1) - \frac{1}{2}\right]\hbar\omega \pm \frac{1}{2}\hbar\Omega_R. \tag{6.8.29}$$

Because there are infinitely many possible n values, there are infinitely many dressed states. These states may be thought of as excited states of the dressed atom, and discrete transitions can take place between these levels, as shown in Figures 6.13 and 6.14. The predicted resonance fluorescence spectrum, shown at lower right in Figure 6.13, is a triplet, in agreement with the results of Section 6.7.

6.8.2 Dressed State Population Dynamics

Using the dressed state formalism we can readily check that the transitions between dressed states depicted in Figure 6.13 have non-zero transition moments and do indeed produce electric dipole radiation that accounts for the Mollow triplet spectrum. Recalling that the atomic dipole operator $\hat{\mu} = \mu(\hat{\sigma}^+ + \hat{\sigma}^-)$ only acts on the atomic part of the atom–field product states, the moment between the upper states of two adjacent dressed state manifolds may be calculated as follows, for example,

$$\begin{aligned}\left\langle D_+(n)\left|e\hat{r}\right|D_+(n+1)\right\rangle &= \mu\left\{\left[\langle 1,n|\sin\theta + \langle 2,n-1|\cos\theta\right]\left(\sigma+\sigma^+\right)\left[\sin\theta\,|1,n+1\rangle + \cos\theta\,|2,n\rangle\right]\right\}\\ &= \mu\left\{\langle n|\,\langle 1|\,(\sin\theta)\sigma^-(\cos\theta)\,|2\rangle\,|n\rangle\right\}\\ &= \frac{1}{2}\mu\sin 2\theta \neq 0.\end{aligned} \tag{6.8.30}$$

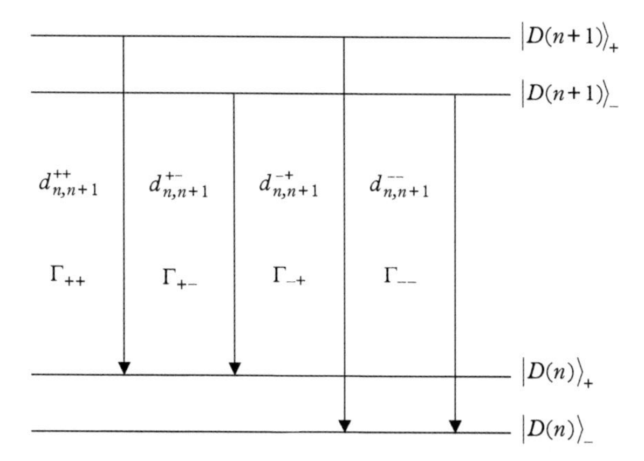

Figure 6.14 *Nomenclature of dipole moments d and transition rates Γ connecting specific adjacent pairs of dressed states during fluorescent transitions.*

Transitions are possible in the dressed state system whenever the dipole moment is non-zero. Since the field has already been incorporated into the electronic structure of the system, radiative transitions take place at a rate Γ_{ij} that is simply proportional to the square of the matrix element of the electric dipole moment between initial and final dressed states D_i and D_j (following Fermi's golden rule):

$$\Gamma_{ij} = \left|\langle D(n_i)|\, e\hat{r}\, |D(n_j)\rangle\right|^2 . \tag{6.8.31}$$

To analyze the transition rates among the dressed states shown in Figure 6.14, which in turn determine the intensities of features in the resonance fluorescence spectrum, all four matrix elements $d^{\pm\pm}_{ij} = \langle D_\pm(n_i)|\, e\hat{r}\, |D_\pm(n_j)\rangle$ are needed. We calculate them explicitly in the following exercises and then proceed to develop rate equations based on knowledge of the relative rates Γ_{ij}.

Exercise: Show that the expected dipole moment $d^{\pm\pm}_{fi} = \langle D_\pm(n_f)|\, e\hat{r}\, |D_\pm(n_i)\rangle$ between initial and final states $|D(n+1)\rangle_+$ and $|D(n)\rangle_+$ differs from that between $|D(n+1)\rangle_+$ and $|D(n)\rangle_-$, and that both have the non-zero values given by:

$$d^{++}_{n,n+1} = \mu\,\{\langle 1,n|\sin\theta + \langle 2,n-1|\cos\theta\}\left(\sigma + \sigma^+\right)\{\sin\theta\,|1,n+1\rangle + \cos\theta\,|2,n\rangle\}$$
$$= \mu \sin\theta\cos\theta, \tag{6.8.32}$$

$$d^{-+}_{n,n+1} = \mu\,\{\langle 1,n|\cos\theta - \langle 2,n-1|\sin\theta\}\left(\sigma + \sigma^+\right)\{\sin\theta\,|1,n+1\rangle + \cos\theta\,|2,n\rangle\}$$
$$= \mu\cos^2\theta. \tag{6.8.33}$$

Exercise: Show that the remaining two transition dipole moments between states $|D(n+1)\rangle_-$ and $|D(n)\rangle_+$ and between $|D(n+1)\rangle_-$ and $|D(n)\rangle_-$ are:

$$d_{n,n+1}^{+-} = -\mu \sin^2\theta. \tag{6.8.34}$$

$$d_{n,n+1}^{--} = -\mu \sin\theta\cos\theta, \tag{6.8.35}$$

The populations in various dressed states can be assessed by calculating diagonal elements of the density matrix $\rho = |D\rangle\langle D|$ and solving rate equations similar to Eqs. (5.1.10) and (5.1.11). Populations of the various dressed states will be denoted by

$$\pi_\pm(n) = \langle D_\pm(n)|\,\rho\,|D_\pm(n)\rangle\,. \tag{6.8.36}$$

Changes in the population of a particular dressed state via one-photon transitions can occur in two ways. Relaxation into a given level involves two transitions down from the upper neighboring manifold and relaxation out of the level takes place by two transitions down to the next neighboring manifold. Hence the population rate equations are very simple:

$$\dot{\pi}_-(n) = -\left(\Gamma_{--} + \Gamma_{+-}\right)\pi_-(n) + \Gamma_{--}\,\pi_-(n+1) + \Gamma_{-+}\,\pi_+(n+1), \tag{6.8.37}$$

$$\dot{\pi}_+(n) = -\left(\Gamma_{-+} + \Gamma_{++}\right)\pi_+(n) + \Gamma_{++}\pi_+(n+1) + \Gamma_{+-}\,\pi_-(n+1). \tag{6.8.38}$$

If we assume the illumination is intense, there is little difference between occupation values of number states labeled with $n-1$, n, or $n+1$. Hence the same may be said of neighboring dressed state manifolds, and we can write $\pi_\pm(n+1) \cong \pi_\pm(n) = p_0(n)\pi_\pm$, where $p_0(n)$ is the distribution of photons among the available modes. Adjacent dressed states connected by one-photon transitions then all obey the same equations:

$$\dot{\pi}_- = -\Gamma_{+-}\pi_- + \Gamma_{-+}\pi_+, \tag{6.8.39}$$

$$\dot{\pi}_+ = -\Gamma_{-+}\pi_+ + \Gamma_{+-}\pi_-. \tag{6.8.40}$$

In steady state, we can solve for the populations themselves by setting $\dot{\pi} = 0$ in Eqs. (6.8.39) and (6.8.40) and making use of the condition

$$\Gamma_{+-}\pi_-(\infty) = \Gamma_{-+}\pi_+(\infty), \tag{6.8.41}$$

together with the closure relation

$$\pi_-(\infty) + \pi_+(\infty) = 1. \tag{6.8.42}$$

The results are

$$\pi_+(\infty) = \frac{\Gamma_{+-}}{\Gamma_{-+} + \Gamma_{+-}} = \frac{\sin^4\theta}{\sin^4\theta + \cos^4\theta}, \tag{6.8.43}$$

$$\pi_-(\infty) = \frac{\Gamma_{-+}}{\Gamma_{-+} + \Gamma_{+-}} = \frac{\cos^4\theta}{\sin^4\theta + \cos^4\theta}. \tag{6.8.44}$$

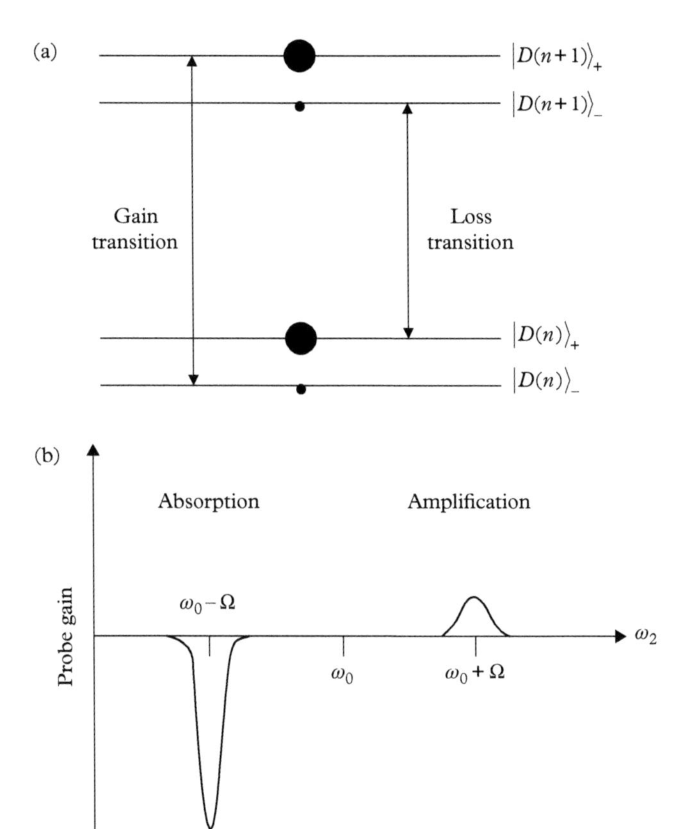

Figure 6.15 *(a) Population differences among dressed states at finite detunings of the pump wave. The size of the filled black circle indicates the relative size of the population. (b) Schematic illustration of the absorption spectrum of a near-resonant probe versus frequency ω_2 of the coupling laser. (After Ref. [6.30].)*

Exactly on resonance ($\theta = \pi/4$) the populations of the upper (+) and lower (–) dressed states are equal. However, for finite detuning of the pump wave the populations may differ (Figure 6.15(a)). In particular, at detunings equal to plus or minus the Rabi frequency, the population difference may be positive or negative. The result is that continuous gain or loss may be experienced by a probe wave passing through the pumped region of a nominally two-level system (Figure 6.15(b)). Because this is a steady-state result, it appears to violate the prescription of thermodynamics that in steady state the excited state population of a two-level system can never exceed that of the ground state. The resolution of this dilemma is that with near resonant excitation the dressed state model no longer consists of just two atom–field levels. It becomes a multi-level system.

Since n is not a good quantum number of the dressed states, the number of photons n in the field is indeterminate. Similarly, it is not known whether the atom is in bare state $|1\rangle$ or $|2\rangle$ at any time, because these states are no longer eigenstates of the coupled system.

Field and atom cannot be considered separate entities near resonance. Transitions of the driven system would normally proceed by stimulated emission and absorption processes, but we have integrated the field with the atom so intimately that we may speak of resonance fluorescence as being the spontaneous emission of dressed atoms.

In summary, near-resonant electromagnetic fields act to couple or "dress" the uncoupled or "bare" states of the atom and field, and effect a significant change in its eigenstates and eigenenergies. The modification of the atomic electronic structure by light is important enough that neither the state of the field (specified purely by the number of photons) nor the state of the atom (specifying which of the original states of the atom is occupied) is any longer well defined. The coupled atom–field system can exhibit optical gain, despite the fact that sustained population inversions of true two-level systems are thermodynamically impossible.

REFERENCES

[6.1] I. Bialynicka-Birula and Z. Bialynicka-Birula, *Phys. Rev. Lett.* 2012, **108**, 140401.
[6.2] M. Sargent, M.O. Scully, and W.E. Lamb, *Laser Physics*. Addison-Wesley, 1974, pp.236–40.
[6.3] J.J. Sakurai, *Advanced Quantum Mechanics*. Menlo Park, CA: Benjamin-Cummings Publ. Co., 1967, Section 2–8.
[6.4] S. Stenholm, *Foundations of Laser Spectroscopy*. New York: J. Wiley & Sons, 1984.
[6.5] A.S. Davydov, *Quantum Mechanics*, 2nd ed. Pergamon Press, 1976, pp.134–5.
[6.6] R.J. Glauber, *Phys. Lett.* 1966, **21**, 650.
[6.7] C. Caves, *Phys. Rev. D* 1981, **23**, 1693.
[6.8] See special issue: Squeezed states of the electromagnetic field. *J. Opt. Soc. Am. B* 1987, **4**.
[6.9] R. Loudon, *The Quantum Theory of Light*, 2nd ed. London: Oxford University Press, 1983; also 3rd ed. London: Oxford University Press, 2000.
[6.10] R. Hanbury-Brown and R.Q. Twiss, *Nature* 1956, **127**, 27.
[6.11] M. Weissbluth, *Photon–Atom Interactions*. Academic Press, 1989.
[6.12] D. Kleppner, *Phys. Rev. Lett.* 1981, **47**, 233.
[6.13] E.M. Purcell, *Phys. Rev.* 1946, **69**, 681.
[6.14] M.O. Scully and M.S. Zubairy, *Quantum Optics*. Cambridge: Cambridge University Press, 1997.
[6.15] P. Lodahl, A.F. van Driel, I.S. Nikolaev, A. Irman, K. Overgagg, D. Vanmaekelbergh, and W.L. Vos, *Nature* 2004, **430**, 654.
[6.16] B.R. Mollow, *Phys. Rev.* 1969, **188**, 1969.
[6.17] P.L. Knight and P.W. Milonni, *Phys. Rep.* 1980, 66, 21.
[6.18] R.E. Grove, F.Y. Wu, and S. Ezekiel, *Phys. Rev. Lett.* 1975, **35**, 1426.
[6.19] F. Schuda, C.R. Stroud, and M. Hercher, *J. Phys. B: At. Mol. Phys.* 1975, 7, L198.
[6.20] W. Hartig, W. Rasmussen, R. Schieder, and H. Walther, *Z. Phys. A* 1976, **278**, 205.
[6.21] A.L. Newstein, *Phys. Rev.* 1968, **167**, 89.
[6.22] A.L. Burshtein, *Sov. Phys. JETP* 1966, **22**, 939.
[6.23] Further discussion can be found in M. Born and E. Wolf, *Principles of Optics*, 7th ed. Cambridge: Cambridge University Press, 1999, pp.566–9.

[6.24] See, for example, A.D. Jackson, *Classical Electrodynamics*, 2nd ed. New York: Wiley, 1975, pp.780–6.
[6.25] W. Heitler, *Quantum Theory of Radiation*, 3rd ed. Oxford: Oxford University Press, 1954.
[6.26] G. Wrigge, I. Gerhardt, J. Hwang, G. Zumofen, and V. Sandoghdar, *Nat. Phys.* 2008, **4**, 60.
[6.27] P.L. Knight, W.A. Molander, and C.R. Stroud Jr., *Phys. Rev. A* 1978, **17**, 1547.
[6.28] P. Meystre and M. Sargent, *Elements of Quantum Optics*, 3rd ed. New York: Springer-Verlag, 1999.
[6.29] C. Cohen-Tannoudji and S. Reynaud, in J.H. Eberly and P. Lambropoulos (eds.), *Multiphoton Processes*. New York: John Wiley & Sons, 1977, pp.103–18.
[6.30] F.Y. Wu, S. Ezekiel, M. Ducloy, and B.R. Mollow, *Phys. Rev. Lett.* 1977, **38**, 1077.

PROBLEMS

6.1. In formal solutions of the Schrödinger equation one sometimes encounters functions of operators that do not commute. Then, identities such as the Weyl relation are useful.

(a) Show that

$$\exp\left(\hat{c} + \hat{d} + \frac{1}{2}\left[\hat{c}, \hat{d}\right]\right) = \exp(\hat{c})\exp(\hat{d})$$

is valid for any pair of operators $\hat{c}$ and $\hat{d}$ which commute with their own commutator according to

$$\left[\hat{c}, \left[\hat{c}, \hat{d}\right]\right] = \left[\hat{d}, \left[\hat{c}, \hat{d}\right]\right] = 0.$$

(b) Use the result of part (a) to show that coherent states can be written in the compact form

$$|\alpha\rangle = \hat{\alpha}\,|0\rangle\,,$$

where $\hat{\alpha} \equiv \exp\left(\alpha\hat{a}^{+} - \alpha * \hat{a}\right)$ transforms the ground state into state $|\alpha\rangle$.

(c) Show that $\hat{\alpha}$ is in fact a displacement operator with the property that

$$\hat{\alpha}^{+}\hat{a}\hat{\alpha} = \hat{a} + \alpha,$$

using the result of part (a) and the commutation relation for field operators $\hat{a}$ and $\hat{a}^{+}$.

6.2. *This problem shows that when an external force acts on a quantized simple harmonic oscillator initially in the ground state, the system can be left in an excited state, with energy determined by the Fourier component $g(\omega)$ of the driving force that is in resonance* *with the unforced or "free" oscillator. The state that is excited in this way is a minimum uncertainty state of constant width that oscillates harmonically incoordinate and momentum space.*

The Hamiltonian of a forced linear harmonic oscillator

$$H = \frac{\rho^2}{2\mu} + \frac{1}{2}\mu\omega^2 q^2 - qF(t) - \rho G(t)$$

can be written in the form

$$H = \hbar\omega\left(a^+ a + \frac{1}{2}\right) + f(t)a + f*(t)a^+,$$

provided we define a complex-valued forcing function

$$f(t) = -\sqrt{\frac{\hbar}{2\mu\omega}}F(t) + i\sqrt{\frac{\hbar\mu\omega}{2}}G(t).$$

(a) Show that the equation of motion for $a(t)$ in the Heisenberg representation is

$$\frac{da(t)}{dt} + i\omega a(t) = -\frac{i}{\hbar}f*(t).$$

(Changes in the excitation level of the oscillator depend on $f*(t)$).

(b) Use a Green's function approach to solve the equation of motion of part (a) for $a(t)$ by adding the appropriate particular solution,

$$a_p(t) = -\frac{i}{\hbar}\int_{-\infty}^{\infty} G(t-t')f*(t')dt',$$

to first one and then the other of the two homogeneous solutions $a_<(t)$ and $a_>(t)$ corresponding to $t<t'$ and $t>t'$, respectively. Write down the two resulting general solutions for $a(t)$—only formal expressions in terms of $a_<(t)$ and $a_>(t)$ are needed. Note that $G(t-t')$ is proportional to $e^{-i\omega(t-t')}$ when $t \neq t'$, but at $t = t'$ integration of the Green's function equation over an interval including t' shows that it must satisfy the condition:

$$\lim_{\varepsilon\to 0}[G(+\varepsilon) - G(-\varepsilon)] = 1,$$

where $\varepsilon \equiv t-t'$. Hence suitable representations for the Green's functions are given by

$$G_<(t-t') = \eta(t-t')e^{-i\omega(t-t')},$$
$$G_>(t-t') = -\eta(t'-t)e^{-i\omega(t-t')},$$

where η is a step function:

$$\eta(x) \equiv \int_{-\infty}^{x} \delta(x')dx' = \begin{cases} 0, & x < 0 \\ 1, & x > 0 \end{cases}.$$

(c) Equating the two solutions from part (b), show that

$$a_>(t) = a_<(t) - \frac{i}{\hbar} g * (\omega),$$

where

$$g * (\omega) \equiv \int_{-\infty}^{\infty} e^{-i\omega(t-t')} f * (t')dt'.$$

(d) Assume that the action of the displacement operator $\hat{\alpha}$ in Problem 6.1 is such that it transforms the field annihilation operator in time according to $a_>(t) = \hat{\alpha}^+ a_< \hat{\alpha}$. Then determine the actual value of the eigenvalue α of the coherent state, to reveal its relationship with the Fourier spectrum of the forcing function.

(e) Show that the number of quanta at frequency ω is zero before the forcing function $f(t)$ is applied, but is proportional to the square of its Fourier component $g(\omega)$ for $t > 0$. To do this, evaluate $\langle 0| \hat{a}_<^+ \hat{a}_< |0\rangle$ and $\langle 0| \hat{a}_>^+ \hat{a}_> |0\rangle$.

6.3. Show that the variables P and Q introduced simply as linear combinations of field amplitudes are indeed the conjugate variables needed for quantum field theory by showing that they obey Hamilton's equations $\dot{Q}_{k\sigma} = \frac{\partial H}{\partial P_{k\sigma}}$ and $\dot{P}_{k\sigma} = -\frac{\partial H}{\partial Q_{k\sigma}}$, where H is expressed explicitly in terms of P and Q.

6.4. Show that unlike the situation for the annihilation operator on which the development of coherent states rests, no eigenstate of the field creation operator exists.

6.5. The momentum density of an electromagnetic field is given by the Poynting vector divided by c^2 (see, for example, Stratton, *Electromagnetic Theory*). Classically, this means the total momentum of the field associated with a volume V in space can be written

$$\bar{G} = \frac{1}{c^2} \int (\bar{E} \times \bar{H}) d^3 r.$$

Show that quantization of the fields in this expression reveals that the momentum of each linearly polarized plane wave is quantized in units of $\hbar\bar{k}$.

6.6. Thermal radiation is quite different from single-mode laser fields, due to dependence of the occupation probability of a state with n photons of frequency ω on the thermodynamic temperature T of the medium (governed by Boltzmann statistics).

(a) From the expression

$$\hat{\rho} = \sum_{\psi} P_{\psi} \,|\psi\rangle\langle\psi|,$$

and the value of P_{ψ} from statistical mechanics for the probability that n photons occupy a mode, show that the radiative density operator for thermal radiation at frequency ω is

$$\hat{\rho} = \left[1 - \exp\left(\frac{-\hbar\omega}{k_B T}\right)\right] \sum_n \exp\left(\frac{-n\hbar\omega}{k_B T}\right) |n\rangle\langle n|.$$

(b) Use the relation

$$\bar{n} = \langle\hat{n}\rangle = Tr\left(\rho a^+ a\right)$$

and the result of part (a) to show that the density operator can be expressed as

$$\hat{\rho} = \sum_n \frac{(\bar{n})^n}{(1+\bar{n})^{1+n}} |n\rangle\langle n|.$$

(c) Also show that the density operator for single-mode thermal radiation can be written in the equivalent form

$$\hat{\rho} = \left\{1 - \exp\left(\frac{-\hbar\omega}{k_B T}\right)\right\} \exp\left(\frac{-\hbar\omega \hat{a}^+ \hat{a}}{k_B T}\right),$$

where the exponential is defined by its usual power-series expansion.

Additional discussion of density operators for different types of radiation field can be found in Ref. [6.9].

6.7. The one-photon coherent state $|\alpha\rangle$ is an eigenstate of the annihilation operator $\hat{a}$. Hence $\hat{a}\,|\alpha\rangle = \alpha\,|\alpha\rangle$. Use this information and the definition of the photon number operator $\hat{n} \equiv \hat{a}^+ \hat{a}^-$ to find

(a) the second-order degree of correlation, $g^{(2)}(\tau) = \dfrac{\langle \hat{a}^+ \hat{a}^+ \hat{a}^- \hat{a}^- \rangle}{\langle \hat{a}^+ \hat{a}^- \rangle^2}$, of the coherent state $|\alpha\rangle$, and the root-mean-square fluctuation, $\Delta n_{rms} \equiv \left(\langle (n - \langle n\rangle)^2 \rangle\right)^{1/2} = \left(\langle n^2\rangle - \langle n\rangle^2\right)^{1/2}$, of the number of photons. Express your answer in terms of $\langle n\rangle$ and note the form of Δn_{rms} that is characteristic of a Poisson process.

(b) Calculate $g^{(2)}(\tau)$ and Δn_{rms} for the Fock state $|n\rangle$. Compare with part (a) to decide which state, $|\alpha\rangle$ or $|n\rangle$, is *noisier*.

6.8. A single-mode number (Fock) state contains exactly n photons.

(a) Calculate the first- and second-order coherence functions, $g^{(1)}$ and $g^{(2)}$, of the state.

(b) As a function of τ, how does each function from part (a) compare with coherences of classical and non-classical sources? (Consider an atom, a thermal source, a "coherent" state $|\alpha\rangle$, and squeezed light in addition to the Fock state. You can simply look up correlation plots for most of these other cases (see, for example, Refs. [6.9] and [6.24] and present comparisons in the form of sketches.)

(c) Which of these sources are non-classical?

6.9. (a) Calculate the magnitude of the quantum degree of first-order coherence $\left|g^{(1)}(\tau)\right|$ of the single mode coherent state $|\alpha\rangle$ given by

$$|\alpha\rangle = \exp(-|\alpha|^2/2)\sum_{n=0}^{\infty}\frac{\alpha^n}{(n!)^{1/2}}|n\rangle,$$

where $|n\rangle$ denotes the single-mode Fock state.

(b) Calculate the root-mean-square fluctuation $\Delta E = [<\hat{E}^2> - <\hat{E}>^2]^{1/2}$ of the electric field operator $\hat{E} = E_0(\hat{a}^- + \hat{a}^+)\sin kz$ for the state

$$|\psi\rangle = \frac{1}{\sqrt{2}}\left[|n\rangle + e^{-i\Omega t}|n+1\rangle\right].$$

6.10. Calculate the root-mean-square fluctuation $\Delta E = [<\hat{E}^2> - <\hat{E}>^2]^{1/2}$ of the vacuum field.

6.11. Solve Eq. (6.7.42) in the low-intensity limit for the first-order quantum coherence $g^{(1)}(\tau)$ of resonance fluorescence.

6.12. "Quantum beats" may be observed in the luminescence of three-level atoms prepared in a coherent superposition state that decays spontaneously. The beats consist of a modulation in the envelope of luminescent decay. In a semi-classical picture, the total emission intensity is simply proportional to the square of the total field radiated by two dipole polarizations comprising the initial tri-level coherence:

$$|E_{TOT}|^2 = \left|E_1 e^{i\omega_1 t} + E_2 e^{i\omega_2 t}\right|^2 \propto E_1 E_2^* e^{i(\omega_1-\omega_2)t} + c.c.$$

The signal intensity should therefore contain a signal at beat frequency $\Delta\omega = \omega_1 - \omega_2$. This seems to provide a convenient method of measuring the level splitting $\Delta\omega$ regardless of whether the splitting is between excited states or ground states ("V-type" or "Λ-type" atoms, respectively—see Figures 1 and 2).

(a) For a V-type atom (Figure 1), write down the semi-classical density matrix equation of motion for $\dot{\rho}_{12}^{(3)}$ to show that there is indeed a radiant polarization on the $1 \leftrightarrow 2$ transition that originates from any coherence $\rho_{32}^{(2)}$ existing between levels 2 and 3. The only non-vanishing matrix elements for the V-type system are $\mu_{12} = e\langle 1|r|2\rangle$ and $\mu_{13} = e\langle 1|r|3\rangle$.

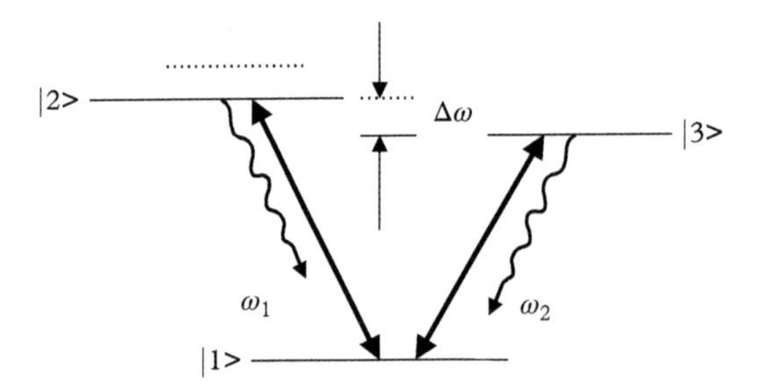

Figure 1 *A V-type energy level scheme. Applied fields at frequencies ω_1 and ω_2 that create the initial coherent superposition state of the atom are shown as bold arrows. Spontaneous decay is indicated by wiggly arrows.*

(b) According to quantum electrodynamics (QED), quantum beats in luminescence arise from tri-level coherence that radiates spontaneously with a signal amplitude of $E_S \propto \langle\psi(t)|\, E_1^-(t)E_2^+(t)\, |\psi(t)\rangle$, where $E_1(t) = E_1^+(t) + E_1^-(t)$ is the field at ω_1 and $E_2(t) = E_2^+(t) + E_2^-(t)$ is the field at ω_2. Note that $E_k^{(+)}(r,t) \equiv |E_k|\, a_k e^{-i\omega_k t} U_k(r)$ and its adjoint are field annihilation and creation operators, respectively. Calculate E_S for V-type atoms using the system wavefunction that comprises the initial superposition state plus states reached by spontaneous emission, namely:

$$|\psi\rangle = |i\rangle|n(\omega_1)\rangle\,|n(\omega_2)\rangle = \sum_i \left[a_i|i\rangle|0\rangle|0\rangle + b_i\,|1\rangle\,|1\rangle\,|0\rangle + c_i\,|1\rangle|0\rangle|1\rangle\right].$$

The factors a_i, b_i, and c_i are occupation probability amplitudes associated with atomic levels $i = 1, 2$, or 3.

(c) Does the QED signal polarization contain the expected beats?

(d) Recalculate E_S for the Λ-type atom depicted in Figure 2, again using a wavefunction which is the initial superposition state plus the states accessed by spontaneous emission, namely $|\psi\rangle = \sum_i [a_i\,|i\rangle\,|0\rangle\,|0\rangle + b_i\,|1\rangle\,|1\rangle\,|0\rangle + c_i\,|3\rangle\,|0\rangle\,|1\rangle]$.

(e) Describe the QED results of parts (b) and (c) in a few words and compare them with the expectation of the semi-classical argument presented at the beginning of this problem. (For an early experiment on quantum beats, see S. Haroche, J.A. Paisner, and A.L. Schawlow, Hyperfine quantum beats observed in Cs vapor under pulsed dye laser excitation. *Phys. Rev. Lett.* 1973, 948.)

6.13. In a system (depicted in the diagram) consisting of three-level atoms in a Λ-configuration, an initial state $|\psi(0)\rangle = C_a(0)\,|a\rangle + C_b(0)\,|b\rangle + C_c(0)\,|c\rangle$ evolves

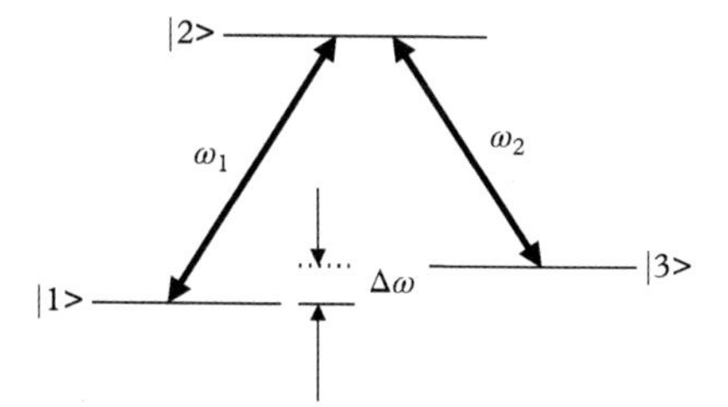

Figure 2 *A Λ-type energy level scheme. The non-vanishing matrix elements in this case are* $\mu_{12} = e\langle 1|r|2\rangle$ *and* $\mu_{23} = e\langle 2|r|3\rangle$. *Again the bold arrows indicate the applied fields that create the initial coherent superposition state of the atom.*

in time under the influence of two driving fields to the exact state $|\psi(t)\rangle = C_a(t)\,|a\rangle + C_b(t)\,|b\rangle + C_c(t)\,|c\rangle$, where

$$C_a(t) = \left[C_a(0)\cos(\Omega t/2) - iC_b(0)\frac{\Omega_{R1}}{\Omega}\sin(\Omega t/2)\ -iC_c(0)\frac{\Omega_{R2}}{\Omega}\sin(\Omega t/2\right]e^{-\gamma t/2},$$

$$\begin{aligned} C_b(t) = &\left\{-iC_a(0)\frac{\Omega_{R1}}{\Omega}\sin(\Omega t/2) + C_b(0)\left[\frac{\Omega_{R2}^2}{\Omega^2} + \frac{\Omega_{R1}^2}{\Omega^2}\cos(\Omega t/2)\right]\right. \\ &\left. + C_c(0)\left[\frac{-\Omega_{R1}\Omega_{R2}}{\Omega^2} + \frac{\Omega_{R1}\Omega_{R2}}{\Omega^2}\cos(\Omega t/2)\right]\right\}e^{-\gamma t/2}, \end{aligned}$$

$$\begin{aligned} C_c(t) = &\left\{-iC_a(0)\frac{\Omega_{R2}}{\Omega}\sin(\Omega t/2) + C_b(0)\left[-\frac{\Omega_{R1}\Omega_{R2}}{\Omega^2} + \frac{\Omega_{R1}\Omega_{R2}}{\Omega^2}\cos(\Omega t/2)\right]\right. \\ &\left. + C_c(0)\left[\frac{\Omega_{R1}^2}{\Omega^2} + \frac{\Omega_{R2}^2}{\Omega^2}\cos(\Omega t/2)\right]\right\}e^{-\gamma t/2}. \end{aligned}$$

Here Ω_{R1} and Ω_{R2} are the Rabi frequencies associated with the resonant fields driving the $|a\rangle \rightarrow |b\rangle$ and $|a\rangle \rightarrow |c\rangle$ transitions, respectively, and $\Omega = \sqrt{\Omega_{R1}^2 + \Omega_{R2}^2}$.

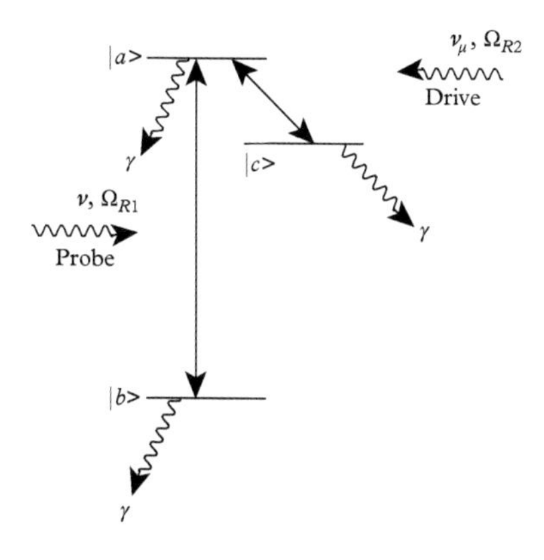

(a) State the main characteristic of a "dark" state.

(b) Assuming $\gamma = 0$, show that atoms prepared in the initial state

$$|\psi(0)\rangle = \frac{\Omega_{R2}}{\Omega}|b\rangle - \frac{\Omega_{R1}}{\Omega}|c\rangle$$

are in a trapped or "uncoupled" state.

(c) Assuming $\gamma = 0$ and that atoms are prepared initially in the ground state, find the probability of absorbing a probe laser photon (Hint: probability of the atom evolving from $|\psi(0)\rangle$ to $|a\rangle$).

(d) What is the condition under which the probability of absorption from the ground state on the $|b\rangle \rightarrow |a\rangle$ transition vanishes, giving electromagnetically induced transparency? (Hint: Use a projection technique to follow the development of excited state probability amplitude.)

6.14. This problem makes use of the expressions for amplitudes $C_a(t)$, $C_b(t)$, and $C_c(t)$ in Problem 6.13 and assumes the initial state of the atom is $|\psi(0)\rangle = |a\rangle$. By simply taking this initial condition into account, and calculating density matrix elements as explicit bilinear products of state amplitudes,

(a) find the polarization $P(t)$ on the $|a\rangle \leftrightarrow |b\rangle$ transition.

(b) Identify the amplitude of the positive frequency component $P(\omega)$ in the macroscopic polarization $P(t) = P(-\omega)e^{i\omega t} + P(\omega)e^{-i\omega t}$ and, for the stated initial condition, calculate the time t_1 at which the gain factor $\Gamma = Im\{P(\omega)\}$ first reaches a maximum.

(c) Find the occupation probabilities of states $|a\rangle$ and $|b\rangle$ and show that at time t_1 (when the gain is maximum) the population in the system is inverted (meaning there is more population in the upper state than in the lower).

(d) Give the time interval, if there is one, over which there exists gain but there is no inversion.

6.15. A two-level atom system is subjected to a strong nearly resonant light field. Verify that the dressed states $|D(n)_+\rangle$ and $|D(n)_-\rangle$ form an orthogonal basis, like the uncoupled product states.

6.16. Using expressions from the dressed state theory of resonance fluorescence for level populations $\pi^{\pm}(\infty)$ and transition rates per atom $\Gamma = |<d>|^2$ between dressed states,

(a) prove that *at exact resonance* the intensity of the central fluorescence component of the Mollow triplet is twice that of either sideband.

(b) "Stimulated" emission rates depend on the difference between initial and final state populations rather than the population of the initial state alone. Show that the rates of stimulated emission and stimulated absorption for one-photon transitions of the dressed atom are different and find their dependence on the mixing angle θ.

(c) Using the result of part (b), consider the gain/loss spectrum that would be measured by a tunable probe wave passing through the dressed atom sample. How many peaks are expected in the spectrum versus detuning, and are they positive peaks or negative?

(d) At exact resonance compare the rates of "stimulated" emission (from part (b)) and "spontaneous" emission (from part (a)) and comment on whether resonance fluorescence should be regarded as the stimulated or spontaneous emission of dressed atoms.

6.17. The quantized version of the interaction Hamiltonian $-\bar{\mu} \cdot \bar{E}$ in the Schrödinger picture is $V = \hbar g\{\hat{a}^{-}\hat{\sigma}^{+} \exp[-i\omega t] + h.c.\}$. Assuming a two-level atom starts in the ground state and interacts with a field containing $n + 1$ photons, calculate steady-state amplitudes for

(a) ρ_{12} and ρ_{21}, separately.

(b) For quantized fields, does one find $\tilde{\rho}_{21} = \tilde{\rho}_{12}^{*}$ or $\tilde{\rho}_{21} \neq \tilde{\rho}_{12}^{*}$? Explain how conservation of energy accounts for the result.

(c) Does this fully quantized interaction Hamiltonian eliminate the unphysical process of spontaneous absorption? Justify.

6.18. The creation and annihilation operators for a single mode of the electromagnetic field are defined in terms of generalized "coordinates" Q and P of the field expansion according to

$$a^{-} = \sqrt{\frac{\omega}{2\hbar}}\left(Q + i\frac{P}{\omega}\right) \text{ and } a^{+} = \sqrt{\frac{\omega}{2\hbar}}\left(Q - i\frac{P}{\omega}\right).$$

(a) Evaluate $\hat{a}^{-}|0\rangle$ in the Q representation and find the ground-state eigenfunction explicitly. (To do this, use operator form $P = -i\hbar(\partial/\partial Q)$ mandated by the commutation relation $[Q, P] = i\hbar$ to evaluate $\langle Q|\hat{a}^{-}|0\rangle$ and find $\langle Q|0\rangle$ from the resulting differential equation.

(b) Next evaluate $\langle Q|\hat{a}^{+}|0\rangle$ in the Q representation and use the result from part (a) to express $\langle Q|\hat{a}^{+}|0\rangle$ in terms of $\langle Q|0\rangle$. Show that the wavefunction of the mode occupied with n quanta is

$$\langle Q|n\rangle = \left(\frac{\omega}{\pi\hbar}\right)^{1/4} \frac{1}{\sqrt{2^{n}n!}} H_{n}\left(\sqrt{\frac{\omega}{\hbar}}Q\right) \exp(-\omega Q^{2}/2\hbar),$$

where the H_n are Hermite polynomials of order n.

(c) Show that wavefunctions of a single mode of the field that contain different numbers of photons are orthogonal.

6.19. The interaction Hamiltonian of a quantized field mode interacting with a two-level atom, in the rotating wave approximation, is

$$H = H_0 + V = \frac{\hbar\omega_0}{2}\sigma_z + \hbar\omega_L a^+ a^- + \hbar g(\sigma^+ a^- + a^+ \sigma^-),$$

where $a^{\pm}$ and $\sigma^{\pm}$ are raising and lowering operators of the field and the atom, respectively.

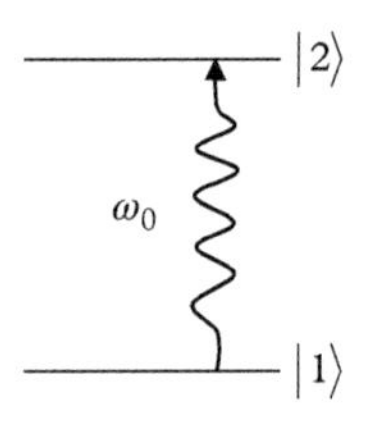

σ_z is one of the Pauli spin matrices, $\hbar\omega_0$ is the atomic energy level separation, and $\hbar\omega_L$ is the energy of incident photons.

(a) Verify by direct substitution that, for an intense field applied on resonance, the states

$$|\pm, n\rangle = \frac{1}{\sqrt{2}}\left(|2, n\rangle \pm |1, n+1\rangle\right)$$

are eigenstates of the combined atom–field system energy and identify the corresponding eigenvalues.

(b) Find the transition electric dipole moment between $|+, n\rangle$ and $|-, n\rangle$.

6.20. Half the energy density of an electromagnetic wave is due to the electric component, as well as half its noise. Even though a Fock state like $|n\rangle$ contains exactly n photons and should therefore be an eigenstate of the field with an energy of exactly $(n + \frac{1}{2})\hbar\omega$, show that

(a) the root-mean-square fluctuation of the modal energy is not zero but equals twice the zero-point energy (i.e., show that $\frac{1}{2}\langle n| (\Delta H_F)^2 |n\rangle^{1/2} \equiv \frac{1}{2}\left\{\langle n| H_F^2 |n\rangle - \langle n| H_F |n\rangle^2\right\} = \frac{1}{2}\hbar\omega$), whereas

(b) the mean square fluctuation of the field (normalized to volume V) depends on n according to $\varepsilon_0 V \langle n| (\Delta E)^2 |n\rangle = \varepsilon_0 V \langle n| E^2 |n\rangle = (n + \frac{1}{2})\hbar\omega$.

(c) Recall that classically the field energy per unit volume is expressible as $H = \varepsilon_0 E^2 + B^2/\mu_0$. Why are the results in parts (a) and (b) different, although both appear to be results for electric field driven fluctuations of the modal energy? Part (a) indicates that only zero-point fluctuations with an average value of $\frac{1}{2}\hbar\omega$ cause uncertainty in the energy of a Fock state, whereas the average noise due to the field itself varies as $|\Delta E|^2 \propto (n + \frac{1}{2})\hbar\omega$ or $|\Delta E| \propto \sqrt{(n + \frac{1}{2})\hbar\omega}$.

6.21. General solutions for the probability amplitudes $c_{2,n}(t)$ and $c_{1,n+1}(t)$ of a two-level atom interacting with a single mode light field (where the atom may be in the excited state with n photons present or in the ground state with $n+1$ photons, respectively) are:

$$c_{2,n}(t) = \left\{ c_{2,n}(0) \left[\cos\left(\frac{\Omega_n t}{2}\right) - \frac{i\Delta}{\Omega_n} \sin\left(\frac{\Omega_n t}{2}\right) \right] - \frac{2ig\sqrt{n+1}}{\Omega_n} c_{1,n+1}(0) \sin\left(\frac{\Omega_n t}{2}\right) \right\} e^{i\Delta t/2}$$

and

$$c_{1,n+1}(t) = \left\{ c_{1,n+1}(0) \left[\cos\left(\frac{\Omega_n t}{2}\right) + \frac{i\Delta}{\Omega_n} \sin\left(\frac{\Omega_n t}{2}\right) \right] - \frac{2ig\sqrt{n+1}}{\Omega_n} c_{2,n}(0) \sin\left(\frac{\Omega_n t}{2}\right) \right\} e^{-i\Delta t/2}.$$

Here, $\Omega_n^2 = \Delta^2 + 4g^2(n+1)$ is the square of the Rabi frequency at detuning Δ for a field containing n photons in the one mode, and g is the transition coupling constant. Assume the atom is initially in the excited state, so that $c_{2,n}(0) = c_n(0)$ and $c_{1,n+1}(0) = 0$, where $c_n(0)$ is the probability amplitude that the field contains n photons.

(a) Write out simplified probability amplitudes for $c_{2,n}(t)$ and $c_{1,n+1}(t)$ using the initial conditions given and then calculate $|c_{2,n}(t)|^2$ and $|c_{1,n}(t)|^2$. Replace the squares of c-numbers with density matrix elements for the probability of the field containing n photons according to the definition $\rho_{nn}(0) \equiv |c_n(0)|^2$.

(b) Calculate the population inversion $W(t)$, given by the excited state population minus the ground state population (from part a) summed over photon number n, and show that it can be rearranged to yield the simple but exact expression

$$W(t) = \sum_{n=0}^{\infty} \rho_{nn}(0) \left[\frac{\Delta^2}{\Omega_n^2} + \frac{4g^2(n+1)}{\Omega_n^2} \cos(\Omega_n t) \right].$$

(c) Determine W for the "vacuum" field and discuss the meaning and/or origin of *vacuum Rabi oscillations*.

6.22. Show that the mean number of photons in a single mode coherent state $|\alpha\rangle$ is different from that in a "displaced" coherent state $\hat{\alpha}\,|\alpha\rangle$, where $\hat{\alpha} \equiv \exp(\alpha\hat{a}^+ - \alpha^*\hat{a}^-)$.

6.23. Using dressed state analysis, find the value of the (resonant) Rabi frequency that corresponds to the onset of absorption saturation of a two-level atom when the light field is *detuned* from resonance ($\Delta \neq 0$)? Base your estimate purely on the condition that the population difference between initial and final states drops by half (i.e., either $\Pi_{D(n)-} - \Pi_{D(n-1)+} = \frac{1}{2}$ or $\Pi_{D(n)+} - \Pi_{D(n-1)-} = \frac{1}{2}$).

6.24. Show that the squared fluctuations from the mean (or variance) of the multimode electromagnetic vacuum field contains electric and magnetic contributions in equal proportion, so that $(\Delta E)^2 = c^2(\Delta B)^2$.

7

Selected Topics and Applications

Here the tools developed in earlier chapters are applied to analyze a few selected topics. They are neither comprehensive nor in any sense the "most important" examples of what has been presented so far in this book. However, they cover surprising results that are close enough to the "research front" that readers can confirm for themselves that problems from many disciplines can be attacked with the basic approaches that have already been introduced. That is, this chapter demonstrates that the now familiar techniques from earlier chapters can be used on challenging new problems. The chosen topics also have the merit of illustrating a few more application areas for which light is an exquisite probe or means of control and for which the density matrix provides tractable analysis.

7.1 Mechanical Effects of Light and Laser Cooling

Light can do more than most of us imagine. It can transfer linear momentum, angular momentum, or energy to material systems. It can also take these things away. Thus it can heat, cool, speed up, slow down, twist, or pull on matter as well as initiate dynamics and chemistry. All that is required is that the light is tuned appropriately with respect to resonant frequencies and has an intensity to place its interaction with matter in an appropriate limit. In this section several surprising mechanical effects of light are described, beginning with physical forces that arise from optical energy gradients in laser beams. Slowing and cooling of atoms as the result of near resonant interactions with light are discussed further in Sections 7.2 and 7.3.

7.1.1 Radiation Pressure, Dipole Forces, and Optical Tweezers

After observing that comet tails always pointed away from the sun, Keppler suggested that light fields exert pressure on matter. Since that time it has been shown that a wave of intensity I produces a force per unit area (or pressure) that is given in classical terms by

$$\frac{F}{A} = \frac{I}{c}. \tag{7.1.1}$$

Lectures on Light. Second Edition. Stephen C. Rand.
© Stephen C. Rand 2016. Published in 2016 by Oxford University Press.

In all earlier sections we ignored center-of-mass motion of atoms interacting with light, as well as spatial variations in the light distribution itself. Although mechanical forces of light are weak, there are micromechanical devices in which they nevertheless lead to important applications [7.1]. To analyze in detail how light can push or pull directly on atoms, we need to account for both these aspects of the problem. This is readily done by including the center-of-mass (kinetic) contribution to the energy in the Hamiltonian of the (two-level) atom,

$$H = \frac{p^2}{2M} + \frac{1}{2}\hbar\omega_0\sigma_z, \tag{7.1.2}$$

and recognizing that both the amplitude and the phase of an inhomogeneous light wave vary in space by writing the interaction Hamiltonian as

$$V(\bar{r}) = -\frac{1}{2}\hbar\Omega(\bar{r})\exp[i(\omega t + \phi(\bar{r}))]|1\rangle\langle 2| + h.c. \tag{7.1.3}$$

The force F exerted by light is best calculated using the Heisenberg equation of motion since it is a property of the field operator itself is of interest. Bearing in mind that the linear momentum operator p commutes with the Hamiltonian H, and that in the coordinate representation $\bar{p} = -i\hbar\bar{\nabla}$, we find

$$\bar{F} = \frac{d\bar{p}}{dt} = \frac{i}{\hbar}[(H+V),\bar{p}] = -\bar{\nabla}V(\bar{r}). \tag{7.1.4}$$

Exercise: Show that upon substitution of Eq. (7.1.3) into Eq. (7.1.4), the force operator for the positive frequency component becomes

$$F = \frac{1}{2}\hbar\Omega(\bar{r})\exp[-i(\omega t + \phi(\bar{r}))]\{\alpha(\bar{r}) - i\beta(\bar{r})\}, \tag{7.1.5}$$

where the real and imaginary parts are given by

$$\bar{\alpha}(\bar{r}) \equiv \frac{\bar{\nabla}\Omega(\bar{r})}{\Omega(\bar{r})} \text{and } \bar{\beta}(\bar{r}) \equiv \bar{\nabla}\phi(\bar{r}). \tag{7.1.6}$$

The expectation value of the force F is given by its trace with the density matrix.

$$\begin{aligned}\langle F\rangle = Tr\{F,\rho\} &= \frac{1}{2}\hbar\Omega(\bar{r})\left\{\bar{\alpha}(\bar{r})(\tilde{\rho}_{12}+\tilde{\rho}_{21}) + i\bar{\beta}(\bar{r})(\tilde{\rho}_{21}-\tilde{\rho}_{12})\right\} \\ &= \frac{1}{2}\hbar\Omega\left\{R_1\bar{\alpha} + R_2\bar{\beta}\right\}\end{aligned} \tag{7.1.7}$$

The first term on the right is purely real and gives the dipole force $\langle F_{dip}\rangle$. Note that it is proportional to the component R_1 of the Bloch vector given by Eq. (3.8.5). The second term is purely imaginary and produces a dissipative force $\langle F_{diss}\rangle$. It is proportional to R_2 of the Bloch vector specified by Eq. (3.8.6). Together these two forces comprise the total force exerted by light on the atom:

$$\langle F\rangle = \langle F_{dip}\rangle + \langle F_{diss}\rangle. \tag{7.1.8}$$

If we make use of the steady-state solution (Eq. (5.1.8)) for the off-diagonal density matrix elements of atoms at rest, the expression for the dipole force term in Eq. (7.1.7) reduces to

$$\langle F_{dip}\rangle = \frac{\hbar}{4}\left(\frac{\Delta}{\Delta^2+\Gamma^2}\right)\bar{\nabla}\Omega^2(\bar{r})[\rho_{11}-\rho_{22}], \tag{7.1.9}$$

since $\Omega\bar{\nabla}\Omega = \bar{\nabla}\Omega^2/2$.

The population difference in Eq. (7.1.9) plays no role far off resonance where this force is commonly applied, since then $\rho_{11}-\rho_{22}\sim 1$. Note too that the dipole force is zero unless there is a gradient of optical intensity. Evidently it is significant only in inhomogeneous fields, such as the focal regions of laser beams (see Figure 7.1). Plane waves do not exert a dipole force, since $\bar{\alpha} = \bar{\nabla}\Omega/\Omega = 0$. Also, the force is "strong-field seeking" for red detunings, since then the force points in the direction of highest intensity. For blue detunings, the opposite is true. The force is repulsive or "weak-field seeking." The result in Eq. (7.1.9) therefore indicates that a strongly focused, red-detuned laser beam can attract atoms or transparent objects to its center, as indicated in Figure 7.1. This is the basis for the manipulation of small particles with laser beams [7.2, 7.3] referred to as "optical tweezers" in biological research [7.4]. Figure 7.2 is intended to illustrate two laser beams trapping particles attached to a molecule in the viewing plane of a microscope.

As already mentioned, a dissipative force accompanies the dipole force. In fact, dissipation is present for each and every plane wave component of the incident beam, because

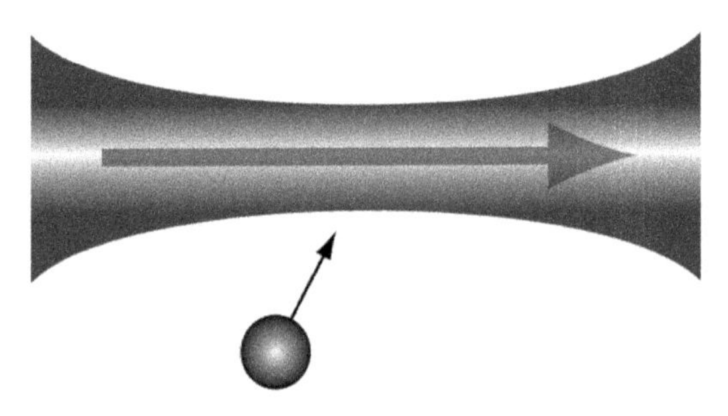

Figure 7.1 *Illustration of a two-level atom being attracted to the high intensity at the focus of a light beam propagating to the right. The light is detuned to the red side of the atomic resonant frequency.*

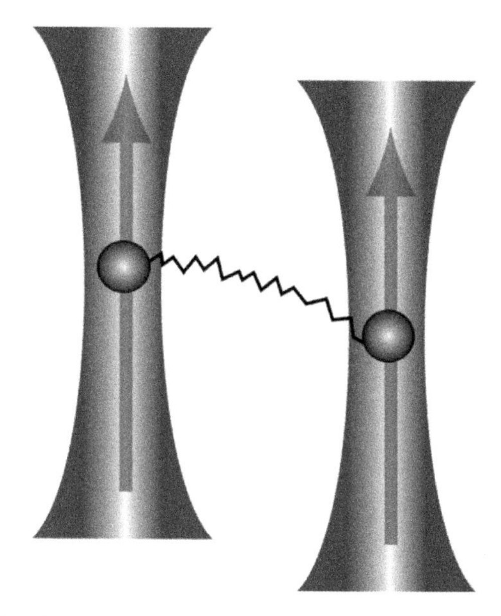

Figure 7.2 *Schematic drawing of two focused laser beams holding silica beads attached to two ends of a macromolecule, permitting it to be moved or stretched for study. By moving the beams individually or in tandem, "optical tweezers" can manipulate the molecule. The beads experience a force that keeps them at the beam foci, as described in the text.*

it does not require a field gradient, only a phase gradient. It originates from the second term in Eq. (7.1.7). Upon evaluation for stationary atoms in a plane wave (i.e., $\bar{\nabla}E_0 = 0$ and $\bar{\nabla}\phi = -k\hat{z}$), the dissipative force yields

$$\langle \bar{F}_{diss}(\mathrm{v}=0)\rangle = \frac{\hbar\Omega\bar{\beta}}{2}\frac{(-\Gamma\Omega)}{(\Delta^2+\Gamma^2)}(\rho_{11}-\rho_{22}) = \frac{\hbar}{2}\frac{k\Gamma\Omega^2\hat{z}}{(\Delta^2+\Gamma^2)}(\rho_{11}-\rho_{22}). \tag{7.1.10}$$

This force is directed along the propagation axis of the wave.

Additional discussion of tweezers can be found in Ref. [7.5]. Trapping with the dipole force using inhomogeneous waves is described further in Refs. [7.6, 7.7]. Use of the dissipative force in Eq. (7.1.10) to cool gases is described next. Approaches to laser cooling based upon alternative concepts are considered in Sections 7.1.3, 7.2, and 7.3.

7.1.2 Laser Cooling via the Doppler Shift

Although the dipole force described in Section 7.1.1 can be used for manipulating or trapping atoms and particles using high-power focused lasers tuned far to the red side of resonances (in regions of spectral transparency), interactions closer to resonance can be effective at more modest powers. The technique described in this section relies on

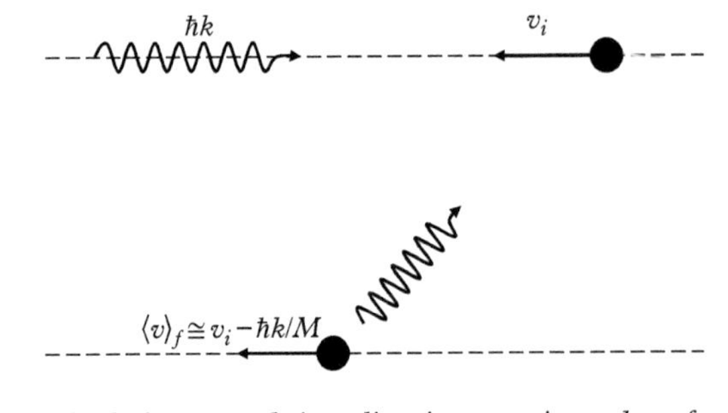

Figure 7.3 *(a) An atom of velocity v travels in a direction opposite to that of an incident beam. (b) Following absorption of a photon, it re-radiates the energy, resulting on average in a slower final velocity, since the emitted photon emerges in a random direction and therefore carries away an average momentum of zero.*

the Doppler effect (Section 3.9) near resonances [7.8] to cool the system, and is usually called Doppler cooling.

In Figure 7.3, a two-level atom is pictured moving at a velocity v opposite that of the light beam, in the $-\hat{z}$ direction. Because of the first-order Doppler shift (Section 3.9), the atom is not in resonance if the light frequency ω is tuned to the rest frame resonant frequency ω_0. However, photons will be resonantly absorbed if the Doppler shift is compensated by detuning to a lower frequency, namely $\omega = \omega_0 - k\mathrm{v}$. For each absorbed photon, the atom then acquires linear momentum in the amount $\hbar k$, opposing its initial momentum. Following absorption, light is re-emitted as spontaneous emission in a random direction that on average carries away no momentum. Hence as the result of absorbing a photon the atom is slowed (on average) by an amount $\Delta \mathrm{v} = \hbar k/M$.

The average emission frequency of the atom is ω_0, since even the Doppler-broadened emission spectrum of an ensemble is centered on the rest frame resonance frequency. Consequently, the average atom radiates more energy than it absorbs after absorbing a photon on the "red" side of resonance. In fact, energy and momentum decrease simultaneously and the atom cools by losing kinetic energy. This is the basis for Doppler laser cooling of gases [7.9] and "anti-Stokes" cooling of solids [7.10].

This argument can be made quantitative by simply extending the result of the last section for dissipative forces on stationary atoms to the case of moving atoms. Consider an atom with positive velocity v that is situated in a wave traveling in the $+\hat{z}$ direction. According to Eq. (5.2.8), the off-diagonal element is then

$$\tilde{\rho}_{21} = \left[\frac{(\Omega/2)}{(\Delta - k\mathrm{v}) - i\Gamma}\right](\rho_{11} - \rho_{22}). \tag{7.1.11}$$

The Doppler shift is small compared to typical linewidths (and optimal detunings as we shall see). That is, $k\mathrm{v} \ll \Gamma, \Delta$. Hence it plays little role in the magnitude of the denominator in Eq. (7.1.11). We find qualitatively new contributions to the force on the atom however by expanding this expression to first order in the shift.

$$\begin{aligned}\tilde{\rho}_{21} &= \frac{i(\Omega/2)}{(\Gamma + i\Delta)} \left(\frac{1}{1 - ikv/[\Gamma + i\Delta]} \right) (\rho_{11} - \rho_{22}) \\ &\cong \frac{i(\Omega/2)(\Gamma - i\Delta)}{\Gamma^2 + \Delta^2} \left\{ 1 + i \left(\frac{\Gamma - i\Delta}{\Gamma^2 + \Delta^2} \right) kv \right\} (\rho_{11} - \rho_{22}).\end{aligned} \tag{7.1.12}$$

For the plane wave under consideration, there is no transverse intensity gradient. Hence $\bar{\alpha} = 0$, and $\bar{\beta} = -k\hat{z}$. The force in Eq. (7.1.7) is therefore simply

$$\langle F \rangle = -\frac{1}{2}\hbar\Omega R_2 k\hat{z}, \tag{7.1.13}$$

where R_2 is the second component of the Bloch vector defined in Eq. (3.8.6) of Chapter 3.

By substituting Eq. (7.1.12) into Eq. (3.8.6) we find that R_2 is given by

$$R_2 = -\left\{ \frac{\Omega\Gamma}{\Delta^2 + \Gamma^2} + \frac{2\Delta\Omega\Gamma kv}{(\Delta^2 + \Gamma^2)^2} \right\} (\rho_{11} - \rho_{22}),$$

and so the expression for the force becomes

$$\langle F \rangle = \langle F_{diss}(v = 0) \rangle + \frac{\Delta\hbar k^2 \Omega^2 \Gamma}{(\Delta^2 + \Gamma^2)^2} (\rho_{11} - \rho_{22}) v\hat{z}. \tag{7.1.14}$$

In this form, it is obvious that the dissipative force on the atom points in the direction of propagation of the light field, regardless of whether the atom is moving or not. The first term on the right side of Eq. (7.1.14) reproduces the result for stationary atoms given by Eq. (7.1.10). The second term is a velocity-dependent force whose sign depends on detuning Δ. For positive ("red") detuning it accelerates atoms traveling in the $(+\hat{z})$-direction and decelerates atoms traveling in the $(-\hat{z})$-direction. However, for any given velocity v, if the detuning is fixed at $\Delta = k\,|v|$, we see that the force is far more effective at deceleration than at acceleration. This is because an atom co-propagating with the light at velocity v sees a detuning of $\Delta = -k\,|v|$ due to the Doppler effect. As a result of this detuning it experiences a greatly reduced force compared to the atom propagating toward the light source, which is in exact resonance.

For negative ("blue") detuning, the forces reverse. Therefore there is great potential for detailed control of the motion of atoms with mechanical forces exerted by light. Red-detuned light is particularly effective, however, as discussed in the introduction of this section, at reducing the energy and momentum of moving atoms simultaneously. This principle is utilized for fast, efficient Doppler cooling in magneto-optic traps, to be described next. In solids, because there is no obvious way to implement the complete Doppler cooling strategy based on moving atoms, optical refrigeration tends to be less efficient [7.11].

7.1.3 Magneto-Optic Trapping

To cool atoms effectively using the Doppler effect, net cooling must be achieved for atoms traveling in either direction along any Cartesian axis. For this purpose, counter-propagating beams along all three axes are needed. Then pairs of red-detuned waves can interact with atoms traveling in positive or negative directions along any given axis. In addition, a central region must be provided where cold atoms experience no force and can collect. Ideally, atoms not in this central region should experience Doppler cooling and a net force driving them toward the origin.

The apparatus depicted in Figure 7.4 uses a combination of magnetic forces and optical trapping to confine and cool atoms in this manner [7.12]. A key ingredient is the introduction of a pair of coils carrying opposing currents, as shown in Figure 7.4(b). This coil configuration gives rise to a magnetic field that increases linearly from the origin. Due to linear Zeeman splitting, states with opposite magnetic projections on either side of the origin experience a magnetic restoring force that points to the center of the trap (see Figure 7.4(a)). However, the restoring force only acts on atoms that reach the excited state through interaction with one of the red detuned beams and thereby develop a magnetic moment. Hence when atoms are slowed and cooled by the standing

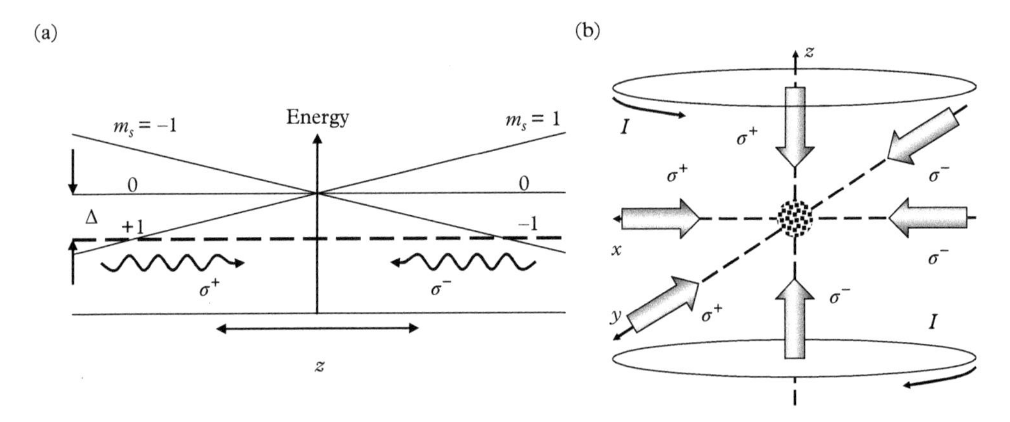

Figure 7.4 *Magneto-optic trap combining Doppler cooling with a magnetic restoring force centered on the origin. (a) For a hypothetical atom with a $S = 0$ ground state and $S = 1$ excited state, the excited substates split as shown in an anti-Helmholtz field. Angular momentum selection rules then stipulate positive (negative) helicity light incident from the left (right) to achieve resonance with left-(right-)going atoms attempting to leave the trap. Hence atoms that would otherwise be lost from the trap are cooled further and may remain trapped. (b) The overall trap geometry showing the coils at top and bottom, and six beams of appropriate polarization to provide cooling on the hypothetical $S = 0 \rightarrow S = 1$ transition.*

wave along each axis, it is possible for them to fall into the capture range of the magnetic trap, whereupon they accumulate at the origin if their velocity is low enough.

The detuning that results in an extremum of the cooling rate by the Doppler method may be readily calculated from Eq. (7.1.14). The stopping force will have a maximum when $\partial\langle F\rangle/\partial\Delta = 0$. The population difference $(\rho_{11} - \rho_{22})$ contains some detuning dependence of its own, particularly as one approaches resonance, but we can estimate the optimal detuning for cooling simply by setting

$$\frac{\partial}{\partial\Delta}\left(\frac{\Delta}{(\Gamma^2+\Delta^2)^2}\right) = 0. \tag{7.1.15}$$

The result is

$$\Delta = \left(\Gamma/\sqrt{3}\right) \approx \Gamma/2. \tag{7.1.16}$$

This relation suggests that the temperature limit of the Doppler cooling method is set by the energy balance between cooling and heating at the detuning specified by Eq. (7.1.16). The lowest temperature achievable by the Doppler cooling method should therefore be

$$k_B T_D = \frac{1}{2}\hbar\Gamma. \tag{7.1.17}$$

Still lower temperatures can be reached using polarization gradient cooling and trapping however, as described next.

7.1.4 Laser Cooling below the Doppler Limit

To cool matter below the Doppler limit of Eq. (7.1.17), new principles must be utilized to improve the cooling rate for slow atoms. An important advance historically was the idea of using polarization gradients in the force law expressed by Eq. (7.1.4) to reach a temperature corresponding to the recoil limit [7.13, 7.14].

The recoil limit is determined by the energy imparted by a single absorbed photon of wavenumber k to an atom of mass M (see Section 3.9 and Ref. [7.15]). This energy equals $\hbar^2k^2/2M$, and so the recoil-limited temperature is

$$k_B T_R = \hbar^2k^2/2M. \tag{7.1.18}$$

Understandably, to reach temperatures lower than the Doppler limit, cooling and trapping mechanisms that do not rely on the Doppler effect and yet still exert viscous damping and cooling must be utilized. Here we shall consider only one polarization gradient technique that provides a higher rate of cooling than Doppler cooling for very slowly moving atoms.

In the previous three sections we analyzed forces arising from spatial gradients associated with Doppler shifts in the light–matter interaction or inhomogeneous light fields.

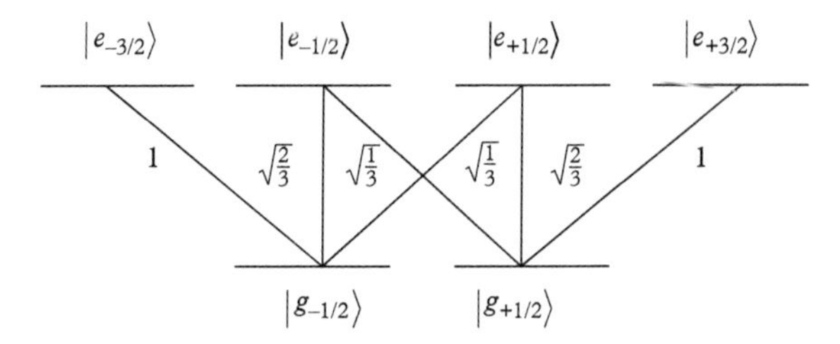

Figure 7.5 *Atomic levels and Clebsch–Gordan (C-G) coefficients for a $J_g = 1/2 \leftrightarrow J_e = 3/2$ transition. Transition probabilities are proportional to the square of the C-G coefficients.*

Spatial variations of the optical polarization can also produce strong forces on atoms. To understand this, consider an atom with the level structure shown in Figure 7.5.

Consider two counter-propagating plane waves with orthogonal linear polarizations (often referred to as a lin $\perp$ lin configuration in a one-dimensional "optical molasses").

Exercise: Show that the total field in a lin $\perp$ lin configuration can be written in the form

$$\bar{E} = \frac{1}{2}E_0 \exp[-i(\omega t - kz)]\{\hat{x} + \hat{y}\exp(-2ikz)\} + c.c.$$
$$= E(z)\exp[-i\omega t] + c.c., \tag{7.1.19}$$

where the amplitude of the positive frequency component is

$$E(z) = \frac{1}{2}E_0(\hat{x} + \hat{y})\cos kz + \frac{1}{2}iE_0(\hat{x} - \hat{y})\sin kz, \tag{7.1.20}$$

and that its ellipticity evolves along z as shown in Figure 7.6.

The optical interaction in a lin $\perp$ lin field configuration is found by substituting the expression (7.1.20) for the field into Eq. (5.3.19) and taking the electric dipole (ED) moment of the transition with a Clebsch–Gordan coefficient of unity to be μ_0. Then we have

$$V(z,t) = -\frac{1}{\sqrt{2}}\hbar\Omega \sin kz\left[|e_{3/2}\rangle\langle g_{1/2}| + \sqrt{\frac{1}{3}}\,|e_{1/2}\rangle\langle g_{-1/2}|\right]\exp(-i\omega t) + h.c. +$$
$$-\frac{1}{\sqrt{2}}\hbar\Omega \cos kz\left[|e_{-3/2}\rangle\langle g_{-1/2}| + \sqrt{\frac{1}{3}}\,|e_{-1/2}\rangle\langle g_{1/2}|\right]\exp(-i\omega t) + h.c., \tag{7.1.21}$$

where $\Omega = \mu_0 E_0/\hbar$, and a shift of origin along z by $\lambda/8$ has been applied to simplify the result. The factor of $1/\sqrt{2}$ that appears in Eq. (7.1.21) takes into account both the Clebsch–Gordon coefficient and the coefficients in the interaction Hamiltonian (see Section 7.3).

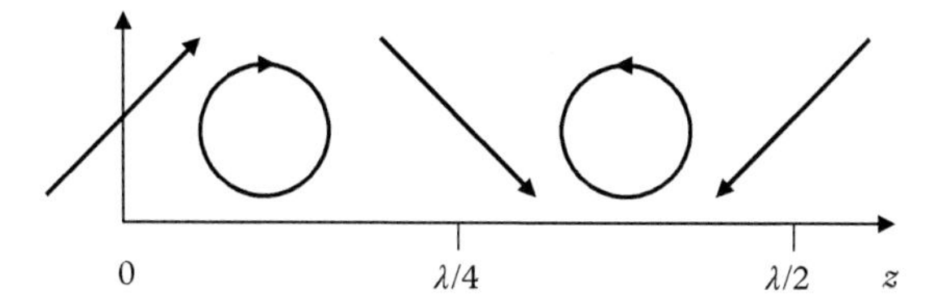

Figure 7.6 *Polarization states of the total optical field in the lin ⊥ lin configuration.*

As before, the average force acting on the atom is given by

$$\langle F\rangle = Tr\{F,\rho\} = Tr\{\bar{\nabla}V,\rho\}$$

$$= \frac{\hbar k\Omega}{\sqrt{2}}\cos kz\left[\langle g_{1/2}|\,\tilde{\rho}\,|e_{3/2}\rangle + \sqrt{\frac{1}{3}}\langle g_{-1/2}|\,\tilde{\rho}\,|e_{1/2}\rangle + c.c.\right] +$$

$$-\frac{\hbar k\Omega}{\sqrt{2}}\sin kz\left[\langle g_{-1/2}|\,\tilde{\rho}\,|e_{-3/2}\rangle + \sqrt{\frac{1}{3}}\langle g_{1/2}|\,\tilde{\rho}\,|e_{-1/2}\rangle + c.c.\right]. \tag{7.1.22}$$

If we assume that atomic motion is slow enough not to mix ground state populations established by optical pumping in different locations of the polarization field (Figure 7.6), we can drop the velocity-dependent term in the transport equations for system coherences. The equations for coherences appearing in Eq. (7.1.22) may then be written

$$i\hbar\langle g_{1/2}|\dot{\rho}\,|e_{3/2}\rangle = -\hbar(\omega_0 + i\Gamma)\langle g_{1/2}|\rho|e_{3/2}\rangle + \langle g_{1/2}|(V\rho - \rho V)|e_{3/2}\rangle, \tag{7.1.23}$$

$$i\hbar\langle g_{-1/2}|\dot{\rho}|e_{1/2}\rangle = -\hbar(\omega_0 + i\Gamma)\langle g_{-1/2}|\rho|e_{1/2}\rangle + \langle g_{-1/2}|(V\rho - \rho V)|e_{1/2}\rangle, \tag{7.1.24}$$

$$i\hbar\langle g_{-1/2}|\dot{\rho}\,|e_{-3/2}\rangle = -\hbar(\omega_0 + i\Gamma)\langle g_{-1/2}|\rho|e_{-3/2}\rangle + \langle g_{-1/2}|(V\rho - \rho V)|e_{-3/2}\rangle, \tag{7.1.25}$$

$$i\hbar\langle g_{1/2}|\dot{\rho}\,|e_{-1/2}\rangle = -\hbar(\omega_0 + i\Gamma)\langle g_{1/2}|\rho|e_{-1/2}\rangle + \langle g_{1/2}|(V\rho - \rho V)|e_{-1/2}\rangle. \tag{7.1.26}$$

Making use of the slowly varying envelope approximation (SVEA), Eqs. (7.1.23)–(7.1.26) become

$$i\hbar(\langle g_{1/2}|\dot{\tilde{\rho}} - i(\Delta + i\Gamma)\tilde{\rho}\,|e_{3/2}\rangle) = -\frac{\hbar\Omega}{\sqrt{2}}\sin kz[\langle e_{3/2}|\rho|e_{3/2}\rangle - \langle g_{1/2}|\rho|g_{1/2}\rangle] +$$

$$-\frac{\hbar\Omega}{\sqrt{2}}\cos kz\sqrt{\frac{1}{3}}\langle e_{-1/2}|\rho|e_{3/2}\rangle, \tag{7.1.27}$$

$$i\hbar\left(\langle g_{-1/2}|\dot{\tilde{\rho}} - i(\Delta + i\Gamma)\tilde{\rho}\,|e_{1/2}\rangle\right) = -\frac{\hbar\Omega}{\sqrt{2}}\sin kz\sqrt{\frac{1}{3}}\left[\langle e_{1/2}|\rho|e_{1/2}\rangle - \langle g_{-1/2}|\rho|g_{-1/2}\rangle\right] + \\ -\frac{\hbar\Omega}{\sqrt{2}}\cos kz\langle e_{-1/2}|\rho|e_{-3/2}\rangle, \tag{7.1.28}$$

$$i\hbar\left(\langle g_{-1/2}|\dot{\tilde{\rho}} - i(\Delta + i\Gamma)\tilde{\rho}\,|e_{-3/2}\rangle\right) = -\frac{\hbar\Omega}{\sqrt{2}}\cos kz\left[\langle e_{-3/2}|\rho|e_{-3/2}\rangle - \langle g_{-1/2}|\rho|g_{-1/2}\rangle\right] + \\ -\frac{\hbar\Omega}{\sqrt{2}}\sin kz\sqrt{\frac{1}{3}}\langle e_{-1/2}|\rho|e_{-3/2}\rangle, \tag{7.1.29}$$

$$i\hbar\left(\langle g_{1/2}|\dot{\tilde{\rho}} - i(\Delta + i\Gamma)\tilde{\rho}\,|e_{-1/2}\rangle\right) = -\frac{\hbar\Omega}{\sqrt{2}}\cos kz\sqrt{\frac{1}{3}}\left[\langle e_{-1/2}|\rho|e_{-1/2}\rangle - \langle g_{1/2}|\rho|g_{1/2}\rangle\right] + \\ -\frac{\hbar\Omega}{\sqrt{2}}\sin kz\langle e_{3/2}|\rho|e_{-1/2}\rangle. \tag{7.1.30}$$

Steady-state solutions may be found by setting $\dot{\tilde{\rho}} = 0$. In the low-intensity limit (characterized by negligible excited state populations and coherences) one finds

$$\langle g_{1/2}|\tilde{\rho}\,|e_{3/2}\rangle = \frac{\Omega}{\sqrt{2}}\frac{1}{\Delta + i\Gamma}\sin kz\langle g_{1/2}|\rho|g_{1/2}\rangle, \tag{7.1.31}$$

$$\langle g_{-1/2}|\tilde{\rho}\,|e_{1/2}\rangle = \frac{\Omega}{\sqrt{6}}\frac{1}{\Delta + i\Gamma}\sin kz\langle g_{-1/2}|\rho|g_{-1/2}\rangle, \tag{7.1.32}$$

$$\langle g_{-1/2}|\tilde{\rho}\,|e_{-3/2}\rangle = \frac{\Omega}{\sqrt{2}}\frac{1}{\Delta + i\Gamma}\cos kz\langle g_{-1/2}|\rho|g_{-1/2}\rangle, \tag{7.1.33}$$

$$\langle g_{1/2}|\tilde{\rho}\,|e_{-1/2}\rangle = \frac{\Omega}{\sqrt{6}}\frac{1}{\Delta + i\Gamma}\cos kz\langle g_{1/2}|\rho|g_{1/2}\rangle. \tag{7.1.34}$$

Substitution of these results into Eq. (7.1.22) for the force leads to

$$\langle F\rangle = \frac{1}{3}\left(\frac{\Omega^2}{\Delta^2 + \Gamma^2}\right)\Delta\hbar k\sin 2kz\left[\langle g_{1/2}|\rho|g_{1/2}\rangle - \langle g_{-1/2}|\rho|g_{-1/2}\rangle\right]. \tag{7.1.35}$$

To complete the evaluation of the force exerted on atoms in this light trap, according to Eq. (7.1.35) we must solve for the (position-dependent) population difference of the two ground state levels. With the compact notation

$$\pi^{(g)}_{\pm 1/2} \equiv \langle g_{\pm 1/2}|\rho|g_{\pm 1/2}\rangle, \tag{7.1.36}$$

$$\pi^{(e)}_{\pm 3/2} \equiv \langle g_{\pm 3/2}|\rho|g_{\pm 3/2}\rangle, \tag{7.1.37}$$

$$\pi^{(e)}_{\pm 1/2} \equiv \langle g_{\pm 1/2}|\rho|g_{\pm 1/2}\rangle, \tag{7.1.38}$$

the ground state population equations may be written

$$\dot{\pi}^{(g)}_{\pm 1/2} = 2\Gamma\left[\pi^{(e)}_{\pm 3/2} + \frac{2}{3}\pi^{(e)}_{\pm 1/2} + \frac{1}{3}\pi^{(e)}_{\mp 1/2}\right] + \left[\frac{i\Omega}{\sqrt{2}}\sin kz\left\langle e_{\pm 3/2}\right|\tilde{\rho}\left|g_{\pm 1/2}\right\rangle \right.$$
$$\left. + \frac{i\Omega}{\sqrt{6}}\cos kz\left\langle e_{\mp 1/2}\right|\tilde{\rho}\left|g_{\pm 1/2}\right\rangle + c.c.\right]. \tag{7.1.39}$$

This equation makes it evident that the task still remains of determining excited state population $\pi^{(e)}$. Steady-state values of $\pi^{(e)}_i$ are readily calculated however from the corresponding transport equations:

$$\dot{\pi}^{(e)}_{3/2} = -2\Gamma\pi^{(e)}_{3/2} + \frac{\Omega}{\sqrt{2}}\sin kz\left[i\left\langle g_{1/2}\right|\tilde{\rho}\left|e_{3/2}\right\rangle + c.c.\right], \tag{7.1.40}$$

$$\dot{\pi}^{(e)}_{1/2} = -2\Gamma\pi^{(e)}_{1/2} + \frac{\Omega}{\sqrt{2}}\sin kz\left[i\sqrt{\frac{1}{3}}\left\langle g_{-1/2}\right|\tilde{\rho}\left|e_{1/2}\right\rangle + c.c.\right] \tag{7.1.41}$$

$$\dot{\pi}^{(e)}_{-1/2} = -2\Gamma\pi^{(e)}_{-1/2} + \frac{\Omega}{\sqrt{2}}\cos kz\left[i\sqrt{\frac{1}{3}}\left\langle g_{1/2}\right|\tilde{\rho}\left|e_{-1/2}\right\rangle + c.c.\right] \tag{7.1.42}$$

$$\dot{\pi}^{(e)}_{-3/2} = -2\Gamma\pi^{(e)}_{-3/2} + \frac{\Omega}{\sqrt{2}}\cos kz\left[i\left\langle g_{-1/2}\right|\tilde{\rho}\left|e_{-3/2}\right\rangle + c.c.\right]. \tag{7.1.43}$$

By inserting the earlier expressions for coherences given by Eqs. (7.1.31)–(7.1.34) into Eqs. (7.1.40)–(7.1.43), the steady-state populations are found to be

$$\pi^{(e)}_{3/2} = \frac{(\Omega^2/2)}{\Delta^2 + \Gamma^2}\pi^{(g)}_{1/2}\sin^2 kz, \tag{7.1.44}$$

$$\pi^{(e)}_{1/2} = \frac{1}{3}\frac{(\Omega^2/2)}{\Delta^2 + \Gamma^2}\pi^{(g)}_{-1/2}\sin^2 kz, \tag{7.1.45}$$

$$\pi^{(e)}_{-1/2} = \frac{1}{3}\frac{(\Omega^2/2)}{\Delta^2 + \Gamma^2}\pi^{(g)}_{1/2}\cos^2 kz, \tag{7.1.46}$$

$$\pi^{(e)}_{-3/2} = \frac{(\Omega^2/2)}{\Delta^2 + \Gamma^2}\pi^{(g)}_{-1/2}\cos^2 kz. \tag{7.1.47}$$

Finally, by inserting Eqs. (7.1.44)–(7.1.47) and (7.1.31)–(7.1.34) into Eq. (7.1.39), we find

$$\dot{\pi}^{(g)}_{1/2} = -\frac{2\Gamma}{9}\left(\frac{(\Omega^2/2)}{\Delta^2 + \Gamma^2}\right)\left[\pi^{(g)}_{1/2}\cos^2 kz - \pi^{(g)}_{-1/2}\sin^2 kz\right]. \tag{7.1.48}$$

Stationary solutions of Eq. (7.1.48) are

$$\pi_{1/2}^{st}(z) = \sin^2 kz, \tag{7.1.49}$$

$$\pi_{-1/2}^{st}(z) = \cos^2 kz, \tag{7.1.50}$$

as shown in Figure 7.7.

When Eqs. (7.1.49) and (7.1.50) are substituted into Eq. (7.1.35), the force on atoms takes the form

$$\langle F\rangle = \frac{1}{3}\left(\frac{\Omega^2}{\Delta^2+\Gamma^2}\right)\Delta\hbar k \sin 2kz\left[\sin^2 kz - \cos^2 kz\right], \tag{7.1.51}$$

which averages to zero over a wavelength. However, slowly moving atoms for which the Doppler shift kv is much smaller than the effective rate constant in Eq. (7.1.48), namely

$$\Gamma' = \tau_p^{-1} \equiv \frac{2\Gamma}{9}\frac{(\Omega^2/2)}{\Delta^2+\Gamma^2}, \tag{7.1.52}$$

alter the populations from the stationary distribution in an important way that yields a large force with a small capture range. As the result of atomic motion over an average (optical pumping) time of τ_p, the population distribution in either ground state decreases by an amount given approximately by

$$\pi_i(z, \mathrm{v}) = \pi_i^{st}(z) - \mathrm{v}\tau_p\frac{d}{dz}\pi_i^{st} + \ldots \tag{7.1.53}$$

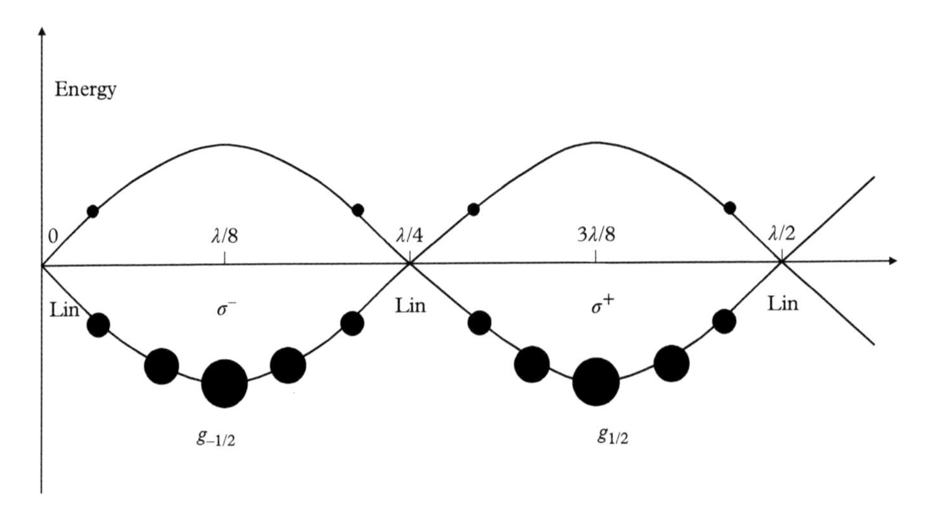

Figure 7.7 *Light-shifted energy levels and steady-state populations (represented by the size of filled circles) in the optical potential of the lin ⊥ lin configuration for positive detuning Δ.*

This adds a velocity-dependent term to the force, namely

$$\langle F\rangle = -\frac{4}{3}\left(\frac{(\Omega^2/2)}{\Delta^2+\Gamma^2}\right)\Delta\hbar k^2 \mathrm{v}\tau_p \left\langle \sin^2 2kz\right\rangle_\lambda = -\left(3\hbar k^2\frac{\Delta}{\Gamma}\right)\mathrm{v}, \tag{7.1.54}$$

which damps atomic motion at a rate that is independent of laser power. The lack of a dependence on power clearly distinguishes this (Sisyphus) laser cooling mechanism from Doppler cooling. Other forms of polarization gradient are discussed in Refs. [7.13, 7.14].

7.2 Dark States and Population Trapping

States that do not interact with light at all often have surprisingly important applications. In the next two sub-sections, "dark" states are introduced and their central role in yet another approach to laser cooling is explored.

7.2.1 Velocity-Selective Coherent Population Trapping

The creation of coherent superpositions of states can substantially alter the interactions of atoms with light. In this section, it is shown that atoms can exist in states that are deliberately decoupled from electromagnetic fields, where they can neither absorb nor emit radiation but are simply "trapped." Ironically, such states can exert an important influence on optical dynamics.

Imagine a three-level system in which circularly polarized light establishes a Λ-configuration, as shown in Figure 7.8. This could be ^{4}He vapor, for example, where $g_\pm$ represents the two Zeeman $m = \pm 1$ sublevels of the 2^3S_1 state and e_0 is the $m = 0$ sublevel of 2^3P_1. In ^{4}He there is of course another ground state sublevel g_0 with $m = 0$, but the transition to the excited state $2^3P_1\,(m = 0)$ is forbidden from this sublevel. Its $3j$ symbol (and Clebsch–Gordon coefficient which is proportional to it) is zero. Hence any initial population in this state will be pumped to g_- and g_+ through excited states e_-

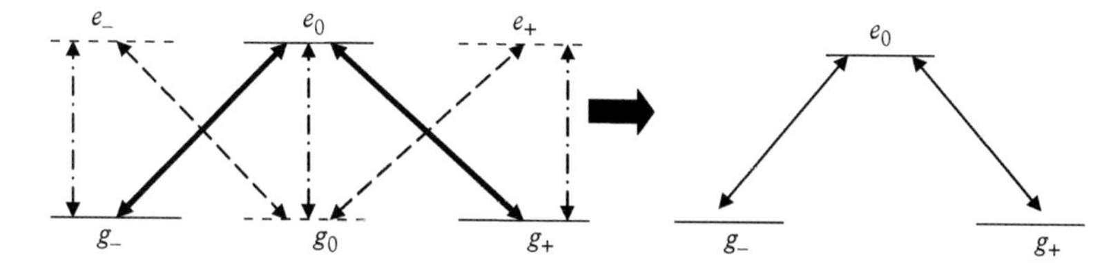

Figure 7.8 *Transitions of a $J = 1 \leftrightarrow J = 1$ system interacting with circularly polarized light beams. The two dashed transitions form a V-system which is not coupled to the Λ-configuration (solid arrows), except through spontaneous emission (dash-dot arrows). Atoms that undergo spontaneous emission contribute to optical pumping, eventually causing all atoms to join the Λ-system.*

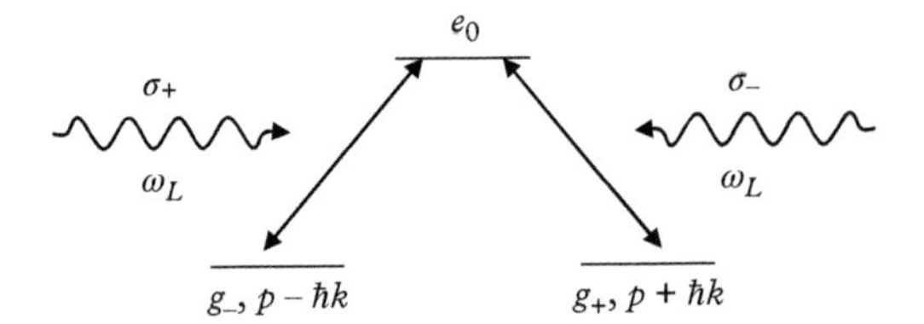

Figure 7.9 *Simplified three-state system that forms a closed family of states interacting only among themselves by absorption and emission processes.*

and e_+ respectively. Thus the g_0 state can be ignored completely, reducing the system to the "family" of states shown in Figure 7.9.

Exercise: Calculate all the $3j$ symbols for transitions shown in Figure 7.8. Show that the transition $|g_0\rangle \leftrightarrow |e_0\rangle$ has zero transition probability.

We shall refer to this family of three states coupled by absorption and emission as

$$\mathcal{F}(p) = \{|e_0, p\rangle, |g_-, p - \hbar k\rangle, |g_+, p + \hbar k\rangle\}. \tag{7.2.1}$$

As long as spontaneous emission and changes of linear momentum are ignored, this is a closed family of states coupled by circularly polarized light. Atoms in the system are taken to be sufficiently cold that the uncertainty in their linear momentum Δp may be much less than the photon momentum ($\Delta p \ll \hbar k$). This causes the atoms to become delocalized since the uncertainty in atomic position obeys a relation of the form $\Delta x \geq \hbar/\Delta p$. The atoms cease to be point-like and it is necessary to consider position z to be an operator ($\hat{z}$), whose expectation values will be governed by the de Broglie wave. Under these conditions the displacement operator $\hat{p} = \exp(\pm ik\hat{z})$ for the atomic momentum, which is conjugate to $\hat{z}$, becomes a function of the $\hat{z}$ operator and restricts changes in momentum to units of $\hbar k$.

The structure of the optical field is unimportant for the topic of velocity-selective population trapping. Hence the laser field will be taken to be classical. As usual, the Hamiltonian is considered to be the sum of two parts:

$$H = H_0 + V, \tag{7.2.2}$$

$$H_0 = \frac{p^2}{2m} + \hbar\omega_0 |e_0\rangle\langle e_0|. \tag{7.2.3}$$

The ground states $|g_+\rangle$ and $|g_-\rangle$ are taken to have the same energy (zero), and the field and the interaction to have the forms given previously by Eqs. (5.3.15) and (5.1.1).

Making use of the scalar definitions

$$\Omega_\pm \equiv \frac{\mu_\pm E_\mp}{\hbar}, \tag{7.2.4}$$

$$\mu_\pm = erC_\pm^{(1)}, \tag{7.2.5}$$

and the selection rules

$$\langle e_0|\mu_+|g_+\rangle = \langle e_0|\mu_-|g_-\rangle = 0, \tag{7.2.6}$$

we can write the perturbation as

$$V = \frac{1}{2}\left[\hbar\Omega_+ |e_0\rangle\langle g_-| \exp(ikz) + \hbar\Omega_- |e_0\rangle\langle g_+| \exp(-ikz)\right] \exp(-i\omega t) + h.c. \tag{7.2.7}$$

It is important to note that, although we shall omit the caret on z, it is to be treated as an *operator* not a continuous variable in Eq. (7.2.7). Hence, $\exp(\pm ikz)$ is a momentum displacement operator, which can be represented by

$$\exp(\pm ikz) = \sum_p |p\rangle\langle p \mp \hbar k|. \tag{7.2.8}$$

$$V = \sum_p \frac{1}{2}\left[\hbar\Omega_+ |e_0, p\rangle\langle g_-, p - \hbar k| + \hbar\Omega_- |e_0, p\rangle\langle g_+, p + \hbar k|\right] \exp(-i\omega t) + h.c. \tag{7.2.9}$$

From this expression for the interaction Hamiltonian, it is immediately apparent that $|e\rangle \equiv |e_0, p\rangle$ is coupled only to $|g_-, p - \hbar k\rangle$ and $|g_+, p + \hbar k\rangle$ optically. Hence this interaction can only induce transitions within the family $\mathcal{F}(p)$. For a fixed value of p, the dynamics are therefore fully describable by a closed set of nine equations.

To write out the evolution equations of the set, we transform to the following slowly varying variables:

$$\tilde{\rho}_{e\pm}(p) = \rho_{e\pm} \exp(i\omega t), \tag{7.2.10}$$

$$\tilde{\rho}_{+-}(p) = \rho_{+-}(p), \tag{7.2.11}$$

$$\tilde{\rho}_{ii}(p) = \rho_{ii}(p). \tag{7.2.12}$$

In Eq. (7.2.12) the index i takes on the values $i = +, -, e$ designating members of the family. Then, the dynamics are completely described by

$$\frac{d}{dt}\tilde{\rho}_{--}(p) = -i\frac{\Omega_+^*}{2}\tilde{\rho}_{e-}(p) + c.c., \tag{7.2.13}$$

$$\frac{d}{dt}\tilde{\rho}_{++}(p) = -i\frac{\Omega_-^*}{2}\tilde{\rho}_{e+}(p) + c.c., \tag{7.2.14}$$

$$\frac{d}{dt}\tilde{\rho}_{ee}(p) = \left[i\frac{\Omega_+^*}{2}\tilde{\rho}_{e-}(p) + i\frac{\Omega_-^*}{2}\tilde{\rho}_{e+}(p)\right] + c.c. \tag{7.2.15}$$

$$\frac{d}{dt}\tilde{\rho}_{e+}(p) = i\left(-\Delta + k\frac{p}{M} + \omega_r\right)\tilde{\rho}_{e+}(p) \tag{7.2.16}$$

$$-i\frac{\Omega_-}{2}\left[\tilde{\rho}_{++}(p) - \tilde{\rho}_{ee}(p)\right] - i\frac{\Omega_+}{2}\tilde{\rho}_{-+}(p),$$

$$\frac{d}{dt}\tilde{\rho}_{e-}(p) = i\left(-\Delta - k\frac{p}{M} + \omega_r\right)\tilde{\rho}_{e-}(p) \tag{7.2.17}$$

$$-i\frac{\Omega_+}{2}\left[\tilde{\rho}_{--}(p) - \tilde{\rho}_{ee}(p)\right] - i\frac{\Omega_-}{2}\tilde{\rho}_{-+}^*(p),$$

$$\frac{d}{dt}\tilde{\rho}_{-+}(p) = -i\frac{\Omega_+^*}{2}\tilde{\rho}_{e+}(p) + i\frac{\Omega_-}{2}\tilde{\rho}_{e-}^*(p) + 2ik\frac{p}{M}\tilde{\rho}_{-+}(p), \tag{7.2.18}$$

together with three complex-conjugate equations for the remaining off-diagonal elements $\tilde{\rho}_{+e} = \tilde{\rho}_{e+}^*$, $\tilde{\rho}_{-e} = \tilde{\rho}_{e-}^*$, and$\tilde{\rho}_{+-} = \tilde{\rho}_{-+}^*$. In Eqs. (7.2.16) and (7.2.17) we have introduced a "recoil" frequency shift $\omega_r \equiv \hbar k^2/2M$ together with the optical wavevector $k = 2\pi/\lambda$. The shift ω_r is relatively unimportant unless $\Delta p \ll \hbar k$, which is exactly the range of interest for matter wave mechanics and laser cooling.

The importance of the evolution equations is that they permit us to analyze how velocity selection can arise during population trapping. For this purpose, consider two special linear combinations of the ground states $|g_+, p + \hbar k\rangle$ and $|g_-, p - \hbar k\rangle$ designated by $|\psi_{NC}(p)\rangle$ and $|\psi_C(p)\rangle$:

$$|\psi_{NC}(p)\rangle = \frac{\Omega_-}{\left(|\Omega_+|^2 + |\Omega_-|^2\right)^{1/2}}|g_-, p - \hbar k\rangle - \frac{\Omega_+}{\left(|\Omega_+|^2 + |\Omega_-|^2\right)^{1/2}}|g_+, p + \hbar k\rangle, \tag{7.2.19}$$

$$|\psi_C(p)\rangle = \frac{\Omega_+^*}{\left(|\Omega_+|^2 + |\Omega_-|^2\right)^{1/2}}|g_-, p - \hbar k\rangle + \frac{\Omega_-^*}{\left(|\Omega_+|^2 + |\Omega_-|^2\right)^{1/2}}|g_+, p + \hbar k\rangle. \tag{7.2.20}$$

The first state $|\psi_{NC}(p)\rangle$ is not coupled radiatively to the excited state $|e_0, p\rangle$ whereas the second state $|\psi_C(p)\rangle$ is.

Exercise: Show that when $\Omega_+ = \Omega_-$ the interaction matrix element of the "non-coupled" state $|\psi_{NC}\rangle$ vanishes:

$$\langle e_0, p|\tilde{V}|\psi_{NC}(p)\rangle = 0. \tag{7.2.21}$$

An atom in the non-coupled state $|\psi_{NC}(p)\rangle$ therefore cannot absorb light and cannot be excited to $|e_0, p\rangle$. On the other hand, the coupled state, represented by $|\psi_C\rangle$, is radiatively coupled to the excited state.

Exercise: Show that the interaction matrix element of the coupled state is non-zero:

$$\langle e_0, p|\tilde{V}|\psi_C(p)\rangle = (\hbar/2)\left(|\Omega_+|^2 + |\Omega_-|^2\right)^{1/2}. \tag{7.2.22}$$

Of main interest to us here is the evolution of population among the available states. To examine trapping dynamics, we next examine the rate of change of population in the non-coupled state. To do this, the time rate of change of the diagonal density matrix element must be evaluated.

$$\frac{d}{dt}\langle\psi_{NC}(p)|\rho\,|\psi_{NC}(p)\rangle = ik\frac{p}{M}\frac{2\Omega_+\Omega_-}{|\Omega_+|^2 + |\Omega_-|^2}\langle\psi_{NC}(p)|\rho|\psi_C(p)\rangle + c.c. \tag{7.2.23}$$

This result shows that the population in $|\psi_{NC}(p)\rangle$ changes only if the atoms are moving. If $p = 0$, the right-hand side of Eq. (7.2.23), which is proportional to p, vanishes. So an atom prepared in $|\psi_{NC}(0)\rangle$ cannot leave this state by free evolution or by the absorption of a photon. Though we have not explicitly considered population change due to spontaneous emission, it is also clear that the atom cannot leave $|\psi_{NC}(0)\rangle$ by this process, since $|\psi_{NC}(0)\rangle$ is a linear combination of two radiatively stable ground states.

Consequently, this situation provides velocity-selective coherent population trapping (VSCPT). If $p \neq 0$, there is a (coherent but non-radiative) coupling proportional to kp/M between $|\psi_{NC}(p)\rangle$ and $|e_0(p)\rangle$. Although the state $|\psi_{NC}(p)\rangle$ cannot evolve radiatively by coupling to the excited state, it also cannot be considered a perfect trap when$p \neq 0$. Excitation by the laser can then still take place after an intermediate transition to$|\psi_C(p)\rangle$. The factor p/M on the right side of Eq. (7.2.23) is the atomic velocity in the excited state for the family $\mathcal{F}(p)$. Hence we note that coherent population trapping in $|\psi_{NC}(p)\rangle$ is velocity selective, since it happens only for $p = 0$.

Exercise: Calculate the matrix element appearing on the right side of Eq. (7.2.23) to show explicitly that when $p \neq 0$ an intermediate transition takes place from $|\psi_{NC}(p)\rangle$ to $|\psi_C(p)\rangle$ and leads indirectly to coupling of the "non-coupled" state with the excited state $|e_0(p)\rangle$.

7.2.2 Laser Cooling via VSCPT

Dark states and VSCPT can be exploited to accomplish laser cooling of atoms by adding two essential ingredients to the discussion of Section 7.2.1. First, a mechanism must exist to couple atoms in families with differing linear momenta (to allow $\mathcal{F}(p) \to \mathcal{F}(0)$). Second, it must be possible to establish the coherent superposition of $|g_+, +\hbar k\rangle$ and$|g_-, -\hbar k\rangle$ that is needed from the outset. That is, a method of producing the non-coupled state $|\psi_{NC}(0)\rangle$ is needed. Fortunately these two ingredients are provided by spontaneous emission and optical pumping, respectively.

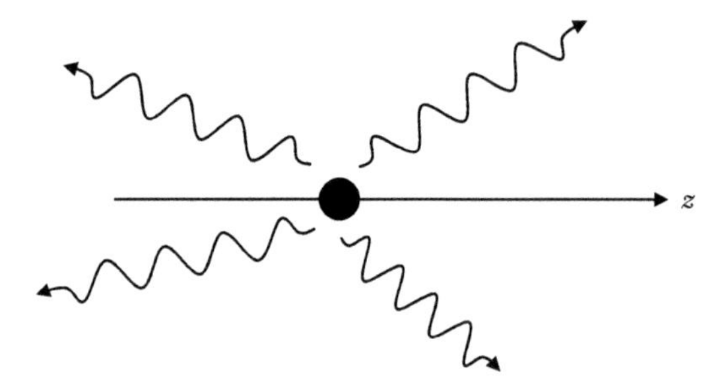

Figure 7.10 *Emission by an atom in the excited state $|e_0(p)\rangle$. The emitted photon directions are arbitrary, so the projection of photon momentum along z has a continuous range between $-\hbar k$ and $+\hbar k$.*

Consider an atom in the excited state $|e_0(p)\rangle$ of the family $\mathcal{F}(p)$. Such an atom can emit a fluorescence photon in any direction, as indicated in Figure 7.10. Suppose that the fluorescence photon has a linear momentum u along $\hat{z}$, where u may have a value between $-\hbar k$ and $+\hbar k$. Due to conservation of momentum, the atomic momentum changes by $-u$. Consequently, it makes a transition from $|e_0(p)\rangle$ to either $|g_+, p-u\rangle$ or $|g_-, p-u\rangle$, or a superposition of these two states. Notice that $|g_+, p-u\rangle$ belongs to the family of states $\mathcal{F}(p-u-\hbar k)$ whereas $|g_-, p-u\rangle$ belongs to $\mathcal{F}(p-u+\hbar k)$.

Spontaneous emission therefore provides a mechanism for atoms of one family to be redistributed to another, according to

$$\mathcal{F}(p) \rightarrow \mathcal{F}\left(p'\right), \tag{7.2.24}$$

where

$$p - 2\hbar k \leq p' \leq p + 2\hbar k. \tag{7.2.25}$$

Notice that the gradual redistribution or "diffusion" of atoms can cause accumulation of atoms in the $\mathcal{F}\,(p=0)$ family, as depicted schematically in Figure 7.11, provided some of the atoms end up in the $|\psi_{NC}(0)\rangle$ state.

Notice however that $|\psi_{NC}(0)\rangle$ is a coherent superposition state. How can such a state be the result of processes involving incoherent dynamics, like spontaneous emission? The answer is that $|\psi_{NC}(0)\rangle$ is the result of a combination of spontaneous emission followed by a selective filtering process. This process is outlined next.

While atoms which diffuse into ground state $|g_+, \hbar k\rangle$ or $|g_-, -\hbar k\rangle$ are in the $\mathcal{F}(p=0)$ family, they are not yet in the trapping state $|\psi_{NC}(0)\rangle$. However, the ground state components contain amplitudes from both $|\psi_{NC}(0)\rangle$ and $|\psi_C(0)\rangle$.

Exercise: Show explicitly that one can write

$$|g_{\pm}, \pm\hbar k\rangle = \mp\frac{1}{\sqrt{2}}\,[|\psi_{NC}(0)\rangle \mp |\psi_C(0)\rangle]. \tag{7.2.26}$$

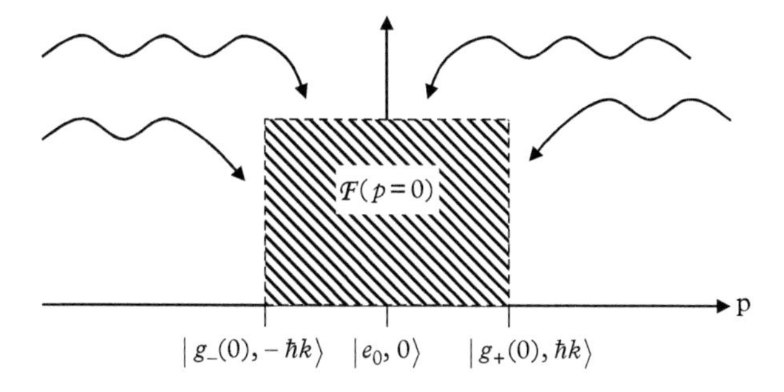

Figure 7.11 *Schematic illustration of the redistribution of atoms among families with different values of linear momentum p, leading to accumulation of low-velocity atoms in the $\mathcal{F}(p = 0)$ family that occupies the shaded region of the diagram.*

Consequently, in the presence of light, the $|\psi_C(0)\rangle$ components of both $|g_+, \hbar k\rangle$ and $|g_-, -\hbar k\rangle$ become involved in new fluorescence cycles, whereas the $|\psi_{NC}(0)\rangle$ component is stable and does not lead to system excitation. Light actively filters the atomic state in favor of the dark, uncoupled state $|\psi_{NC}(0)\rangle$.

In summary, "diffusion" of atoms in momentum space causes rapid redistribution of atoms among the available momentum families, including $\mathcal{F}(p=0)$. Once in the $\mathcal{F}(p=0)$ family, atoms undergo a self-filtering process driven by the applied light field that generates a dark state. Cold atoms then remain cold, while others join $\mathcal{F}(p=0)$. Curiously, although the atoms are left in a state in which the average momentum is much less than $\hbar k$, the result of a measurement of p can only be $\pm\hbar k$, since the act of measuring p produces an eigenstate of the linear momentum operator. Since $|\psi_{NC}(p)\rangle$ is not an eigenstate of operator $\hat{p}$, the result of many *measurements* of its momentum yields a distribution like that in Figure 7.12 when the effective temperature descends below the recoil limit.

VSCPT is an effective method for cooling gas samples below the recoil limit. Additional reading can be found in Refs. [7.9, 7.16, 7.17]. The development of such

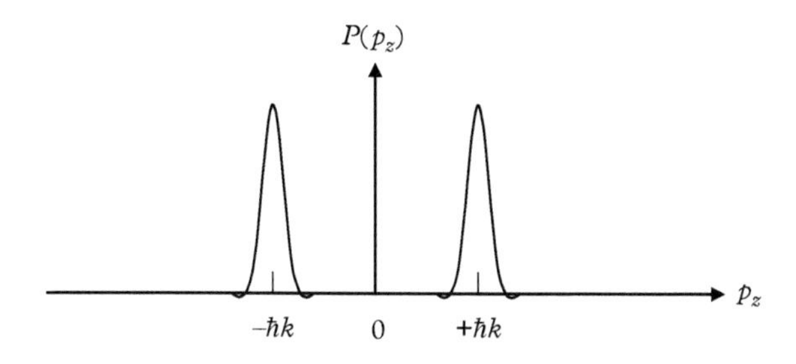

Figure 7.12 *Schematic diagram of the probability distribution of linear momentum measurements in a sample of atoms cooled by VSCPT below the recoil limit. (After [7.17].)*

techniques was an important step on the road to achieving Bose–Einstein condensation of neutral atoms [7.18–7.20].

7.3 Coherent Population Transfer

7.3.1 Rapid Adiabatic Passage

Next we examine the application of three-level coherence to population transfer in a system driven by two fields. The solution to the problem of transferring population in the most efficient way from state $|1\rangle$ to $|3\rangle$ of the system depicted in Figure 7.13 proves to be interesting, because it prescribes a counterintuitive procedure that utilizes adiabatic passage, a process of fundamental importance in quantum mechanics. It also generalizes the concept of coherent population trapping to the case of time-varying fields, since the incident light in this case is a two-color pulsed field.

In this application, a coherent two-photon method is sought to transfer population between levels 1 and 3. There are many reasons why population transfer of this type might be useful. A few are mentioned at the end of the section. The symmetry of the states is assumed to be such that ED transitions are allowed between $1 \leftrightarrow 2$ and $2 \leftrightarrow 3$, but not between $1 \leftrightarrow 3$. That is, $\mu_{13} = 0$. This is essential if the population is to remain in level 3 for any significant period of time after its transfer.

The two applied light fields are assumed to have frequencies close to the transition frequencies ω_{12} and ω_{23}. They are detuned by small amounts (Δ_{12} and Δ_{23}) from their respective one-photon transitions to minimize spontaneous emission losses, as shown in Figure 7.13. An effective Hamiltonian, H_{eff}, that incorporates both fields using the rotating wave approximation (RWA) can be constructed to simplify the search for eigenstates of this problem [7.21–7.24]. Here however we shall proceed by first writing out the exact equations of motion for $\rho(t)$ in terms of the full interaction Hamiltonian $H(t)$ and then proceeding to identify a simplified matrix $H_{eff}(t)$ that reproduces the equations for the slowly varying amplitudes, according to

$$i\hbar\dot{\tilde{\rho}} = \left[H_{eff}(t), \tilde{\rho}\right]. \tag{7.3.1}$$

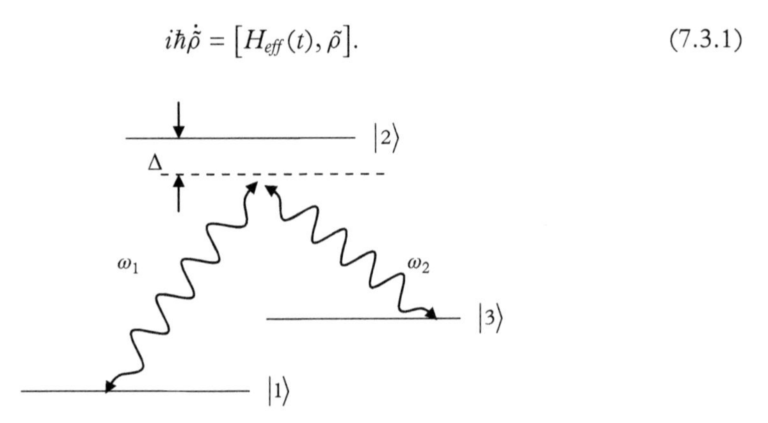

Figure 7.13 *Three-level model for the analysis of coherent population transfer by adiabatic passage.*

This procedure will identify the effective Hamiltonian as having the matrix form

$$H_{eff} = -\frac{1}{2}\hbar \begin{bmatrix} 0 & \Omega_1(t) & 0 \\ \Omega_1(t) & 2\Delta(t) & \Omega_2(t) \\ 0 & \Omega_2(t) & 0 \end{bmatrix}, \tag{7.3.2}$$

where $\Delta(t)$ is the one-photon detuning and $\Omega_n(t) = \mu_n E_n(t)/\hbar$ are the resonant Rabi frequencies of the two fields ($n = 1, 2$). A discussion of effective Hamiltonians can be found in Appendix I.

The interaction is of the form

$$V(t) = -\frac{1}{2}\left[\mu_{12}E_1(t)\exp(i\omega_1 t) + c.c.\right] - \frac{1}{2}\left[\mu_{23}E_2(t)\exp(i\omega_2 t) + c.c.\right] \tag{7.3.3}$$

$$= -\frac{1}{2}\left[\hbar\Omega_1(t)\exp(i\omega_1 t) + c.c.\right] - \frac{1}{2}\left[\hbar\Omega_2(t)\exp(i\omega_2 t) + c.c.\right], \tag{7.3.4}$$

where $\Omega_1(t) \equiv \mu_{12}E_1(t)/\hbar$ and $\Omega_2(t) \equiv \mu_{23}E_2(t)/\hbar$. The equations of motion are therefore

$$i\hbar\dot{\rho}_{11} = -\frac{1}{2}\hbar\Omega_1(t)\exp(i\omega_1 t)\rho_{21} + \frac{1}{2}\rho_{12}\hbar\Omega_1^*(t)\exp(-i\omega_1 t), \tag{7.3.5}$$

$$i\hbar\dot{\rho}_{22} = -\frac{1}{2}\hbar\Omega_1^*(t)\exp(-i\omega_1 t)\rho_{12} + \frac{1}{2}\rho_{21}\hbar\Omega_1(t)\exp(i\omega_1 t)$$

$$-\frac{1}{2}\hbar\Omega_2(t)\exp(-i\omega_2 t)\rho_{32} + \frac{1}{2}\rho_{23}\hbar\Omega_2^*(t)\exp(i\omega_2 t), \tag{7.3.6}$$

$$i\hbar\dot{\rho}_{33} = -\frac{1}{2}\hbar\Omega_2^*(t)\exp(i\omega_2 t)\rho_{23} + \frac{1}{2}\rho_{32}\hbar\Omega_2(t)\exp(-i\omega_2 t), \tag{7.3.7}$$

$$i\hbar\dot{\rho}_{12} = -\hbar\omega_{12}\rho_{12} - \frac{1}{2}\hbar\Omega_1(t)\exp(i\omega_1 t)(\rho_{22} - \rho_{11}) + \frac{1}{2}\rho_{13}\hbar\Omega_2^*(t)\exp(i\omega_2 t), \tag{7.3.8}$$

$$i\hbar\dot{\rho}_{23} = \hbar\omega_{23}\rho_{23} - \frac{1}{2}\hbar\Omega_2(t)\exp(-i\omega_2 t)(\rho_{33} - \rho_{22}) - \frac{1}{2}\hbar\Omega_1^*(t)\exp(-i\omega_1 t)\rho_{13}, \tag{7.3.9}$$

$$i\hbar\dot{\rho}_{13} = -\hbar\omega_{13}\rho_{13} - \frac{1}{2}\hbar\Omega_1(t)\exp(i\omega_1 t)\rho_{23} + \frac{1}{2}\rho_{12}\hbar\Omega_2(t)\exp(-i\omega_2 t). \tag{7.3.10}$$

These six equations are augmented by three for the conjugate coherences $\rho_{21} = \rho_{12}^*$, $\rho_{31} = \rho_{13}^*$, and $\rho_{32} = \rho_{23}^*$. Since the total population may be assumed to be constant, the closure condition $\rho_{11} + \rho_{22} + \rho_{33} = 1$ can be used to reduce the number of independent elements of the density matrix from nine (in a three-level system) to eight. (For an interesting discussion of the relationship between these eight density matrix elements and Gell-Mann's SU(3) generators, isospin, and hypercharge of particle physics, see Ref. [7.25]).

We begin by solving Eqs. (7.3.8)–(7.3.10) for the steady-state system coherences. Using the SVEA, one can anticipate solutions based on the substitutions

$$\rho_{12} = \tilde{\rho}_{12} \exp(i\omega_1 t) \rightarrow \dot{\rho}_{12} = \dot{\tilde{\rho}}_{12} \exp(i\omega_1 t) + i\omega_1 \tilde{\rho}_{12} \exp(i\omega_1 t), \tag{7.3.11}$$

$$\rho_{13} = \tilde{\rho}_{13} \exp(i\omega_{13} t) \rightarrow \dot{\rho}_{13} = \dot{\tilde{\rho}}_{13} \exp(i\omega_{13} t) + i\omega_{13} \tilde{\rho}_{13} \exp(i\omega_{13} t), \tag{7.3.12}$$

$$\rho_{23} = \tilde{\rho}_{23} \exp(-i\omega_2 t) \rightarrow \dot{\rho}_{23} = \dot{\tilde{\rho}}_{23} \exp(-i\omega_2 t) - i\omega_2 \tilde{\rho}_{23} \exp(-i\omega_2 t), \tag{7.3.13}$$

where $\omega_{13} \equiv \omega_{12} - \omega_{23}$ (and equals $\omega_1 - \omega_2$ at two-photon resonance). Starting with Eq. (7.3.11), this yields

$$\begin{aligned} i\hbar \left[\dot{\tilde{\rho}}_{12} \exp(i\omega_1 t) + i\omega_1 \tilde{\rho}_{12} \exp(i\omega_1 t) \right] &= -\hbar\omega_{12} \tilde{\rho}_{12} \exp(i\omega_1 t) + \\ &\quad -\frac{1}{2}\hbar\Omega_1(t) \exp(i\omega_2 t) \left(\rho_{22}^{(0)} - \rho_{11}^{(0)} \right) + \frac{1}{2}\hbar\Omega_2^*(t) \exp(i\omega_2 t) \tilde{\rho}_{23} \exp(i\omega_{23} t) \\ &= -\hbar\omega_{12} \tilde{\rho}_{12} \exp(i\omega_1 t) - \frac{1}{2}\hbar\Omega_1(t) \exp(i\omega_1 t) \left(\rho_{22}^{(0)} - \rho_{11}^{(0)} \right) \\ &\quad + \frac{1}{2}\hbar\Omega_2^*(t) \tilde{\rho}_{13} \exp(i\omega_1 t). \end{aligned} \tag{7.3.14}$$

By introducing the definitions

$$\Delta_{12} \equiv \omega_{12} - \omega_1, \tag{7.3.15}$$

$$\Delta_{23} \equiv \omega_{23} - \omega_2, \tag{7.3.16}$$

this and subsequent equations of motion for the coherences can be simplified by assuming that the two-photon resonance condition ($\Delta = -\Delta_{12} = \Delta_{23}$) is met:

$$\begin{aligned} i\hbar\dot{\tilde{\rho}}_{12} &= \hbar\Delta\tilde{\rho}_{12} - \frac{1}{2}\hbar\Omega_1(t) \left[\rho_{22}^{(0)} - \rho_{11}^{(0)} \right] + \frac{1}{2}\hbar\Omega_2^*(t)\tilde{\rho}_{13}, \\ \dot{\tilde{\rho}}_{12} &= -i\Delta\tilde{\rho}_{12} + \frac{1}{2}\Omega_1(t) \left[\rho_{22}^{(0)} - \rho_{11}^{(0)} \right] - \frac{1}{2}\hbar\Omega_2^*(t)\tilde{\rho}_{13}, \end{aligned} \tag{7.3.17}$$

$$\begin{aligned} i\hbar \left[\dot{\tilde{\rho}}_{23} \exp(-i\omega_2 t) - i\omega_2 \tilde{\rho}_{23} \exp(-i\omega_2 t) \right] &= \hbar\omega_{23} \tilde{\rho}_{23} \exp(-i\omega_2 t) \\ &\quad -\frac{1}{2}\hbar\Omega_2(t) \exp(-i\omega_2 t) \left[\rho_{33}^{(0)} - \rho_{22}^{(0)} \right] - \frac{1}{2}\hbar\Omega_1^*(t) \exp(-i\omega_1 t) \tilde{\rho}_{13} \exp(i\omega_{13} t), \\ \dot{\tilde{\rho}}_{23} &= -i\Delta\tilde{\rho}_{23} + \frac{i}{2}\Omega_2(t) \left[\rho_{33}^{(0)} - \rho_{22}^{(0)} \right] + \frac{i}{2}\Omega_1^*(t)\tilde{\rho}_{13}, \end{aligned} \tag{7.3.18}$$

$$i\hbar\left[\dot{\tilde{\rho}}_{13}\exp(i\omega_{13}t) + i\omega_{13}\tilde{\rho}_{13}\exp(i\omega_{13}t)\right] = -\hbar\omega_{13}\tilde{\rho}_{13}\exp(i\omega_{13}t) +$$

$$-\frac{1}{2}\hbar\Omega_1(t)\exp(i\omega_1 t)\tilde{\rho}_{23}\exp(-i\omega_2 t) + \frac{1}{2}\hbar\Omega_2(t)\tilde{\rho}_{12}\exp(-i\omega_2 t)\exp(i\omega_1 t),$$

$$\dot{\tilde{\rho}}_{13} = \frac{i}{2}\Omega_1(t)\tilde{\rho}_{23} - \frac{i}{2}\Omega_2(t)\tilde{\rho}_{12}. \tag{7.3.19}$$

Exercise: Show that Eqs. (7.3.5)–(7.3.7) and (7.3.17)–(7.3.19) are reproduced by the matrix equation of motion for $\tilde{\rho}$ in terms of the effective Hamiltonian. That is, show that in agreement with Eqs. (7.3.1) and (7.3.2),

$$i\hbar\dot{\tilde{\rho}} = -\frac{\hbar}{2}\begin{bmatrix} 0 & \Omega_1 & 0 \\ \Omega_1 & 2\Delta & \Omega_2 \\ 0 & \Omega_2 & 0 \end{bmatrix}\begin{bmatrix} \rho_{11} & \tilde{\rho}_{12} & \tilde{\rho}_{13} \\ \tilde{\rho}_{21} & \rho_{22} & \tilde{\rho}_{23} \\ \tilde{\rho}_{31} & \tilde{\rho}_{32} & \rho_{33} \end{bmatrix} + \frac{\hbar}{2}\begin{bmatrix} \rho_{11} & \tilde{\rho}_{12} & \tilde{\rho}_{13} \\ \tilde{\rho}_{21} & \rho_{22} & \tilde{\rho}_{23} \\ \tilde{\rho}_{31} & \tilde{\rho}_{32} & \rho_{33} \end{bmatrix}\begin{bmatrix} 0 & \Omega_1 & 0 \\ \Omega_1 & 2\Delta & \Omega_2 \\ 0 & \Omega_2 & 0 \end{bmatrix}. \tag{7.3.20}$$

Once the effective Hamiltonian has been written in matrix form, it is straightforward to solve for its eigenvalues λ_i and eigenvectors $\bar{W}_i$ by examining the determinant in the secular equation

$$\left[H_{eff} - \lambda\bar{\bar{U}}\right]\bar{W} = 0, \tag{7.3.21}$$

where $\bar{\bar{U}}$ represents the unit matrix.

Exercise: Show that the three eigenvalues of Eq. (7.3.21) are given by

$$\lambda_1 = 0, \tag{7.3.22}$$

$$\lambda_\pm = \Delta(t) \pm \sqrt{\Delta^2 + \Omega_1^2 + \Omega_2^2}. \tag{7.3.23}$$

By introducing a "mixing" angle defined by

$$\tan\theta(t) \equiv \Omega_1(t)/\Omega_2(t), \tag{7.3.24}$$

it may easily be verified that the eigenvector corresponding to the first eigenvalue in Eq. (7.3.22) is given by

$$|W_1(t)\rangle = \cos\theta(t)\,|1\rangle - \sin\theta(t)\,|3\rangle\,, \tag{7.3.25}$$

independent of Δ. This particular state vector does not evolve temporally on the timescale set by our choice of interaction reference frame, because its eigenvalue is zero.

Notice that as the ratio of the field amplitudes is varied, the mixing angle $\theta(t)$ changes. $\theta(t)$ goes to the limiting value in which complete transfer of population occurs from state $|1\rangle$ to state $|3\rangle$ provided the ratio of pulse amplitudes follows the prescription

$$[\Omega_1(t)/\Omega_2(t)]_{t=-\infty} \to 0 \tag{7.3.26}$$

and

$$[\Omega_1(t)/\Omega_2(t)]_{t=+\infty} \to \infty. \tag{7.3.27}$$

The first limit ensures that the system starts in state $|1\rangle$, since $|\langle 1|W_1(t)\rangle|^2_{t=-\infty} = 1$. The second limit ensures that the final state is $|3\rangle$, since $|\langle 3|W_1(t)\rangle|^2_{t=+\infty} = 1$. Notice that the limiting values of the mixing angles that are required for complete transfer of population require the second transition to precede the first. Hence the optimal pulse interaction corresponds to the counterintuitive sequence of pulses illustrated in Figure 7.14. The prescription for optimal transfer of population seems non-intuitive only because there is initially no population in state 3 to respond in the usual way to the pulse at ω_2. Nevertheless the system does respond by establishing a coherence at this frequency—a charge oscillation that is not accompanied by an atomic transition—that can mediate a two-photon transition if ground state atoms are excited within the dephasing time of the ω_2 transition.

It is easy to see now that $|W_1(t)\rangle$ is a coherently trapped state even when θ is time dependent. This stems from the fact that

$$H_{eff}|W_1(t)\rangle = 0, \tag{7.3.28}$$

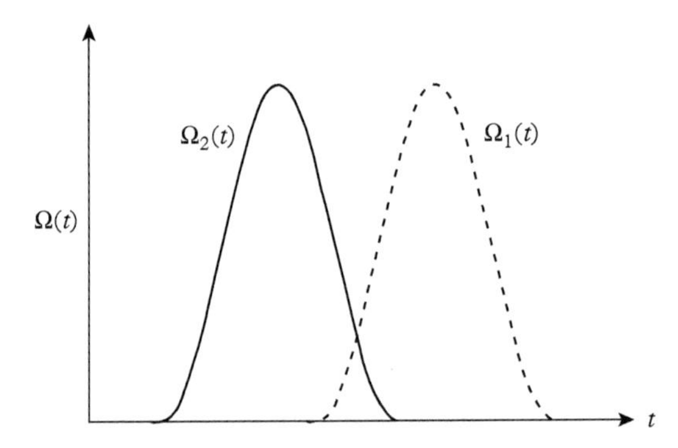

Figure 7.14 *Illustration of a suitable pulse sequence for adiabatic transfer of population from level 1 to 3. Notice that pulse 2 is applied before pulse 1.*

regardless of the values of t, $\Omega_1(t)$, $\Omega_2(t)$, and $\Delta(t)$. On the basis of the Schrödinger equation, Eq. (7.3.28) can obviously also be written in the form

$$i\hbar\frac{\partial}{\partial t}|W_1(t)\rangle = 0, \tag{7.3.29}$$

showing explicitly that, contrary to the notational implication, the state $|W_1(t)\rangle$ is essentially independent of time. The probability amplitude $|W_1\rangle$ is constant.

Moreover, any component of a more general state of the system $\psi(t)$ that has the form of $|W_1(t)\rangle$ is trapped. This is easily shown by projecting the Schrödinger equation onto state $\langle W_1(t)|$. This procedure yields

$$\langle W_1(t)|\, i\hbar\frac{\partial}{\partial t}|\psi(t)\rangle = \langle W_1(t)|\, H_{eff}(t)|\psi(t)\rangle\,. \tag{7.3.30}$$

Because $\langle W_1(t)|\, H_{eff} = 0$, or because $|W_1(t)\rangle$ is time independent according to Eq. (7.3.29) implying that the left side of Eq. (7.3.30) can be written as $i\hbar\frac{\partial}{\partial t}\langle W_1(0)|\psi(t)\rangle$, we find

$$\frac{\partial}{\partial t}\langle W_1(0)|\psi(t)\rangle = 0, \tag{7.3.31}$$

$$\langle W_1(0)\,|\psi(t)\rangle = \text{constant}. \tag{7.3.32}$$

To conclude this section, we mention a few applications. Using coherent population transfer, a population inversion of level 3 with respect to level 1 can be produced. If level 3 were a metastable level with a small ED moment for transitions to the ground state, this method of transfer could produce optical gain at frequency ω_{13} with minimal losses associated with spontaneous decay of atoms in excited state 2. Also, one can imagine storing a bit of information in the quantum bit (qubit) comprised of the "upper" and "lower" states (states 3 and 1 respectively) of the system. Qubits are discussed further in Section 7.8.

Coherent population transfer is also useful for laser cooling as described in the next sub-section, and for chemistry in state-selective reactions (see Ref. [7.26]). For further discussion of adiabatic passage see Ref. [7.24].

7.3.2 Laser Cooling of Solids

The topic of laser cooling in earlier sections began with observations of the mechanical effects of radiation pressure on atomic motion in gases. Historically, optical refrigeration of solids was similarly preceded by experiments to reduce the mechanical vibrations of mirrors along the axes of Fabry–Perot resonators. This was accomplished by monitoring excursions from resonance and using the error signal to control mirror position with variations of light pressure applied to the back side of one of the cavity mirrors. An early demonstration was reported in Ref. [7.27]. However, such an approach does not lead to refrigeration because it does not reduce internal modes of vibration irreversibly. To refrigerate a solid, the light field must couple to one or more internal modes *and* reduce entropy of the sample through radiative transport [7.28].

In gases the kinetic energy of moving atoms can be exploited to shift them preferentially into resonance with incident light via the Doppler effect [7.9]. Thus light can be applied to slow down fast-moving atoms selectively. In condensed matter however, free translational motion is absent. As a consequence, laser cooling experiments in crystals or glasses must exploit different principles. In particular they must address the energy and momentum of collective excitations of the medium if they are to provide temperature control in isolated environments like vacuum, which is only possible through radiative transport.

The earliest demonstration of optical refrigeration of a solid relied on inducing anti-Stokes fluorescence by simply tuning the wavelength of incident light to the "red" side of an impurity resonance [7.10]. Consistent with the implications of Figure 6.12, the absorption of photons on the low-energy side of a resonance is followed on average by photon emission at the center frequency of the transition. In this circumstance an overall loss of energy is imposed even on stationary atoms (Figure 7.15). Hence this general approach has met with considerable success in the cooling of rare earth solids [7.29], liquids [7.30], semiconductors [7.31], and internal acoustic modes of optomechanical resonators [7.32].

The efficiency of anti-Stokes cooling can easily be estimated. In the two-level system of Figure 7.15, energy $h\nu$ is added to the system for each absorbed photon. Energy $h\nu_{fl}$ is removed by each fluorescent photon. Hence if these two steps account for all system dynamics, then absorbed input power of P_{abs} produces a cooling power of

$$P_c = \left\{\left(\frac{\nu_{fl}}{\nu}\right) - 1\right\} P_{abs}, \tag{7.3.33}$$

and the cooling efficiency would be $\eta \equiv P_c/P_{abs} = (\nu_{fl}/\nu_L) - 1$. However, in the presence of non-radiative relaxation, which is commonplace in rare earth solids, the radiative

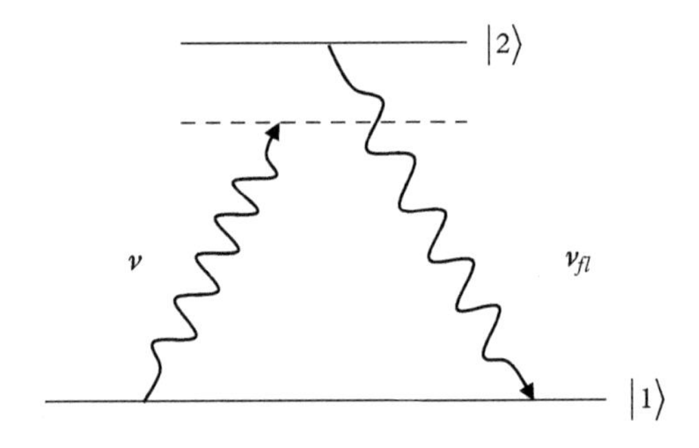

Figure 7.15 *Anti-Stokes fluorescence emission in a two-level system.*

quantum efficiency is lowered. Radiative and non-radiative decay at rates W_r and W_{nr} then contribute a quantum efficiency factor of the form $\eta_{ext} = W_r/(W_r + W_{nr})$. When corrected for fluorescence escape efficiency η_e, which is the fraction of emitted photons that escape the sample, this external quantum efficiency factor becomes

$$\eta_{ext} = \eta_e W_r/(W_r + W_{nr}). \tag{7.3.34}$$

Additionally, only a fraction of the absorbed photons result in excitation of coolant atoms. The rest are absorbed by background impurities. With the absorption coefficient of the coolant atoms designated by α_c and that of the background by α_b, the proportion of absorbed photons that contribute to cooling is

$$\eta_{abs} = \frac{\alpha_c}{(\alpha_c + \alpha_b)}. \tag{7.3.35}$$

If the assumption is made that phonon-mediated absorption is proportional to the occupation probability of the phonon mode, namely $\langle n \rangle = (\exp[\hbar\Omega/k_B T] - 1)^{-1}$, the efficiency becomes $\eta_{abs} = \alpha_c \langle n \rangle / (\alpha_c \langle n \rangle + \alpha_b)$. Then at high temperatures ($k_B T \gg \hbar\Omega$), η_{abs} varies little with temperature when the background absorption is low ($\alpha_b/\alpha_c \ll 1$). On the other hand, at low temperatures, as $\langle n \rangle \rightarrow 0$, η_{abs} decreases in proportion to $\langle n \rangle$. Hence the absorption efficiency drops dramatically through the important cryogenic region. The overall cooling is determined by the product of the various efficiency factors:

$$\eta_c = \eta_{ext}\eta_{abs}\left(\frac{\nu_{fl}}{\nu}\right) - 1. \tag{7.3.36}$$

From Eqs. (7.3.33) and (7.3.36) it is easy to show that the cooling power density is $P_c/V = (\eta_{abs}\eta_{ext}(\nu_{fl}/\nu) - 1)\alpha_{tot}I_0$, where $\alpha_{tot} \equiv \alpha_c + \alpha_b$. Here I_0 is the incident intensity and V is the irradiated volume. If the bandwidth of the light source equals the full linewidth 2Γ of the optical transition, and the density of atoms in the interaction volume is N/V, the energy loss rate per atom is

$$R = \eta_c \alpha_{tot} I_0 (V/N). \tag{7.3.37}$$

The condition for net cooling can be expressed in a simple way based on Eq. (7.3.36), since η_c must exceed zero. By assuming that the laser is optimally detuned by the phonon frequency Ω, we can rewrite the numerator using $h\nu_{fl} - h\nu = \hbar\Omega$. Furthermore the average phonon energy is $\hbar\Omega \cong k_B T$ near the boundary of the classical regime [7.33]. Hence the approximate condition for laser cooling ($\eta_c > 0$) becomes

$$\eta_{ext}\eta_{abs} > 1 - \frac{k_B T}{h\nu}. \tag{7.3.38}$$

An important limitation of refrigeration based on anti-Stokes fluorescence emerges from this condition by determining the temperature at which the cooling efficiency drops to zero ($\eta_c = 0$). By solving Eq. (7.3.38) as an equality, it is found that a bound exists for the lowest attainable temperature. The minimum temperature, $T_{\min}$, that can be reached with anti-Stokes cooling found from Eq. (7.3.38) is

$$T_{\min} = \frac{h\nu}{k_B}\left(\frac{(\alpha_b/\alpha_c) + (1 - \eta_{ext})}{1 + (\alpha_b/\alpha_c)}\right). \tag{7.3.39}$$

According to Eq. (7.3.39), if the external quantum efficiency is high ($\eta_{ext} \cong 1$), the temperature limit for laser cooling by anti-Stokes emission is primarily proportional to the frequency of light and the ratio of background to resonant absorption. The dependence on these two factors arises because heat input to the system is proportional to incident photon energy and absorptive heating efficiency. Only when the background absorption coefficient vanishes is it possible to reach arbitrarily low temperatures. For the moderate levels of background absorption encountered in practice in solids, this places a severe constraint on the lowest temperatures that can be reached by laser cooling.

Exercise: Show that the anti-Stokes cooling limit at a wavelength of 1 μm in a system with background absorption coefficient $\alpha_b = \alpha_c/1000$ is $T_{\min} = 145.3$ K when the external quantum efficiency is $\eta_{ext} = 0.90$. Assume $\langle n \rangle = 1$.

The lower bound on attainable temperatures can be interpreted in another way. It arises because the cooling rate is proportional to phonon occupation, a factor which diminishes as temperature decreases. Eventually the rate of anti-Stokes cooling drops below the rate of heating from background impurity absorption and cooling is no longer possible. For three-dimensional refrigeration of solids to be as effective as laser cooling of gases, techniques capable of maintaining high rates of cooling through the cryogenic temperature range are clearly needed.

In the remainder of this section two approaches are considered that conserve momentum in the overall light–matter interaction, to see if they offer improvement over anti-Stokes fluorescence cooling. Judging from the successes of laser cooling of gases, cooling schemes that conserve momentum and energy simultaneously during optical interactions are the most efficient. Yet momentum is not taken into account in the anti-Stokes fluorescence method. Other techniques may offer the capability of balancing the momentum and energy requirements more consistently to accommodate phonon dispersion (Figure 7.16).

In gases the link between linear momentum and the absorption of a photon is simple. For each photon absorbed, the linear momentum of an approaching atom decreases by an average of $\hbar k$. Translational momentum is thereby reduced, provided fluorescence randomizes the direction of re-emitted photons. In solids things are not so simple because free translation of atoms cannot take place. Atomic displacements are oscillatory. Motion is conveyed in collective waves. Momentum is distributed. Nevertheless, charge motions associated with departures of individual atoms from their equilibrium positions have Fourier components with well-defined amplitude, momentum, and energy [7.34]. Moreover, Doppler-shifted interactions with these phonon modes are well known to

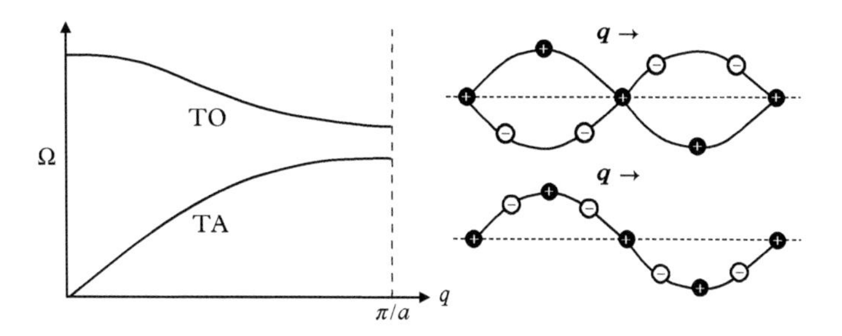

Figure 7.16 *(a) Dispersion curves illustrating the variation of phonon frequency Ω with wavevector q for transverse acoustic (TA) and optic (TO) modes. The lattice constant is a. (b) Particle displacements in a diatomic linear lattice for TO (upper) and TA (lower) modes of the same wavelength.*

take place in acousto-optic (Brillouin) scattering [7.33]. So analogies to Doppler cooling methods in gases are useful. In the remainder of this section we therefore formulate the problem of light interacting with acoustic and optical modes of vibration in terms of interaction Hamiltonians and equations of motion that display momentum explicitly. Two recent developments are reviewed. Detailed consideration is given to laser cooling by stimulated Raman scattering because in theory it furnishes a cooling rate that is independent of temperature.

The energy of interaction between light and a solid derives from the electron–phonon interaction in which the force of light displaces ions from equilibrium by an amount $\delta\bar{R}(t)$. This displacement can be written as a Fourier expansion in terms of normal modes of polarization $\hat{\varepsilon}_{k\lambda}$ ($\lambda = 1, 2, 3$), amplitude $Q_{q\lambda}$, wavevector $\bar{q}$, and frequency $\Omega_{q\lambda}$. For an ion situated at equilibrium position $\bar{R}_0$ in a solid, undergoing motion described by a single phonon mode, the displacement consists of only one Fourier term:

$$\delta\bar{R}(t) = (NM)^{-1/2}Q(\Omega, t)\hat{\varepsilon}\exp[i\bar{q}\cdot\bar{R}_0]. \tag{7.3.40}$$

N is the total number of particles. M is the particle mass. The energy of interaction is given by force times distance, or

$$H_I = -\bar{\nabla}V\cdot\delta\bar{R}. \tag{7.3.41}$$

The quantized form of the interaction Hamiltonian H_I depends on the coupling mechanism between light and the vibrational mode of interest, and the cases of acoustic and optical vibrational modes are different. Hence they are considered separately in the following paragraphs.

For transverse *acoustic modes*, which are propagating density fluctuations, the light–matter interaction is determined by electrostrictive deformations of the medium because transverse optical fields cannot couple directly to longitudinal waves. The energy dependence of acoustic and optical modes on wavevector is shown in Fig. 7.16. To generate pressure waves, a deformation potential [7.35] is needed of the form

$$\bar{\nabla}V = -\gamma\bar{\nabla}\left(\bar{E}_1 \cdot \bar{E}_2^*\right) = -\gamma q E_1 E_2^* \hat{\varepsilon}. \tag{7.3.42}$$

Here

$$E_1(z,t) = \xi_k\left(\hat{a}_1^- e^{-i(\omega_1 t - kz)} + \hat{a}_1^+ e^{i(\omega_1 t - kz)}\right) \tag{7.3.43}$$

and

$$E_2^*(z,t) = \xi_{k+q}\left(\hat{a}_2^+ e^{i(\omega_2 t - [k+q]z)} + \hat{a}_2^- e^{-i(\omega_2 t - [k+q]z)}\right) \tag{7.3.44}$$

are quantized pump and anti-Stokes fields, $\xi_k = \sqrt{\hbar\omega_k/2\varepsilon_0 V}$ is the field per photon, and $\gamma \equiv [\rho(\partial\varepsilon/\partial\rho)]_{\rho_0}$ is the electrostrictive constant relating changes of permittivity to changes of density [7.36]. The operator for normalized mode amplitude is related to the phonon creation and annihilation operators $\hat{b}^+$ and $\hat{b}^-$ by

$$\hat{Q}(\Omega,t) = \left(\frac{\hbar}{2\Omega}\right)^{1/2}\left(\hat{b}_q^- \exp[-i\Omega t] + \hat{b}_q^+ \exp[i\Omega t]\right). \tag{7.3.45}$$

We omit the subscript λ customarily used to label the polarization of the mode because we consider only one type of mode at a time here. When Eq. (7.3.40) is combined with Eqs. (7.3.42)–(7.3.45) in Eq. (7.3.41) the interaction Hamiltonian becomes

$$\begin{aligned} H_I &= -(NM)^{-1/2}\left(\hat{\varepsilon} \cdot \bar{\nabla}V\right) Q(\Omega,t) \exp[i\bar{q} \cdot \bar{R}_0] \\ &= \hbar f\left(\hat{b}_q^- a_{k+q}^+ \hat{a}_k^- + h.c.\right). \end{aligned} \tag{7.3.46}$$

In Eq. (7.3.46) an interaction strength $f \equiv (2\hbar NM\Omega)^{-1/2} V\gamma q\xi_k\xi_{k+q}$ has been introduced. V is the volume of interaction. Conservation of momentum in the interaction is illustrated in Figure 7.17.

The Hamiltonian can be used to calculate temporal changes in the vibrational energy (or temperature) of the medium. To appreciate the connection between H_I and vibrational energy or temperature, note that the conjugate momentum of the vibrational amplitude $\hat{Q}(\Omega,t)$ is

$$\hat{P}(\Omega,t) = -i\left(\frac{\hbar\Omega}{2}\right)^{1/2}\left(\hat{b}_q^- \exp[-i\Omega t] - \hat{b}_q^+ \exp[i\Omega t]\right). \tag{7.3.47}$$

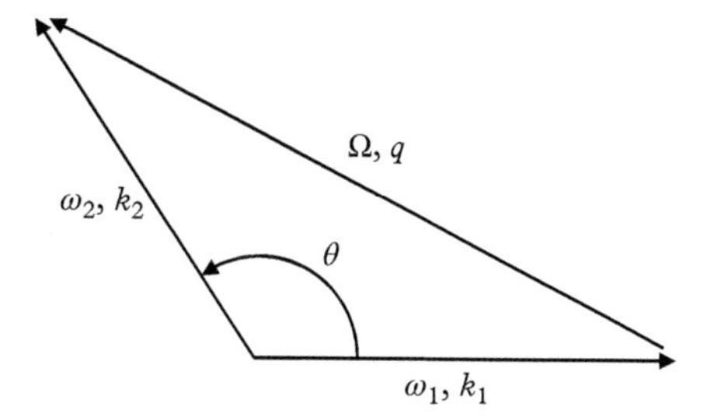

Figure 7.17 *Wavevector diagram for phonon scattering through angle θ. Phonon frequency Ω and wavevector $\bar{q} = \bar{k}_2 - \bar{k}_1$ conserve energy and linear momentum in the wave interaction involving optical fields $\bar{E}_1(\omega_1)$ and $\bar{E}_2(\omega_2)$.*

When $\hat{Q}$ and $\hat{P}$ are as previously defined, the single mode Hamiltonian specifying vibrational energy reduces to a familiar form, as already outlined at the beginning of Chapter 6:

$$H_q = \frac{1}{2}\left(\hat{P}_q^+\hat{P}_q + \Omega_q^2\hat{Q}_q^+\hat{Q}_q\right) = \frac{1}{2}\hbar\Omega_q\left(\hat{b}_q^+\hat{b}_q^- + h.c.\right). \tag{7.3.48}$$

Taking all modes into account (see Problem 6.5 for a related summation), the total energy is found to be

$$H = \sum_q H_q = \sum_q \hbar\Omega_q\left(\hat{b}_q^+\hat{b}_q^- + \frac{1}{2}\right). \tag{7.3.49}$$

Since $\langle\hat{n}_q\rangle = \left\langle\hat{b}_q^+\hat{b}_q^-\right\rangle$ is the number of phonons in mode q, the summation in Eq. (7.3.49) describes the total vibrational kinetic energy of the medium as a weighted sum of the individual mode energies which in turn determines the sample temperature.

With the interaction Hamiltonian of Eq. (7.3.46) in hand, optically driven dynamics of two optical waves interacting with an acoustic mode can be analyzed. For example, the change of phonon occupation with time can be calculated using the Heisenberg equation of motion

$$\begin{aligned}\dot{\hat{n}}_b &= (i/\hbar)[H_0 + H_{\text{int}}, \hat{n}] \\ &= (i/\hbar)\left\{\hbar f\left(\hat{b}^-\hat{a}_2^+\hat{a}_1^- + h.c.\right), \hat{b}^+\hat{b}^-\right\} \\ &= if\left\{\hat{b}^-\hat{a}_2^+\hat{a}_1^-\left(\hat{b}^+\hat{b}^-\right) - \hat{b}^+\left(\hat{b}^-\hat{b}^+\right)\hat{a}_2^-\hat{a}_1^+\right\}.\end{aligned} \tag{7.3.50}$$

The static Hamiltonian $H_0 = \hbar\omega_1\hat{a}_1^+\hat{a}_1^- + \hbar\omega_2\hat{a}_2^+\hat{a}_2^- + \hbar\Omega\hat{b}^+\hat{b}^-$ plays no role in dynamics. Hence both terms on the right of Eq. (7.3.50) are proportional to f, the strength of the optical interaction. The first term determines the rate of annihilation of phonons while the second is the rate of creation. We can show this explicitly by solving for the temporal behavior of $\hat{b}^+$ and $\hat{b}^-$, again using the Heisenberg picture:

$$(d\hat{b}^+/dt) = i\Omega\hat{b}^+ + if\hat{b}^-\hat{a}_1^-\hat{a}_2^+\hat{b}^+ - if\hat{b}^+\hat{a}_1^+\hat{a}_2^-\hat{b}^- - \Gamma_b\hat{b}^+, \tag{7.3.51}$$

$$(d\hat{b}^-/dt) = -i\Omega\hat{b}^- - if\hat{b}^+\hat{a}_1^+\hat{a}_2^-\hat{b}^- + if\hat{b}^-\hat{a}_1^-\hat{a}_2^+\hat{b}^+ - \Gamma_b\hat{b}^-. \tag{7.3.52}$$

Making use of the SVEA, steady-state solutions for the operator amplitudes are simply

$$\hat{b}^+ = \left\{if\hat{b}^-\hat{a}_1^-\hat{a}_2^+\hat{b}^+ + h.c.\right\}/\Gamma_b, \tag{7.3.53}$$

$$\hat{b}^- = \left\{-if\hat{b}^+\hat{a}_1^+\hat{a}_2^-\hat{b}^- + h.c.\right\}/\Gamma_b. \tag{7.3.54}$$

The net rate of change of phonon occupation in the state $|n_1\rangle\,|n_2\rangle\,|n_b\rangle$ is determined by the balance of annihilation versus creation. By substituting Eqs. (7.3.53) and (7.3.54) in Eq. (7.3.50) to determine the expectation value of $\dot{\hat{n}}_b$, one finds

$$\begin{aligned}\left\langle\dot{\hat{n}}_b\right\rangle &= if\langle n_b|\langle n_1|\langle n_2|\left\{-if\hat{b}^+\hat{a}_1^+\hat{a}_2^-\hat{b}^- + h.c.\right\}\hat{a}_2^+\hat{a}_1^-\left(\hat{b}^+\hat{b}^-\right)|n_2\rangle\,|n_1\rangle\,|n_b\rangle/\Gamma_b \\ &\quad -if\langle n_b|\langle n_1|\langle n_2|\left\{if\hat{b}^-\hat{a}_1^-\hat{a}_2^+\hat{b}^+ + h.c.\right\}\hat{a}_2^-\hat{a}_1^+\left(\hat{b}^-\hat{b}^+\right)|n_2\rangle|n_1\rangle|n_b\rangle/\Gamma_b.\end{aligned} \tag{7.3.55}$$

Upon evaluation of the matrix elements we obtain the rate equation for phonon occupation:

$$\dot{n}_b = -(f^2/\Gamma_b)n_1(n_2+1)n_b + (f^2/\Gamma_b)(n_1+1)n_2(n_b+1). \tag{7.3.56}$$

The first term on the right side of Eq. (7.3.56) describes cooling while the second describes heating. Possible detuning of the modes has been ignored. The phonon annihilation rate for resonant Brillouin scattering in the interaction volume is therefore the coefficient of the first term in Eq. (7.3.56):

$$\gamma_B = (f^2/\Gamma_b)n_1(n_2+1)n_b. \tag{7.3.57}$$

The rate R of energy loss by the system must take into account the phonon frequency:

$$R = \gamma_B\hbar\Omega = (f^2/\Gamma_b)n_1(n_2+1)n_b\hbar\Omega. \tag{7.3.58}$$

Laser cooling is realized if conversion of an optical pump wave to an anti-Stokes wave prevails over its conversion to a Stokes-shifted wave. Brillouin cooling of a mechanical mode was first demonstrated in this way by simultaneously matching the momentum and energy difference of wavevectors and frequencies of the pump and anti-Stokes waves in a micro-resonator to the acoustic mode [7.32]. To do this the pump and anti-Stokes waves were simultaneously enhanced in the resonator while the Stokes component of Brillouin scattering was suppressed through wavevector mismatch.

Next, stimulated Raman scattering is considered for the cooling of solids. The general theory of stimulated scattering [7.37, 7.38] was developed in the 1960s and despite the complexity of equations coupling the various mode amplitudes Louisell [7.39] pointed out that the interaction energy is just the added electromagnetic energy resulting from a change in the permittivity ε, or $H_I = \frac{1}{2}\int \delta\varepsilon E^2$. This perspective facilitates the analysis of both Brillouin and Raman scattering since it is straightforward to write $\delta\varepsilon$ to reflect one interaction or the other. In the last section the dielectric fluctuation in Brillouin scattering was written $\delta\varepsilon = -\gamma(\partial Q/\partial x)$ to describe electrostriction by the incident light. In this section $\delta\varepsilon$ arises from dielectric fluctuations due to transverse charge motion caused directly by the transverse electric field of the light. We shall proceed as before nevertheless, writing down the interaction Hamiltonian using Eq. (7.3.41), since the deformation potential approach is equivalent to that of Louisell.

For transverse *optical modes*, the net optical electric field can act directly on charged ions at the difference frequency to create or destroy optical phonons. In this case the potential is just the optical energy density. Hence the gradient of the potential in Eq. (7.3.41) is $\bar{\nabla}V = \varepsilon_0\bar{\nabla}(\bar{E}_1 \cdot \bar{E}_2^*)\hat{\varepsilon}$, and the single mode interaction Hamiltonian is

$$\begin{aligned} H_I &= \left(\frac{\hbar\varepsilon_0^2}{2\Omega NM}\right)^{1/2} \xi_k\xi_{k+q}\left(\hat{a}_1^-\hat{a}_2^+\hat{b}^- + h.c.\right). \\ &= \hbar f'\left(\hat{a}_1^-\hat{a}_2^+\hat{b}^- + h.c.\right). \end{aligned} \tag{7.3.59}$$

Here the coupling strength in the interaction volume V has been redefined as $f' \equiv -(2\hbar NM\Omega)^{-1/2}\varepsilon_0 Vq\xi_k\xi_{k+q}$. By careful selection of the frequencies and wavevectors of two optical fields, the energy and momentum can be conserved for optical modes just as for acoustic modes.

Two-photon Raman transitions (Figure 7.18) can couple to optical modes of vibration purely through anti-Stokes scattering. Consequently, they can be utilized for efficient cooling. Early analysis and experiments on Raman cooling in gases may be found in Refs. [7.40–7.42]. Here we follow Ref. [7.43] to explain how the Raman technique can be adapted to cooling solids and to estimate the cooling rate. The potential advantage of applying adiabatic rapid passage to this technique is also considered. The fact that the cooling rate is temperature independent through the cryogenic range is emphasized.

With pulsed excitation, the two fields at ω_1 and ω_2 in Figure 7.18 cause the system population to evolve in the direction indicated by wiggly arrows. The light drives population from state 1 to state 3 at a rate determined by the one- and two-photon detunings Δ and δ. The $1 \rightarrow 2$ and $2 \rightarrow 3$ transitions are assumed to be electric-dipole allowed. The selection rules therefore dictate that state 3 is a long-lived storage state. For prolonged excitation, population builds up in level 3 and the reverse process begins to take place at a comparable rate. For the two photon detuning shown in the figure, a small energy deficit ($\hbar\delta$) and momentum reduction ($2\hbar k$) is incurred for each Raman

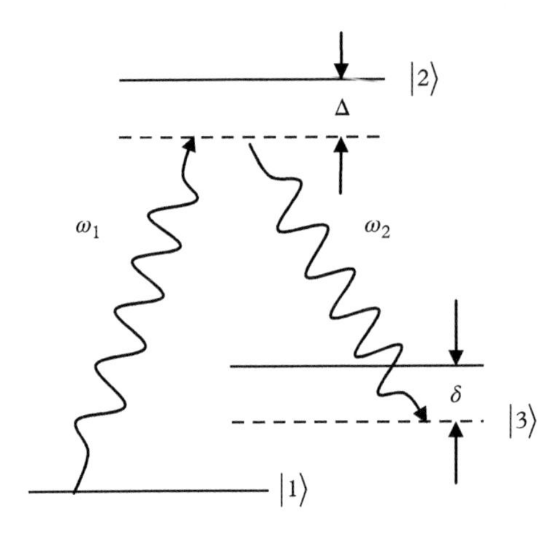

Figure 7.18 *A three-level system undergoing a two-photon Raman process.* $\Delta = \omega_{12} - \omega$ *is the one-photon detuning and* $\delta = \omega_{13} - (\omega_1 - \omega_2)$ *is the two-photon detuning.*

cycle, provided the pulses are short enough to avoid substantial buildup of population in state 3 [7.44, 7.45]. Atoms with a velocity centered at $v = 2\hbar k/m$ lose kinetic energy. By emptying state 3 periodically with a fast, random process like fluorescence, changes in the energy and momentum distributions become irreversible through an increase in the entropy of the outgoing radiation field [7.46, 7.47]. Hence the ensemble of atoms can be progressively cooled.

To estimate the stimulated Raman transition rate for the $1 \rightarrow 3$ transition it is instructive to assess the importance of "forward" and "backward" pumping processes between states 1 and 3 (Figures 7.18 and 7.19, respectively). It is easily shown that the net rate of change of population is zero in steady state because of the balance of these two processes, which provide cooling and heating respectively. One can also show that the transition rate

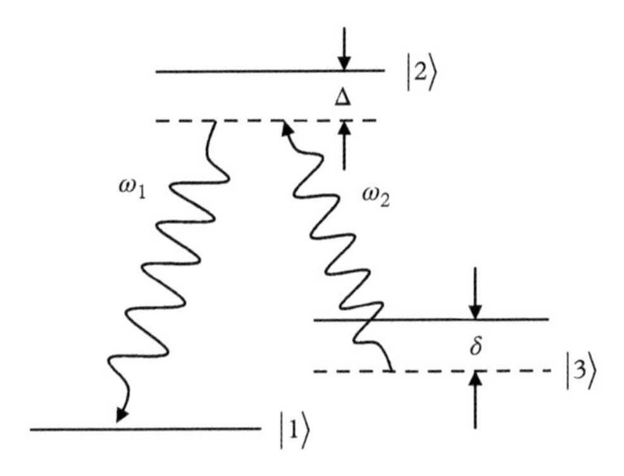

Figure 7.19 *Raman interaction of fields* $E_1(\omega_1)$ *and* $E_2(\omega_2)$ *in time-reversed ("backward") sequence.*

for a single process is given by the square of an off-diagonal (coherence) element of the density matrix that may be evaluated straightforwardly to estimate the cooling rate.

The net rate of population change in excited state 3 is just $\dot{\rho}_{33} \equiv d\rho_{33}/dt$. In the weak field limit the wavefunction is dominated by its ground state component, which varies little with time. Hence we may assume $\rho_{11} = c_1 c_1^* \cong 1$ and insert $\langle 1| c_1 c_1^* |1\rangle \cong 1$ into the defining expression for $\dot{\rho}_{33}$ in order to calculate the transition rate:

$$\dot{\rho}_{33} = \frac{d}{dt}\left[\langle 3|\psi\rangle\langle\psi|3\rangle\right] \cong \frac{d}{dt}\left[\langle 3|\, c_3(t)|3\rangle\langle 1|\, c_1^* c_1\, |1\rangle\langle 3| c_3^*(t)|3\rangle\right] = \dot{\rho}_{31}\rho_{13} + \rho_{31}\dot{\rho}_{13}. \tag{7.3.60}$$

Since $\rho_{31} = \rho_{13}^*$ and $|\rho_{31}| = |\rho_{13}|$, one finds $\dot{\rho}_{31} = \dot{\rho}_{13}^* = -i\omega_{13}\rho_{31}$ and $\dot{\rho}_{13} = i\omega_{13}\rho_{13}$. The result is that the two contributions to $\dot{\rho}_{33}$ given in Eq. (7.3.60) are equal and opposite. Under steady-state conditions the occupation of state 3 therefore does not change:

$$\dot{\rho}_{33} = -i\omega_{13}\,|\rho_{13}|^2 + i\omega_{13}\,|\rho_{13}|^2 = 0. \tag{7.3.61}$$

The rate at which level 3 population increases is determined by the positive term in Eq. (7.3.61), assuming one vibrational quantum occupies the mode. The transition rate per atom per unit (angular frequency) excitation bandwidth that leads to cooling is therefore

$$|\dot{\rho}_{33}/\omega_{13}| = |\rho_{13}|^2\,. \tag{7.3.62}$$

To determine the rate itself from the field-driven coherence in Eq. (7.3.62), the source bandwidth must be specified. Taking this to equal the full transition linewidth $2\Gamma_{13}$ in the following discussion, it remains only to calculate the matrix element ρ_{13} explicitly.

For quasi, steady-state fields, the interaction Hamiltonian can be written

$$V = \left(-\frac{1}{2}\hbar\Omega_{12}\exp(i\omega_1 t) + c.c.\right) + \left(-\frac{1}{2}\hbar\Omega_{23}\exp(i\omega_2 t) + c.c.\right). \tag{7.3.63}$$

Contributions to state 3 dynamics other than the one depicted in Figure 7.18 will be ignored owing to their large detunings from resonance. The first-order coherence is therefore given by the standard perturbation result (Appendix G):

$$\tilde{\rho}_{12}^{(1)} = \left(\frac{\Omega_{12}/2}{\Delta_1 + i\Gamma_{12}}\right)\left(\rho_{11}^{(0)} - \rho_{22}^{(0)}\right). \tag{7.3.64}$$

The second-order coherence established by the forward Raman process is obtained by solving the second-order transport equation (Appendix D), which is

$$i\hbar\dot{\rho}_{13}^{(2)} = \hbar(\omega_1 - \omega_3)\rho_{13}^{(2)} + \left[V^{(1)}, \rho^{(1)}\right]_{13} - i\hbar\Gamma_{13}\rho_{13}^{(2)}. \tag{7.3.65}$$

The two-photon coherence oscillates at the frequency of the (combined) driving fields. So if the SVEA is adopted, the form of ρ_{13} is

$$\rho_{13}^{(2)}(t) = \tilde{\rho}_{13}^{(2)} \exp[i(\omega_1 - \omega_2)t]. \tag{7.3.66}$$

By substituting Eqs. (7.3.63), (7.3.64), and (7.3.66) into Eq. (7.3.65) and grouping terms with the same time dependence one finds the solution for the coherence to be

$$\tilde{\rho}_{13}^{(2)} = \left\{ \left(\frac{\Omega_{12}\Omega_{23}^*/4}{[\Delta_1 + i\Gamma_{12}][\Delta_2 + i\Gamma_{13}]} \right) \left(\rho_{11}^{(0)} - \rho_{22}^{(0)} \right) \right\}. \tag{7.3.67}$$

The second denominator incorporates the two-photon detuning $\Delta_2 \equiv \omega_{31} - (\omega_1 - \omega_2)$. Using this result in Eq. (7.3.62), the Raman transition rate per atom is obtained with an appropriate assumption regarding excitation bandwidth. If the excitation matches the bandwidth of the Raman transition ($2\Gamma_{13}$), the number of phonon annihilations per second per coolant atom is

$$\gamma_R = 2\Gamma_{13} <n> \left| \tilde{\rho}_{13}^{(2)} \right|^2 = 2\Gamma_{13}^{-1} <n> |\langle f | M | i \rangle|^2, \tag{7.3.68}$$

where adjustment has been made for the number of phonons occupying the mode, using the Planck distribution $\langle n \rangle = (\exp[\hbar\Omega/k_B T] - 1)^{-1}$. In this model the implicit summation over intermediate states in the matrix element $\langle f | M | i \rangle$ reduces to a single term:

$$\langle f | M | i \rangle = \left[\frac{\langle 3 | \bar{\mu}^{(e)} \cdot \hat{e}_2 | 2 \rangle \langle 2 | \bar{\mu}^{(e)} \cdot \hat{e}_1 | 1 \rangle}{[\Delta_1 + i\Gamma_{12}][\Delta_2 + i\Gamma_{13}]} \right] \frac{E(\omega_1)E * (-\omega_2)}{4\hbar^2}. \tag{7.3.69}$$

The rate of energy loss is therefore

$$R = \gamma_R \hbar\Omega = 2\Gamma_{13}^{-1} \langle n \rangle |\langle f | M | i \rangle|^2 \hbar\Omega. \tag{7.3.70}$$

This rate applies to each atom in the interaction volume because it was calculated using the density matrix which is normalized to the number of coolant ions.

To realize this cooling rate in practice, the optical interaction must be made irreversible. To this point, the dynamics we have described are coherent and proceed as readily in the "backward" direction as the "forward" direction. We must ensure that the second term in Eq. (7.3.60) describing reverse transitions is negligible. Also the overall efficiency may benefit from interaction with more than one mode at a time. So we briefly consider these aspects of Raman cooling next with the aid of Figure 7.20.

Improvement in the Raman cooling rate can be anticipated by interacting with a range of vibrational modes during each cooling cycle. More than one mode could be addressed for example by rapidly scanning the two-photon detuning δ in discrete steps [7.44] or by using some form of adiabatic passage [7.46]. Reversal of the temporal order of two fixed frequency pulses was shown in Section 7.3.1 to transfer population

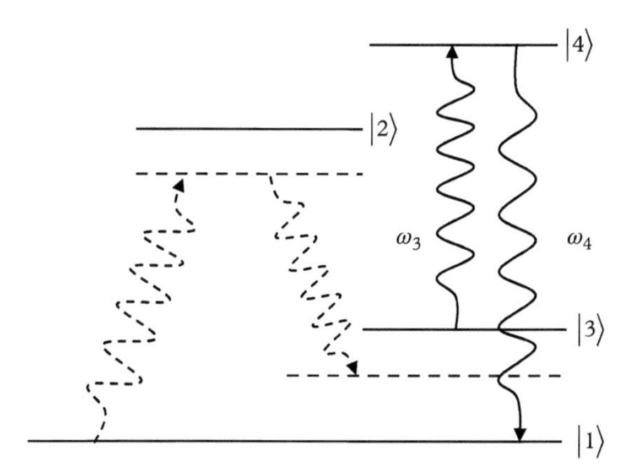

Figure 7.20 *The coherent Raman process (dashed arrows) must be followed by random emission for there to be net cooling. Here, spontaneous emission is induced at ω_4 with a pump source at frequency ω_3 that returns population from state 3 to state 1.*

between two discrete states with nearly unit efficiency at a fixed detuning. For rapid cooling over a broad range of phonon frequencies, rapid adiabatic passage is a superior method in which the excitation detuning is swept from one side of Raman resonance to the other adiabatically. This approach can be implemented very simply by imposing a "chirp" on $\delta(t)$. By sweeping the frequency rapidly from one side of resonance to the other, all phonons with frequencies between the endpoints $\pm\delta_{\text{max}}$ of the chirp are cooled at once.

Refrigeration requires irreversibility. The coherent (reversible) Raman process outlined must therefore be combined with a random process that increases the entropy of outgoing radiation [7.47, 7.48] without decreasing efficiency. For this purpose fluorescence on an allowed transition at frequency ω_4 can be induced by a strong "repump" beam (at frequency ω_3) as illustrated schematically in Figures 7.20 and 7.21. To avoid lowering the rate of cooling set by anti-Stokes Raman transitions, the return of excited state population to the ground state should be fast. Provided state 3 is emptied at a rate faster than the Raman excitation process which fills it, and no extra heat is generated in the process, efficient cooling will be maintained.

The need to avoid heat generation during the repump cycle affects how closely the repump beam may be tuned to the nearest resonance. The criterion for optimizing one-photon detuning in Raman cooling of solids is different from that for gases because the configuration of neighboring atoms surrounding a coolant atom changes as the dipole forms during transition of a coolant atom from a low- to a high-energy state. That is, the equilibrium positions of neighbors change in response to the absorption process. Phonons are unavoidably generated by relaxation of this non-equilibrium configuration unless the light is detuned on the red side of resonance by more than the phonon energy. In Figure 7.21, detuning requirements are illustrated in the context of energy levels of Ce^{3+} ions. Given the electronic structure of Ce^{3+}, the laser must be detuned by an

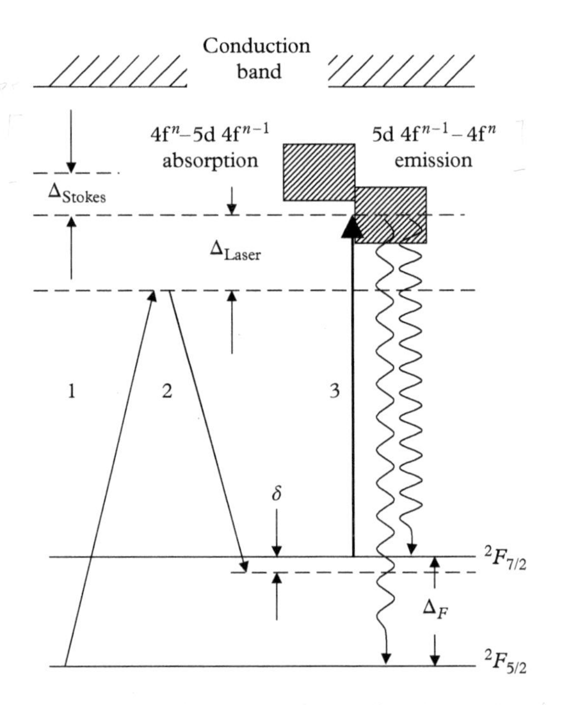

Figure 7.21 *Energy levels and detunings involved in Raman laser cooling of a three-level coolant atom in a solid. The example is based on the electronic structure of* Ce^{3+}.

amount equal to or greater than the Stokes shift of the longest wavelength $4f^{n-1}5d \leftrightarrow 4f^n$ transition. For Ce^{3+} this means

$$\Delta_{Laser} \geq \Delta_{Stokes} + \Delta_F, \tag{7.3.71}$$

where Δ_F is the energy separating the $^2F_{5/2}$ and $^2F_{7/2}$ energy levels. This choice ensures that absorbed photons have less than the energy required to generate either a vibrationally excited upper state or a vibrationally excited ground state. In the Ce^{3+} system, cooling ought to maximize when both the Raman and repump beams are detuned by $\Delta_{Laser} = \Delta_{Stokes} + \Delta_F$. In practice, smaller detunings might improve the cooling if the transition rate from the 5d state to $^2F_{7/2}$ is less than that to $^2F_{5/2}$ as it is in Ce:YAG [7.49]. Despite the reduction in cooling rate that large detunings incur in general, high overall cooling rates can still be expected when the radiative lifetime of the excited state is short, and the homogeneous linewidth of the transition is exceptionally broad, as for the 5d state of Ce^{3+} [7.43].

To conclude this chapter we emphasize two key advantages of Raman cooling of solids, namely that the cooling rate is temperature independent and that the cooling process is divided into two main steps involving different levels, thereby permitting separate optimization of absorption and fluorescence in principle. At low temperatures, the phonon occupation factor in Eq. (7.3.68) drops below one and varies as $<n> \cong \exp(-\hbar\Omega/k_B T)$. In this non-classical regime, any cooling rate that is solely proportional to $<n>$ falls exponentially as temperature decreases. Indeed γ_R is proportional to $<n>$, just like the rate for anti-Stokes fluorescence cooling. However, the phonon-broadened linewidth Γ_{13} of the Raman transition is also strongly temperature dependent and offsets the decline with a matching proportionality to $<n>$ of its own. In the following model there is complete cancellation of the factor of $<n>$ in Eq. (7.3.68) at two-photon resonance. Since the transition to state 3 is detuned by the phonon frequency below the terminal electronic state (see Figure 7.18), the relevant resonant condition is $\Delta_2 = 0$.

While other mechanisms may contribute, we assume here that the Raman linewidth is governed primarily by a two-phonon Orbach process [7.50]. Thus it can be modeled as $\Gamma_{13}(T) \cong \Gamma_0 \left(1 + \langle n \rangle \left[\Gamma_0'/\Gamma_0\right]\right)$. In this expression Γ_0 is the inhomogeneous linewidth at zero temperature and Γ_0' is the linewidth constant for the Orbach term. When $\Gamma_0' > \Gamma_0$ the linewidth reduces to $\Gamma_{13}(T) \cong <n>\Gamma_0'$. Upon substitution into Eq. (7.3.68) with zero two-photon detuning ($\Delta_2 = 0$), the rate of phonon annihilation per atom becomes constant:

$$\gamma_R = \frac{2}{\Gamma_0'} \left|\langle f| M' |i\rangle\right|^2 . \tag{7.3.72}$$

In this expression the modified matrix element M' appears because of the cancellation between $<n>$ and two-photon linewidth factors in the model outlined. Its definition is

$$\langle f| M' |i\rangle = \left[\frac{\langle 3|\,\bar{\mu}^{(e)} \cdot \hat{e}_2\,|2\rangle\langle 2|\,\bar{\mu}^{(e)} \cdot \hat{e}_1\,|1\rangle}{[\Delta_1 + i\Gamma_{12}]}\right] \frac{E(\omega_1)E*(-\omega_2)}{4\hbar^2}. \tag{7.3.73}$$

The energy loss rate per atom is

$$R = \gamma_R \hbar\Omega = \frac{2}{\Gamma_0'} \left|\langle f| M' |i\rangle\right|^2 \hbar\Omega. \tag{7.3.74}$$

This is similarly expected to be temperature independent over the entire range $k_B T > \hbar\Gamma_0$. Unlike acoustic modes, optical phonons have very little dispersion (Figure 7.16). Consequently, the frequency of the dominant optical phonon does not vary significantly and R remains constant over a wide range of temperature.

The one-dimensional Raman interaction described in this section involves a phonon near the center of the Brillouin zone propagating along a single optical axis. However, zone center phonons cannot participate directly in the collisional processes that establish thermal equilibrium within the phonon distribution. Hence it is natural to inquire as to how effectively a mode that is cooled irreversibly can equilibrate with the reservoir of other phonons in the solid. Restoration of equilibrium in solids is governed by

Umklapp processes, phonon collisions requiring a reciprocal lattice vector $\bar{G}$ to conserve momentum [7.51]. Zone center phonons have low wavevectors, so when two phonons interact with a third via an Umklapp process, there is no way to satisfy the relation $\bar{k}_1 + \bar{k}_2 = \bar{k}_3 + \bar{G}$ with a wavevector $\bar{G}$ that spans the entire Brillouin zone. To achieve uniform cooling of even a single mode, phonons propagating along orthogonal axes should be addressed, and this calls for the introduction of two more sets of counter-propagating Raman beam pairs to cover all three orthogonal space axes. Additionally, beam switching along each axis is needed to interrogate forward- and backward-traveling phonons of the same frequency. Fast directional switching of each beam pair can be implemented with a Pockel's cell that reverses the propagation direction of the ω_1 and ω_2 beams simultaneously within the sample [7.44] on a timescale much shorter than the lifetime of the $^2F_{7/2}$ shelving state.

While uniform cooling of one mode can be assured by such procedures, thermal equilibration of the sample as a whole relies on Umklapp processes that involve phonons of wavevector $\bar{G}$. Although the occupation probability of such phonons drops exponentially with decreasing temperature, the thermal conductivity which is proportional to phonon mean free path actually rises through most of the cryogenic range. Hence sample equilibration times drop until temperatures around 10–20 K are reached. Then the mean free path becomes limited by sample dimensions and thermal conductivity drops as T^3 due to its proportionality to specific heat [7.51]. The consequence of this is that fast thermal equilibration can be anticipated in Raman laser-cooled solids at all but the very lowest ($T < 10$ K) temperatures.

7.4 Coherent Transverse Optical Magnetism

The optical magnetic field generally plays a negligible role in light–matter interactions. This is attributable to the fact that the Lorentz force $q(\bar{v} \times \bar{B})$ exerted by the optical B field is much weaker than the force of E on charges moving at low speeds ($v \ll c$). For a plane wave in free space this seems to be a foregone conclusion in view of the fixed relationship between the field amplitudes ($B = E/c$, where c is the speed of light). Nevertheless strong magnetic interactions can be induced by light in nominally "non-magnetic" materials [7.52] at intensities far below the relativistic intensity threshold ($\ll 10^{18}$ W/cm^2). Light scattered at 90° with respect to a laser beam passing through a "non-magnetic" sample of liquid CCl_4 is shown in Figure 7.22, for example. It reveals an intense cross-polarized component due to magnetic dipole (MD) scattering. In this section a dressed state approach is applied to understand how light induces enhanced optical magnetism through the exertion of magnetic torque on molecules. This nonlinear optical phenomenon is shown to be second order in the applied fields, in agreement with the data of Figure 7.15(b). The analysis provides an exactly solvable model for intense interactions between light and molecules that includes both the optical electric and magnetic fields.

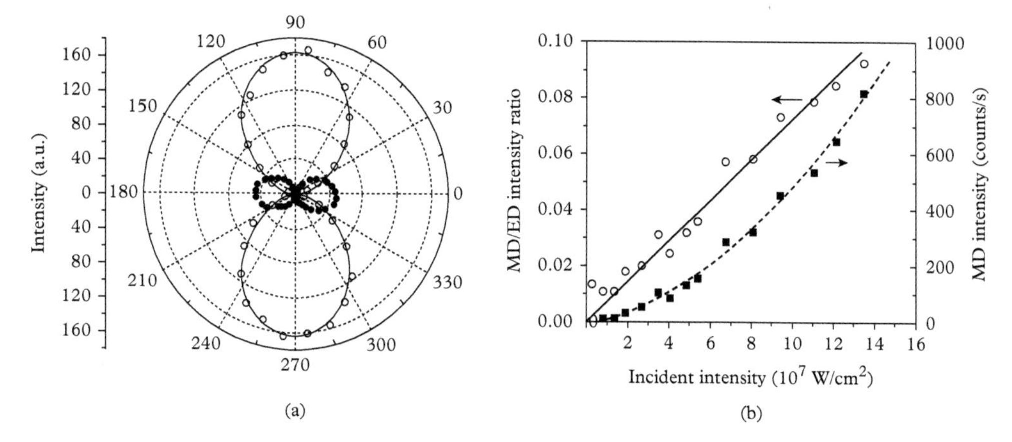

Figure 7.22 *(a) Radiation patterns of polarized light scattered at* 90° *in liquid* CCl_4. *Open circles were measured with the analyzer parallel to the incident electric field at* 10^8 *W/cm*^{2}, *whereas filled circles are for an analyzer orientation perpendicular to the incident electric field. (b) The intensity of scattered MD radiation versus incident intensity in* CCl_4, *together with its ratio with respect to ED scattering (after [7.52]).*

The filled circles in Figure 7.22(a) show the radiation pattern of a simple dipole oriented perpendicular to the polarization of Rayleigh scattering (open circles). This scattered light from a sample of spherical top molecules, which are neither chiral nor polar, is produced by moments that are orthogonal to the induced electric dipoles in the medium and can only arise from magnetization at the optical frequency. That is, it arises from magnetic dipoles induced by the optical magnetic field. Treatment of the time-dependence of this nonlinear signal is beyond the scope of this book, since it involves parametric resonance and optical instabilities [7.53]. However, steady-state analysis provides an understanding of the basic phenomenon and the reader can consult existing literature regarding these advanced topics [7.54, 7.55]. In this section we shall merely show that strongly enhanced, transverse magnetization can be induced at optical frequencies through a process driven jointly by the E and B components of light acting on bound electrons.

In this treatment only orbital and rotational angular momentum of a nominally two-level molecule are taken into account using a dressed state approach (see Section 6.8). Spin is ignored. The total angular momentum of the system is then a sum of the internal (orbital) and external (rotational) momenta ($J = L + O$) that are coupled by the action of magnetic torque on excited molecules [7.56, 7.57]. The coupling of L and O makes it possible to mix angular momentum into the ground state, enabling a nearly resonant magnetic transition at the optical frequency. The magnetic field also breaks the inversion symmetry of the medium [7.58, 7.59]. The medium becomes parity-time (P T) symmetric. As a result, transverse (radiant) magnetization becomes *allowed* even in centrosymmetric media, in contradiction to traditional symmetry analysis [7.36, 7.59].

The model we consider is that of a single electron bound to a homonuclear diatomic molecule, a rigid rotor with a one-photon resonance at frequency ω_0. The quantization axis is chosen to lie along $\hat{x}$, parallel to both the electric field and the axis of the molecule. Linearly polarized light of frequency ω propagates along $\hat{z}$, impinging on the molecule with a detuning of $\Delta \equiv \omega_0 - \omega$. The ground electronic state is taken to be ${}^1\Sigma_g^+$, the excited state ${}^1\Pi_u$, analogous to S and P states of atomic physics. Orbital angular momentum is specified by the eigenvalue of L and its projection m_l (or Λ) on the axis. The notation $|\alpha L m_l\rangle$ is used to denote uncoupled electronic states, with $\alpha = 1, 2$ specifying the principal quantum number. The basis states support an ED transition from $L = 0$ to $L = 1$ followed by an MD transition from $m_l = 0$ to $m_l = -1$, as in Figure 7.23.

The RWA is made for both the ED and MD interactions. The basis set is chosen to comprise only three electronic states: $|100\rangle, |210\rangle, |21-1\rangle$, although there is no a priori reason to exclude $|211\rangle$. Exclusion of $|211\rangle$ has the benefit of decreasing the induced magnetic moment by only 30% while permitting an analytic solution that explains magnetic enhancement. (Exact numerical results that include $|211\rangle$ in the basis are shown in Figure 7.25). Molecular rotational states are written $|Om_o\rangle$ and similarly comprise only the three states: $|00\rangle$, $|10\rangle$, and $|11\rangle$. The optical field is a single-mode Fock state $|n\rangle$, where n denotes the photon occupation number. The molecule–field states form an uncoupled product basis in which the full Hamiltonian may be readily diagonalized to evaluate eigenstates dressed by both E and B and dipole moments induced at frequencies 0 and ω.

To simplify notation, the three uncoupled product states are written as $|1\rangle \equiv |100\rangle|00\rangle|n\rangle, |2\rangle \equiv |210\rangle|10\rangle|n-1\rangle$, and $|3\rangle \equiv |21-1\rangle|11\rangle|n\rangle$. These states are eigenstates of the molecule–field Hamiltonian given by

$$\hat{H}_{mf} = \hat{H}_{mol} + \hat{H}_{field} = (\hbar\omega_0/2)\hat{\sigma}_z + \hat{O}^2/2I + \hbar\omega\hat{a}^+\hat{a}^-, \tag{7.4.1}$$

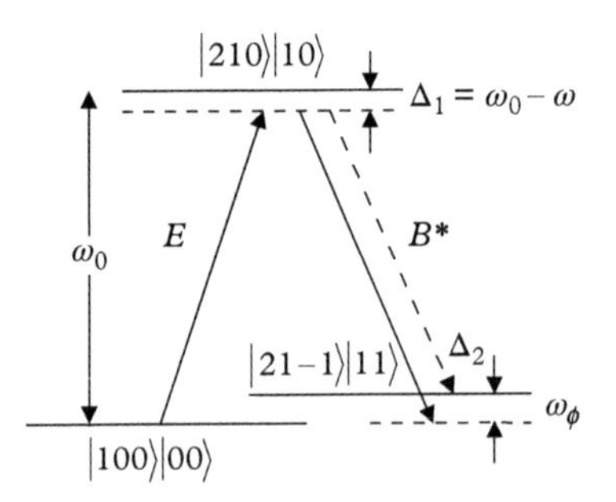

Figure 7.23 *Energy levels and transitions in a nominally two-level molecule undergoing a two-photon magneto-electric transition. The transition on the left is mediated by the electric field according to standard selection rules. The transition on the right is mediated by magnetic torque that mixes angular momentum into the dressed ground state, making possible a resonant or nearly resonant transition at the optical frequency.*

with eigenenergies $E_i (i = 1, 2, 3)$ defined by $\hat{H}_{mf}|i\rangle = E_i|i\rangle$. $\hat{O}^2/2I$ designates the kinetic energy of rotation perpendicular to the internuclear axis with moment of inertia I. Basis state energies are:

$$E_1 = -\frac{\hbar\omega_0}{2} + n\hbar\omega, \tag{7.4.2}$$

$$E_2 = E_1 + \hbar\Delta = \frac{\hbar\omega_0}{2} + (n-1)\hbar\omega, \tag{7.4.3}$$

$$E_3 = E_2 - \hbar\Delta + \hbar\omega_\phi = -\frac{\hbar\omega_0}{2} + \hbar\omega_\phi + n\hbar\omega. \tag{7.4.4}$$

In Eq. (7.4.1), $\hat{\sigma}_z$ is a Pauli spin operator. $\hat{a}^+$ and $\hat{a}^-$ are the customary raising and lowering operators of the single mode field. The sign of the first term in Eq. (7.4.4) reflects the assignment of state 3 to the electronic ground state. This energy is the result of transfer of orbital angular momentum to rotational angular momentum through magnetic torque. The torque interaction swaps the internal angular momentum (i.e., the excited state kinetic energy) for external rotation of the molecule. This has the unexpected consequence of returning the molecule to the electronic ground state [7.60]. Note that state 3 has rotational energy $\hbar\omega_\phi \equiv \hbar^2/I$ consistent with a two-photon interaction that terminates in a rotationally excited ground-state sublevel.

As depicted in Figure 7.23, the *EB** process has a small detuning from two-photon resonance. In the case of monochromatic light the detuning is equal to the rotational frequency of the molecule, which is much smaller than the optical frequency ($\Delta_2 = \omega_\phi \ll \omega$). At low intensities the second step in the process, the magnetic transition, would be expected to take place at zero frequency among degenerate magnetic sublevels in the excited state via the interaction Hamiltonian $-\bar{\mu} \cdot \bar{B}$. However this overlooks the possibility of transfer of angular momentum between the internal and external rotational degrees of freedom of the molecule at moderately high intensities. When torque is included [7.58], the Hamiltonian of the magnetic transition is $\hat{H}_{\text{int}}^{(m)} = \hbar f \hat{L}'_- \hat{O}'_+ \hat{a}^+ + h.c.$ Primes on the raising and lowering operators for orbital ($\hat{L}'_\pm \equiv \hat{L}_\pm/\hbar$) and rotational angular momentum ($\hat{O}'_\pm \equiv \hat{O}_\pm/\hbar$) indicate division by $\hbar$. Their action on angular momentum states follows the usual prescription (see Appendix H), which is $\hat{J}_\pm|\alpha jm\rangle = \sqrt{(j \mp m)(j \pm m + 1)}\hbar\,|\alpha jm \pm 1\rangle$. The total interaction becomes a sum of electric and magnetic terms:

$$\hat{H}_{\text{int}} = \hat{H}_{\text{int}}^{(e)} + \hat{H}_{\text{int}}^{(m)} = \hbar g(\hat{\sigma}^+\hat{a}^- + h.c.) + \left(\hbar f \hat{L}'_- \hat{O}'_+ \hat{a}^+ + h.c.\right). \tag{7.4.5}$$

The electric and magnetic interaction strengths are $\hbar g \equiv -\mu^{(e)}\xi$ and $\hbar f \equiv i\mu_{eff}^{(m)}\xi/c$, respectively [7.58], where $\xi \equiv \sqrt{\hbar\omega/2\varepsilon_0 V}$ is the electric field per photon and $\mu_{eff}^{(m)} \equiv (\omega_0/\omega_c)(e\hbar/2m_e)$. Note that the effective magnetic moment $\mu_{eff}^{(m)}$ is much larger than the orbital moment $e\hbar/2m_e$. The full Hamiltonian can now be written as $\hat{H} = \hat{H}_{mol} + \hat{H}_{field} + \hat{H}_{\text{int}}$, and the corresponding eigenvalue equation $\hat{H}\,|D\rangle = E_D\,|D\rangle$ is

$$\begin{pmatrix} E_3 & 2\hbar f\sqrt{n} & 0 \\ 2\hbar f^*\sqrt{n} & E_2 & \hbar g\sqrt{n} \\ 0 & \hbar g^*\sqrt{n} & E_1 \end{pmatrix} \begin{pmatrix} c_i \\ b_i \\ a_i \end{pmatrix} = E_{Di} \begin{pmatrix} c_i \\ b_i \\ a_i \end{pmatrix}. \tag{7.4.6}$$

Equation (7.4.6) may be solved to yield "doubly dressed" eigenvalues E_{Di} and eigenstates $|D_i\rangle$ $(i = 1, 2, 3)$ by setting the secular determinant equal to zero.

$$\begin{vmatrix} E_3 - E_{Di} & 2\hbar f\sqrt{n} & 0 \\ 2\hbar f^*\sqrt{n} & E_2 - E_{Di} & \hbar g\sqrt{n} \\ 0 & \hbar g^*\sqrt{n} & E_1 - E_{Di} \end{vmatrix} = 0. \tag{7.4.7}$$

This equation has the form

$$y^3 + py^2 + qy + r = 0, \tag{7.4.8}$$

where $y = E_{Di}$ are dressed-state eigenvalues and the coefficients are

$$p \equiv -(E_1 + E_2 + E_3), \tag{7.4.9}$$

$$q \equiv \left(E_1E_2 + E_2E_3 + E_3E_1 - 4n\hbar^2\,|f|^2 - n\hbar^2 g^2\right), \tag{7.4.10}$$

$$r \equiv \left(-E_1E_2E_3 + 4n\hbar^2\,|f|^2\,E_1 + n\hbar^2 g^2 E_3\right). \tag{7.4.11}$$

In order to write down analytic solutions in a compact form, a further substitution $y = x + p/3$ may be made so that the eigenvalue equation becomes $x^3 + ax + b = 0$ with new coefficients:

$$\begin{aligned} a &= \frac{1}{3}\left(3q - p^2\right) \\ &= \left(E_1E_2 + E_2E_3 + E_3E_1 - 4n\hbar^2\,|f|^2 - n\hbar^2 g^2\right) - \frac{1}{3}\left(E_1 + E_2 + E_3\right)^2, \end{aligned} \tag{7.4.12}$$

$$\begin{aligned} b &= \frac{1}{27}\left(2p^3 + 27r - 9pq\right) \\ &= \frac{2}{27}\left(E_1 + E_2 + E_3\right)^3 + \left(-E_1E_2E_3 + 4n\hbar^2\,|f|^2\,E_1 + n\hbar^2 g^2 E_3\right) \\ &\quad + \frac{1}{3}\left(E_1 + E_2 + E_3\right)\left(E_1E_2 + E_2E_3 + E_3E_1 - 4n\hbar^2\,|f|^2 - n\hbar^2 g^2\right). \end{aligned} \tag{7.4.13}$$

The solutions are then expressible as

$$x_k = 2\sqrt{-\frac{a}{3}}\cos\left(\varphi - k\frac{2\pi}{3}\right), \quad k = 0, 1, 2, \tag{7.4.14}$$

where the phase angle φ is defined by

$$\varphi = \frac{1}{3}\cos^{-1}\left(\frac{3b}{2a}\sqrt{\frac{-3}{a}}\right). \tag{7.4.15}$$

Energy eigenvalues for the doubly dressed eigenstates are

$$E_{D1} = 2\sqrt{-\frac{a}{3}}\cos\left(\varphi - \frac{4\pi}{3}\right) + \frac{1}{3}(E_1 + E_2 + E_3), \tag{7.4.16}$$

$$E_{D2} = 2\sqrt{-\frac{a}{3}}\cos\left(\varphi - \frac{2\pi}{3}\right) + \frac{1}{3}(E_1 + E_2 + E_3), \tag{7.4.17}$$

$$E_{D3} = 2\sqrt{-\frac{a}{3}}\cos(\varphi) + \frac{1}{3}(E_1 + E_2 + E_3). \tag{7.4.18}$$

The subscripts of the eigenenergies have been chosen so that E_{D1} and E_{D2} reduce to those of a two-level atom dressed by the electric field alone. E_{D3} is the additional eigenenergy introduced when the magnetic field is included. These eigenvalues may be substituted back into Eq. (7.4.6) to find sets of coefficients $\{a_i, b_i, c_i\}$ for the three dressed eigenstates ($i = 1, 2, 3$):

$$|D_i(n)\rangle = a_i|100\rangle|00\rangle|n\rangle + b_i|210\rangle|10\rangle|n-1\rangle + c_i|21-1\rangle|11\rangle|n\rangle. \tag{7.4.19}$$

The coefficients a_i, b_i, and c_i must satisfy standard normalization requirements:

$$1 = |a_i|^2 + |b_i|^2 + |c_i|^2. \tag{7.4.20}$$

Written in terms of the eigenenergies E_{Di} of Eqs. (7.4.16)–(7.4.18), the eigenstates acquire the form

$$|D_i(n)\rangle = \frac{1}{\Xi_i}\left[\left(\frac{(E_{Di}-E_2)}{\hbar g\sqrt{n}} - \frac{4\hbar|f|^2\sqrt{n}}{g(E_{Di}-E3)}\right)|1\rangle + (1)|2\rangle + \left(\frac{2\hbar|f|^2\sqrt{n}}{(E_{Di}-E_3)}\right)|3\rangle\right], \tag{7.4.21}$$

where normalization is provided by the dimensionless factor

$$\Xi_i = \sqrt{\left(\frac{(E_{Di}-E_2)}{\hbar g\sqrt{n}} - \frac{4\hbar|f|^2\sqrt{n}}{g(E_{Di}-E3)}\right)^2 + 1^2 + \left(\frac{2\hbar|f|^2\sqrt{n}}{(E_{Di}-E_3)}\right)^2}, \tag{7.4.22}$$

and the expansion coefficients are

$$a_i = \frac{1}{\Xi_i}\left(\frac{(E_{Di} - E_2)}{\hbar g\sqrt{n}} - \frac{4\hbar\,|f|^2\sqrt{n}}{g(E_{Di} - E3)}\right), \tag{7.4.23}$$

$$b_i = \frac{1}{\Xi_i}, \tag{7.4.24}$$

$$c_i = \frac{1}{\Xi_i}\left(\frac{2\hbar\,|f|^2\sqrt{n}}{(E_{Di} - E_3)}\right). \tag{7.4.25}$$

Exercise: In the limit of negligible magnetic coupling ($f \to 0$), verify that the eigenstates of Eq. (7.4.21) reduce to the quasi-eigenstates of an atom dressed by the electric field alone [7.61].

Once the eigenstates have been determined, dipole moments between components of these doubly dressed states may be readily evaluated. Examination of the admixtures coupled by arrows in Figure 7.24 reveals that two nonlinear dipoles appear that satisfy standard selection rules. The first is a nonlinear *ED* moment $p_z^{(2)}(0)$ that is longitudinally polarized. It lies perpendicular to the quantization axis, and couples the ground state admixtures $|21-1\rangle$ and $|100\rangle$, as indicated by the curved double-headed arrow in Figure 7.24. The moment of main interest here though is the transverse magnetization whose expectation value in terms of density matrix elements is

$$\begin{aligned}\langle\hat{m}\rangle &= Tr\left(\mu^{(m)}, \tilde{\rho}\right) \\ &= \left(\mu_{21}^{(m)}\tilde{\rho}_{12} + \mu_{31}^{(m)}\tilde{\rho}_{13} + \mu_{32}^{(m)}\tilde{\rho}_{23}\right) + c.c.\end{aligned} \tag{7.4.26}$$

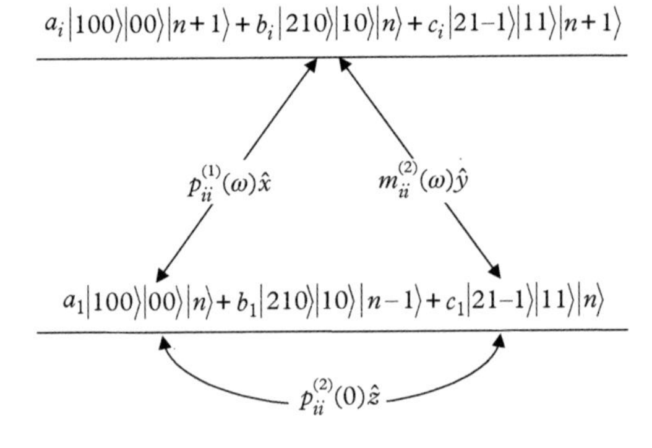

Figure 7.24 *Diagram of three dipole moments formed by strong excitation of a nominally two-level molecule during a two-photon EB* process.* $p^{(1)}(\omega)\hat{x}$ *is the linear ED polarization along the quantization axis.* $p^{(2)}(0)\hat{z}$ *and* $m^{(2)}(\omega)\hat{y}$ *are nonlinear rectification and magnetization moments oriented along* $\hat{z}$ *and* $\hat{y}$*, respectively.*

Two contributions to $\langle \hat{m} \rangle$ are non-zero. One is second order, and is measured as the y-projection of a magneto-electric (mixed, two-photon) moment. This contribution accounts for the intensity dependence of the data in Figure 7.22. The other is third order and accounts for saturation of the magnetization. These terms in the trace may be written explicitly as

$$\langle \hat{m} \rangle = \langle \hat{m}^{(2)} \rangle + \langle \hat{m}^{(3)} \rangle = \left([\mu_{13}^{me}]_y \tilde{\rho}_{31}^{(2)} + \mu_{23}^{(m)} \rho_{32}^{(3)} \right) + c.c. \tag{7.4.27}$$

The off-diagonal density matrix elements that appear in Eq. (7.4.27) can be anticipated following the perturbation procedure outlined in Appendix D. They are given by

$$\tilde{\rho}_{13}^{(2)} = \frac{\tilde{\rho}_{12}^{(1)} \tilde{V}_{23}^{(1)}}{\Delta_{13} + i\Gamma_{13}}, \tag{7.4.28}$$

where $\Delta_{13} \equiv (\omega_1 - \omega_3) - \omega$, and

$$\tilde{\rho}_{23}^{(3)} = \frac{V_{21}^{(1)} \tilde{\rho}_{13}^{(2)}}{\Delta_{23} + i\Gamma_{23}}, \tag{7.4.29}$$

where $\Delta_{23} \equiv (\omega_2 - \omega_3) - \omega$. Guided by these expressions, it is straightforward to revert to the dressed state treatment by identifying the coefficients that contribute to the magnetic moment. The following replacements are made: $\tilde{\rho}_{12}^{(1)} \leftrightarrow a_i b_i^*$, $\tilde{\rho}_{13}^{(2)} \leftrightarrow a_i c_i^*$, $\tilde{V}_{23}^{(1)} \mu_{31}^{(me)} / (\Delta_{13} + i\Gamma_{13}) \leftrightarrow \langle D_i | \mu_{21}^{(e)} \xi \mu_{eff}^{(m)} \hat{L}'_- \hat{O}'_+ \hat{\sigma}_+ | D_i \rangle$, and $\tilde{V}_{21}^{(1)} \mu_{32}^{(m)} / (\Delta_{23} + i\Gamma_{23}) \leftrightarrow \langle D_i | \mu_{21}^{(e)} \xi 2\mu_{eff}^{(m)} \hat{\sigma}_+ | D_i \rangle$. With these substitutions in Eq. (7.4.27) the two contributions to the magnetic moment become

$$\begin{aligned}
\langle \hat{m}^{(2)} \rangle &= \left\{ a_i b_i^* \langle D_i | \mu_{12}^{(e)} \xi \mu_{eff}^{(m)} \hat{L}'_- \hat{O}'_+ \hat{\sigma}_+ | D_i \rangle + h.c. \right\}_y \\
&= a_i b_i^* \langle n | \langle 11 | \langle 21,-1 | c_i^* \mu_{eff}^{(m)} \hat{L}'_- \hat{O}' + b_i | 210 \rangle | 10 \rangle | n \rangle + h.c. \\
&= 2 a_i b_i^* b_i c_i^* \mu_{eff}^{(m)} \langle 11 | \langle 21,-1 | 21,-1 \rangle | 11 \rangle + c.c. \\
&= 2c\mu_{12}^{(e)} \left(a_i b_i^* b_i c_i^* + c.c. \right)
\end{aligned} \tag{7.4.30}$$

and

$$\begin{aligned}
\langle \hat{m}^{(3)} \rangle &= \left\{ a_i c_i^* \langle D_i | \mu_{21}^{(e)} \xi 2\mu_{eff}^{(m)} \hat{\sigma}_+ | D_i \rangle + h.c. \right\}_y \\
&= 2 a_i c_i^* \langle n | \langle 10 | \langle 210 | b_i^* \mu_{eff}^{(m)} \hat{\sigma}_+ a_i | 100 \rangle | 00 \rangle | n \rangle + h.c. \\
&= 2 a_i^* b_i a_i c_i^* \mu_{eff}^{(m)} \langle 10 | \langle 210 | 210 \rangle | 10 \rangle + c.c. \\
&= 2c\mu_{12}^{(e)} \left(a_i^* b_i a_i c_i^* + c.c. \right).
\end{aligned} \tag{7.4.31}$$

In the last steps of (7.4.30) and (7.4.31) the effective magnetic moment has been replaced by the upper bound for the enhanced magnetic moment $\mu_{eff}^{(m)} = c\mu_{12}^{(e)}/2$, which equals the value derived in Ref. [7.52] using a classical argument. Importantly, all terms contributing to the magnetic moments are proportional to the electric polarization $\tilde{\rho}_{12}^{(1)} \leftrightarrow a_i b_i^*$. This reflects the need to set charges in motion before they can be deflected by the Lorentz force to form a magnetic moment.

Exercise: Show that when the MD moment is maximally enhanced (i.e., $\mu_{eff}^{(m)} = c\mu_{12}^{(e)}/2$), the magnetic and electric interaction strengths are equal. That is, show that $2\,|f| = g$.

Enhancement of the magnetic moment is due to the increase in area enclosed by charge motion following the transfer of orbital angular momentum to molecular rotation [7.59]. The cross-sectional area of molecules is generally much bigger than that of a localized excited state orbital. Since magnetic moments scale with the area encompassed by circulating current, the rotational moment is therefore enhanced with respect to the orbital moment. The measured moment is a summation of Eqs. (7.4.30) and (7.4.31) that treats terms from different initial dressed states as part of an incoherent sum:

$$\langle \hat{m}(\omega) \rangle = -2\mu_{eff}^{(m)} \left\{ \sum_{j=1}^{3} \left(a_j b_j^* a_j c_j^* + c.c. + a_j b_j^* b_j c_j^* + c.c. \right)^2 \right\}^{1/2}. \tag{7.4.32}$$

The magnetization given by Eq. (7.4.32) is plotted versus photon number and rotational frequency of the molecule in Figure 7.25. Several things can be noted in the figure. First the magnetization is quadratic at low intensities and cubic at high intensities. Second, in the region where the cubic term dominates, the moment saturates at a level equal to the first-order electric polarization.

For moderate input fields, the torque dynamics responsible for magnetic enhancement take place on an ultrafast timescale. We can confirm this by evaluating the torque itself exerted about an axis perpendicular to the internuclear axis of the molecule. Torque obeys an equation governing rotary motion in the same way that the force law $\hat{F} = d\hat{p}/dt$ governs linear motion. In the Heisenberg representation this is

$$\hat{T} = \frac{d\hat{L}}{dt} = \frac{i}{\hbar}\left[H_{int}^{(m)}, \hat{L} \right]. \tag{7.4.33}$$

Using a semi-classical approach we can use Eq. (7.4.33) to determine the expected value of torque around the z-axis from its trace with the density matrix:

$$\langle \hat{T}_z \rangle = \hat{z}\frac{i}{\hbar}\left\{ -\tilde{\rho}_{23} \langle 3 \left| L_z \hbar f L'_- O'_+ a^+ \right| 2 \rangle + \tilde{\rho}_{32} \langle 2 \left| \hbar f^* L'_+ O'_- a^- L_z \right| 3 \rangle \right\} = 2i\sqrt{n}\left\{ f\tilde{\rho}_{23} - f^*\tilde{\rho}_{32} \right\} \hat{z}. \tag{7.4.34}$$

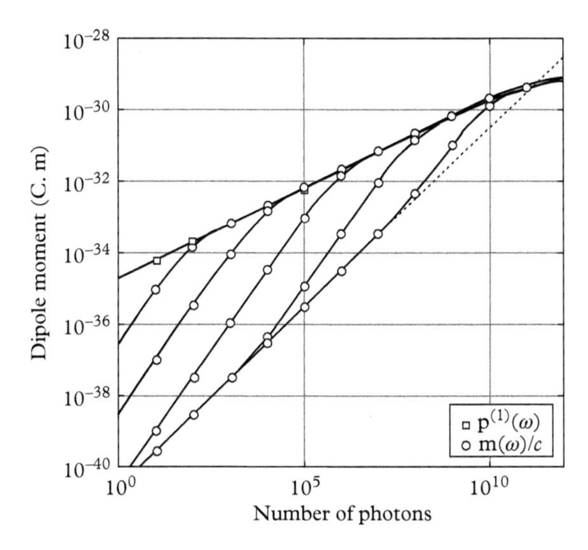

Figure 7.25 *Plot of the MD moment (divided by c for comparison with the ED polarization) plotted versus photon number for rotational frequencies of* $\omega_\varphi/\omega_0 = 10^{-11}, 10^{-9}, 10^{-7}, 10^{-5}, 10^{-3}$ *(left to right). These numerical results include the fourth basis state* $|211\rangle$ *discussed in the text, and reveal that nonlinear magnetic moments can be as large as ED moments at optical frequencies.*

The right hand side of Eq. (7.4.34) is real. This expression shows that a finite torque exists during the magnetic transition. Since there are no atoms positioned on the axis of rotation perpendicular to the internuclear axis in the diatomic model, the orbital angular momentum (L) can only be exchanged for end-over-end rotation of the molecule (O). L and O undergo a flip-flop interaction to conserve total angular momentum as torque is applied. The coupling of L, O, and B then conserves energy with a very small defect. Transfer of the internal energy and momentum of the excited state to an optical quantum plus rotational motion results in a two-photon detuning of only $\omega-(\omega-\omega_\phi) = \omega_\phi$, where $\omega_\phi \ll \omega$. Thus energy and angular momentum are conserved simultaneously during the transition.

The time interval τ in which the orbital angular momentum is reduced by $\hbar$ as it is converted to rotational motion can easily be estimated. First we remind ourselves that the effective moment $\mu_{eff}^{(m)}$ determining f in Eq. (7.4.34) is substantially larger than the orbital magnetic moment. Physically, this is because the area encompassed by rotational motion is larger than the initial orbital motion. But angular momentum before torque is applied is the same as after it has acted. So $I_{\hat{L}}\omega_0 = I_{\hat{O}}\omega_\phi$. Since the moment of inertia is proportional to area in diatomic molecules, the consequence is that there should be an enhancement of the magnetic moment of at least $\mu_{eff}^{(m)}/\mu_{\hat{L}}^{(m)} \propto \omega_0/\omega_\phi > 10^3$ due to the torque dynamics. For the purpose of estimation we use the upper bound of the enhanced moment $\mu_{eff}^{(m)} = c\mu_{12}^{(e)}/2$ and the high-intensity value of the off-diagonal density matrix element $\tilde{\rho}_{23} \approx \frac{1}{2}$ to obtain $\tau = \hbar/\mu_0^{(e)}E \approx 125\,\text{fs}$. This is the transfer time for an optical field of $E \approx 10^8$V/m. This short timescale indicates that, at the field strengths and pulse durations of published experiments, the dynamic torque interaction goes to completion during single ultrafast excitation pulses.

Transverse coherent magnetism driven by a combination of the E and B fields of light offers a mechanism for enhancing magnetization and controlling the permeability $\mu^{(m)}$ of a wide variety of materials with light. Near resonances it introduces magnetic dispersion that depends on the intensity of light. Nonlinear beam-beam interactions become possible in right-angle geometries for novel ultrafast switching. Rough estimates indicate that it may be possible to induce magnetic fields of $B \approx 1$–10 T in dielectrics or semiconductors illuminated with intense light below the band edge. So, optical magnetism could be useful for preserving spin orientation during electron transport over extended distances. Finally, there are proposals to use finely spaced magnetization structures to induce negative refractive index behavior in the propagation of de Broglie waves for atom optics [7.62]. Optical magnetization potentially provides a way of creating magnetic structures for many applications without permanent magnets or coils.

7.5 Electromagnetically Induced Transparency

An important manifestation of the tri-level coherence covered in Chapter 5 is electromagnetically induced transparency or EIT. This phenomenon [7.63, 7.64] provides a way of making opaque media transparent at the very center of an absorption transition where loss would ordinarily be at its peak. It can also provide a means of altering the refractive index of optical media, making it useful in nonlinear optics where the objective is generally to arrange for waves of different frequencies to propagate at the same speed so that they can exchange energy coherently and produce waves at new frequencies efficiently. With EIT, two or more waves can indeed be made to see the same index. Because the transparency in EIT is achieved on resonance, this process simultaneously maximizes the resonant enhancement of multiwave-mixing processes. References to these and other applications are mentioned at the end of this section.

EIT typically utilizes a pump that has a long enough wavelength to avoid inter-band absorption, so that it can penetrate the medium and interfere with the probe wave to render its absorption negligible (Figure 7.26). As usual, we begin by writing down a system Hamiltonian consisting of terms for the unperturbed atom and the optical interaction:

$$H = H_0 + V, \tag{7.5.1}$$

$$H_0 = \hbar\omega_2 |2\rangle\langle 2| + \hbar\omega_1 |1\rangle\langle 1| + \hbar\omega_3 |3\rangle\langle 3|, \tag{7.5.2}$$

$$V = -(\hbar/2)\,[\Omega_1 \exp(i\omega_1 t)\,|2\rangle\langle 1| + \Omega_2 \exp(i\omega_2 t)\,|2\rangle\,\langle 3|] + h.c. \tag{7.5.3}$$

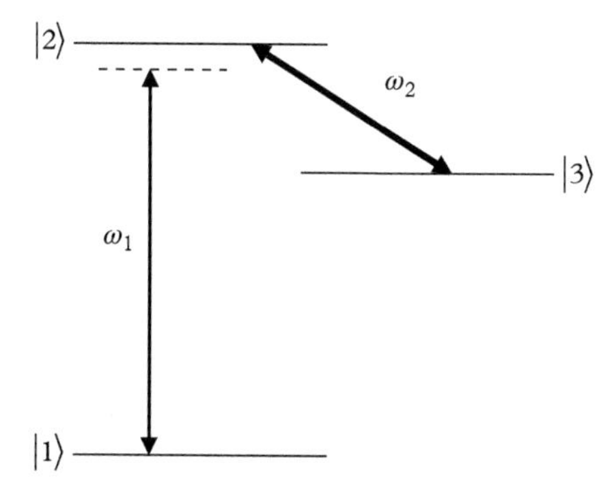

Figure 7.26 *Three-level system subjected to a strong pump wave at $\omega_2 = \omega_{23}$ and probed for absorption at frequency ω_1 in the vicinity of the ground state resonance at frequency ω_{12}.*

Note that V includes two waves, the probe (Ω_1) and the pump (Ω_2). The equations of motion for the off-diagonal matrix elements are then given by

$$\dot{\rho}_{21} = -(i\omega_{21} + \Gamma_{21})\rho_{21} + \frac{i}{2}\Omega_1^* \exp(-i\omega_1 t)\,(\rho_{11} - \rho_{22}) + \frac{i}{2}\Omega_2^* \exp(-i\omega_2 t)\rho_{31}, \tag{7.5.4}$$

$$\dot{\rho}_{31} = -(i\omega_{31} + \Gamma_{31})\rho_{31} + \frac{i}{2}\Omega_1^* \exp(-i\omega_1 t)\rho_{32} + \frac{i}{2}\Omega_2 \exp(i\omega_2 t)\rho_{21}, \tag{7.5.5}$$

$$\dot{\rho}_{23} = -(i\omega_{23} + \Gamma_{23})\rho_{23} + \frac{i}{2}\Omega_2^* \exp(-i\omega_2 t)\,(\rho_{33} - \rho_{22}) + \frac{i}{2}\Omega_1^* \exp(-i\omega_1 t)\rho_{13}. \tag{7.5.6}$$

Because the coherent driving field that couples $|2\rangle$ and $|3\rangle$ is presumed to be both intense and resonant ($\omega_2 = \omega_{23}$), we adopt a procedure which is based on Eqs. (7.5.4)–(7.5.6) but treats the Ω_2 interaction exactly and the Ω_1 interaction as a perturbation. It focuses on two key coherences, namely ρ_{21} and ρ_{31}, by dropping the second term on the right of Eq. (7.5.5). Since a weak probe wave cannot generate significant coherence between states 1 and 3 to first order, this term (which provides the only coupling to ρ_{23}) may be dropped.

The initial conditions are assumed to be that the atom resides in the ground state and there is no pre-established coherence in the system:

$$\rho_{11}^{(0)} = 1, \tag{7.5.7}$$

$$\rho_{22}^{(0)} = \rho_{33}^{(0)} = \rho_{32}^{(0)} = 0. \tag{7.5.8}$$

With these assumptions and the substitutions

$$\rho_{21} = \tilde{\rho}_{21} \exp(-i\omega_1 t), \tag{7.5.9}$$

$$\rho_{31} = \tilde{\rho}_{31} \exp[-i(\omega_1 - \omega_2)t], \tag{7.5.10}$$

one finds that the three equations for coherences in the system reduce to just two:

$$\dot{\tilde{\rho}}_{21} = -(\Gamma_{21} + i\Delta)\tilde{\rho}_{21} + \frac{i}{2}\Omega_1^* + \frac{i}{2}\Omega_2^*\tilde{\rho}_{31}, \tag{7.5.11}$$

$$\dot{\tilde{\rho}}_{31} = -(\Gamma_{31} + i\Delta)\tilde{\rho}_{31} + \frac{i}{2}\Omega_2\tilde{\rho}_{21}, \tag{7.5.12}$$

where $\Delta = \omega_{21} - \omega_1$.

Equations (7.5.11) and (7.5.12) may be readily solved in steady state by setting $\dot{\tilde{\rho}}_{21} = \dot{\tilde{\rho}}_{31} = 0$, as in earlier chapters. The microscopic polarization at the probe frequency is found to be

$$\rho_{21} = \frac{\left(i\Omega_1^*/2\right)(\Gamma_{31} + i\Delta)}{\left[(\Gamma_{21} + i\Delta)(\Gamma_{31} + i\Delta) + |\Omega_2|^2/4\right]}\exp(-i\omega_1 t). \tag{7.5.13}$$

For a sample of atomic density N, the macroscopic polarization P is given by

$$\begin{aligned} P &= N(\mu_{21}\rho_{12} + \mu_{12}\rho_{21}) \\ &= \frac{\left(Ni\mu_{12}\Omega_1^*/2\right)(\Gamma_{31} + i\Delta)}{\left[(\Gamma_{21} + i\Delta)(\Gamma_{31} + i\Delta) + |\Omega_2|^2/4\right]}\exp(-i\omega_1 t) + c.c. \end{aligned} \tag{7.5.14}$$

P may also be written in terms of the electric susceptibility $\chi = \chi' + i\chi''$ as

$$P = \frac{1}{2}\varepsilon_0 E_1\left[\chi(\omega)e^{-i\omega t} + \chi(-\omega)e^{i\omega t}\right]. \tag{7.5.15}$$

To determine the real and imaginary components of the susceptibility, Eqs. (7.4.14) and (7.5.15) must be compared term by term. The results are

$$\chi' = \left(\frac{N|\mu_{21}|^2}{\varepsilon_0\hbar}\right)\frac{\Delta\left[\Gamma_{31}(\Gamma_{21} + \Gamma_{31}) + (\Delta^2 - \Gamma_{21}\Gamma_{31} - |\Omega_2|^2/4)\right]}{\left(\Delta^2 - \Gamma_{21}\Gamma_{31} - |\Omega_2|^2/4\right)^2 + \Delta^2(\Gamma_{21} + \Gamma_{31})^2}, \tag{7.5.16}$$

$$\chi'' = \left(\frac{N|\mu_{21}|^2}{\varepsilon_0\hbar}\right)\frac{\left[\Delta^2(\Gamma_{21} + \Gamma_{31}) - \Gamma_{31}\left(\Delta^2 - \Gamma_{21}\Gamma_{31} - |\Omega_2|^2/4\right)\right]}{\left(\Delta^2 - \Gamma_{21}\Gamma_{31} - |\Omega_2|^2/4\right)^2 + \Delta^2(\Gamma_{21} + \Gamma_{31})^2}. \tag{7.5.17}$$

A plot of χ'' (see Figure 7.27) versus detuning reveals that as the result of strong pumping at ω_2 where the sample is assumed to be transparent, absorption at the *probe* frequency develops a local minimum at line center ($\Delta = 0$). The residual absorption on resonance is determined by

$$\chi'' = \left(\frac{N|\mu_{21}|^2}{\varepsilon_0\hbar}\right)\left[\frac{\Gamma_{31}}{\Gamma_{21}\Gamma_{31} + |\Omega_2|^2/4}\right], \tag{7.5.18}$$

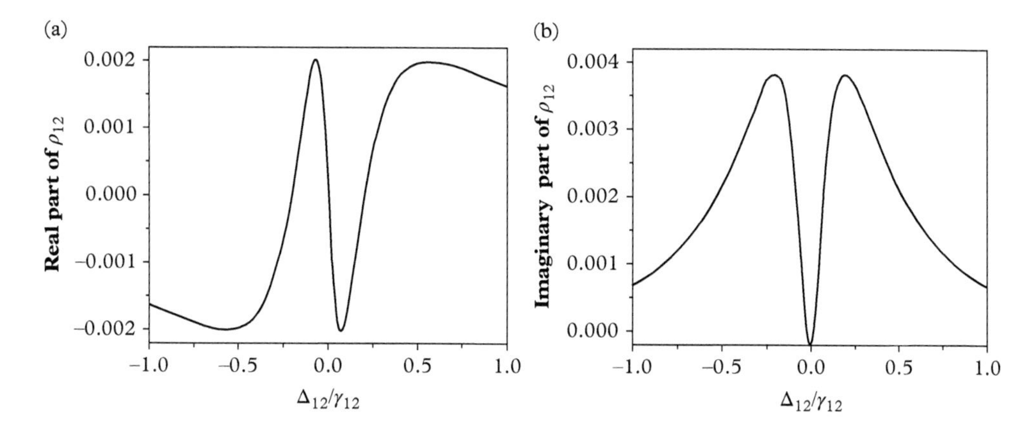

Figure 7.27 *Plots of (a) the real part and (b) the imaginary part of the polarization versus detuning at probe frequency ω_1 in a Λ-system that is strongly driven on the second transition at frequency ω_2. These curves correspond to the dispersive and absorptive components of the polarization and curve (b) shows the induced transparency on resonance.*

and approaches zero asymptotically when the Rabi frequency Ω_2 exceeds the decay rate $\Gamma_{23} = \sqrt{\Gamma_{21}\Gamma_{31}}$ on the pump transition.

EIT is related to Fano interference and can operate at arbitrarily low pump intensities (see Section 5.4.3 on quantum interference). However it is not the only way to induce transparency in an opaque medium. Another method, distinct from EIT, is based on Rabi splitting and may be understood on the basis of the simple physical picture presented in Figure 7.28. In the Λ-system shown, an intense coupling laser at frequency ω_2 modulates energy levels 2 and 3, even if there is no population in these levels. Both levels therefore undergo Rabi splitting, as shown in the figure. The upper level of the probe transition is displaced from its original position, thereby reducing the absorption at the

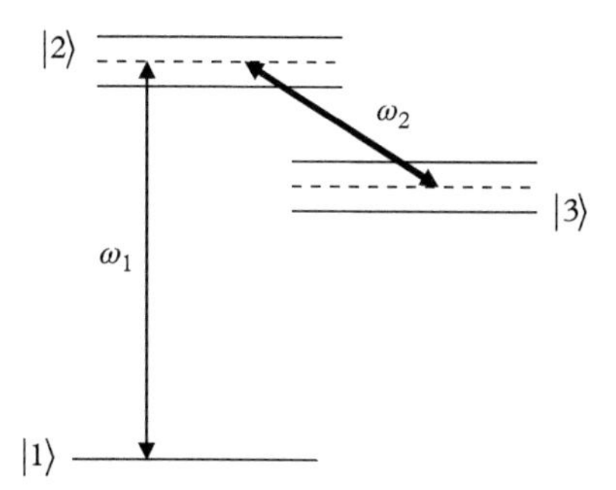

Figure 7.28 *Physical picture of induced transparency due to Rabi splitting on the strongly driven coupling transition at ω_2. In the strong pump limit, transparency is achieved by shifting the final state of the probe transition off resonance. However, high pump intensity is not necessary for transparency, as discussed earlier in Section 5.4.3.*

probe center frequency. Clearly this requires a high-power pump wave to reduce the probe losses at the resonance frequency, and it is this feature which distinguishes it from true EIT [7.65].

Notice in Figure 7.27 that the slope of the index profile is very steep at the resonant frequency under EIT conditions. The large dispersion near resonance is useful for changing the speed at which light propagates through the system, in addition to altering its losses. In particular, light can be slowed, stopped, or stored using EIT [7.66]. Usually we think of the speed of light in a medium as a fixed property which exhibits only small variations with wavelength. For example the group velocity, which is the speed at which information travels through a medium, is completely determined by refractive index properties. Its definition is $v_g \equiv \partial\omega/\partial k$ and it is equal to

$$v_g = c/(n(\omega) + \omega[\partial n(\omega)/\partial\omega]), \tag{7.5.19}$$

where c is the speed of light in vacuum. However, in the presence of EIT, for $n > 1$ and $\omega[\partial n(\omega)/\partial\omega] \gg 1$, the group velocity is reduced to much less than c. This reduction has now been confirmed in many published experiments and is useful for delay lines in signal processing [7.67].

To conclude this section a number of intersections with other topics should be mentioned. In avalanche upconversion systems, the dispersion $\partial n/\partial\omega$ of the EIT feature can be reversed [7.68] compared to that in Figure 7.27(a) (where $\partial n/\partial\omega > 0$). Hence so-called "superluminal" velocities may be reached in this and other situations through the use of EIT. However, the bandwidth of such effects is extremely limited, so signal and energy velocities never exceed the speed of light in vacuum. EIT does provide a method for transmitting light through highly absorbing media [7.63], slowing and storing light [7.69], enhancing nonlinear optical interactions [7.64, 7.70], producing low-loss media with high refractive indices [7.71], and enabling lasing without inversion [7.72]. Some of these applications are discussed in Ref. [7.73] and the phenomenon of EIT can be observed at photon energies in the hard x-ray region [7.74]. For electromagnetically induced absorption, see Ref. [7.75]. Continuous tuning of some systems between EIT and electromagnetically induced absorption is possible [7.76]. Classical analogs of EIT are described in Ref. [7.77].

7.6 Squeezed Light

In Chapter 6 the concept of squeezed states was introduced and some of the properties of these special coherent states were examined. Here, the challenge of finding a way to create a squeezed state of an electromagnetic wave is considered, together with potential applications.

To get started, a squeezing operator $\hat{S}$ is identified that can mathematically modify the statistics of the field operators $\hat{a}, \hat{a}^+$ in just the right way to generate a squeezed state of the light field, instead of an ordinary coherent state. After confirming that the squeezing operator works, we try to interpret it to identify experimental processes that allow manipulation of light fields in the same way as the squeezing operator.

The squeezing operation affects the noise properties of a system. So it can be anticipated that the operator $\hat{S}$ involves changes in occupation of the modes contributing to the field. Moreover it should be a nonlinear function of the field creation and annihilation operators, because as we have seen noise is redistributed between quadratures of each cycle of light at the fundamental frequency. To have low losses the operator should also be unitary.

Consider the operator

$$\hat{S}(z) = \exp\left[\left(z(\hat{a}^-)^2 - z^*(\hat{a}^+)^2\right)/2\right], \tag{7.6.1}$$

where $z = r\exp(-i\theta)$ is referred to as the complex squeeze parameter. Equation (7.6.1) has the form $\hat{S}(z) = \exp(\hat{A})$, where the argument

$$\hat{A} = \left(z(\hat{a}^-)^2 - z^*(\hat{a}^+)^2\right)/2 \tag{7.6.2}$$

is Hermitian. $\hat{S}(z)$ is therefore unitary:

$$\hat{S}^+(z) = \hat{S}^{-1}(z). \tag{7.6.3}$$

The electromagnetic field consists of an expansion in terms of the creation and annihilation operators. If we consider just a single mode, the transformed annihilation operator is

$$\begin{aligned}\hat{t}^- &\equiv \hat{S}(z)\hat{a}^-\hat{S}^+(z) = \exp(\hat{A})\hat{a}^-\exp(-\hat{a}) \\ &= \hat{a}^- + \left[\hat{A}, \hat{a}^-\right] + \frac{1}{2!}\left[\hat{A}\left[\hat{A}, \hat{a}^-\right]\right] + \frac{1}{3!}\left[\hat{A}\left[\hat{A}\left[\hat{A}, \hat{a}^-\right]\right]\right] + \ldots \end{aligned}\tag{7.6.4}$$

Since $[\hat{a}^-, \hat{a}^+] = 1$, this can rewritten as

$$\begin{aligned}\hat{t}^- &= \hat{a}^- + z^*\hat{a}^+ \frac{1}{2!}|z|^2\hat{a}^- + \frac{1}{3!}|z|^2 z^*\hat{a}^+ + \frac{1}{4!}|z|^4\hat{a}^- + \ldots \\ &= \hat{a}^-\left(1 + \frac{1}{2!}r^2 + \frac{1}{4!}r^4 + \ldots\right) + \hat{a}^+\exp(i\theta)\left(r + \frac{1}{3!}r^3 + \frac{1}{5!}r^5 + \ldots\right) \\ &= \hat{a}^-\cosh(r) + \hat{a}^+\exp(i\theta)\sinh(r), \end{aligned}\tag{7.6.5}$$

$$\hat{t}^+ = \hat{a}^+\cosh(r) + \hat{a}^-\exp(-i\theta)\sinh(r). \tag{7.6.6}$$

By introducing definitions

$$\mu \equiv \cosh(r), \tag{7.6.7}$$

$$\nu \equiv \exp(i\theta)\sinh(r), \tag{7.6.8}$$

the transformed annihilation operator for a single mode can be written

$$\hat{t}^- = \mu^* \hat{a}^- - \nu \hat{a}^+, \tag{7.6.9}$$

where

$$|\mu|^2 - |\nu|^2 = 1. \tag{7.6.10}$$

This operator can be decomposed into quadrature components

$$\hat{t}_1 = \frac{i}{\sqrt{2}} \left(\hat{t}^+ + \hat{t}^- \right) = \hat{a}_1^- \exp(r) \tag{7.6.11}$$

and

$$\hat{t}_2 = \frac{i}{\sqrt{2}} \left(\hat{t}^+ - \hat{t}^- \right) = \hat{a}_2^- \exp(-r). \tag{7.6.12}$$

Exercise: Show that the quadrature operators in Eqs. (7.6.11) and (7.6.12) have a commutator given by

$$\left[\hat{t}_1, \hat{t}_2 \right] = i/2. \tag{7.6.13}$$

Based on Appendix B, Eq. (B.20) and Eq. (7.6.13), the uncertainty product associated with the quadrature operators $\hat{t}_1$ and $\hat{t}_2$ is

$$\left\langle (\Delta t_1)^2 \right\rangle^{1/2} \left\langle (\Delta t_2)^2 \right\rangle^{1/2} = 1/4. \tag{7.6.14}$$

Any operator for which the standard (root-mean-square) deviation is below the minimum for an ordinary coherent state, namely

$$\left\langle (\Delta t_i)^2 \right\rangle^{1/2} < 1/2, \ (i = 1, \ 2), \tag{7.6.15}$$

is called a squeezed state. Consequently it should be possible for either quadrature $\hat{t}_1$ or $\hat{t}_2$ in Eq. (7.6.14) to exhibit squeezing.

How can light with the squeezed property (7.6.15) be produced experimentally? For guidance on this question, note that application of the squeeze operator $\hat{S}$ to the optical field operator $\hat{a}^-$ generates a squeezed coherent state operator $\hat{t}^-$. This suggests that we compare the form of $\hat{S}$ with the interaction Hamiltonians $\hat{V}$ for various possible optical interactions, particularly nonlinear parametric processes involving the simultaneous creation or annihilation of pairs of photons. The reason for this is that Eq. (7.6.2) contains nonlinear combinations of the field operators not found in the simple one-photon interaction Hamiltonian (6.2.10). So it is natural to search for quadratic (or higher order) processes that have a form like the quadratic operator terms that appear in the argument $\hat{A}$ of the squeezing operator in Eq. (7.6.2).

A suitable Hamiltonian in this regard is the one for parametric down conversion involving three modes. In parametric conversion, an intense pump wave (mode 3) splits into signal and idler waves (modes 1 and 2) via the nonlinear susceptibility $\chi^{(2)}(\omega_3 = \omega_1 + \omega_2)$ of the medium. The interaction destroys one pump photon of frequency ω_3 and creates one photon in each of the modes 1 and 2. The inverse (Hermitian conjugate) process also takes place, giving rise to second harmonic generation. The interaction energy can therefore be written as

$$\hat{V} = -\varepsilon_0 \left[\chi^{(2)} \hat{E}_1^- \hat{E}_2^- \right] \cdot \hat{E}_3^+ + h.c. \tag{7.6.16}$$

For high conversion efficiency, the pump wave is normally intense. Hence it can be treated as a classical wave of amplitude $E_3^+ = E_0$. Moreover we can choose to make the undepleted pump approximation [7.78] so that the pump intensity is simply constant throughout the sample. When the modes are degenerate ($a_1 = a_2$), and the substitution $f \equiv (-\varepsilon_0 \chi^{(2)}/2V)\sqrt{\omega_1 \omega_2 / \varepsilon_1 \varepsilon_2} E_0$ is made to incorporate constant pump intensity, Eq. (7.6.16) reduces to

$$\hat{V} = \hbar f[(\hat{a}^+)^2 + (\hat{a}^-)^2]. \tag{7.6.17}$$

Notice that this has the same form as the operator $\hat{A}$ in Eq. (7.6.2) for which squeezed states are the eigenstates. Parametric down conversion and second harmonic generation can therefore generate squeezed states of light. When the states or modes in question are unoccupied, squeezing modifies the fluctuations driving spontaneous processes, producing what is known as squeezed vacuum (Figure 7.29).

Squeezed light offers a means of communicating securely and improving optical interferometry [7.79]. The "quiet" quadrature of squeezed light can be used for example to transmit information at levels below the usual shot-noise limit in communication channels. The precision of interferometric measurements can also be improved when squeezed light is used because noise invariably limits the precision with which fringe positions can be determined. In general, the inherent uncertainty of measurements made using squeezed light can be made less than that of the "unsqueezed" world in which we live. Methods for the production and use of squeezed states of light for applications of various kinds are described or referenced by Mandel and Wolf [7.80].

Finally, we note that squeezing is not limited to states of light. The currents in common electronic components like resistors can be squeezed. Squeezing can also be applied to matter waves—such as the de Broglie waves associated with particles. Even the fluctuations of modes of the vacuum field can be squeezed, as illustrated in Figure 7.29.

Exercise: Evaluate the squeezed number operator $\hat{t}^+\hat{t}^-$ in the vacuum state $|0\rangle$. (a) Show that $\langle 0| \hat{t}^+\hat{t}^- |0\rangle = \sinh^2 r$. (b) Is it possible for the squeezed, single-mode vacuum state to contain an average of one photon or more?

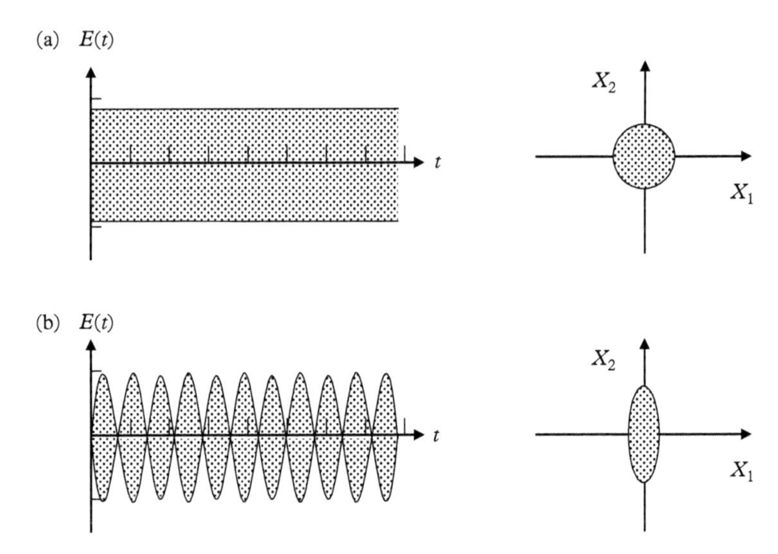

Figure 7.29 *Illustration of the fluctuations of the electric field $E(t)$ of vacuum. Left: Time dependence of the fluctuations in a single mode of (a) unsqueezed vacuum, and (b) amplitude-squeezed vacuum. Right: Loci of the tip of the electric field vector at a single point in time for the corresponding cases.*

7.7 Cavity Quantum Electrodynamics

An important example of the quantized reservoir theory of Section 6.3 is provided by the damping of a light field by atoms passing through a single-mode field region in an optical cavity. A reservoir of excitable atoms can damp the field intensity exponentially in the same way that spontaneous emission damps an excited atom. In the three subsections that follow this problem is analyzed first by ignoring spontaneous emission effects, then by including natural decay in the weak atom–field coupling limit, and finally by treating the interesting case of strong atom–field coupling and its effect on spontaneous decay, which is the realm of cavity quantum electrodynamics (QED).

7.7.1 Damping of an Optical Field by Two-Level Atoms

Consider a lossless linear optical cavity, as shown in Figure 7.30. In the figure, two-level atoms pass through a light field confined to a single cavity mode $|n\rangle$. They interact with the field through absorption and stimulated emission for a time τ. Spontaneous emission is ignored. The state of the atom is described by

$$\rho_A = \begin{bmatrix} \rho_{22} & 0 \\ 0 & \rho_{11} \end{bmatrix} \tag{7.7.1}$$

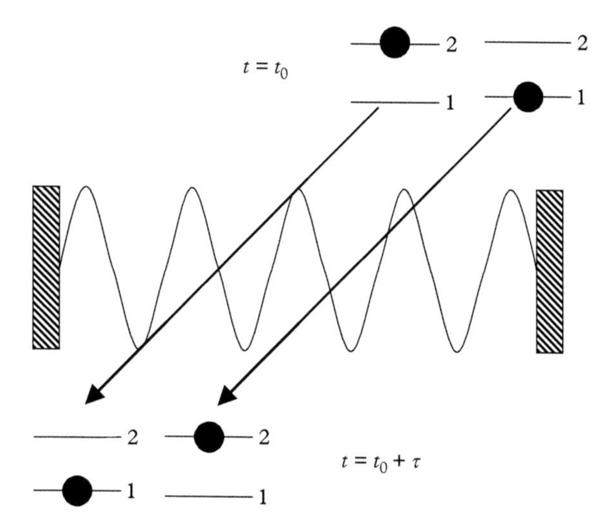

Figure 7.30 *An atomic beam of two-level atoms passing through a single-mode cavity, acting as a reservoir for decay of the optical simple harmonic oscillator.*

and the state of the field has a density operator

$$\rho_B = |n\rangle\langle n| \,. \tag{7.7.2}$$

In a frame rotating at the cavity frequency Ω, which is also the optical frequency in this case, the interaction between the atoms and the field is given by

$$V = \hbar g \begin{bmatrix} 0 & \sigma^+ a^- \exp[-i(\Omega - \omega_0)t] \\ \sigma^- a^+ \exp[+i(\Omega - \omega_0)t] & 0 \end{bmatrix}. \tag{7.7.3}$$

Using the equations of motion derived earlier for the temporal evolution of subsystems A and B, only with A and B interchanged, we write

$$\begin{aligned}\dot{\rho}_B(t) &= -\frac{1}{\hbar^2}\int_{t_0}^{t} dt'\, Tr_A[V(t)V(t')\rho_B(t')\rho_A(t_0) - V(t)\rho_B(t')\rho_A(t_0)V(t')] + h.c. \\ &= -\frac{1}{\hbar^2}\int_{t_0}^{t} dt' \left\{\exp[-i(\Omega-\omega_0)(t-t')][V_{12}V_{21}\rho_{11}(t_0)\rho_B(t') - V_{12}\rho_{22}(t_0)\rho_B(t')V_{21}]\right\} + h.c.\end{aligned} \tag{7.7.4}$$

On resonance ($\Omega = \omega_0$) the argument of the exponential function is zero, and if we ignore the small change that takes place in ρ_B during the short transit time $\tau = t - t_0$ of each atom passing through the beam, the integration yields

$$\dot{\rho}_B(t) = -g^2\tau[(a^+a^-\rho_B - a^-\rho_B a^+)\rho_{11} + (\rho_B a^- a^+ - a^+\rho_B(t)a^-)\rho_{22}] + h.c. \tag{7.7.5}$$

If atoms pass through the beam at a rate r, the number of atoms in the beam at any time is $r\tau$. By identifying the absorption and emission rates per photon as $R_1 = 2r\tau^2 g^2 \rho_{11}$ and $R_2 = 2r\tau^2 g^2 \rho_{22}$, respectively, the equation of motion for decay of the intensity in the cavity mode becomes

$$\dot{\rho}_B(t) = -\frac{R_1}{2}\left[a^+a^-\rho_B(t) - a^-\rho_B(t)a^+\right] - \frac{R_2}{2}\left[\rho_B(t)a^-a^+ - a^+\rho_B(t)a^-\right] + h.c. \tag{7.7.6}$$

This result has the same form as the equation for atomic decay through coupling to a reservoir of simple harmonic oscillators. Hence it is not surprising that when one evaluates the rate of change of the probability that the mode contains n photons, we find a similar result:

$$\begin{aligned}\langle n|\dot{\rho}_B(t)|n\rangle &= \left\{-\frac{R_1}{2}[\langle n|a^+a\rho_B(t)|n\rangle - \langle n|a\rho_B(t)a^+|n\rangle]\right.\\ &\quad \left.-\frac{R_2}{2}\left[\langle n|\rho_B(t)aa^+|n\rangle - \langle n|a^+\rho_B(t)a|n\rangle\right]\right\} + h.c.\\ &= -R_1[n\langle n|\rho_B|n\rangle - (n+1)\langle n+1|\rho_B|n+1\rangle]\\ &\quad - R_2[(n+1)\langle n+1|\rho_B|n+1\rangle - n\langle n|\rho_B|n\rangle].\end{aligned} \tag{7.7.7}$$

The occupation of the cavity mode evidently depends on the balance between absorption of cavity photons by ground state atoms and stimulated emission into the cavity mode by excited atoms passing through the beam. For example, if there are no excited atoms passing through the beam ($R_2 = 0$), then exponential decay of the intensity will ensue. This can be seen by substituting $\rho_B = |n\rangle\langle n|$ in Eq. (7.7.7), whereupon one finds that the total number of photons in the beam is governed by the equation $\dot{n}(t) = (-R_1 + R_2)n(t)$, with the solution

$$n(t) = n(0)\exp[(-R_1 + R_2)t]. \tag{7.7.8}$$

7.7.2 Weak Coupling Regime

Real atom–cavity interactions must take into account losses from spontaneous emission by atoms in the cavity, and loss of photons from the cavity mode through scattering or finite reflectivity of the mirrors, as well as the strength g of atom–cavity coupling (Figure 7.31). Here we consider both types of loss explicitly in the weak coupling limit.

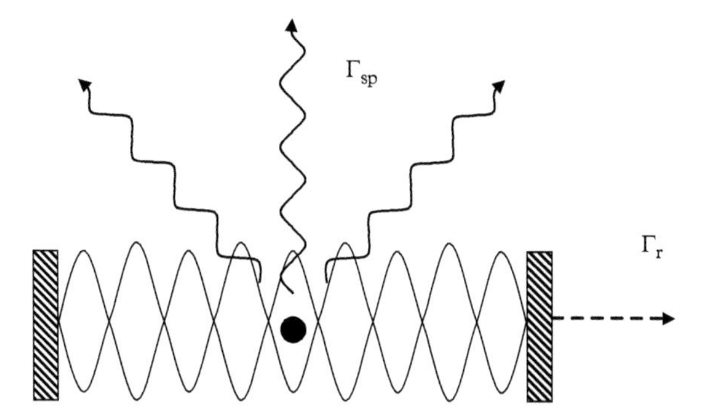

Figure 7.31 *A stationary atom interacting with photons in a cavity. Losses due to spontaneous emission and finite mirror transmission occur with loss rates of* Γ_{sp} *and* Γ_{r}*, respectively.*

First, recall that in the presence of dissipation (through coupling of the system to a reservoir) the equation of motion separates into an Hermitian portion that depends on the Hamiltonian and a non-Hermitian part $L(\rho)$ that describes dissipation. The quantum relaxation term $L(\rho)$ was considered in Section 6.6, and must be of the so-called Lindblad form

$$L(\rho) = -\frac{1}{2}\sum_i \left(\hat{A}_i^+\hat{A}_i\rho + \rho\hat{A}_i^+\hat{A}_i\right) + \sum_i \hat{A}_i\rho\hat{A}_i^+, \tag{7.7.9}$$

if the description is to preserve trace ($Tr(\rho) = 1$). The last term in $L(\rho)$, called the quantum jump Liouvillian, is discussed further at the end of this section.

In a coupled atom–cavity system there are two additive sources of loss. We treat both as couplings to thermal reservoirs, modeled as before as continua of simple harmonic oscillators. Using the Markoff approximation, the equation of motion for the reduced matrix element of the atom–cavity system becomes

$$\begin{aligned}\dot{\rho} = &-\frac{i}{\hbar}[H_0, \rho] - \frac{i}{\hbar}[V, \rho] - \frac{1}{2}\Gamma_{sp}[(\sigma^+\sigma^-\rho - \sigma^-\rho\sigma^+) + h.c.] \\ &-\frac{1}{2}\Gamma_r[(a^+a^-\rho - a^-\rho a^+) + h.c.].\end{aligned} \tag{7.7.10}$$

The interaction Hamiltonian is given by

$$V = \hbar g[a^+\sigma^- \exp(i\Omega_c t) + h.c.], \tag{7.7.11}$$

where Ω_c is the cavity mode resonant frequency. Now we choose a particularly simple set of states with which to explore the dynamics.

If there is initially only one excitation in the system, there are exactly three states to consider. Two of the available $|atom, field\rangle$ product states are the "one-quantum" states $|2\rangle \equiv |2,0\rangle$ and $|1\rangle \equiv |1,1\rangle$ that are directly coupled by the driving field interaction.

The third state $|0\rangle \equiv |1,0\rangle$ results from the irreversible loss of a photon through the mirror, and is coupled to $|1\rangle \equiv |1,1\rangle$ by loss of cavity photons through the mirrors and to $|2\rangle \equiv |2,0\rangle$ by spontaneous emission.

Exercise: Show from Eq. (7.7.10) that the equations of motion are

$$i\hbar\dot{\rho}_{00} = i\hbar\Gamma_r\rho_{11} + i\hbar\Gamma_{sp}\rho_{22}, \tag{7.7.12}$$

$$i\hbar\dot{\rho}_{11} = V_{12}\rho_{21} - \rho_{12}V_{21} - i\hbar\Gamma_r\rho_{11}, \tag{7.7.13}$$

$$i\hbar\dot{\rho}_{22} = -V_{12}\rho_{21} + \rho_{12}V_{21} - i\hbar\Gamma_{sp}\rho_{22}, \tag{7.7.14}$$

$$i\hbar\dot{\rho}_{12} = -\hbar\omega_0\rho_{12} + V_{12}(\rho_{22} - \rho_{11}) - (i\hbar/2)(\Gamma_{sp} + \Gamma_r)\rho_{12}. \tag{7.7.15}$$

We proceed to solve these equations by choosing the coherence judiciously to be

$$\rho_{12}(t) = \tilde{\rho}_{12}\exp(i\Omega_c t), \tag{7.7.16}$$

whereupon the time derivative becomes

$$\dot{\rho}_{12}(t) = \dot{\tilde{\rho}}_{12}\exp(i\Omega_c t) + i\Omega_c\tilde{\rho}_{12}\exp(i\Omega_c t). \tag{7.7.17}$$

Substituting Eq. (7.7.17) into Eq. (7.7.15), one finds

$$i\hbar\dot{\tilde{\rho}}_{12} = -\hbar\Delta\tilde{\rho}_{12} + \tilde{V}_{12}(\rho_{22} - \rho_{11}) - (i\hbar/2)(\Gamma_{sp} + \Gamma_r)\tilde{\rho}_{12}, \tag{7.7.18}$$

where $\Delta \equiv \omega_0 - \Omega_c$. In order to integrate Eq. (7.7.18), it is helpful to introduce an integrating factor via the substitution $\tilde{\rho}_{12} = \rho'_{12}\exp(i\Delta - \Gamma_\phi)t$, where $\Gamma_\phi \equiv (\Gamma_{sp} + \Gamma_r)/2$. Equation (7.7.18) then becomes

$$\dot{\rho}'_{12}\exp(i\Delta - \Gamma_\phi)t = -ig(\rho_{22} - \rho_{11}), \tag{7.7.19}$$

since $\tilde{V}_{12} = \hbar g$. Formal integration yields

$$\rho'_{12} = -ig\int_0^t [\rho_{22}(t') - \rho_{11}(t')]\exp[-(i\Delta - \Gamma_\phi)t']dt'. \tag{7.7.20}$$

Next consider the "bad" cavity limit of Eq. (7.7.20) in which $\Gamma_r \gg g, \Gamma_{sp}$. Assume that the excited state population is slowly varying for times $t \gg \Gamma_r^{-1}$, and that the ground-state population remains low because of rapid decay to the reservoir. Then the population difference can be removed from the integral:

$$\rho'_{12} = -ig(\rho_{22}(t) - \rho_{11}(t))\int_0^t \exp\left[-\left(i\Delta - \frac{1}{2}\Gamma_r\right)t'\right]dt'. \tag{7.7.21}$$

As a result, for times longer than the inverse cavity loss rate, the integral can easily be evaluated.

$$\tilde{\rho}_{12}(t) = \left(\frac{-ig}{-i\Delta + \Gamma_r/2}\right)(\rho_{22}(t) - \rho_{11}(t)). \tag{7.7.22}$$

Equations for the populations can be obtained by substituting Eq. (7.7.22) into Eqs. (7.7.13) and (7.7.14). The substitution yields

$$\dot{\rho}_{11} = g^2\left[\frac{1}{-i\Delta + \Gamma_r/2} + c.c.\right](\rho_{11} - \rho_{22}) - \Gamma_r\rho_{11}, \tag{7.7.23}$$

$$\dot{\rho}_{22} = -g^2\left[\frac{1}{-i\Delta + \Gamma_r/2} + c.c.\right](\rho_{11} - \rho_{22}) - \Gamma_{sp}\rho_{22}. \tag{7.7.24}$$

According to these equations the temporal evolution of ρ_{11} and ρ_{22} is governed by two different complex rates, which we take to be α and β. To ensure real populations, solutions are assumed to have the form

$$\rho_{22}(t) = \rho_2(0)\exp[(\alpha + \alpha*)t] \tag{7.7.25}$$

and

$$\rho_{11}(t) = \rho_1(0)\exp[(\beta + \beta*)t], \tag{7.7.26}$$

where α and β are yet to be determined. Finally, substitution of Eqs. (7.7.25) and (7.7.26) into Eq. (7.7.24) furnishes an expression for the decay constant of the excited state:

$$\alpha = -\frac{g^2(i\Delta + \Gamma_r/2)}{\Delta^2 + (\Gamma_r/2)^2} - \frac{1}{2}\Gamma_{sp}, \tag{7.7.27}$$

$$\alpha + \alpha^* = -\frac{g^2\Gamma_r}{\Delta^2 + (\Gamma_r/2)^2} - \Gamma_{sp}. \tag{7.7.28}$$

The excited state of the atom decays exponentially, according to

$$\rho_{22}(t) = \exp[(\alpha + \alpha*)t]. \tag{7.7.29}$$

However, note that the decay rate is altered by the detuning, reservoir coupling, and quality of the cavity as specified by Eq. (7.7.27). This solution corresponds to the weak coupling limit in which the cavity modifies the decay rate, but does not qualitatively change the dynamics, due to its low quality factor. When the atom–field coupling increases, the effective atom–field coupling may be enhanced to the point that the dynamics do change qualitatively. Oscillatory behavior is then observed as shown in the next section.

7.7.3 Strong Coupling Regime

Equation (7.7.27) is not the most general solution to the equation of motion of the coupled atom–field problem. In the earlier treatment of weak coupling, the population in the lower state was assumed to decay rapidly to the reservoir, so that ρ_{11} played little role in the solution of Eq. (7.7.20). An exact approach must recognize that strong atom–cavity interaction leads to strong coupling of the ground and excited state that manifests itself as a splitting of modes at resonance, in a manner reminiscent of the Rabi splitting of atomic emission in resonance fluorescence. This is the strong coupling regime of cavity QED [7.81].

Unlike the situation in the "bad" cavity limit of the last section, the ground state plays an important role when $g \gg \Gamma_r, \Gamma_{sp}$. Then, according to Eqs. (7.7.13) and (7.7.14), both $\rho_{11}(t)$ and $\rho_{22}(t)$ are dominated by the coherences $\rho_{12} = c_1^* c_2$ and $\rho_{21} = c_1 c_2^*$. A photon is present only when the atom is in the ground state, so reservoir couplings are different in the two atomic states. Hence it may be anticipated that only two complex frequencies α_+ and α_- will determine the time development of probability amplitudes in the system (c_1 and c_2, respectively). Hence combination frequencies $\alpha_+^* + \alpha_-$ and $\alpha_+ + \alpha_-^*$ will govern the evolution of products of these amplitudes such as $\rho_{21} = c_1^* c_2$ and $\rho_{12} = c_1 c_2^*$ which determine the evolution of diagonal elements of the density matrix. Based on this argument, the general form of the populations may be taken to be

$$\rho_{22}(t) = \rho_{2a}\,e^{(\alpha_+ + \alpha_+^*)t} + \rho_{2b}\,e^{(\alpha_+ + \alpha_-^*)t} + \rho_{2b}^*\,e^{(\alpha_+^* + \alpha_-)t} + \rho_{2c}\,e^{(\alpha_- + \alpha_-^*)t}, \tag{7.7.30}$$

$$\rho_{11}(t) = \rho_{1a}\,e^{(\alpha_+ + \alpha_+^*)t} + \rho_{1b}\,e^{(\alpha_+ + \alpha_-^*)t} + \rho_{1b}^*\,e^{(\alpha_+^* + \alpha_-)t} + \rho_{1c}\,e^{(\alpha_- + \alpha_-^*)t}, \tag{7.7.31}$$

where α_+ and α_- are to be determined.

In principle, one can proceed to analyze strong coupling by finding ρ_{12}' with the use of an exact evaluation of Eq. (7.7.20), rather than with the earlier approximation of slowly varying populations. This can be done by inserting Eqs. (7.7.30) and (7.7.31) into Eq. (7.7.20) prior to performing the integration. However, this approach leads to cubic equations for the exponents. It is preferable to proceed in a simpler way by solving the equations of motion for the probability amplitudes themselves that compose the density matrix elements. These equations are

$$\dot{C}_2(t) = -(\Gamma_{sp}/2)C_2(t) - igC_1(t), \tag{7.7.32}$$

$$\dot{C}_1(t) = (i\Delta - \Gamma_r/2)C_1(t) - igC_2(t). \tag{7.7.33}$$

Just like the populations, the amplitudes $C_2(t)$ and $C_1(t)$ will evolve with two complex frequencies α_+ and α_-. Hence they are assumed to have the forms

$$C_2(t) = C_{2+}\exp(\alpha_+ t) + C_{2-}\exp(\alpha_- t), \tag{7.7.34}$$

and

$$C_1(t) = C_{1+}\exp(\alpha_+ t) + C_{1-}\exp(\alpha_- t). \tag{7.7.35}$$

Upon substitution of Eqs. (7.7.34) and (7.7.35) into Eqs. (7.7.32) and (7.7.33), and collection of terms in $\exp(\alpha_+ t)$ or $\exp(\alpha_- t)$, the differential equations for the amplitudes reduce to coupled algebraic equations:

$$[\alpha_\pm + (\Gamma_{sp}/2)]C_{2\pm} = -igC_{1\pm}, \tag{7.7.36}$$

$$[\alpha_\pm - i\Delta + (\Gamma_r/2)]C_{1\pm} = -igC_{2\pm}. \tag{7.7.37}$$

These are readily solved to give

$$\alpha_\pm = -\frac{1}{2}(\Gamma_\phi - i\Delta) \pm \frac{1}{2}\sqrt{(\Gamma_\phi - i\Delta)^2 - 4\left(g^2 - \frac{1}{2}i\Delta\Gamma_{sp} + \frac{1}{4}\Gamma_r\Gamma_{sp}\right)}. \tag{7.7.38}$$

Applying the conditions of the strong coupling limit ($g \gg \Gamma_r, \Gamma_{sp}$) this result reduces to

$$\alpha_\pm = -\frac{1}{2}(\Gamma_\phi - i\Delta) \pm ig. \tag{7.7.39}$$

In the strong coupling regime the decay of the atom develops Rabi-like oscillations. Any photon emitted into the cavity mode in this "good" cavity limit is likely to be re-absorbed by the atom before exiting the cavity. The atom is therefore driven by its own emission in this limit, and exhibits the behavior shown in Figure 7.32, referred to as

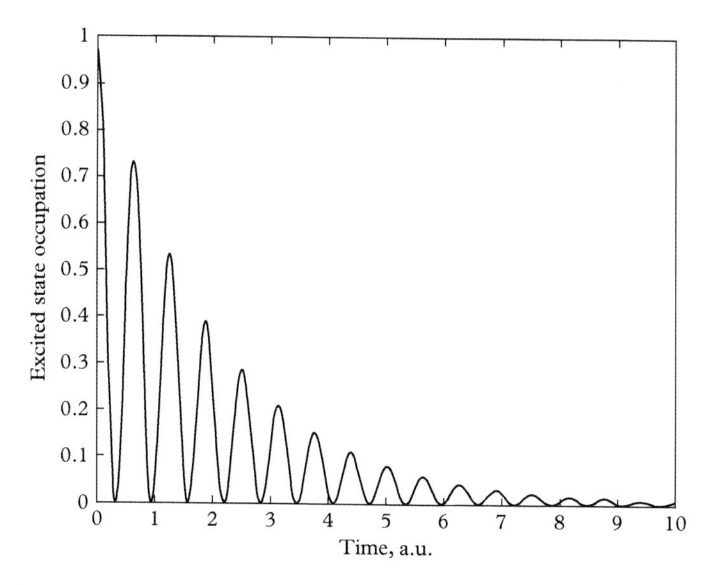

Figure 7.32 *Oscillations in the excited state population decay of an atom interacting with a resonant cavity in the strong coupling regime. Since electric field fluctuations of the vacuum are responsible for spontaneous emission, these are called vacuum Rabi oscillations. The envelope of these oscillations corresponds to the exponential decay encountered in the weak coupling regime. Parameter values:* $\Gamma_\phi = 0.25$, $g = 5$, *and* $\Delta = 0$.

vacuum Rabi oscillations. The absorption of a probe wave shows only two transitions, however, split by the Rabi frequency $2g$. The three distinguishable transition frequencies of our earlier dressed atom picture of the resonance fluorescence of atoms are no longer observed, because in the present case only the ground state has a photon present that can dress the atom.

As the cavity is tuned through resonance, the Rabi sidebands in the probe absorption spectrum are found to exchange intensity. First principally one and then the other transition is excited (Figure 7.33). For large positive detunings, the symmetric dressed state has the character of bare state $|1,1\rangle$ and an energy like that of the bare excited atom. Hence it is "atom like." The character of the anti-symmetric state is primarily that of $|2,0\rangle$ and it has an energy determined by the cavity frequency. It is therefore deemed "cavity like." As the detuning is varied from positive to negative, these states evolve from mostly atom like to mostly cavity like, and vice versa, as one can show with dressed state analysis. At zero detuning, where the atom–field interaction is strongest because of cavity enhancement, the transitions have mixed character and undergo an avoided crossing. The intensity exchange through the avoided crossing region is a signature of the strong coupling regime of cavity QED that has been observed in many contexts (see, for example, [7.82] and [7.83]).

Superposition states, including those of cavity quantum electrodynamics, are important in the field of quantum information science. They are the basis for quantum computational speed-up proposals, quantum encryption, and secure quantum communication [7.84]. When it is possible to separate the components of superposition

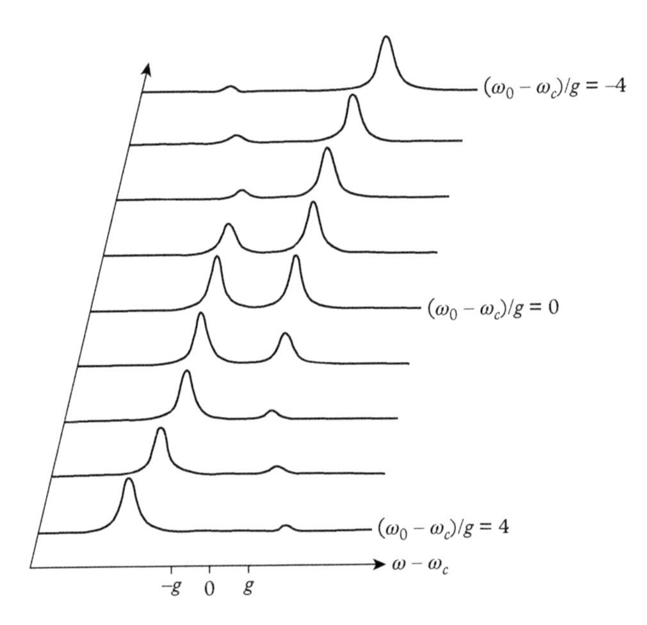

Figure 7.33 *Illustration of the intensity exchange between features at the avoided crossing that characterizes the probe absorption spectrum of strongly coupled cavity QED.*

states physically while still retaining the coherent phase information, they are said to be *entangled.* Transport of quantum state information to remote locations then becomes possible. The strongly coupled field–atom system of this section is an example of entanglement that is potentially suitable for such applications, since the atom and light propagate in different directions following their interaction (see Ref. [7.85] for an experimental realization). While cavity enhancement and entanglement are important ingredients in some quantum information applications, neither is essential however [7.86]. There are many other approaches to the creation, transport and manipulation of entangled states [7.87], which can include entanglement by polarization in the emission from light-emitting diodes [7.88]. This topic is explored further in Problems 5.1 and 7.5, and in Section 7.8.

This section has analyzed the interaction of atoms with light for atom–field couplings that are sufficiently enhanced by a cavity that the Rabi frequency exceeds the decoherence rate. This is the regime of strong coupling, which corresponds to the condition $\Gamma \ll \Omega_R$. Generally, the optical carrier frequency greatly exceeds the Rabi frequency ($\Omega_R \ll \omega$). However one can imagine even stronger coupling, characterized by $\Omega_R \approx \omega$, in which photon exchange occurs on the timescale of the optical period itself. Under such extreme conditions the RWA breaks down, and interactions become non-adiabatic. Efforts to extend strong cavity QED to this regime of ultra-strong light–matter interactions have been undertaken in experiments based on intersubband cavity polaritons [7.89].

7.8 Quantum Information Processing

7.8.1 Introduction

Modern computers perform computations by combining digital signals at the moment they appear at the input or by using "remembered" results to obtain deterministic outcomes. That is, they either combine signals actually present at the input or they apply sequential logic to a series of results obtained at earlier times. These operations can also be mixed, but it is noteworthy that the first process requires no memory whereas the second relies on prior results, yet all classical computations can be analyzed as a deterministic sequence of such operations. In a classical binary computer where information is represented in terms of two-state "bits" (with values of either 0 or 1), calculations become a sequence of operations performed on one bit or two bits at a time. One-bit operations typically perform arithmetic and two-bit operations assign the extra bit to a control function that applies a "remembered" result to determine whether the one-bit outcome should be inverted or not, thereby enabling logic operations.

Quantum information processing has followed this established approach to classical computing and information processing, but paradoxically it offers improvements based on the non-deterministic character of quantum mechanics. Computations in quantum mechanics are probabilistic rather than deterministic, and it is not obvious that information processing can be improved by relaxing the certainty of an outcome. Nevertheless, the introduction of quantum mechanical concepts and uncertainty does hold unexpected benefits for machine-based computation and information processing. In this section the origin of computational speed-up for factoring large numbers and accelerating search and optimization problems is explored, and ways of implementing quantum information processing with optical science are considered.

The heart of quantum information processing is the use of superposition states to exploit delocalized cross-correlations that are absent from classical operations. As previously mentioned, quantum correlations hold hidden benefits for the treatment of certain types of computational problem. Entanglement also has advantages for other topics, such as quantum cryptography, superdense coding and quantum key distribution for secure communication, and teleportation, but these topics are too specialized for our present purposes. The reader interested in these topics is referred to the bibliography [7.90]. Here we confine our attention to examples of how entangled states can be produced experimentally, how optical transformations are performed, and what relevance the density matrix has to the subject.

7.8.2 Classical Logic Circuits and Computation

Classical operations are performed by electronic "gates," the building blocks of circuits that process information or perform logical calculations. Common gates have names like AND, OR, XOR, and NOT, and are simple to construct from basic circuit elements [7.91]. These gates execute completely predictable operations that can be described in a table specifying their outputs for all possible inputs. Symbols for the Inverter or NOT gate and the AND gate are given in Figure 7.34, together with truth tables of their operation. The NOT gate is a circuit that acts on one bit by simply inverting the input. Although many other types of gate are available and offer various design advantages, the importance of the NOT and AND gates is that together they constitute a universal set of gates with which any classical computation can be performed [7.92, 7.93].

A typical objective of information processing might be to sort a list of digitally encoded signals that represent names and place them in alphabetic order. Or, it might be to find the shortest path between two cities based on available road and map information. Regardless of the objective of the processing, there is a strict limit on the number of operations that can be executed in a specified period of time by a particular machine. Each gate operation requires a circuit element and takes time. So there are problems that are impractical to solve with available time or resources. Limited resources render some problems effectively unsolvable. It is this limitation that accounts for the great interest in quantum information processing, which promises to reduce the number of operations required to analyze important classes of problem, thereby rendering some "unsolvable" problems "solvable."

Quantum computation is expected to be significantly faster than classical computation, but how many fewer operations than a classical processing algorithm would a quantum calculation actually take? The answer to this question seems to be that although speed-up depends on the calculation under consideration, a quantum computer

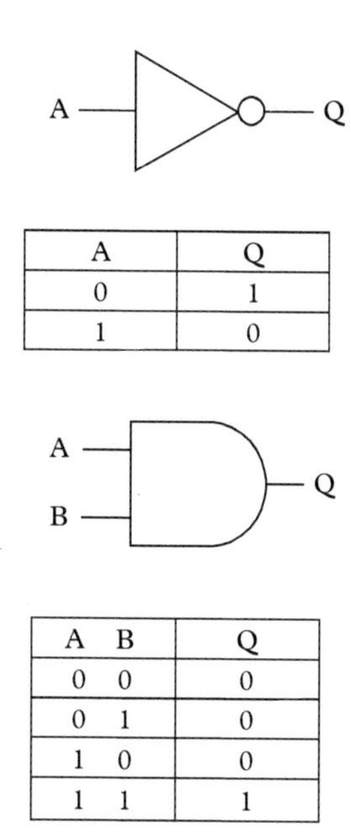

A	Q
0	1
1	0

A B	Q
0 0	0
0 1	0
1 0	0
1 1	1

Figure 7.34 *Examples of classical computational gates and corresponding truth tables. (a) A one-bit NOT gate (top), and a (b) two-bit AND gate (bottom).*

may exceed even the theoretical constraints of perfect "Turing machine" models [7.94], based on the analyses of Deutsch and Shor in recent decades [7.95, 7.96]. For example, to factor a large number represented by n bits, a classical calculation requires a number of operations that is exponential in n (i.e., $\exp\{O(n^{1/3}\log^{2/3} n)\}$), whereas a quantum computation can finish in a time that is polynomial in the logarithm of n (i.e., $\exp\{O(n^2 \log n \log \cdot \log n)\}$). This means that quantum algorithms are predicted to be exponentially faster for example at factorization than their classical counterparts [7.97]. For the much broader class of search and optimization problems the advantage of quantum over classical computations is predicted to be quadratic in n [7.98], so this too is very significant for large-scale problems.

7.8.3 Quantum Bits and Quantum Logic Gates

The most distinctive feature of quantum computation is the storage and processing of superposition states [7.99]. Hence its most fundamental elements are the quantum bits (qubits) and unitary operations that create and process superposition states. Quantum bits (or qubits) are typically formed from two-level quantum systems rather than the binary bits of an electronic circuit. For example, two energy levels $|0\rangle$ and $|1\rangle$ of a two-level atom or molecule might be chosen to realize a quantum bit. A means must also exist however to create and transform linear combinations of the basis states in arbitrary superposition states. For example, $|0\rangle$ must be transformable into $\alpha|0\rangle + \beta|1\rangle$, and $|1\rangle$must be transformable into $\gamma|0\rangle + \delta|1\rangle$. This requires a logic gate operation represented by the unitary 2×2 matrix $\begin{bmatrix} \delta & \beta \\ \gamma & \alpha \end{bmatrix}$ which produces these combinations by acting on $|0\rangle = \begin{bmatrix} 0 \\ 1 \end{bmatrix}$ or $|1\rangle = \begin{bmatrix} 1 \\ 0 \end{bmatrix}$ while preserving trace.

The qubit itself is really to be thought of as a normalized vector $|\psi\rangle = a|0\rangle + b|1\rangle$, with the normalization constraint $|a|^2 + |b|^2 = 1$. If $|\psi\rangle$ represents the state of a two-level atom with an allowed optical transition, any desired series of transformations can be accomplished using near resonant light pulse sequences, as discussed in Chapter 4. The application of pulses on timescales much less than the coherence decay time causes unitary rotations in the Hilbert space of the atom. The result of a sequence of transformations can also be "read out" by measuring components of the Bloch vector. Readout is usually accomplished by coupling the superposition state to an empty fluorescent level whose subsequent population (and fluorescent emission intensity) is made proportional to the square of a coefficient of the superposition. But wherein lies the big difference between qubits and classical bits? The answer is that the state of the system $|\psi\rangle = a|0\rangle + b|1\rangle$ can take on a limitless number of values. Although only two states are involved in specifying $|\psi\rangle$, just as for a classical bit, the coefficients a and b can assume any continuous value between 0 and 1. So in principle a two-level system can store or represent limitless information instead of just one bit.

Given the potential advantages of qubits, is it possible to perform arbitrary logic operations with them? To answer this question we can remind ourselves of the fact that 2×2 Pauli matrices provide a complete basis for the description of two-level systems and are also generators of rotations. They perform rotations $R_{\hat{n}}(\theta)$ through an angle θ about an arbitrary axis $\hat{n}$ in the Hilbert space of two-level atoms (see Problem 2.8). Hence they cover the entire space of two-state operations or two-state logic gates based on qubits. The function of all single qubit operators is either to create a superposition state by assigning a complex phase to an input or to perform a rotation. Consequently the most general single qubit operator has the form

$$\hat{U} = \exp(i\alpha)\hat{R}_{\hat{n}}(\theta), \tag{7.8.1}$$

which does both of these operations. The Pauli matrices have forms that are not necessarily easy to implement optically however, and minimizing the number of components in a quantum computer is always important. So it is helpful to point out a useful decomposition of this most general operator.

It can be shown that the Hadamard (H) and $\pi/8$ (T) gates whose matrix representations are given in Eq. (7.8.2) can approximate the functionality of a single qubit with arbitrarily good accuracy [7.90]. Moreover they are simple to build. To these gates, though not strictly necessary for quantum logic algorithms, we shall add the phase (S) gate. This is because we plan to analyze an implementation of a quantum gate that incorporates it shortly. More importantly, the S gate makes the representation of single qubit operations by H and T gates fault tolerant (Ref. [7.90], p. 197).

$$H = \frac{1}{\sqrt{2}}\begin{bmatrix} 1 & 1 \\ 1 & -1 \end{bmatrix}; \; T = \exp(i\pi/8)\begin{bmatrix} \exp(-i\pi/8) & 0 \\ 0 & \exp(i\pi/8) \end{bmatrix}; \; S = \begin{bmatrix} 1 & 0 \\ 0 & i \end{bmatrix}. \quad (7.8.2)$$

The matrices of Eq. (7.8.2) represent H, T, and S gates and are some of the mathematical ingredients related to qubits.

But what physical systems have the superposition states required to construct qubits? Fortunately many quantum systems can potentially serve as individual qubits. For instance nuclear spins, Josephson junctions, pairs of energy levels in atoms, and orthogonal polarization states of light are all candidates for binary qubit systems and there are many other possibilities. Because there are so many ways to create superposition states, let us move ahead to discuss coupled qubits, which constitute the main technical challenge in building a quantum computer. This is an essential step because single qubits are insufficient by themselves to perform quantum computations. We need to discuss specific examples of single and double qubits.

Just as in classical computing, quantum computation is always expressible as a sequence of one- and two-bit gate operations [7.84]. Hence a quantum two-bit gate is required for general quantum computation and it needs to complete a universal set of gates that can implement any sequence of instructions. The C-NOT gate is widely regarded as the standard two-bit gate for quantum information processing. It passes the input states directly to the output when the control bit is in the 0 state but switches the output state if the control is 1. The output Q that results from all possible inputs A and B is given by the truth table in Figure 7.35. The combination of the C-NOT and any one-bit quantum gate like the NOT provides a universal set from which any quantum computational sequence can be constructed [7.100, 7.101].

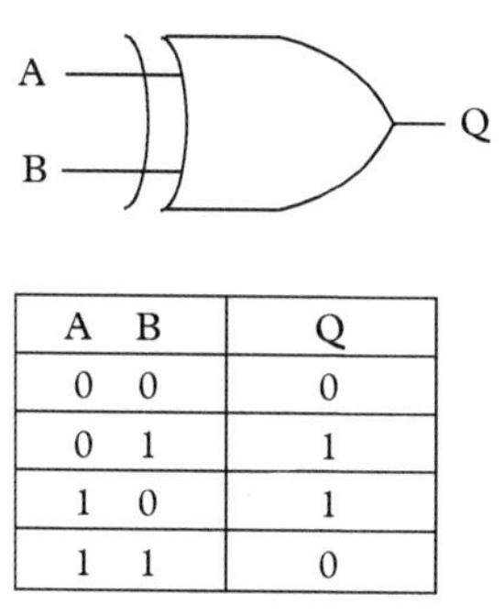

A	B	Q
0	0	0
0	1	1
1	0	1
1	1	0

Figure 7.35 *The C-NOT gate and its truth table. Any quantum computation can be expressed as a sequence of this two-bit quantum gate and a one-bit qubit gate like the NOT in Figure 7.34.*

The decomposition of an arbitrary quantum computation into a sequence of C-NOT and single qubit operations is the subject of Gray code analysis. This topic will not be covered here, but the interested reader can find a full discussion and examples in Ref. [7.102].

Just as for single-bit gates there is an important distinction between classical and quantum two-bit gates. To qualify as a quantum two-bit gate, a C-NOT processor must handle inputs that are *entangled*. Entanglement implies that an arbitrary state of the coupled qubits can be written as $|\psi\rangle = a|0\rangle|1\rangle \pm b|1\rangle|0\rangle$. This is a superposition of product states with the same energy that is referred to as a *dual-rail representation*. If the two input signals are not entangled, that is if they are not strongly coupled and able to form superpositions of the basis states, the gate is classical. Just as a single qubit must be capable of storing a superposition state, a quantum double bit (two-qubit) must store superpositions of its product basis states.

7.8.4 Realization of Quantum Gates

In the discussion in Section 7.8.3, the combination of single qubit and C-NOT gate operations was identified as a universal set of computing elements for quantum computation. Additionally it was pointed out that single qubit operations can be approximated with arbitrarily good precision by Hadamard and $\pi/8$ gates. Hence any quantum "circuit" can be well approximated by a sequence of Hadamard, $\pi/8$, and C-NOT gates [7.103]. We now turn to practical considerations as to how to construct basic quantum gates.

Much of this book has been devoted to understanding coherences that can be established by irradiating two-level and three-level atoms with light. Hence the idea of using simple quantum systems to create superposition states should by now be quite familiar. However, a specific realization of a qubit is illustrated in Figure 7.36 which highlights two states of a $^9\text{Be}^+$ ion that can serve our purpose and will subsequently allow discussion of a two-bit gate as well. First, single-bit gate operation consists of the application of

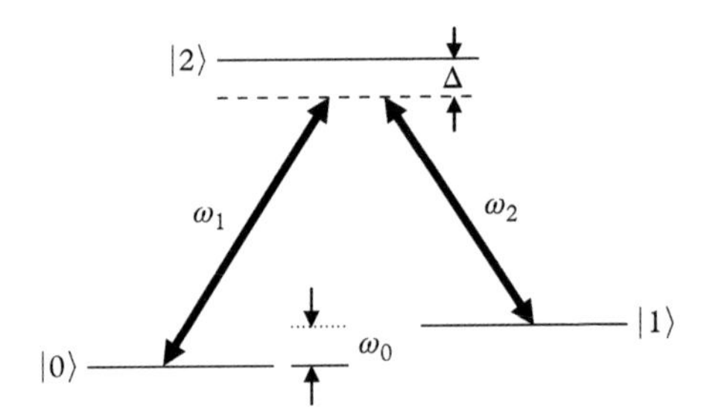

Figure 7.36 *Example of a single qubit consisting of a three-level system with transfer between 0 and 1 states accomplished by a two-photon (Raman) interaction. State* $|1\rangle$ *is long-lived, both to be useful as a latched state and to render the two-photon transition allowed. For the specific case of a* $^9\text{Be}^+$ *ion, the states in question are* $|0\rangle \equiv {}^2S_{1/2}(F = 2; m_F = 2)$, $|1\rangle \equiv {}^2S_{1/2}(F = 1; m_F = 1)$, *and* $|2\rangle \equiv {}^2P_{1/2}$.

a Raman pulse pair that causes a transition, either $|0\rangle \rightarrow |1\rangle$ or $|1\rangle \rightarrow |0\rangle$ depending on which state is occupied at the outset. The two optical pulses are applied simultaneously, so there is no directionality to the gate operation that originates from pulse timing. If the qubit is 0 at first, the state is flipped to $|1\rangle$ by application of a Raman π-pulse. If the qubit is 1 at first, the state is flipped to $|0\rangle$ by the π-pulse. Operation of the gate therefore satisfies the truth table of the NOT gate in Figure 7.34 and can also support the creation of arbitrary superposition states if excited with a pulse having an area between 0 and π. If the ion is sufficiently isolated from the environment, and the two-photon transition sufficiently detuned from resonance with intermediate state $|2\rangle$ so as to avoid spontaneous emission, eigenstates or superposition states can also be stored in this qubit indefinitely (in principle).

Next we consider the more complex problem of realizing a two-bit quantum gate. This problem contains a new challenge, since not only does each bit now have to support superposition states but the bits have to be strongly coupled in a linear combination of product states. That is, there must exist the possibility of entanglement *between the bits.*

To understand why entanglement is a challenge for construction of a quantum computer, consider the following. A two-level atom can act as a C-NOT gate that satisfies the truth table in Figure 7.35 and yet *fails* to yield a quantum two-bit gate. In Problem 7.15 the presence or absence of a π phase shift is shown to control whether a two-pulse sequence excites the atom or not. This operation is equivalent to two applications of the S phase gate in Eq. (7.82). If we consider the phase shift conditions $|A(\pi)\rangle = 0$ and $|A(0)\rangle = 1$ to represent the binary states of the control bit A and atomic states $|0\rangle$ and $|1\rangle$ to represent states of the target bit B, then the result of applying pulses to the system satisfies the truth table for C-NOT. When the control phase shift of π is applied, the atom does not change state (regardless of whether it is initially in the ground state or the excited state). On the other hand, when a phase shift of zero is applied, the output states are flipped. So the atom together with the phase shifting interactions act as a C-NOT gate. Unfortunately, the control bit is not a qubit and cannot form superposition states with the atom. Hence the gate as a whole is merely an unnecessarily complicated way to implement classical logic.

The first demonstration of a C-NOT gate that incorporated entanglement of the separate bits comprising the quantum register as well as C-NOT functionality was provided using optically controlled dynamics of a laser-cooled $^9Be^+$ ion [7.104]. Figure 7.37 shows the experimental quantum bits consisting of two-level excitations between electronic and vibrational states of the ion. The vibrational levels were provided by quantized motion of the cold atom in the harmonic potential of an ion trap. The electronic levels were hyperfine levels of the ion. The available states of this two-bit system are shown schematically in the figure. As for the single-bit gate described previously, transitions were controlled with stimulated Raman pulse pairs tuned to one of the three frequencies ω_0, ω_B (blue sideband), and ω_R (red sideband) shown in the figure. To prepare one of the four eigenstates, for instance $|0\rangle|v = 1\rangle$, the ions were first laser-cooled to $|0\rangle|v = 0\rangle$ and then excited with a π-pulse at ω_B followed by a second π-pulse at ω_0. This caused the ion+vibron system to cycle through the sequence of coupled states $|0\rangle|v = 0\rangle \rightarrow |1\rangle|v = 1\rangle \rightarrow |0\rangle|v = 1\rangle$ to the desired eigenstate.

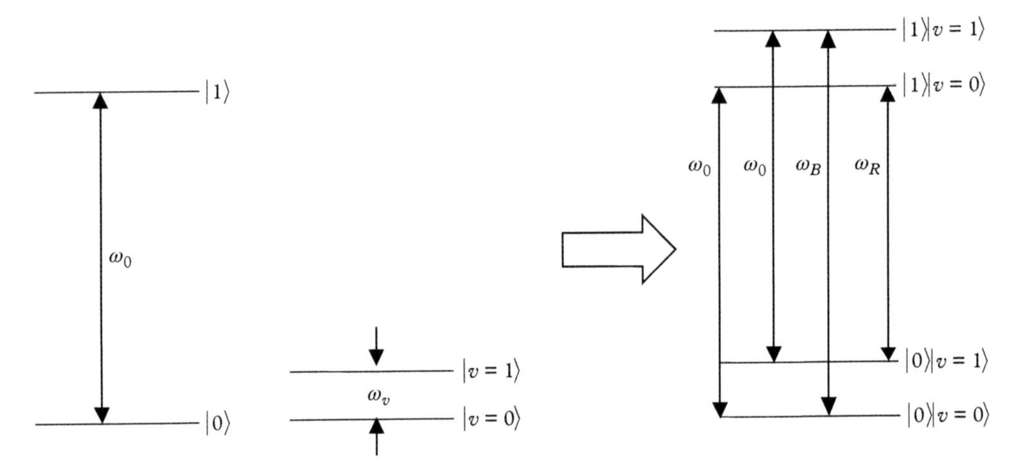

Figure 7.37 *Electronic and vibrational qubits (left) of a two-qubit system comprised of a single* $^9\mathrm{Be}^+$ *ion (right) localized in an rf-ion trap. Four transitions take place between the two-qubit product states, at three frequencies* ω_0, ω_B *(blue sideband), and* ω_R *(red sideband).*

Confirmation that the quantum register had been prepared in the desired eigenstate was provided by reading out the probability that the electronic (target) bit was in the $|0\rangle$ state and independently measuring the probability that the vibrational (control) bit was in the $|v = 1\rangle$ state. In the case of the first electronic bit, this readout was accomplished by irradiating the ion with σ^+-polarized light that coupled only the $|0\rangle|v=0\rangle$ and $|0\rangle|v=1\rangle$ levels to the $^2P_{3/2}$ fluorescent state, as shown in Figure 7.38, and collecting fluorescence. Optical selection rules ($\Delta F = 0, \pm 1$ and $\Delta m_F = +1$) prevented

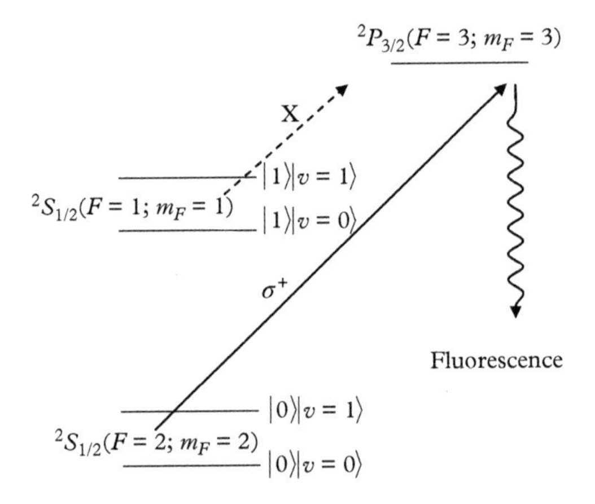

Figure 7.38 *Application of a circularly polarized readout pulse to determine whether the electronic state of the quantum register is 0 or 1.*

coupling of the fluorescent state to the $|1\rangle|v=0\rangle$ and $|1\rangle|v=1\rangle$ levels altogether, so only the probability amplitude of the $|0\rangle$ component was monitored. To separately measure the vibrational state, the ion was prepared a second time in the same input state by the transition sequence $|0\rangle|v=0\rangle \rightarrow |1\rangle|v=1\rangle \rightarrow |0\rangle|v=1\rangle$ followed by application of a π-pulse.

If the first measurement showed the ion to be in the $|0\rangle$ state, the π-pulse was applied on the red sideband at frequency ω_R to empty the $|0\rangle|v=1\rangle$ state. Thus when σ^+-polarized light was applied subsequently, the presence or absence of fluorescence indicated a vibrational state of $|v=0\rangle$ or $|v=1\rangle$, respectively. If the first measurement instead found the ion in the $|1\rangle$ state, the π-pulse was applied at the blue sideband frequency ω_R to empty the $|0\rangle|v=0\rangle$ state. Thus when σ^+-polarized light was applied, the presence or absence of fluorescence in this case indicated a vibrational state of $|v=1\rangle$ or $|v=0\rangle$, respectively. This time the second fluorescence measurement read out the second bit as 1 or 0. This procedure conveniently mapped the vibrational state onto the electronic degrees of freedom and verified reliable state preparation.

Once the protocols were established for preparing and reading out arbitrary states, the operation of this quantum logic element could be tested. The C-NOT operation was implemented with a fixed sequence of pulses applied to a $^9Be^+$ ion prepared in various states. Using the readout procedures described previously, the experimental outcomes were then checked for agreement with those specified by the truth table of Figure 7.35. The three pulses applied to operate the C-NOT gate were:

(i) $\pi/2$ Ramsey pulse at frequency ω_0;

(ii) 2π pulse at an auxiliary, blue sideband frequency ω' to cause a complete Rabi spin cycle $|1\rangle\,|v=1\rangle \rightarrow |aux\rangle\,|v=0\rangle \rightarrow -\,|1\rangle\,|v=1\rangle$ that negates the electron state;

(iii) $\pi/2$ Ramsey pulse at frequency ω_0 with a π phase shift.

Entangled states containing the $|v=0\rangle$ state of the vibrational qubit do not change sign as the result of control step (ii). The two Ramsey pulses simply cancel and cause no change. However, any $|1\rangle\,|v=1\rangle$ component of the input state is negated by the auxiliary transition, because the Ramsey pulses add constructively to form a π pulse that flips the target qubit. These specialized pulse effects are explored using the Bloch vector model in Problem 7.16 at the end of the chapter. The direct effect of a pulse of area A on an atomic wavefunction $|\psi\rangle$ is provided by the transformation $|\psi'\rangle = \{\cos(A/2) + i\sin(A/2)\}\,|\psi\rangle$, which for a 2π pulse yields $|\psi'\rangle = \{\cos(\pi) + i\sin(\pi)\}|\psi\rangle$ or $|\psi'\rangle = -|\psi\rangle$.

To implement C-NOT operation, the three-pulse sequence given by (i)–(iii) can simply be applied to any input state of the quantum register. When the control qubit is 0, the gate preserves the state of the target qubit. When the control qubit is 1, the gate flips the target qubit. Thus the output reproduces the C-NOT truth table with two entangled qubits and can succeed in performing unitary transformations on one quantum system conditioned by the quantum state of another. Thus the $^9Be^+$ ion system can realize the basic operations necessary to construct a quantum computer.

7.8.5 Fidelity of Gate Operations

In practice, imperfections in the apparatus controlling the qubits or the presence of pure dephasing or spontaneous emission can compromise the performance of any gate. Obviously, rapid decoherence processes destroy superposition states altogether, and systems with long coherence times are desirable. The challenge of creating, measuring and maintaining the entanglement of multiple qubits becomes more severe as the computational basis set is enlarged [7.105], so it is important to define what is meant by the fidelity of operation.

Basic operation of quantum gate can be checked using the experimental procedures already outlined. Since reliable computation is the ultimate goal of the exercise however, it is important to introduce a reliable quantitative measurement of how well a gate conforms to its intended function. This calls for evaluation of coherences which is the domain of the density matrix ρ. Consequently it should come as no surprise that the metric of gate performance, called the fidelity, is defined in terms of it. The fidelity evaluates how well the initial coherence is preserved throughout an operation by calculating the modulus of the probability that the final state is merely a rotated version of the initial (or reference) state:

$$F = \sqrt{\langle \psi_{ref} | \tilde{\rho}_{final}(t) | \psi_{ref} \rangle}. \tag{7.8.3}$$

If the final state is indistinguishable from the reference state for example, then clearly $F = 1$. If the projection of the final state on ψ_{ref} yields $F = 0$, then the fidelity of the gate is zero. In other words, the gate failed. The deleterious effect that spontaneous emission has on the fidelity of any readout operation is explored in Problem 7.5.

REFERENCES

[7.1] T. Carmon, H. Rokhsari, L. Yang, T.J. Kippenberg, and K.J. Vahala, *Phys. Rev. Lett.* 2005, **94**, 223902.
[7.2] J.P. Gordon and A. Ashkin, *Phys. Rev. A* 1980, **21**, 1606.
[7.3] A. Ashkin, J.M. Dziedzic, J.E. Bjorkholm, and S. Chu, *Opt. Lett.* 1986, **11**, 288.
[7.4] A. Ashkin and J.M. Dziedzic, *Science* 1987, **235**, 1517.
[7.5] S. Chu, *Rev. Mod. Phys.* 1998, **70**, No. 3.
[7.6] J. Dalibard and C. Cohen-Tannoudji, *J. Opt. Soc. Am. B* 1985, **2**, 1707.
[7.7] S. Chu, J.E. Bjorkholm, A. Ashkin, and A. Cable, *Phys. Rev. Lett.* 1986, 57, 314.
[7.8] T. Hansch and A.L. Schawlow, *Opt. Commun.* 1975, **13**, 68.
[7.9] See, for example, H. Metcalf and P. van der Straten, *Laser Cooling and Trapping*. New York: Springer-Verlag, 1999.

[7.10] R.I. Epstein, M.I. Buchwald, B.C. Edwards, T.R. Gosnell, and C.E. Mungan, *Nature* 1995, **377**, 500.
[7.11] J. Thiede, J. Distel, S.R. Greenfield, and R.I. Epstein, *Appl. Phys. Lett.* 2005, **86**, 154107.
[7.12] E.L. Raab, M. Prentiss, A. Cable, S. Chu, and D.E. Pritchard, *Phys. Rev. Lett.* 1987, **59**, 2631.
[7.13] J. Dalibard and C. Cohen-Tannoudji, *J. Opt. Soc. Am. B* 1989, **6**, 2023.
[7.14] P.J. Ungar, D.S. Weiss, E. Riis, and S. Chu, *J. Opt. Soc. Am. B* 1989, **6**, 2058.
[7.15] J.L. Hall, C.J. Borde, and K. Uehara, *Phys. Rev. Lett.* 1976, **37**, 1339.
[7.16] M. Kasevich and S. Chu, *Phys. Rev. Lett.* 1992, **69**, 1741.
[7.17] A. Aspect, E. Arimondo, R. Kaiser, N. Vansteenkiste, and C. Cohen-Tannoudji, *J. Opt. Soc. Am. B* 1989, **6**, 2112.
[7.18] M.H. Anderson, J.R. Ensher, M.R. Matthews, C.E. Wieman, and E.A. Cornell, *Science* 1995, **269**, 198.
[7.19] L. Pitaevskii and S. Stringari, *Bose–Einstein Condensation.* Oxford: Clarendon Press, 2003.
[7.20] C.A. Regal, M. Geiner, and D.S. Lin, *Phys. Rev. Lett.* 2004, **92**, 040403.
[7.21] B.W. Shore, *The Theory of Atomic Excitation.* New York: Wiley, 1990.
[7.22] J. Oreg, G. Hazak, and J.H. Eberly, *Phys. Rev. A* 1985, **32**, 2776.
[7.23] J.R. Kuklinski, U. Gaubatz, F.T. Hioe, and K. Bergmann, *Phys. Rev. A* 1989, **40**, 6741.
[7.24] M. P. Fewell, B. W. Shore, and K. Bergmann, *Aust. J. Phys.* 1997, **50**, 281.
[7.25] F. Hioe, *Phys. Rev. A* 1983, **28**, 879.
[7.26] J. Gong and S.A. Rice, *J. Chem. Phys.* 2004, **120**, 5117.
[7.27] P.F. Cohadon, A. Heidmann, and M. Pinard, *Phys. Rev. Lett.* 1999, **83**, 3174.
[7.28] X.L. Ruan, S.C. Rand, and M. Kaviany, *Phys. Rev. B* 2007, **75**, 214304.
[7.29] M. Sheik-Bahaie and R.I. Epstein, *Nat. Photonics* 2007, **1**, 693.
[7.30] J.L. Clark, P.F. Miller, and G. Rumbles, *J. Phys. Chem. A* 1998, **102**, 4428.
[7.31] J. Zhang, D. Li, R. Chen, and Q. Xiong, *Nature* 2013, **493**, 504.
[7.32] G. Bahl, M. Tomes, F. Marquardt, and T. Carmon, *Nat. Phys.* 2012, **8**, 203.
[7.33] C. Kittel, *Introduction to Solid State Physics,* 2nd ed. New York: John Wiley & Sons, 1956.
[7.34] B. Donovan and J.F. Angress, *Lattice Vibrations*. London: Chapman and Hall, 1971.
[7.35] D. Pines, *Elementary Excitations in Solids.* New York: W.A. Benjamin, Inc., 1964, p.232.
[7.36] Y.R. Shen, *The Principles of Nonlinear Optics*. New York: John Wiley & Sons, 1984, p. 192.
[7.37] Y.R. Shen and N. Bloembergen, *Phys. Rev.* 1965, **137**, 1787
[7.38] A. Yariv, *IEEE J. Quantum Electron.* 1965, **1**, 28.
[7.39] W.H. Louisell, *Coupled Modes and Parametric Electronics.* New York: Wiley, 1960.
[7.40] P.E. Toschek, *Ann. Phys. (Fr.)* 1985, **10**, 761.
[7.41] M. Lindberg and J. Javanainen, *J. Opt. Soc. Am. B* 1986, **3**, 1008.
[7.42] M. Kasevich and S. Chu, *Phys. Rev. Lett.* 1992, **69**, 1741.
[7.43] S.C. Rand, *J. Lumin.* 2013, **133**, 10.
[7.44] J. Reichel, O. Morice, G.M. Tino, and C. Salomon, *Europhys. Lett.* 1994, **28**, 477.
[7.45] J. Reichel, F. Bardou, M. Ben Dahan, E. Peik, S.C. Rand, C. Salomon, and C. Cohen-Tannoudji, *Phys. Rev. Lett.* 1995, **75**, 4575.
[7.46] A. Kuhn, H. Perrin, W. Hansel, and C. Salomon, in K. Burnett (ed.), *OSA TOPS on Ultracold Atoms and BEC*, Vol. 7. Optical Society of America, 1997, pp.58–65.
[7.47] P. Pringsheim, *Z. Phys. A* 1929, **57**, 739.
[7.48] L. Landau, *J. Phys. (Moscow)* 1946, **10**, 503.
[7.49] S.P. Feofilov, A.B. Kulinkin, T. Gacoin, G. Mialon, G. Dantelle, R.S. Meltzer, and C. Dujardin, *J. Lumin.* 2012, **132**, 3082.

[7.50] See A. Abragam and B. Bleaney, *Electron Paramagnetic Resonance of Transition Ions*. New York: Dover Publications, 1970, pp.560–2.
[7.51] See, for example, C. Kittel, *Introduction to Solid State Physics,* 4th ed. New York: Wiley, 1971, p.228.
[7.52] S.C. Rand, W.M. Fisher, and S.L. Oliveira, *J. Opt. Soc. Am. B* 2008, **25**, 1106.
[7.53] B.Y. Zeldovich, *Phys. Uspekhi* 2008, **51**, 465.
[7.54] See, for example, A.H. Nayfeh, *Perturbation Methods*. New York: John Wiley & Sons, 1973.
[7.55] R. Di Leonardo, G. Ruocco, J. Leach, M.J. Padgett, A.J. Wright, J.M. Girkin, D.R. Burnham, and D. McGloin, *Phys. Rev. Lett.* 2007, **99**, 010601.
[7.56] C.H. Townes and A.L. Schawlow, *Microwave Spectroscopy*. Toronto: Dover Publications, 1975, p.178.
[7.57] G. Herzberg, *Spectra of Diatomic Molecules*, 2nd ed. New York: Van Nostrand Reinhold, 1950, p.129.
[7.58] A.A. Fisher, E. Dreyer, A. Chakrabarty, and S.C. Rand, Opt. Express **24**, 26055(2016).
[7.59] E. Dreyer, A.A. Fisher, G. Smail, P. Anisimov, and S.C. Rand, Opt. Express **26**, 17755 (2018).
[7.60] A.A. Fisher, E. Dreyer, A. Chakrabarty, and S.C. Rand, Opt. Express **24**, 26064(2016).
[7.61] C. Cohen-Tannoudji and S. Reynaud, in J. Eberly and P. Lambropoulos (eds.), *Multiphoton Processes*, New York: John Wiley & Sons, 1977, pp.103–18.
[7.62] J. Baudon, M. Hamamda, J. Grucker, M. Boustimi, F. Perales, G. Dutier, and M. Ducloy, *Phys. Rev. Lett.* 2009, **102**, 140403.
[7.63] K.-J. Boller, A. Imamoglu, and S.E. Harris, *Phys. Rev. Lett.* 1991, **66**, 2593.
[7.64] K. Hakuta, L. Marmet, and B.P. Stoicheff, *Phys. Rev. Lett.* 1991, **66**, 596.
[7.65] P.M. Anisimov, J.P. Dowling, and B.C. Sanders, *Phys. Rev. Lett.* 2011, **107**, 163604.
[7.66] L.V. Hau, S.E. Harris, and Z. Dutton, *Nature* 1999, **397**, 594.
[7.67] M.F. Yanik and S. Fan, Nat. Phys. 2007, **3**, 372; Z. Shi, R.W. Boyd, R.M. Camacho, P.V. Vudyasetu, and J.C. Howell, *Phys. Rev. Lett.* 2007, **99**, 240801.
[7.68] Q. Shu and S.C. Rand, *Phys. Rev. B* 1997, **55**, 8776.
[7.69] A.S. Zibrov, A.B. Matsko, O. Kocharovskaya, Y.V. Rostovtsev, G.R. Welch, and M.O. Scully, *Phys. Rev. Lett.* 2002, **88**, 103601.
[7.70] K. Hakuta, M. Suzuki, M. Katsuragawa, and J.Z. Li, *Phys. Rev. Lett.* 1997, **79**, 209.
[7.71] M.O. Scully, *Phys. Rev. Lett.* 1991, **67**, 1855; M. Fleischauer, C.H. Kreitel, M.O. Scully, C. Su, B.T. Ulrich, and S.-Y. Zhu, *Phys. Rev. A* 1992, **46**, 1468.
[7.72] A. Nottlemann, C. Peters, and W. Lange, *Phys. Rev. Lett.* 1993, **70**, 1783; E.S. Fry, X. Li, D. Nikonov, G.G. Padmabandu, M.O. Scully, A.V. Smith, F.K. Tittel, C. Wang, S.R. Wilkinson, and S.-Y. Zhu, *Phys. Rev. Lett.* 1993, **70**, 3235; W.E. van der Veer, R.J.J. van Diest, A. Donszelmann, and H.B. van Linden van den Heuvell, *Phys. Rev. Lett.* 1993, **70**, 3243.
[7.73] M.O. Scully and M.S. Zubairy, *Quantum Optics*. Cambridge: Cambridge University Press, 1997.
[7.74] R. Rohlsberger, H.-C. Wille, K. Schlage, and B. Sahoo, *Nature* 2012, **482**, 199.
[7.75] A. Lezama, S. Barreiro, and A.A. Akulshin, *Phys. Rev. A* 1999, **59**, 4732.
[7.76] H. Ian, Y.X. Liu, and F. Nori, *Phys. Rev. A* 2010, **81**, 063823.
[7.77] C.L.G. Alzar, M.A.G. Martinez, and P. Nussenzveig, *Am. J. Phys.* 2002, **70**, 37.
[7.78] A. Yariv, *Quantum Electronics*, 3rd ed. Wiley, 2003.
[7.79] H.P. Yuen and J.H. Shapiro, *IEEE Trans. Inf. Theory* 1980, **26**, 78.
[7.80] L. Mandel and E. Wolf, *Optical Coherence and Quantum Optics*. Cambridge: Cambridge University Press, 1995.

[7.81] P. Berman (ed.), *Cavity Quantum Electrodynamics*. New York: Academic Press, 1994.

[7.82] C. Weisbuch, M. Nishioka, A. Ishikawa, and Y. Arakawa, *Phys. Rev. Lett.* 1992, **69**, 3314.

[7.83] J. McKeever, A. Boca, A.D. Boozer, J.R. Buck, and H.J. Kimble, *Nature* **425**, 6955.

[7.84] M.A. Nielson and I.L. Chuang, *Quantum Computation and Quantum Information*. Cambridge: Cambridge University Press, 2000.

[7.85] Q.A. Turchette, C.J. Hood, W. Lange, H. Mabuchi, and H.J. Kimble, *Phys. Rev. Lett.* 1995, 75, 4710.

[7.86] P. Knight, *Science* 2000, **287**, 441.

[7.87] Supplementary reading on advanced quantum information experimentation using single atoms, quantum dots, ions, color centers, and nuclear spins may be found in: B.B. Blinov, D.L. Moehring, L.-M. Duan, and C. Monroe, *Nature* 2004, **428**, 153; X. Li, Y. Wu, D.G. Steel, D. Gammon, T.H. Stievater, D.S. Katzer, D. Park, C. Piermarocchi, and L.J. Sham, *Science* 2003, **301**, 809; R.G. Brewer, R.G. DeVoe, and R. Kallenbach, *Phys. Rev. A* 1992, **46**, R6781; F. Jelezko, T. Gaebel, I. Popa, M. Domhan, A. Gruber, and J. Wrachtrup, *Phys. Rev. Lett.* 2004, **93**, 130501; I.L. Chuang, N. Gershenfeld, and M.G. Kubinec, *Phys. Rev. Lett.* 1998, **18**, 3408.

[7.88] C.L. Salter, R.M. Stevenson, I. Farrer, C.A. Nicoll, D.A. Ritchie, and A.J. Shields, *Nature* 2010, **465**, 594.

[7.89] C. Ciuti, G. Bastard, and I. Carusotto, *Phys. Rev. B* 2005, **72**, 115303; A.A. Anappara, S. De Liberato, A. Tredicucci, C. Ciuti, G. Biasiol, L. Sorba, and F. Beltram, *Phys. Rev. B* 2009, **79**, R201303; G. Gunter, A.A. Anappara, J. Hees, A. Sell, G. Biasial, L. Sorba, S. De Liberato, C. Ciuti, A. Tredicucci, A. Leitenstorfer, and R. Huber, *Nature* 2009, **458**, 178.

[7.90] M.A. Nielsen and I.L. Chuang, *Quantum Computation and Quantum Information*. Cambridge: Cambridge University Press, 2000, p.196.

[7.91] See, for example, P. Horowitz and W. Hill, *The Art of Electronics*. Cambridge: Cambridge University Press, 1980.

[7.92] C.E. Shannon, *Trans. Am. Inst. Electr. Eng.* 1938, **57**, 713.

[7.93] W. Wernick, *Trans. Am. Math. Soc.* 1942, **51**, 11732.

[7.94] A.M. Turing, *Proc. Lond. Math. Soc.* 1936, 42, 230.

[7.95] D. Deutsch, *Proc. R. Soc. Lond. A* 1985, **400**, 97.

[7.96] P.W. Shor, in *Proceedings of the 35th Annual Symposium on Foundations of Computer Science*. Los Alamitos, CA: IEEE Press, 1994.

[7.97] A. Ekert and R. Jozsa, *Rev. Mod. Phys.* 1996, **68**, 733.

[7.98] L.K. Grover, *Phys. Rev. Lett.* 1997, **79**, 325.

[7.99] R.P. Feynman, *Int. J. Theor. Phys.* 1982, **21**, 467.

[7.100] D.P. DiVincenzo, *Phys. Rev. A* 1995, **51**, 1015.

[7.101] A. Barenco, D. Deutsch, A. Ekert, and R. Jozsa, *Phys. Rev. Lett.* 1995, **74**, 4083.

[7.102] A. Barenco, C.H. Bennett, R. Cleve, D.P. Vincenzo, N. Margolus, P. Schor, T. Sleator, J. Smolin, and H. Weinfurter, *Phys. Rev. A* 1995, **52**, 3457.

[7.103] M.A. Nielsen and I.L. Chuang, *Quantum Computation and Quantum Information*. Cambridge: Cambridge University Press, 2000, pp.188–202.

[7.104] C. Monroe, D.M. Meekhof, B.E. King, W.M. Itano, and D.J. Wineland, *Phys. Rev. Lett.* 1995, 75, 4714.

[7.105] T. Monz, P. Schindler, J.T. Barreiro, M. Chwalla, D. Nigg, W.A. Coish, M. Harlander, W. Hansel, M. Hennrich, and R. Blatt, *Phys. Rev. Lett.* 2011, **106**, 130506.

[7.106] J.-T. Shen and S. Fanhui, *Phys. Rev. A* 2009, **79**, 023837.

PROBLEMS

7.1. Verify the values of the Clebsch–Gordan coefficients shown in Figure 7.5 for a $J_g = 1/2 \leftrightarrow J_e = 3/2$ transition.

7.2. Calculate the seven Clebsch–Gordan coefficients for a $J_g = 1 \leftrightarrow J_e = 1$ transition.

7.3. Calculate the 15 Clebsch–Gordan coefficients that determine the relative transition probabilities between hyperfine levels of sodium on the $F = 2 \rightarrow F' = 3$ transition used in the first experimental observations of the Mollow triplet.

7.4. *Statistics of squeezed states*:

(a) Using the squeezed field operator $\hat{t} \equiv \hat{S}\hat{a}\hat{S}^{+} = \hat{a}^{-} \cosh r + \hat{a}^{+} e^{i\theta} \sinh r$ (and its Hermitian conjugate), show that the mean square of the number of photons in a squeezed, single-mode vacuum state $|0\rangle$ is $\langle 0|\, \hat{t}^{+}\hat{t}^{+}\hat{t}\hat{t}\, |0\rangle = 3 \sinh^4 r + \sinh^2 r$.

(Note: $\hat{S}^{+}\hat{S} = 1$, and r and θ are the magnitude and phase of the complex squeeze parameter $z = re^{i\theta}$ that describes how strong the squeezing is. For large $|z|$ there is strong squeezing and for $|z| = 0$ or $r = 0$ there is no squeezing at all).

(b) Use the result of part (a) and the exercise in Section 7.6 to show that the quantum degree of second-order coherence for a squeezed single-mode vacuum state is

$$g^{(2)}(0) = \frac{\langle a^{+} a^{+} \hat{a}\hat{a} \rangle}{\langle \hat{a}^{+} \hat{a}^{-} \rangle^2} = 3 + \frac{1}{\langle n \rangle}$$

Squeezed vacuum therefore shows photon bunching rather than antibunching.

7.5. *In this problem the practical issue of preserving entanglement and the need to avoid spontaneous emission are explored.*

Consider a two-level atom that is coupled to a single mode field in a cavity. The atom may be in state $|\uparrow\rangle$ or $|\downarrow\rangle$, while the field is in $|+\rangle$ or $|-\rangle$, depending on whether the atom has added a photon to the field by undergoing a transition $|\uparrow\rangle \rightarrow |\downarrow\rangle$ or has removed a photon by absorption. The only other process possible is that the atom spontaneously decays to the ground state without adding a photon to the mode (the emitted photon escapes from the cavity). There are thus three states in the product basis: $|\uparrow\rangle|-\rangle$, $|\downarrow\rangle|+\rangle$, and $|\downarrow\rangle|-\rangle$, and the full density matrix is 3×3 instead of 2×2. That is,

$$\tilde{\rho} = \begin{bmatrix} \rho_{\uparrow-,\uparrow-} & \rho_{\uparrow-,\downarrow+} & \rho_{\uparrow-,\downarrow-} \\ \rho_{\downarrow+,\uparrow-} & \rho_{\downarrow+,\downarrow+} & \rho_{\downarrow+,\downarrow-} \\ \rho_{\downarrow-,\uparrow-} & \rho_{\downarrow-,\downarrow+} & \rho_{\downarrow-,\downarrow-} \end{bmatrix}.$$

Note however that if spontaneous emission is absent, the density matrix elements in the 2 × 2 sub-matrix

$$\tilde{\rho}_{sub} = \begin{bmatrix} \rho_{\uparrow -,\uparrow -} & \rho_{\uparrow -,\downarrow +} \\ \rho_{\downarrow +,\uparrow -} & \rho_{\downarrow +,\downarrow +} \end{bmatrix}$$

fully describe the system. On resonance, elements of $\tilde{\rho}_{sub}$ can be found from the Bloch vector $\bar{R}(t)$:

$$\bar{R}(t) = \begin{bmatrix} V(t) \\ W(t) \end{bmatrix} = \overleftrightarrow{M}\bar{R}_i(0) = \begin{bmatrix} \cos\theta & -\sin\theta \\ \sin\theta & \cos\theta \end{bmatrix} \begin{bmatrix} V(0) \\ W(0) \end{bmatrix},$$

using standard relations $\tilde{\rho}_{\uparrow -,\downarrow +} = -iV/2$, $\tilde{\rho}_{\downarrow +,\uparrow -} = \tilde{\rho}*_{\uparrow -,\downarrow +}$, and $W = \rho_{\uparrow -,\uparrow -} - \rho_{\downarrow +,\downarrow +}$. In the absence of spontaneous emission, one also has $\rho_{\uparrow -,\uparrow -} + \rho_{\downarrow +,\downarrow +} = 1$. Also, without spontaneous emission the remaining elements of $\tilde{\rho}$ are zero since they do not conserve energy in the coupled atom–cavity system.

(a) A fully entangled "initial" state is produced in the 2 × 2 sub-space of $\tilde{\rho}_{sub}$ by applying an ultrashort $\pi/2$ pulse to the $|\downarrow\rangle|+\rangle$ state. The initial Bloch vector *after pulse 1* is

$$\bar{R}(t_1 = 0) = \begin{bmatrix} 1 \\ 0 \end{bmatrix}. \tag{7.10.1}$$

Calculate the Bloch vectors $\bar{R}(t_2)$ and $\bar{R}(t_3)$ following a second $\pi/2$ pulse at time t_2 that converts the excitation to a pure excited state population, and a third pulse of *arbitrary* area θ at time t_3, respectively. Omit precession and dephasing between the pulses

(b) Determine all elements of the 3 × 3 density matrix after the third pulse (in the *absence* of spontaneous emission).

(c) Show that the probabilities of finding the system in states $|\uparrow\rangle|-\rangle$ and $|\downarrow\rangle|+\rangle$ are exactly anti-correlated for arbitrary θ by calculating the expectation values of $\tilde{\rho}$ for these states. Show that when one is maximum the other is always minimum.

(d) Calculate the "fidelity (F)" of entanglement in the final state when $\theta = -\pi/2$. (Fidelity $F = \sqrt{\langle\psi_1(0)|\,\tilde{\rho}_{final}(t)\,|\psi_1(0)\rangle}$ measures the degree of overlap with the initial reference state $\psi_1(0) = \frac{1}{\sqrt{2}}\begin{bmatrix} 1 \\ 1 \end{bmatrix}$ to see how much of the original entanglement is recovered after transformation.) To find F, a rotation must be applied independently to the off-diagonal terms of $\tilde{\rho}_{sub}(\theta = -\pi/2)$ according to the prescription

$$\tilde{\rho}_{final} = \tilde{\rho}'_{sub}\left(-\frac{\pi}{2}, \phi\right) = \begin{bmatrix} \rho_{\uparrow -,\uparrow -} & \rho_{\uparrow -,\downarrow +}(e^{i\phi}) \\ \rho_{\downarrow +,\uparrow -}(e^{-i\phi}) & \rho_{\downarrow +,\downarrow +} \end{bmatrix}.$$

Simply multiply your off-diagonal entries by $e^{\pm i\phi}$, as indicated, evaluate F, and give its value for $\phi = \pi/2$.
(Note: In practice the phase angle ϕ is varied to map out the magnitude of residual polarization terms, converting coherences to populations so that they can be measured.)

(e) Suppose that between pulses 2 and 3 a spontaneous emission event occurs and the atom drifts into a null of the standing wave in the cavity. Write down the full density matrix just prior to pulse 3 and determine the fidelity of entanglement as before, following the $\theta = -\pi/2$ pulse. Compare with the result of part (d).

7.6. (a) In laser cooling, position z must be considered an operator instead of a scalar under certain circumstances. Briefly, why is this so and what is the implication for the eigenvalues of z when the spread in linear momentum Δp of the atoms becomes less than the photon momentum $\hbar k$?

(b) Using the expressions for the interaction V of ^{4}He atoms with circularly polarized light on the $2^3S_1\,(m=\pm 1)$–$2^3P_1\,(m=0)$ transitions, consider the coherent superposition states $|\psi_{NC}(p)\rangle$ and $|\psi_C(p)\rangle$, and show explicitly that the NC state does not couple to the excited state whereas the C state does. To do this, calculate matrix elements $\langle e_0, p|V|\,\psi_{NC}(p)\rangle$ and $\langle e_0, p|V|\,\psi_C(p)\rangle$, and comment.

7.7. (a) Calculate the root-mean-square fluctuations of the electric field operator $E = \frac{1}{\sqrt{2}}(a + a^+)\sin(kz - \omega t)$ in the following two distinct states:

$$|\psi_1\rangle = |n\rangle \text{ and } |\psi_2\rangle = e^{i(kz-\omega t)}\,|n-3\rangle + |n-2\rangle + e^{i(kz-\omega t)}\,|n+2\rangle + |n+3\rangle\,.$$

(b) Show that in state $|\psi_1\rangle$, noise is a minimum only at times when the amplitude of the field goes to zero as the result of its sinusoidal motion (i.e., when $kz = \omega t$). This shows that individual photon number states exhibit noise that is proportional to the instantaneous field amplitude.

(c) Show that for the simple superposition state $|\psi_2\rangle$, the noise can be zero at another time during each period by finding a *non-zero* phase angle $\phi = kz - \omega t$ for which the root-mean-square field is zero. Assume n is large.

(Note: The field noise at this second instant is lower than the noise in the case described at this same phase angle. This appears to show that simple linear superpositions of number states can exhibit amplitude squeezing. However, after proper normalization of $|\psi_2\rangle$, no real solution for the second phase value can be found. Try to find one! Only more complicated superpositions of number states (coherent states) can exhibit squeezing when properly normalized.)

7.8. On resonance ($\Delta = 0$), the second-order degree of coherence for a quantum light source is given by the expression

$$g^{(2)}_{atom}(\tau) = \left[1 - \exp\left(\frac{-3\beta\tau}{2}\right)\left(\cos(\Omega''\tau) + \frac{3\beta}{2\Omega''}\sin(\Omega''\tau)\right)\right],$$

where $\Omega'' \equiv \sqrt{\Omega^2 - (\beta^2/4)}$, $\Omega \equiv \mu E/\hbar$, and $\beta \equiv \gamma_{sp}/2$.

(a) What specific value of $g^{(2)}(\tau)$ is *uniquely* characteristic of emission from a quantum light source consisting of *a single atom?* Give a one sentence explanation of the mechanism that allows an atom to exhibit this special value.

(b) Suppose that $g^{(2)}(\tau)$ is measured for an ensemble of *two uncorrelated atoms*. Each atom emits individually in accord with $g^{(2)}_{atom}$. However, when a measurement of $g^{(2)}(\tau)$ is made based on the total intensity from both atoms, one atom may emit at a time when the other cannot, thereby altering the value of $g^{(2)}_{ensemble}$ with respect to $g^{(2)}_{atom}$. Estimate $g^{(2)}_{ensemble}(0)$ for the two-atom ensemble in the *weak field limit* by approximating it with $g^{(2)}(\tau)$ for one atom at a delay τ equal to its most probable time of emission. (This is equivalent to measuring the unwanted contribution of the second atom when trying to measure $g^{(2)}_{atom}(0)$ of the first alone.)

(c) Formulate a criterion that ensures that a single molecule experiment was actually performed with a single molecule.

7.9. A two-level atom is coupled to a cavity with a coupling strength g, and the cavity in turn is coupled to a single-mode optical fiber waveguide, as depicted in the diagram. Light of frequency ω impinges from the left at a velocity v_g and may be reflected or transmitted in the coupling region as indicated by arrows. The atomic resonant frequency is ω_0 and the cavity resonant frequency is ω_c. The quantities τ_a and τ_c are atomic and cavity lifetimes respectively, and V is the coupling strength of the waveguide–cavity interaction. Exact solutions for the transmitted and reflected amplitudes [7.106] are found to be:

$$t = \frac{(\omega - \omega_0 + (i/\tau_a))(\omega - \omega_c + (i/\tau_c)) - g^2}{(\omega - \omega_0 + (i/\tau_a))(\omega - \omega_c + (i/\tau_c)) + (iV^2/v_g)) - g^2},$$

$$r = \frac{-(\omega - \omega_0 + (i/\tau_a))(iV^2/v_g)}{(\omega - \omega_0 + (i/\tau_a))(\omega - \omega_c + (i/\tau_c)) + (iV^2/v_g)) - g^2}.$$

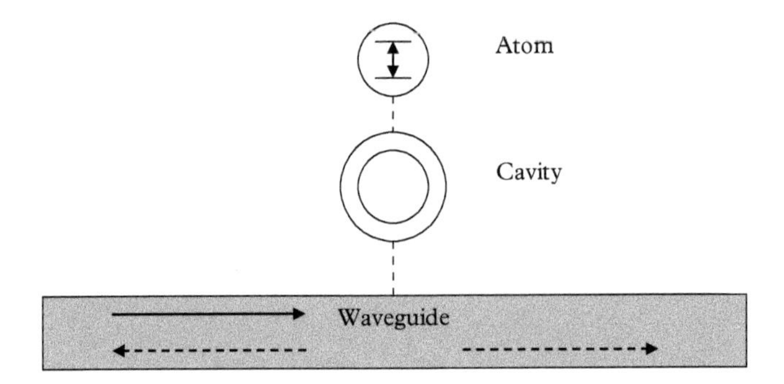

(a) Ignoring dissipation processes altogether, show that when the light frequency is chosen so that the guide is detuned from the cavity and atomic resonances, there is no reflection.

(b) Ignoring dissipation when the atom and the light in the guide are both tuned to the cavity frequency, show that there is still no reflection.

(c) Ignoring dissipation but considering the case when the coupling coefficient between the light and the cavity is small ($V^2 \ll v_g$), determine a condition that yields total reflection of the light.

(d) When dissipation is *included*, show from the expressions for r, t that the overall interaction conserves energy in the weak coupling limit, regardless of detuning.

(e) Comment on whether frequency detuning in this structure could theoretically provide a transmission switch for *individual photons* or not.

7.10. *Cavity quantum relaxation theory:*

In the simplest model for cavity QED, the atom/field product basis states are taken to be $|2\rangle \equiv |2\rangle|0\rangle$, $|1\rangle \equiv |1\rangle|1\rangle$, and $|0\rangle \equiv |1\rangle|0\rangle$ and the density matrix equation of motion is

$$\dot{\rho} = -\frac{i}{\hbar}[H_0, \rho] - \frac{i}{\hbar}[V, \rho] - \frac{\Gamma_{sp}}{2}\left(\sigma^+\sigma^-\rho + \rho\sigma^+\sigma^- - 2\sigma^-\rho\sigma^+\right) + \dots$$
$$\dots - \frac{\Gamma_r}{2}\left(a^+a^-\rho + \rho a^+a^- - 2a^-\rho a^+\right).$$

The off-diagonal matrix element ρ_{12} follows the slowly varying amplitude approximation and therefore oscillates at the optical frequency (identical to the cavity frequency Ω_c in this model). That is, you may assume that $\rho_{12} = \tilde{\rho}_{12}\exp(i\Omega_c t)$.

(a) Calculate the explicit form of the equation of motion for population of the excited state $|2\rangle \equiv |2\rangle|0\rangle$. That is, write out the equation of motion for $\dot{\rho}_{22}$ using the given master equation expression, evaluating all the matrix elements on the *right-hand side* of the equation. Determine whether the "jump" Liouvillian plays a role in the dynamics of the excited state.

(b) Calculate the explicit form of the equation of motion for the reservoir state $|0\rangle \equiv |1\rangle|0\rangle$. That is, write out the equation of motion for $\dot{\rho}_{00}$, and evaluate the matrix elements on the *right-hand side* of the equation. Show again explicitly whether the "jump" Liouvillian contributes to dynamics of the reservoir state or not.

7.11. The mechanical force of light on a two-level atom comprises a dipole force that depends on the intensity gradient and resonant radiation pressure that is usually called the dissipative force.

(a) Find the ratio of the dipole to radiation pressure forces in the limit of a very high-intensity gradient, on the order of $\nabla\Omega^2 \approx k\Omega^2$, and describe what conditions of detuning favor one force over the other.

(b) The dipole force is conservative. By integrating the force expression, find the corresponding potential energy U_{dip} and show that in the limit of very large detuning $(\Delta \gg \Gamma, \Omega)$ it is approximately equal to the light shift in the ground state: $U_{dip} = \hbar\Omega^2/4\Delta$.

7.12. In a one-dimensional lin $\perp$ lin configuration for laser cooling with polarization gradients, localization of atoms occurs at very low temperatures* due to the formation of potential wells by the light shift (see Figure 7.7). Cold atoms can become trapped in the lowest levels of the optical potentials and therefore become simple harmonic oscillators with quantized motion. The position coordinate of a trapped atom becomes the operator $\hat{z} = z_0(\hat{b}^+ + \hat{b}^-)$, where $z_0 \equiv \sqrt{\frac{\hbar}{2m\omega_k}}$ and ω_k is the frequency of the mode.

(a) Find the root-mean-square position of such an atom as a function of the quantum number n specifying its vibrational energy level.

(b) If the width of the well is taken to be $\lambda/2$ at an optical wavelength of 780 nm, estimate the highest value of n that a localized Rb atom ($m = 1.419 \times 10^{-25}$ kg) can have when the light intensity is adjusted to give a level separation of 100 kHz.

(c) How deep is the potential well (in Joules) under these conditions?

(d) Estimate the maximum temperature a collection of Rb could have and still be confined to such a well.

(*Further reading: P.S. Jessen et al., *Phys. Rev. Lett.* 1992, **69**, 49.)

7.13. From the solutions for probability amplitudes and transition frequencies in the strong coupling regime of cavity QED, calculate the off-diagonal element of the density matrix.

(a) Show that when the atom is prepared in the excited state ($C_1(0) = 0$ and $C_2(0) = 1$) the coherence amplitude is

$$\tilde{\rho}_{12} = -4C_{1+}C^*_{2-}\exp(-\Gamma_\phi t)\sin^2(gt).$$

(b) Find $\tilde{\rho}_{12}$ when the system is prepared in the ground state with $C_1(0) = 1$ and $C_2(0) = 0$.

7.14. A simple model of cavity QED involves a single quantum of excitation mediating transitions between three available $|atom\rangle|field\rangle$ product states. Two of the states, $|2\rangle \equiv |2{,}0\rangle$ and $|1\rangle \equiv |1{,}1\rangle$, have the same total energy. The third state, $|0\rangle \equiv |1{,}0\rangle$, results from the irreversible loss of a photon from the atom–cavity system, either through the mirrors or by spontaneous emission. The relevant master equation, including decay terms of the Lindblad form is

$$\dot{\rho} = -\frac{i}{\hbar}[H_0, \rho] - \frac{i}{\hbar}[V, \rho] - \frac{1}{2}\Gamma_{sp}[(\sigma^+\sigma^-\rho - \sigma^-\rho\sigma^+) + h.c.]$$

$$-\frac{1}{2}\Gamma_r[(a^+a^-\rho - a^-\rho a^+) + h.c.],$$

where $V = \hbar g[a^+\sigma^- \exp(i\Omega_c t) + h.c.]$ and Ω_c is the cavity mode frequency. Evaluate the explicit contributions, if any, of decay terms in the cavity QED equations of $\dot{\rho}_{00}$ and $\dot{\rho}_{11}$.

7.15. In the weak coupling limit of cavity QED, a high-quality cavity enhances the emission rate of atoms within it. To relate the enhancement to the conventional cavity quality factor Q one can introduce it into Eq. (7.7.27) through its definition in terms of cavity loss rate $\Gamma_r = \Omega_c/Q$. The square of the atom–field coupling strength g is also proportional to the square of the dipole matrix element $\mu^2 = \frac{3\pi\varepsilon_0\hbar c^3}{\Omega_c^3}\Gamma_{sp}$.

(a) Show that near resonance the excited state decay constant is enhanced beyond the natural decay rate to a value of $\Gamma_{sp}\left\{1 + \frac{3\lambda^3 Q}{4\pi^2 V}\right\}$.

(b) Show that far off resonance (for $\Delta \sim \Omega_c$) the decay constant is $\Gamma_{sp}\left\{1 + \frac{3\lambda^3}{16\pi^2 VQ}\right\}$, so that under this condition high Q cavities have little effect on atomic decay.

7.16. Two resonant, ultrashort pulses with areas of $\pi/2$ are separated in time by a short delay $\tau \ll T_2$ during which FID takes place. A phase shift of δ is applied during the second pulse. Ignoring decay during the FID period between pulses, use the Bloch vector model to show that interference between these *Ramsey pulses* (see Problem 4.15) causes

(a) complete excitation to the excited state when $\delta = 0$, and

(b) cancellation of absorption when $\delta = \pi$.

(c) Show that a phase-shifted, resonant $\pi/2$ pulse with $\delta = \pi$ has the same evolution matrix as the resonant pulse sequence π, $\pi/2$ with no interpulse delay.

(d) Show that a resonant π pulse negates the coherence represented by Bloch vector $\begin{bmatrix} 0 \\ 1 \\ 0 \end{bmatrix}$.

(e) Show that a resonant 2π pulse negates an arbitrary superposition state.

(For an application of these results in quantum computation, see Ref. [7.100].)

7.17. Effective Hamiltonians can be helpful in solving for the temporal dynamics of optical interactions.

(a) Verify that replacement of the Hamiltonian for cavity QED in the weak coupling limit by the H_{eff} in Eq. (I.6) of Appendix I leads to the expression for coherence given by Eq. (7.7.18).

(b) Similarly, verify that substitution of the effective Hamiltonian of Eq. (7.3.2) into the master equation Eq. (7.3.1) yields Eqs. (7.3.5)–(7.3.7) and Eqs. (7.3.17)–(7.3.19).

7.18. The rate of change of photon number in laser cooling is governed by the commutation properties of the three operators $\partial/\partial t$, $\hat{a}^+$, and $\hat{a}^-$.

(a) For arbitrary operators $\hat{p}_1$, $\hat{p}_2$, and $\hat{p}_3$ show that

$$\left[\hat{p}_1, \hat{p}_2\hat{p}_3\right] = \left[\hat{p}_1, \hat{p}_2\right]\hat{p}_3 + \hat{p}_2\left[\hat{p}_1, \hat{p}_3\right]$$

and

$$\left[\hat{p}_1\hat{p}_2, \hat{p}_3\right] = \hat{p}_1\left[\hat{p}_2, \hat{p}_3\right] + \left[\hat{p}_1, \hat{p}_3\right]\hat{p}_2.$$

(b) Use these results and $[\hat{a}^-, \hat{a}^+] = 1$ to prove that

$$\left(\frac{\partial \hat{a}^-}{\partial t}\right)\hat{a}^+ = \left(\frac{\partial \hat{a}^+}{\partial t}\right)\hat{a}^-.$$

7.19. A good model for the operation of a quantum logic gate is a simple 50:50 beam splitter, which presents each incident photon with an equal probability of being reflected or transmitted.

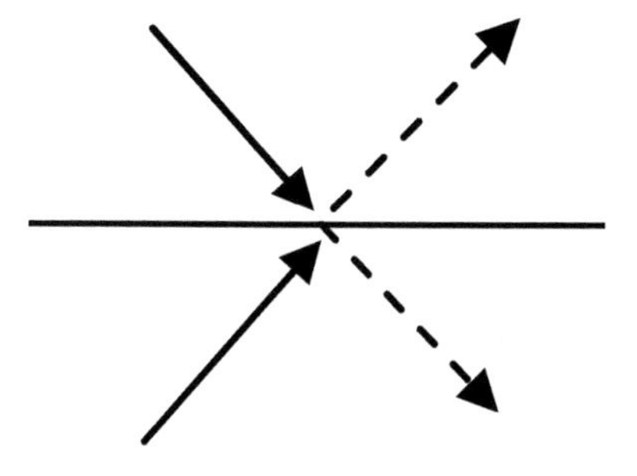

The splitter mixes incident modes a and b to give outputs c and d by the unitary transformation $\begin{pmatrix} c \\ d \end{pmatrix} = \frac{1}{\sqrt{2}}\begin{bmatrix} 1 & 1 \\ 1 & -1 \end{bmatrix}\begin{pmatrix} a \\ b \end{pmatrix}$, which ensures they are orthogonal superpositions with integer occupations of the input modes. Let the input modes be Fock states with creation operators that obey the relation $\hat{a}^+\hat{b}^+\,|0\rangle|0\rangle = |1\rangle|1\rangle$ and show that the output $\hat{c}^+\hat{d}^+\,|0\rangle|0\rangle$ consists only of $|2\rangle|0\rangle$ or $|0\rangle|2\rangle$ in equal proportion, never yielding a $|1\rangle|1\rangle$ state. Thus the photons always exit together, whatever the port.

(This result is called the Hong–Ou–Mandel effect and is a two-photon interference effect in quantum optics. See C.K. Hong, Z.Y. Ou, and L. Mandel, *Phys. Rev. Lett.* 1987, **59**, 2044.)

7.20. In Sisyphus laser cooling the potential wells formed by the light shifts of orthogonal, counter-propagating, linearly polarized waves are separated by a constant distance equal to a quarter wavelength.

(a) Show that if the relative polarization angle of the beams is not 90° but θ, the intensity decomposes into $\hat{\varepsilon}_+$ and $\hat{\varepsilon}_-$ components in a circular basis, with distributions describable by $I_+ = I_0\cos^2(kz - \theta/2)$ and $I_- = I_0\cos^2(kz + \theta/2)$ whose relative positioning can be controlled by varying θ.

(b) Find the separation $(z_+ - z_-)$ of adjacent wells in terms of θ and λ.

Appendices

Appendix A
Expectation Values

In quantum mechanics, operators appear in place of the continuous variables from classical mechanics and their commutation properties must be taken into account when predicting the outcome of measurements. For an operator $\hat{O}$ that in general does not commute with wavefunction ψ, the average value of $\hat{O}$ expected in a series of measurements could in principle be defined in two ways:

$$\langle \hat{O} \rangle = \int \hat{O}\, \psi^* \psi\, d^3r, \tag{A.1}$$

or

$$\langle \hat{O} \rangle = \int \psi^* \hat{O} \psi\, d^3r. \tag{A.2}$$

We can judge whether Eq. (A.1) or (A.2) is correct by testing the outcome of each expression against Schrödinger's wave equation (1.3.5). As an example, consider the Hamiltonian operator $\hat{H}$. If we ignore operator commutation rules we can imagine two possible expressions for the first moment, namely

$$\langle \hat{H} \rangle = i\hbar \int \frac{\partial}{\partial t} (\psi^* \psi) d^3r, \tag{A.3}$$

$$\langle \hat{H} \rangle = i\hbar \int \psi^* \frac{\partial}{\partial t} \psi\, dx. \tag{A.4}$$

The correct expression for $\langle \hat{H} \rangle$ must however be fully consistent with Eq. (1.3.5). If we multiply through $\hat{H}\psi = i\hbar \frac{\partial \psi}{\partial t}$ by ψ^* and integrate over volume, we find

$$\langle \hat{H} \rangle = \int \psi^* \hat{H} \psi\, d^3r = i\hbar \int \psi^* \frac{\partial \psi}{\partial t} d^3r, \tag{A.5}$$

in agreement with definition (A.2) and Eq. (1.3.11). If instead we operate with $\hat{H}$ on $\psi^*\psi$, we find

$$\int \hat{H} \psi^* \psi\, d^3r = i\hbar \int \frac{\partial}{\partial t} (\psi^* \psi)\, d^3r = i\hbar \int \left(\psi^* \frac{\partial}{\partial t} \psi + \psi \frac{\partial}{\partial t} \psi^* \right) d^3r. \tag{A.6}$$

Evaluation of the last integral for the simplest case in which ψ is an eigenstate with eigenvalue E yields $2E$. The outcome $2E$ clearly contradicts the starting assumption and is therefore inconsistent with Schrödinger's equation. Only the second definition yields the correct result, uniquely establishing Eq. (A.2) as the definition of expectation value. Overall, this illustrates the importance of commutation properties of operator expressions in quantum mechanics.

Appendix B
The Heisenberg Uncertainty Principle

To derive the Heisenberg uncertainty principle analytically for specific operators such as x and p_x, the following argument can be used. The squared standard deviations of position and momentum observables are

$$\left\langle \Delta x^2 \right\rangle = \int \psi^* \left(x - <x>\right)^2 \psi \, dV, \tag{B.1}$$

$$\left\langle \Delta p_x^2 \right\rangle = \int \psi^* \left(-i\hbar \frac{\partial}{\partial x} - \langle p_x \rangle\right)^2 \psi \, dV. \tag{B.2}$$

Taking the average values to be zero ($\langle x \rangle = \langle p_x \rangle = 0$), we find

$$\left\langle \Delta p_x^2 \right\rangle \left\langle \Delta x^2 \right\rangle = \int \psi^* \left(-\hbar^2 \frac{\partial^2}{\partial x^2}\right) \psi \, dr \int \psi^* x^2 \psi \, dV. \tag{B.3}$$

Integration by parts permits us to write

$$\int \psi^* \frac{\partial^2 \psi}{\partial x^2} dV = -\int \frac{\partial \psi^*}{\partial x} \frac{\partial \psi}{\partial x} dV, \tag{B.4}$$

$$\left\langle \Delta p_x^2 \right\rangle \left\langle \Delta x^2 \right\rangle = \hbar^2 \int \frac{\partial \psi^*}{\partial x} \frac{\partial \psi}{\partial x} dV \int \psi^* x^2 \psi \, dV. \tag{B.5}$$

Now, using the Schwartz inequality, we obtain

$$\int ff^* dV \int g^* g dV \geq \left[\frac{1}{2}\left(\int fg^* dV + \int gf^* dV\right)\right]^2. \tag{B.6}$$

Setting $f \equiv \partial\psi/\partial x$ and $g \equiv x\psi$ in this expression, one finds

$$\begin{aligned}
\left\langle \Delta p_x^2 \right\rangle \left\langle \Delta x^2 \right\rangle &\geq \frac{\hbar^2}{4}\left(\int \frac{\partial \psi}{\partial x} x \psi^* dV + \int x\psi \frac{\partial \psi^*}{\partial x} dV\right)^2 \\
&\geq \frac{\hbar^2}{4}\left(\int x \frac{\partial}{\partial x}\left(\psi\psi^*\right) dV\right)^2 = \frac{\hbar^2}{4} \\
&\geq \frac{\hbar^2}{4}.
\end{aligned} \tag{B.7}$$

The final integral is again evaluated by parts and is equal to −1. Defining the uncertainties in momentum and position as $\Delta p_x \equiv \left\langle \Delta p_x^2 \right\rangle^{1/2}$ and $\Delta x \equiv \left\langle \Delta x^2 \right\rangle^{1/2}$, we obtain

$$\Delta p_x \Delta x \geq \frac{\hbar}{2}. \tag{B.8}$$

The operators x and p_x provide one example of two Hermitian operators $\hat{A}$ and $\hat{B}$ with a commutator that is Hermitian and non-zero. Note that this results in an uncertainty principle between $\hat{A}$ and $\hat{B}$. Next, this result is extended to the general case.

Consider two Hermitian operators $\hat{A}$ and $\hat{B}$ with a Hermitian commutator

$$[\hat{A}, \hat{B}] = i\hat{C}. \tag{B.9}$$

It is shown subsequently that the standard deviations ΔA and ΔB have the product

$$\Delta A \Delta B \geq \frac{1}{2} \left| \langle C \rangle \right|. \tag{B.10}$$

The standard deviations (uncertainties) in $\hat{A}$ and $\hat{B}$ are defined as

$$\Delta A \equiv \hat{A} - \langle A \rangle, \tag{B.11}$$

$$\Delta B \equiv \hat{B} - \langle B \rangle; \tag{B.12}$$

hence observed values of their squares are

$$\left\langle (\Delta A)^2 \right\rangle = \langle \psi \Delta A | \Delta A \psi \rangle = \| \Delta A \psi \|^2 \tag{B.13}$$

and

$$\left\langle (\Delta B)^2 \right\rangle = \langle \psi \Delta B | \Delta B \psi \rangle = \| \Delta B \psi \|^2 . \tag{B.14}$$

The Schwartz inequality is equivalent to the inequality expressing the fact that the inner product of two vector operators, namely $\langle \psi \Delta A | \, \Delta B \psi \rangle$, is less than the product of their lengths, given by $\| \Delta A \psi \| \, \| \Delta B \psi \|$. That is,

$$\| \Delta A \psi \|^2 \, \| \Delta B \psi \|^2 \geq \left| \langle \psi \Delta A | \Delta B \psi \rangle \right|^2 , \tag{B.15}$$

or

$$(\Delta A)^2 (\Delta B)^2 \geq \left| \langle \psi \Delta A | \, \Delta B \psi \rangle \right|^2 = \left| \langle \psi \, | \, \Delta A \Delta B \, | \psi \rangle \right|^2 . \tag{B.16}$$

The last step is due to the Hermiticity of ΔA. The product $\Delta A \Delta B$ on the right-hand side of this result can be re-expressed in terms of the commutator of $\hat{A}$ and $\hat{B}$, because any operator can be written as a linear combination of two Hermitian operators:

$$\begin{aligned}\Delta A \Delta B &= \frac{1}{2}(\Delta A \Delta B + \Delta B \Delta A) + \frac{1}{2}[\Delta A, \Delta B] \\ &= \frac{1}{2}(\Delta A \Delta B + \Delta B \Delta A) + \frac{1}{2}[A, B] \\ &= \hat{D} + \frac{i}{2}\hat{C}.\end{aligned} \tag{B.17}$$

Hence

$$(\Delta A)^2 (\Delta B)^2 \geq \left| \langle \psi | \hat{D} + \frac{i}{2}\hat{C} | \psi \rangle \right|^2 = \left| \langle D \rangle + \frac{i}{2}\langle C \rangle \right|^2 . \tag{B.18}$$

Because $\hat{D}$ and $\hat{C}$ are Hermitian, their expectation values are real. Consequently, the squared expression on the right-hand side of Eq. (B.18) is simply

$$(\Delta A)^2 (\Delta B)^2 \geq |\langle D \rangle|^2 + \frac{1}{4}|\langle C \rangle|^2 \geq \frac{1}{4}|\langle C \rangle|^2 . \tag{B.19}$$

The uncertainties in A and B therefore satisfy the general relation

$$(\Delta A)(\Delta B) \geq \frac{1}{2}|\langle C \rangle| . \tag{B.20}$$

Appendix C
The Classical Hamiltonian of Electromagnetic Interactions

The non-relativistic force on a charged particle of velocity $\bar{v}$ is the Lorentz force

$$\bar{F} = q\left[\bar{E} + (\bar{v} \times \bar{B})\right]. \tag{C.1}$$

On the basis of the vector identity

$$\bar{\nabla} \cdot (\bar{\nabla} \times \bar{A}) = 0, \tag{C.2}$$

where $\bar{A}$ is an arbitrary vector potential, we conclude from the Maxwell equation $\bar{\nabla} \cdot \bar{B} = 0$ that $\bar{B}$ can always be written in the form

$$\bar{B} = \bar{\nabla} \times \bar{A}, \tag{C.3}$$

where $\bar{A}$ is a vector potential.

Similarly, applying Eq. (C.3) in the Maxwell equation $\bar{\nabla} \times \bar{E} = -\partial \bar{B}/\partial t$, we see that it can be written as

$$\bar{\nabla} \times \bar{E} + \frac{\partial}{\partial t}\left(\bar{\nabla} \times \bar{A}\right) = 0, \tag{C.4}$$

or by reversing the order of derivatives, as

$$\bar{\nabla} \times \left(\bar{E} + \frac{\partial \bar{A}}{\partial t}\right) = 0. \tag{C.5}$$

Making use of a second vector identity, namely

$$\bar{\nabla} \times \bar{\nabla}\phi = 0, \tag{C.6}$$

where ϕ is a scalar potential, the electric field can clearly always be written as

$$\bar{E} = -\bar{\nabla}\phi - \frac{\partial \bar{A}}{\partial t}. \tag{C.7}$$

The expression for the Lorentz force becomes

$$\bar{F} = q\left[-\bar{\nabla}\phi - \frac{\partial \bar{A}}{\partial t} + \left(\bar{\mathrm{v}} \times \left[\bar{\nabla} \times \bar{A}\right]\right)\right]. \tag{C.8}$$

Using the replacements $\frac{d\bar{A}}{dt} = \frac{\partial \bar{A}}{\partial t} + \left(\bar{\mathrm{v}} \cdot \bar{\nabla}\right)\bar{A}$ and $\bar{\mathrm{v}} \times \left(\bar{\nabla} \times \bar{A}\right) = \bar{\nabla}\left(\bar{\mathrm{v}} \cdot \bar{A}\right) - \left(\bar{\mathrm{v}} \cdot \bar{\nabla}\right)\bar{A}$, we find

$$\begin{aligned}
\bar{F} &= q\left[-\bar{\nabla}\phi + \bar{\nabla}\left(\bar{\mathrm{v}} \cdot \bar{A}\right) - \frac{\partial \bar{A}}{\partial t} - \left(\bar{\mathrm{v}} \cdot \bar{\nabla}\right)\bar{A}\right] \\
&= q\left[-\bar{\nabla}\left(\phi - \bar{\mathrm{v}} \cdot \bar{A}\right) - \frac{d\bar{A}}{dt}\right] \\
&= q\left[-\bar{\nabla}\left(\phi - \bar{\mathrm{v}} \cdot \bar{A}\right) - \frac{d}{dt}\bar{\nabla}_{\mathrm{v}}\left(\bar{\mathrm{v}} \cdot \bar{A}\right)\right]
\end{aligned} \tag{C.9}$$

Since the scalar potential ϕ does not depend on velocity, the ith component of the force on a particle may be written as

$$F_i = q\left[-\frac{\partial U}{\partial x_i} - \frac{d}{dt}\frac{\partial U}{\partial \mathrm{v}_i}\right], \tag{C.10}$$

where $i = x, y, z$ and $U \equiv q\left(\phi - \bar{A} \cdot \bar{\mathrm{v}}\right)$. Here U is a generalized potential function from which we obtain the Lagrangian in traditional, non-relativistic mechanics:

$$L = T - U = \frac{m\mathrm{v}^2}{2} - q\phi + q\bar{A} \cdot \bar{\mathrm{v}}. \tag{C.11}$$

The Lagrangian in turn furnishes the Hamiltonian:

$$H = \sum_i \dot{x}_i \frac{\partial L}{\partial \dot{x}_i} - L = \frac{1}{2}m\mathrm{v}^2 + q\,\phi, \tag{C.12}$$

which, when expressed in terms of the canonical momentum $p_i = \partial L/\partial \dot{x}_i = m\mathrm{v}_i + qA_i$, yields

$$H = \frac{1}{2m}\left(\bar{p} - q\bar{A}\right) \cdot \left(\bar{p} - q\bar{A}\right) + q\phi. \tag{C.13}$$

The Hamiltonian in Eq. (C.13) can be converted to a form that is much more convenient for describing the system in terms of the applied fields E and B, to separate out multipole contributions, and for eventual quantization [C.1]. To make the conversion, we note that Eqs. (C.3) and (C.7) can be satisfied if $\phi(\bar{r}, t)$ and $\bar{A}(\bar{r}, t)$ are expanded in Taylor's series representations as

$$\phi(\bar{r}, t) = -\bar{r} \cdot \bar{E}(0, t) - \frac{1}{2}\bar{r}\bar{r} : (\bar{\nabla}\bar{E}(0, t)) + \ldots \tag{C.14}$$

and

$$\bar{A}(\bar{r},t) = \frac{1}{2}\bar{B}(r) \times \bar{r} + \frac{1}{3}\bar{r}\cdot\bar{\nabla}\bar{B}(0,t) \times \bar{r} + \ldots. \tag{C.15}$$

These expansions constitute a specific choice of gauge consistent with Taylor expansions of the fields themselves:

$$\bar{E}(\bar{r},t) = \bar{E}(0,t) + \bar{r}\cdot(\bar{\nabla}\bar{E}(0,t) + \ldots, \tag{C.16}$$

$$\bar{B}(\bar{r},t) = \bar{B}(0,t) + \bar{r}\cdot(\bar{\nabla}\bar{B}(0,t) + \ldots. \tag{C.17}$$

By substituting Eqs. (C.14) and (C.15) into Eq. (C.13), one finds that the interaction Hamiltonian assumes the form

$$H = -\bar{\mu}^{(e)}\cdot\bar{E} - \bar{\mu}^{(m)}\cdot\bar{B}, \tag{C.18}$$

if we retain only the leading electric and magnetic dipole terms, with moments $\bar{\mu}^{(e)} = e\bar{r}$ (pointing from the negative to the positive charge) and $\bar{\mu}^{(m)} = \frac{1}{2}e\bar{r}\times\bar{v}$.

REFERENCE

[C.1] L.D. Barron and C.G. Gray, *J. Phys. A: Math., Nucl. Gen.* 1973, **6**, 59.

Appendix D
Stationary and Time-Dependent Perturbation Theory

A. Stationary Perturbation Theory

Assume that the system Hamiltonian can be written as the sum of two parts, one of which is the unperturbed part ($\hat{H}_0$) and the second of which is an interaction $\hat{V}$.

$$\hat{H} = \hat{H}_0 + \lambda\hat{V} \tag{D.1}$$

λ is a perturbation parameter that merely keeps track of the presence and order of perturbation. When $\lambda = 0$ the perturbation is removed, and $\hat{H} = \hat{H}_0$. When $\lambda = 1$ the perturbation is present, and $\hat{H} = \hat{H}_0 + \hat{V}$.

Eigenvalues and eigenfunctions of the unperturbed system may be found by solving the energy equation $\hat{H}_0\phi_n = E_n^0\phi_n$. When the interaction is included, this requires solution of the equation

$$\left(\hat{H}_0 + \lambda\hat{V}\right)\psi = E\psi. \tag{D.2}$$

Solutions can readily be found when there is no degeneracy, that is, when all energy levels are widely separated, by using an approximate method. For this purpose, an expansion based on the complete set of eigenfunctions ϕ_n of the unperturbed problem may be used.

$$\psi = \sum_n a_n\phi_n. \tag{D.3}$$

Inserting this expansion into Eq. (D.2), one obtains

$$\left(E - E_m^0\right)a_n = \lambda\sum_n V_{mn}a_n, \tag{D.4}$$

where $V_{mn} \equiv \left\langle\phi_m\middle|\hat{V}\middle|\phi_n\right\rangle$.

Because of the perturbation, the energies and eigenfunctions of the system change. To determine the corrections to the energy and eigenfunction of the lth stationary state, a scheme of successive approximations for E_l and a_n is used:

$$E_l = E_l^{(0)} + \lambda E_l^{(1)} + \lambda^2 E_l^{(2)} + \ldots, \tag{D.5}$$

$$a_m = a_m^{(0)} + \lambda a_m^{(1)} + \lambda^2 a_m^{(2)} + \ldots. \tag{D.6}$$

If we assume the system starts in state l, we have $a_m^{(0)} = \delta_{ml}$, so that

$$a_m = \delta_{ml} + \lambda a_m^{(1)} + \lambda^2 a_m^{(2)} + \ldots \tag{D.7}$$

Substituting Eqs. (D.7) and (D.5) into Eq. (D.4), one finds

$$\left[E_l^0 - E_m^0 + \lambda E_l^{(1)} + \lambda^2 E_l^{(2)} + \ldots\right]\left[\delta_{ml} + \lambda a_m^{(1)} + \lambda^2 a_m^{(2)} + \ldots\right] \\ = \lambda \sum_n V_{mn}\left[\delta_{nl} + \lambda a_n^{(1)} + \ldots\right]. \tag{D.8}$$

If $m = l$, then comparison of terms of the same order in λ yields

$$E_l^{(1)} = V_{ll}, \\ E_l^{(2)} + E_l^{(1)} a_l^{(1)} = \sum_n V_{\text{ln}} a_n^{(1)} \ldots. \tag{D.9}$$

If $m \neq l$ the result is

$$a_m^{(1)}\left(E_l^0 - E_m^0\right) = V_{ml}, \\ E_l^{(1)} a_m^{(1)} + \left(E_l^0 - E_m^0\right) a_m^{(2)} = \sum_n V_{mn} a_n^{(1)} \ldots. \tag{D.10}$$

By setting $\lambda = 1$, the first-order approximation for the energy of the perturbed state l is found to be

$$E_l = E_l^0 + \lambda E_l^{(1)} = E_l^0 + V_{ll}. \tag{D.11}$$

Then, using Eqs. (D.9) and (D.10) in Eq. (D.3), the first-order wavefunction becomes

$$\psi_l = \phi_l + \lambda a_l^{(1)} \phi_l + \sum_{m \neq l} \frac{V_{ml}}{E_l^0 - E_m^0} \phi_m. \tag{D.12}$$

In order that this expression yields the initial condition $\psi_l = \phi_l$ when $V_{m\ell} = 0$ after setting $\lambda = 1$, the coefficient in Eq. (D.12) must have the value $a_l^{(1)} = 0$. Hence

$$\psi_l = \phi_l + \sum_{m \neq l} \frac{V_{ml}}{E_l^0 - E_m^0} \phi_m. \tag{D.13}$$

In second order we find

$$E_l^{(2)} = \sum_{m \neq l} \frac{V_{l,n} V_{nl}}{E_l^0 - E_n^0} \phi_m \tag{D.14}$$

and

$$E_l = E_l^0 + V_{ll} + \sum_{n \neq l} \frac{|V_{l,n}|^2}{E_l^0 - E_n^0} \phi_m. \tag{D.15}$$

From this expression it follows that the second-order correction to the ground-state energy of any system is *always negative* (when $E_\ell^0 < E_n^0$).

Perturbation theory is only valid if the expansion series converges. For this to take place, each correction must be smaller than the preceding one. That is, the condition

$$|H_{l,n}| = |V_{l,n}| \ll \left|E_l^0 - E_n^0\right| \tag{D.16}$$

must be fulfilled for all $m \neq l$.

Hence a necessary condition for validity of the perturbation method is that the distance between any given level and all other levels of the unperturbed problem must be large compared to the magnitude of the change in energy caused by the perturbation. Naturally this entails that the level ℓ cannot be degenerate, since otherwise the energy difference in the unperturbed problem would vanish.

Perturbation theory is also valid only if both the eigenfunctions and eigenvalues of $\hat{H}$ change continuously into those of the operator H_0 as $\lambda \to 0$. Sometimes this requirement is not satisfied when the character of the solution changes as the perturbation parameter vanishes. An example is provided by the evolution of a system from one with a discrete spectrum into one with a continuous spectrum [D.1]. Consider a Hamiltonian with potential energy

$$U(x) = \frac{1}{2}\mu^2\omega^2 x^2 + \lambda x^3. \tag{D.17}$$

For $\lambda = 0$, this is a harmonic oscillator with a discrete spectrum $E_n^0 \left(n + \frac{1}{2}\right)\hbar\omega$, as is well known. If λ is not zero however, we find that the condition

$$\lambda \left|\left\langle m \left| x^3 n \right|\right\rangle\right| \ll \left|E_m^0 - E_n^0\right| = \hbar\omega|m - n| \tag{D.18}$$

is always satisfied. For any non-vanishing value of λ, the spectrum of the perturbed Hamiltonian is then continuous! For large, negative values of x, the potential energy becomes lower than the total energy of the particle, so it can pass through the potential barrier in the direction of $x = -\infty$ and not return. This means that while the condition $|V_{lm}| \ll \left|E_l^0 - E_m^0\right|$ is necessary for perturbation theory to be applicable, it is not sufficient in itself to guarantee validity of the result.

B. Time-Dependent Perturbation Theory

In time-dependent perturbation theory, the eigenstates of H_0 are taken to be the basis states of the problem, and the task of calculating the wavefunction simplifies to one of determining the time-varying superposition of these states caused by the interaction $V(t)$. This approach ignores the possibility that the basis states themselves are altered by the "perturbing" interaction $V(t)$, but for weak, non-resonant interactions this is acceptable.

A perturbation parameter λ is used to keep track of the order in which each successive contribution to the approximate description arises.

$$H = H_0 + \lambda V(t), \tag{D.19}$$

$$\psi(\bar{r}, t) = \sum_n \left[C_n^{(0)} + \lambda C_n^{(1)} + \lambda^2 C_n^{(2)} + \ldots\right] U_n(\bar{r}) \exp(-i\omega_n t) \tag{D.20}$$

These expansions may be substituted into the Schrödinger equation $i\hbar \frac{\partial \psi}{\partial t} = H\psi$ to obtain

$$\begin{aligned} i\hbar \frac{\partial}{\partial t} \sum_n \left[C_n^{(0)}(t) + \lambda C_n^{(1)}(t) + \ldots\right] U_n(\bar{r}) \exp(-i\omega_n t) = \\ = H_0 \sum_n \left[C_n^{(0)}(t) + \lambda C_n^{(1)}(t) + \ldots\right] U_n(\bar{r}) \exp(-i\omega_n t) + \\ + \lambda V \sum_n \left[C_n^{(0)}(t) + \lambda C_n^{(1)}(t) + \ldots\right] U_n(\bar{r}) \exp(-i\omega_n t), \end{aligned} \tag{D.21}$$

$$\begin{aligned} i\hbar \sum_n \left[\dot{C}_n^{(0)}(t) + \lambda \dot{C}_n^{(1)}(t) + \ldots\right] U_n(\bar{r}) \exp(-i\omega_n t) + \hbar\omega_n \psi \\ = H_0 \psi + \lambda V \sum_n \left[C_n^{(0)}(t) + \lambda C_n^{(1)}(t) + \ldots\right] U_n(\bar{r}) \exp(-i\omega_n t) \end{aligned} \tag{D.22}$$

Note that $H_0 \psi = \hbar\omega_n \psi$. Hence, after multiplying both sides by $U_n(\bar{r}) * \exp(i\omega_n t)$, integrating over all space, and making use of the orthonormal properties of the eigenstates, one finds

$$i\hbar \left[\dot{C}_k^{(0)}(t) + \lambda \dot{C}_k^{(1)}(t) + \ldots\right] = \sum_n \langle k|V|n\rangle [\lambda C_n^{(0)}(t) + \lambda^2 C_n^{(1)}(t) + \ldots] \exp(-i[\omega_n - \omega_k]t). \tag{D.23}$$

This yields

$$\left[\dot{C}_k^{(0)} + \lambda \dot{C}_k^{(1)} + \ldots\right] = \sum_n \frac{V_{kn}}{i\hbar} \left[\lambda C_n^{(0)} + \lambda^2 C_n^{(1)} + \ldots\right] \exp(-i[\omega_n - \omega_k]t). \tag{D.24}$$

Equating coefficients of various powers of λ:

$$\lambda^{(0)} : \dot{C}_k^{(0)} = 0, \tag{D.25}$$

$$\lambda^{(1)} : \dot{C}_k^{(1)} = \frac{V_{kn}}{i\hbar} \dot{C}_n^{(0)} \exp(-i[\omega_n - \omega_k]t), \tag{D.26}$$

$$\lambda^{(2)} : \dot{C}_k^{(2)} = \frac{V_{kn}}{i\hbar} \dot{C}_n^{(1)} \exp(-i[\omega_n - \omega_k]t). \tag{D.27}$$

By solving the equations for each order, an expression for the wavefunction valid to any desired degree of approximation may be obtained.

By solving the first-order equations for a sinusoidal perturbation, such as an optical field, an expression known as Fermi's golden rule can be derived for the transition rate between states that is valid for short times. Equation (D.25) shows that the zero-order coefficients $C_k^{(0)}$ are constant in time. We assume that all zero-order coefficients are zero except one with $k = m$, so that prior to the onset of the perturbation the system is in an eigenstate of the unperturbed Hamiltonian. Integration of Eq. (D.25) then yields

$$C_k^{(1)}(t) = \frac{1}{i\hbar} \int_{-\infty}^{t} V_{km}(t') \exp(-i[\omega_m - \omega_k]t')dt'. \tag{D.28}$$

The constant of integration has been set equal to zero so that $C_k^{(1)}$ itself is zero at $t = -\infty$. Note that if V is of finite duration, the probability amplitude of state $k(k \neq m)$ following the perturbation is proportional to the Fourier component of the perturbation between state k and m (see also Problem 6.2).

Next we assume the perturbation is harmonic in time and is turned on at $t = 0$ and off at $t = t_0$. Thus

$$V_{km}(t') = 2V_{km} \cos(\omega t') \tag{D.29}$$

and

$$C_k^{(1)}(t \geq t_0) = -\frac{V_{km}}{\hbar} \left\{ \frac{\exp(i[\omega_{km} + \omega]t_0) - 1}{\omega_{km} + \omega} + \frac{\exp(i[\omega_{km} - \omega]t_0)}{\omega_{km} - \omega} \right\}. \tag{D.30}$$

Close to optical resonances, where $\omega_{km} - \omega \approx 0$), the second term in Eq. (D.30) is dominant. Hence one can simplify the calculation by making what is known as the rotating wave approximation (RWA) in which the first term is ignored. Then

$$\left| C_k^{(1)}(t \geq t_0 \right|^2 = \left| \frac{V_{km}}{\hbar} \right|^2 t_0^2 \frac{\sin^2 \left[\frac{1}{2}(\omega_{km} - \omega)t_0 \right]}{\left[\frac{1}{2}(\omega_{km} - \omega)t_0 \right]^2}. \tag{D.31}$$

The factor of 2 inserted into Eq. (D.29) ensures that in the RWA the physically relevant matrix element is V_{km} rather than $V_{km}/2$ (see Ref. [D.2]). If we denote the transition probability per unit time as rate Γ_{km}, we now find

$$\Gamma_{km} = \frac{1}{t_0} \int \left| C_k^{(1)}(t \geq t_0) \right|^2 \rho(\omega_k) d\omega_k, \tag{D.32}$$

where $\rho(\omega_k)d\omega_k$ is the number of final states with frequencies between ω_k and $\omega_k + d\omega_k$. $\rho(\omega_k)$ is referred to as the density of final frequency states. If we change variables to $x \equiv \frac{1}{2}(\omega_{km} - \omega)t_0$, the integration is readily performed and yields

$$\Gamma_{km} = (2\pi/\hbar^2) \left| V_{km} \right|^2 \rho(\omega_k). \tag{D.33}$$

For a monochromatic field, the perturbation is $V = V(\omega_0)$. Assuming the transition is between discrete states, whereupon the density of final states can be written as $\rho(\omega) = \delta(\omega - \omega_0)$, Eq. (D.33) simplifies to the expression

$$\Gamma(\omega) = \left(2\pi/\hbar^2\right) |V(\omega_0)|^2 \, \delta(\omega - \omega_0), \tag{D.34}$$

where Γ is expressed in s^{-1}. If the linewidth of the light source covers a range $d\omega$ much narrower than the atomic lineshape given by $g(\omega) = \Gamma/[(\omega_0 - \omega)^2 + \Gamma^2]$, then the rate becomes

$$\Gamma(\omega) = \left(2\pi/\hbar^2\right) |V(\omega_0)|^2 \, g(\omega) d\omega. \tag{D.35}$$

Calculations of transition rates based on Fermi's golden rule in Eq. (D.34) give results in agreement with those of Section 3.5 upon specification of the corresponding density of states available in the atom that overlap the spectral density of the optical field.

C. Perturbation Theory with the Density Matrix

The approach outlined in Section B of this Appendix can also be applied to time-dependent perturbation analysis of the density matrix. In this case, the expressions for the density matrix and the interaction are written

$$\rho = \rho^{(0)} + \lambda \rho^{(1)} + \lambda^2 \rho^{(2)} + \lambda^3 \rho^{(3)} + \ldots, \tag{D.36}$$

$$V = V^{(0)} + \lambda V^{(1)} + \lambda^2 V^{(2)} + \lambda^3 V^{(3)} + \ldots, \tag{D.37}$$

and upon substitution into the equation of motion,

$$i\hbar \frac{\partial \rho}{\partial t} = [H_0, \rho] + [V, \rho] + i\hbar \frac{\partial \rho}{\partial t} |_{relax} \,, \tag{D.38}$$

one obtains

$$\begin{aligned} i\hbar \frac{\partial}{\partial t} \left(\rho^{(0)} + \lambda \rho^{(1)} + \lambda^2 \rho^{(2)} + \lambda^3 \rho^{(3)} + \ldots\right) = \\ = \left[H_0, \rho^{(0)} + \lambda \rho^{(1)} + \lambda^2 \rho^{(2)} + \lambda^3 \rho^{(3)} + \ldots\right] \\ + \left[V^{(0)} + \lambda V^{(1)} + \lambda^2 V^{(2)} + \lambda^3 V^{(3)} + \ldots, \rho^{(0)} + \lambda \rho^{(1)} + \lambda^2 \rho^{(2)} + \lambda^3 \rho^{(3)} + \ldots\right] \end{aligned} \tag{D.39}$$

Equating coefficients of corresponding orders of λ on the left and right side of Eq. (D.39), one finds:

zeroth order $\lambda^{(0)}$:

$$i\hbar \frac{\partial}{\partial t} \rho^{(0)} = \left[H_0, \rho^{(0)}\right] + \left[V^{(0)}, \rho^{(0)}\right] + i\hbar \frac{\partial}{\partial t} \rho^{(0)} |_{relax} \,, \tag{D.40}$$

first order $\lambda^{(1)}$:

$$i\hbar\frac{\partial}{\partial t}\rho^{(1)} = \left[H_0, \rho^{(1)}\right] + \left[V^{(0)}, \rho^{(1)}\right] + \left[V^{(1)}, \rho^{(0)}\right] + i\hbar\frac{\partial}{\partial t}\rho^{(1)}\,|_{relax}\,, \tag{D.41}$$

second order λ^2 :

$$i\hbar\frac{\partial}{\partial t}\rho^{(2)} = \left[H_0, \rho^{(2)}\right] + \left[V^{(0)}, \rho^{(2)}\right] + \left[V^{(1)}, \rho^{(1)}\right] + \left[V^{(2)}, \rho^{(0)}\right] + i\hbar\frac{\partial}{\partial t}\rho^{(2)}\,|_{relax}\,, \tag{D.42}$$

third order λ^3 :

$$\begin{aligned} i\hbar\frac{\partial}{\partial t}\rho^{(3)} &= \left[H_0, \rho^{(3)}\right] + \left[V^{(0)}, \rho^{(3)}\right] + \left[V^{(1)}, \rho^{(2)}\right] + \left[V^{(2)}, \rho^{(1)}\right] \\ &\quad + \left[V^{(3)}, \rho^{(0)}\right] + i\hbar\frac{\partial}{\partial t}\rho^{(3)}\,|_{relax}\,, \end{aligned} \tag{D.43}$$

and so on for higher orders.

In practice, one finds that matrix elements for populations are zero in odd orders and non-zero in even orders of perturbation. This corresponds to the simple physical fact that interactions begin with populations in eigenstates and coherence can only be established by the action of an applied field. Thus one order of perturbation is required to establish polarization. Only after a superposition state has been established that includes a new state can the distribution of population change. Consequently, population change appears in the next order of perturbation. The procedure for solving the equations of motion (D.40)–(D.43) begins with the determination of first-order coherences—the off-diagonal elements of the density matrix—and then proceeds to populations. Note that once the contributions $\rho^{(0)}, \rho^{(1)}, \rho^{(2)}, \rho^{(3)} \ldots$ to the desired order are determined, they must be added according to Eq. (D.36) to find the entire density matrix $\rho(t)$ itself to that order, after setting $\lambda = 1$.

REFERENCES

[D.1] A.S. Davydov, *Quantum Mechanics*, 2nd ed. Oxford: Pergamon Press, 1976, p. 189.
[D.2] L.I. Schiff, *Quantum Mechanics*, 3rd ed. New York: McGraw-Hill, 1968, p. 283.

Appendix E
Second Quantization of Fermions

With few exceptions, the systems considered in examples and problems of this book are assumed to be bosonic in character. That is, they are assumed not to have occupation probabilities dominated by spin-related considerations. Their spectra generally have large numbers of energy levels and their statistics are governed by the Bose–Einstein distribution, unlike fermionic systems that have a very small number of energy levels and obey Fermi–Dirac statistics (to describe the latter we shall follow the discussion in Ref. [E.1]). This of course ignores the important role played by the Pauli exclusion principle in many systems of practical importance, such as semiconductors, as well as the possibility of converting bosonic systems into fermionic systems and vice versa by altering experimental conditions. At the end of Chapter 3, an example of the conversion of a multi-level atom into a two-level atom was considered, and other examples can be found in the literature on Bose–Einstein condensates near Feshbach resonances [E.2]. Here we briefly consider some of the implications of the exclusion principle that were omitted from earlier discussions.

Fermions obey population statistics that are different from those of bosons, and this difference is reflected in the commutation relations for fermionic operators. Fermionic operators *anti-commute*. For example, the annihilation and creation operators of a fermionic state labeled with index k, namely $\hat{\alpha}_k$ and $\hat{\alpha}_k^+$, obey the commutation rules:

$$\begin{aligned} \left[\hat{\alpha}_k, \hat{\alpha}_\ell\right]_+ &= \left[\hat{\alpha}_k^+, \hat{\alpha}_l^+\right]_+ = 0, \\ \left[\hat{\alpha}_k, \hat{\alpha}_\ell^+\right]_+ &= \delta_{k\ell}, \end{aligned} \tag{E.1}$$

where the symbol $[\ldots]_+$ implies the sum $[\hat{\alpha}, \hat{\beta}]_+ \equiv \hat{\alpha}\hat{\beta} + \hat{\beta}\hat{\alpha}$. Other than electrons, what systems are fermionic, and do they exhibit familiar features? On the basis of Eq. (E.1) and our earlier representation of two-level atoms by anti-commuting Pauli matrices, we may infer that two-level atoms are fermionic as one example. On the other hand, electromagnetic fields and simple harmonic oscillators, both of which are characterized by an infinite number of equally spaced energy levels, are bosonic. In this appendix we discuss differences in the way wavefunctions are constructed for fermionic and bosonic systems, while emphasizing that their system energies share a simple feature in the limit of weak interactions.

Define a fermionic number operator $\hat{N}_k = \hat{\alpha}_k^+ \hat{\alpha}_k$ and note that according to Eq. (E.1)

$$\hat{N}_k^2 = \hat{\alpha}_k^+ \hat{\alpha}_k \hat{\alpha}_k^+ \hat{\alpha}_k = \hat{\alpha}_k^+ \left(1 - \hat{\alpha}_k^+ \hat{\alpha}_k\right) \hat{\alpha}_k = \hat{\alpha}_k^+ \hat{\alpha}_k = \hat{N}_k. \tag{E.2}$$

If $\hat{N}_k$ has eigenvalues n_k, then obviously

$$n_k^2 = n_k \rightarrow n_k = 0, 1. \tag{E.3}$$

For a given state k, the operator $\hat{N}$ can be represented by the matrix

$$\hat{N} = \begin{bmatrix} 0 & 0 \\ 0 & 1 \end{bmatrix}, \tag{E.4}$$

when there is no energy level degeneracy. Suitable matrices for $\hat{\alpha}, \hat{\alpha}^+$ which are consistent with the commutation relations are

$$\hat{\alpha} = \begin{bmatrix} 0 & 1 \\ 0 & 0 \end{bmatrix},$$

$$\hat{\alpha}^+ = \begin{bmatrix} 0 & 0 \\ 1 & 0 \end{bmatrix}. \tag{E.5}$$

The two possible states of the system are

$$|0\rangle = \begin{bmatrix} 1 \\ 0 \end{bmatrix},$$

$$|1\rangle = \begin{bmatrix} 0 \\ 1 \end{bmatrix}, \tag{E.6}$$

and it is easily verified that $\hat{\alpha}$ and $\hat{\alpha}^+$ play the roles of destruction and creation operators respectively. That is,

$$\hat{\alpha}\,|\,n >= n\,|\,1-n >; \quad \hat{\alpha}^+\,|\,n >= (1-n)\,|\,1-n > \tag{E.7}$$

where $n = 0, 1$.

To describe a complete system of fermions we must introduce the exclusion principle by permitting the sign of the wavefunction to be altered by the action of $\hat{\alpha}, \hat{\alpha}^+$. Order the states of the system in an arbitrary but definite way, for example, $1, 2, \ldots k$. Then permit $\hat{\alpha}_k, \hat{\alpha}_k^+$ to act as before, except that the sign will be determined by whether the kth state is preceded in the assumed order by an even or odd number of occupied states.

$$\hat{\alpha}_k\,|\,n_1, \cdots n_k, \cdots >= \theta_k n_k\,|\,n_1, \cdots 1-n_k, \cdots >,$$

$$\hat{\alpha}_k^+\,|\,n_1, \cdots n_k, \cdots >= \theta_k\,(1-n_k)\,|\,n_1, \cdots 1-n_k, \cdots >, \tag{E.8}$$

$$\theta_k \equiv (-)^{\nu_k}; \nu_k \equiv \sum_{j=1}^{k-1} n_j.$$

ν_k is the number of states, preceding state k, which are occupied. For a particular single particle state k of the system, it is easy to show that

$$\begin{aligned}
\hat{\alpha}_k\hat{\alpha}_k^+\,|\cdots n_k\cdots &>= (1-n_k)^2\,\theta_k^2\,|\cdots n_k\cdots > \\
&= (1-n_k)\,|\cdots n_k\cdots >, \\
\hat{\alpha}_k^+\hat{\alpha}_k\,|\cdots n_k\cdots &>= n_k\,|\cdots n_k\cdots >, \\
\hat{\alpha}_k\hat{\alpha}_k\,|\cdots n_k\cdots &>= n_k\,(1-n_k)\,|\cdots n_k\cdots >= 0, \\
\hat{\alpha}_k^+\hat{\alpha}_k^+\,|\cdots n_k\cdots &>= \alpha_k^+\theta\,(1-n_k)\,|\cdot(1-n_k)\cdots > \\
&= \theta^2\,(1-n_k)\,n_k\,|\cdots n_k\cdots > \\
&= 0.
\end{aligned} \tag{E.9}$$

From these relations it is easily seen that the commutation relations are satisfied for $k = \ell$. For example,

$$\left[\hat{\alpha}_k \hat{\alpha}_k^+\right] = 1 - n_k + n_k = 1. \tag{E.10}$$

For $k \neq \ell$ $(k > \ell)$:

$$\begin{aligned}
\hat{\alpha}_\ell \hat{\alpha}_k | \cdots n_\ell \cdots n_k \cdots > &= \theta_k n_k \hat{\alpha}_\ell | \cdots n_\ell \cdots | 1 - n_k \cdots > \\
&= \theta_k \theta_\ell n_k n_\ell | \cdots 1 - n_\ell \cdots 1 - n_k \cdots >, \\
\hat{\alpha}_k \hat{\alpha}_\ell | \cdots n_\ell \cdots n_k \cdots > &= \theta_\ell n_\ell \hat{\alpha}_k | \cdots 1 - n_\ell \cdots n_k \cdots > \\
&= \theta_{k-1} \theta_\ell n_\ell n_k | \cdots 1 - n_\ell \cdots 1 - n_\ell \cdots 1 - n_k \cdots > .
\end{aligned} \tag{E.11}$$

Hence $\hat{\alpha}_\ell \hat{\alpha}_k + \hat{\alpha}_k \hat{\alpha}_\ell \Rightarrow (-1)^{k+\ell} n_k n_\ell + (-)^{k+\ell-1} n_k n_\ell = 0$. Remaining commutation relations are satisfied in the same way.

Notice from this that the result of applying fermi operators $\hat{\alpha}_k \hat{\alpha}_k^+$ depends not only on occupation number k but also on the occupation numbers of all preceding states. Thus $\hat{\alpha}_k$ and $\hat{\alpha}_\ell$ are not entirely independent. This means that the order of processes in fermionic systems is even more important to the outcome of interactions than it is for bosonic systems. It should therefore not be taken for granted in a fermionic system that the application of a sum of single particle operators to the system wavefunction yields a sum of single particle energies. Let us test whether this is true in the perturbative limit of a fermionic system that has extremely weak interactions.

To the extent that particles can be viewed as being distinct, the total system Hamiltonian in the configuration representation can be written

$$\hat{H}(\xi, \xi_2, \cdots \xi_n) = \sum_{i=1}^{N} \hat{H}(\xi_i), \tag{E.12}$$

where ξ_i gives the spatial and spin coordinates of each of the N fermions. Here $\hat{H}(\xi_i)$ is the single particle Hamiltonian for the particle at coordinate ξ_i. Single particle states $\phi_s(\xi_i)$ and their eigenvalues ε_s are determined by

$$\left[\hat{H}(\xi) - \varepsilon_s\right] \phi_s(\xi) = 0. \tag{E.13}$$

To see what the form of the total system Hamiltonian is in the occupation number representation, we introduce field operators $\hat{\psi}$ defined by

$$\hat{\psi}(\xi, t) = \sum\nolimits_s \hat{\alpha}_s \phi_s(\xi) \exp(-i\omega_s t), \tag{E.14}$$

where $\omega_s \equiv \varepsilon_s / \hbar$. The single particle states are assumed to form a complete set. Hence the total Hamiltonian operator may be changed into the occupation representation using the transformation (compare Eq. (2.4.34) for discrete states)

$$\hat{H} = \int \hat{\psi}^+(\xi) \hat{H}(\xi) \hat{\psi}(\xi) d\xi, \tag{E.15}$$

where the integration over coordinates implicitly includes a sum over spin variables σ. At a fixed time t it may readily be shown that

$$\left[\hat{\psi}(\xi'), \hat{\psi}^+(\xi)\right] = \sum \phi_s(\xi')\phi_\ell^*(\xi)\left[\hat{\alpha}_s, \hat{\alpha}_\ell^+\right]_+ = \delta(\xi' - \xi), \tag{E.16}$$

where

$$\delta(\xi' - \xi) = \delta_{\sigma\sigma'}\delta(\vec{r}' - \vec{r}). \tag{E.17}$$

Similarly

$$\left[\hat{\psi}(\xi'), \hat{\psi}(\xi)\right]_+ = \left[\hat{\psi}^+(\xi'), \psi^+(\xi)\right]_+ = 0. \tag{E.18}$$

Substituting these expressions into Eq. (E.15), one finds

$$\hat{H} = \sum_s \varepsilon_s \hat{\alpha}_s^+ \hat{\alpha}_s = \sum_s \hat{n}_s \varepsilon_s. \tag{E.19}$$

This result shows that the energy of a system of non-interacting fermions is just the sum of the individual particle energies, a familiar result that holds for purely bosonic systems also. In interacting systems of fermions, a similar decomposition of the system into a simple sum of (renormalized) individual particle energies can be performed only for extremely weak interactions that can be taken into account via an effective field. For this to be true, the fermions have to be widely spaced. For close particles and strong interactions, the concept of single particle states and single particle energies breaks down completely. Equations (E.12) and (E.19) are no longer valid in this limit. In order to ensure that no two fermions have the same wavefunction (occupy the same state) in the strong-coupling limit, many-body techniques are required for analysis.

REFERENCES

[E.1] A.S. Davydov, *Quantum Mechanics*, 2nd ed. Oxford: Pergamon Press, 1976, p. 362.
[E.2] See, for example, C.A. Regal, M. Geiner, and D.S. Lin, *Phys. Rev. Lett.* 2004, **92**, 040403.

Appendix F
Frequency Shifts and Decay Due to Reservoir Coupling

A single parameter such as the decay constant γ introduced in Chapter 3 is typically thought to account for a single process, such as the rate of population decay in two-level atoms. However, here we show on the basis of perturbation theory that as many as four fundamentally different physical processes are linked to the decay constants in our description of atomic dynamics.

In Section 6.3 an exponential form was derived for the population decay of a two-level system via spontaneous emission. This justified the assumed form of phenomenological decay terms in Chapter 3, where perturbation theory was used to solve for the probability amplitudes of a two-level system undergoing radiative decay after being prepared in the excited state:

$$C_2^{(1)}(t) = \exp(-\gamma_2 t/2) \tag{F.1}$$

for times short compared to lifetime γ_1^{-1} but long compared to field periods. The process of emission from $|2\rangle$ to $|1\rangle$ also excites one of many radiation modes of the electromagnetic field. These modes are labeled by index λ. Then, setting $(\Omega_{21})_\lambda \equiv \mu_{21} E_\lambda/\hbar$ in the rate equations of Chapter 3, one finds

$$(\dot{C}_1)_\lambda = \sum_\lambda \frac{i}{2}(\Omega_{21})_\lambda \exp(-i[\omega_0 - \omega_\lambda]t)(C_2)_\lambda, \tag{F.2}$$

$$(\dot{C}_2)_\lambda = \sum_\lambda \frac{i}{2}(\Omega_{12})_\lambda \exp(i[\omega_0 - \omega_\lambda]t)(C_1)_\lambda, \tag{F.3}$$

by ignoring the effect of decay at short times. Substitution of Eq. (F.1) into Eq. (F.2) then yields

$$(\dot{C}_1)_\lambda = \sum_\lambda \frac{i}{2}(\Omega_{21})_\lambda \exp(-i[\omega_0 - \omega_\lambda]t - \gamma_2 t/2). \tag{F.4}$$

This is easily solved to give

$$\left(C_1^{(1)}\right)_\lambda(t) = \sum_\lambda \frac{i}{2}(\Omega_{21})_\lambda \left[\frac{\exp(-i[\omega_0 - \omega_\lambda]t - \gamma_2 t/2) - 1}{-i[\omega_0 - \omega_\lambda] - \frac{1}{2}(\gamma_2 - \gamma_1)}\right]. \tag{F.5}$$

Now we use the first-order perturbation results to extract a general form of γ_2. Upon substitution of Eqs. (F.1) and (F.5) into Eq. (F.3), we find

$$-\frac{\gamma_2}{2}\exp(-\gamma_2 t/2) = \sum_\lambda (i/2)^2 \, |(\Omega_{12})_\lambda|^2 \left[\frac{\exp(-\gamma_2 t/2) - \exp(i[\omega_0 - \omega_\lambda]t)}{-i[\omega_0 - \omega_\lambda] - \frac{1}{2}(\gamma_2 - \gamma_1)}\right]$$

or

$$\gamma_2 = \frac{1}{2}\sum_\lambda |(\Omega_{12})_\lambda|^2 \left[\frac{1 - \exp(i[\omega_0 - \omega_\lambda]t + \gamma_2 t/2)}{-i[\omega_0 - \omega_\lambda] - \frac{1}{2}(\gamma_2 - \gamma_1)}\right]. \tag{F.6}$$

Replacing the summation over modes with an integral over the density of modes $D(\omega) = V\omega^2/\pi^2c^3$ available per unit frequency interval according to

$$\sum_\lambda \rightarrow \int D(\omega)d\omega, \tag{F.7}$$

Eq. (F.6) becomes

$$\gamma_2 = \frac{1}{2}\int |\Omega_{12}(\omega)|^2 \left[\frac{1 - \exp(i[\omega_0 - \omega]t + \gamma_2 t/2)}{-i[\omega_0 - \omega] - \frac{1}{2}(\gamma_2 - \gamma_1)}\right] D(\omega)d\omega. \tag{F.8}$$

To make the expression for γ_2 more compact we can define the factor

$$f(\omega) \equiv \frac{1}{2}|\Omega_{12}(\omega)|^2 D(\omega). \tag{F.9}$$

In terms of $f(\omega)$, the decay constant becomes

$$\gamma_2 = i\int_\omega f(\omega) \left[\frac{1 - \exp(i[\omega_0 - \omega]t + \gamma_2 t/2)}{[\omega_0 - \omega] - \frac{i}{2}(\gamma_2 - \gamma_1)}\right] d\omega. \tag{F.10}$$

If $(\gamma_2 - \gamma_1) \ll \omega$ the denominator in Eq. (F.10) reduces to $\sim [\omega_0 - \omega]^{-1}$, and at short times the portion of the integrand in square brackets has the limiting form

$$\mathrm{Lim}_{t\rightarrow 0}\left[\frac{1 - \exp(i[\omega_0 - \omega]t + \gamma_2 t/2)}{[\omega_0 - \omega] - \frac{i}{2}(\gamma_2 - \gamma_1)}\right] \approx \left[\frac{1 - \cos(\omega_0 - \omega)t}{(\omega_0 - \omega)}\right] + i\left[\frac{\sin(\omega_0 - \omega)t}{(\omega_0 - \omega)}\right] \tag{F.11}$$

With this simplification, and the values for the integrals over all frequencies of $\cos \Delta t/\Delta$ and $\sin \Delta t/\Delta$ which are zero and $-\pi$ respectively since $\Delta \equiv \omega_0 - \omega \propto -\omega$, the square bracket in Eq. (F.10) becomes

$$P\left\{\frac{1}{\omega_0 - \omega}\right\} - i\pi\delta(\omega_0 - \omega),$$

where P denotes the principal part. The decay constant thereby acquires the form

$$\gamma_2 = i\int_{\omega} \left\{P\left[\frac{1}{\omega_0 - \omega}\right] - i\pi\delta(\omega_0 - \omega)\right\} f(\omega)d\omega. \tag{F.12}$$

This development demonstrates (with only first-order perturbation theory) that finite lifetime, and the coupling to radiation modes that causes it, leads to a separation of γ_2 into real and imaginary parts. Thus

$$\gamma_2 = \gamma_r + i\gamma_{im}, \tag{F.13}$$

where the associated expressions are given by

$$\gamma_{im} \propto \frac{1}{2}P\left\{\int \frac{|\Omega_{12}(\omega)|^2}{\omega_0 - \omega} D(\omega)d\omega\right\}, \tag{F.14}$$

$$\gamma_r \propto \frac{\pi}{2} D(\omega_0)\,|\Omega_{12}(\omega_0)|^2. \tag{F.15}$$

The decay processes in Eqs. (F.14) and (F.15) are both induced by real optical fields, as is apparent by their mutual dependencies on $|\Omega_{12}|^2$. Now we are in a position to gain some perspective on the purely quantum mechanical effects that result from quantization of the radiation field by taking into account that the total electric field includes the vacuum state which does not depend on mode occupation number n_λ, and a part which does depend on n_λ. By recognizing that, according to Sections 6.2 and 6.3, fluctuations of the vacuum field can induce spontaneous decay in much the same manner that the photon field induces stimulated relaxation, we see that the imaginary and real rate constants split into two additional parts corresponding to spontaneous and stimulated processes. That is,

$$\gamma_2 = \gamma_r^{(spont)} + \gamma_r^{(induced)} + i\gamma_{im}^{(spont)} + i\gamma_{im}^{(induced)} \tag{F.16}$$

The physical effects of relaxation are reflected in the frequency and time dependence of the emitted radiation. That is, the field leaving the atom has the form

$$\begin{aligned} E &\propto \exp(-i\omega_0 t)\cdot\exp(-\gamma_2 t) \\ &= \exp\left[-i\left(\omega_0 + \gamma_{im}^{(spont)} + \gamma_{im}^{(induced)}\right)t\right]\cdot\exp\left[-\left(\gamma_r^{(spont)} + \gamma_r^{(induced)}\right)t\right]. \end{aligned} \tag{F.17}$$

As a result of atom–field coupling four important processes take place, namely spontaneous decay, stimulated decay, the Lamb shift, and the AC Stark shift. This is illustrated in Figure F.1.

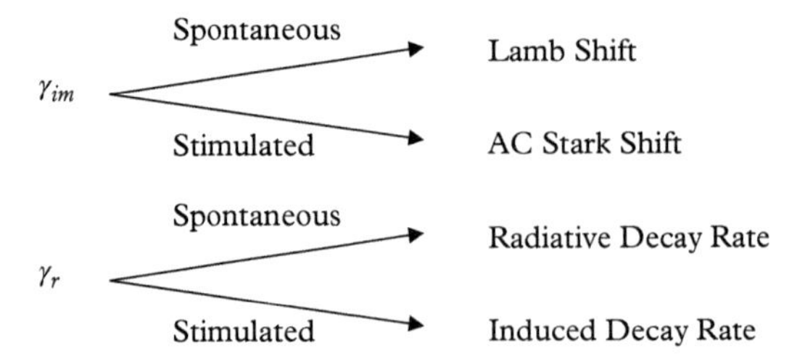

Figure F.1 *Four fundamental, distinguishable physical processes that are intimately linked to field-driven decay of atoms.*

Appendix G
Solving for Off-Diagonal Density Matrix Elements

Off-diagonal elements of the density matrix describe charge oscillations initiated by applied fields. Hence the full temporal evolution of their amplitudes reflects transient build-up (or decay) of an oscillation prior to the establishment of any steady-state amplitude. The frequency of the oscillation is ω, as determined by the driving field. Hence the general form of the solution is

$$\rho_{ij}(t) = \tilde{\rho}_{ij}(t)\exp(i\omega t), \tag{G.1}$$

where i, j specify the initial and final states of the transition closest in frequency to ω and $\tilde{\rho}_{ij}$ is the slowly varying amplitude of the polarization.

The time derivative of ρ_{ij} is

$$\dot{\rho}_{ij} = \dot{\tilde{\rho}}_{ij}\exp(i\omega t) + i\omega\tilde{\rho}_{ij}\exp(i\omega t). \tag{G.2}$$

If the assumption is made that $\dot{\tilde{\rho}}_{ij} = 0$, then substitution of Eq. (G.2) into the equation of motion yields an algebraic equation that may be readily solved to find the steady-state solution. More generally $\dot{\tilde{\rho}}_{ij} \neq 0$, and the equation of motion must be integrated to find a solution, as follows.

Take the interaction to be

$$V_{ij}(t) = \frac{1}{2}\left[\tilde{V}_{ij}(t)\exp(i\omega t) + \tilde{V}^{*}_{ij}(t)\exp(-i\omega t)\right]. \tag{G.3}$$

Substituting Eqs. (G.3) and (G.2) into the equation of motion for a two-level system, as an example, one finds

$$\dot{\tilde{\rho}}_{12} = i\Delta\tilde{\rho}_{12} + \left(\tilde{V}_{12}/2i\hbar\right)(\rho_{22} - \rho_{11}) - \Gamma_{12}\tilde{\rho}_{12}, \tag{G.4}$$

where $\Delta \equiv \omega_0 - \omega$. Next, use is made of an integrating factor through the substitution of

$$\tilde{\rho}_{12} = \tilde{\rho}'_{12}\exp[(i\Delta - \Gamma_{12})t]. \tag{G.5}$$

This yields

$$\dot{\tilde{\rho}}'_{12}\exp[(i\Delta - \Gamma_{12})t] + (i\Delta - \Gamma_{12})\tilde{\rho}'_{12}\exp[(i\Delta - \Gamma_{12})t] =$$
$$= (i\Delta - \Gamma_{12})\tilde{\rho}'_{12}\exp[(i\Delta - \Gamma_{12})t] + \left(\tilde{V}_{12}/2i\hbar\right)(\rho_{22} - \rho_{11}),$$

or

$$\dot{\tilde{\rho}}_{12}' = (\tilde{V}_{12}/2i\hbar)(\rho_{22} - \rho_{11})\exp[-(i\Delta - \Gamma_{12})t], \tag{G.6}$$

The next step is to integrate over time between appropriate limits, for example, 0 and t. One finds

$$\tilde{\rho}_{12}' = \frac{-i}{2\hbar}\int_0^t \tilde{V}_{12}(t')(\rho_{22}(t') - \rho_{11}(t'))\exp\left[-(i\Delta - \Gamma_{12})t'\right]dt'. \tag{G.7}$$

If the interaction starts at time zero, as assumed in Eq. (G.7), then the integral must be performed by taking the transient buildup of the envelope function $\tilde{V}_{12}(t')$ into account. Since $\tilde{V}_{12}(t')$ itself cannot be approximated by a constant over the interval of integration, it cannot be removed from the integral. The steady-state solution, identical to that obtained by solving Eq. (G.4) algebraically with the assumption $\dot{\tilde{\rho}}_{12} = 0$, may be found directly after setting the lower limit of integration equal to $t' = -\infty$ and integrating by parts:

$$\begin{aligned}\int_{-\infty}^{t} &\tilde{V}_{12}(t')\left(\rho_{22}(t') - \rho_{11}(t')\right)\exp\left[-(i\Delta - \Gamma_{12})t'\right]dt' \\ &= \left[\frac{\tilde{V}_{12}(t')(\rho_{22}(t') - \rho_{11}(t'))\exp\left[-(i\Delta - \Gamma_{12})t'\right]}{-(i\Delta - \Gamma_{12})}\right]_{-\infty}^{t} \\ &- \int_{-\infty}^{t}\left(\frac{d}{dt'}\tilde{V}_{12}(t')(\rho_{22}(t') - \rho_{11}(t'))\right)\frac{\exp\left[-(i\Delta - \Gamma_{12})t'\right]}{-(i\Delta - \Gamma_{12})}dt'.\end{aligned} \tag{G.8}$$

Under steady-state conditions ($\dot{\rho}_{11} = \dot{\rho}_{22} = 0$), assuming the interaction is turned on adiabatically at early times so that $d\tilde{V}_{12}(t')/dt' \cong 0$, the second integral on the right is zero. One therefore finds the result

$$\begin{aligned}\frac{-i}{2\hbar}\int_{-\infty}^{t} &\tilde{V}_{12}(t')(\rho_{22}(t') - \rho_{11}(t'))\exp\left[-(i\Delta - \Gamma_{12})t'\right]dt' \\ &= \frac{i}{2\hbar}\left[\frac{\tilde{V}_{12}(t)\exp[-(i\Delta - \Gamma_{12})t]}{(-i\Delta + \Gamma_{12})}\right](\rho_{11}(t) - \rho_{22}(t))\end{aligned} \tag{G.9}$$

and

$$\tilde{\rho}_{12}'(t) = \frac{i\tilde{V}_{12}(t)}{2\hbar(-i\Delta + \Gamma_{12})}(\rho_{11}(t) - \rho_{22}(t))\exp[-(i\Delta - \Gamma_{12})t]. \tag{G.10}$$

Using Eq. (G.10) in Eq. (G.5), the steady-state value for the slowly varying polarization amplitude is found to be

$$\tilde{\rho}_{12} = \frac{-\tilde{V}_{12}}{2\hbar(\Delta + i\Gamma_{12})}(\rho_{11} - \rho_{22}) = \frac{(\Omega_{12}/2)}{(\Delta + i\Gamma_{12})}(\rho_{11} - \rho_{22}), \tag{G.11}$$

and the final result for the off-diagonal coherence (polarization) is therefore

$$\rho_{12}(t) = \left(\frac{\Omega_{12}/2}{\Delta + i\Gamma_{12}}\right)(\rho_{11}(t) - \rho_{22}(t))\exp(i\omega t). \tag{G.12}$$

This result is the same as that obtained by substituting Eq. (G.2) into the equation of motion for ρ_{12} after setting $\dot{\tilde{\rho}}_{12} = 0$. Determinations of ρ_{12} that include transient behavior associated with a rapid onset of the interaction must include an evaluation of the second term on the right of Eq. (G.8), since then $d\tilde{V}_{12}(t')/dt' \neq 0$. Similarly, population oscillations or rapid population dynamics require a full evaluation of Eq. (G.8). A useful alternative is to assume the interaction appears instantaneously at time $t = 0$, and to evaluate the integral using Eq. (G.7). However, for steady-state conditions we note that the result in Eq. (G.12) agrees with the result obtained in Chapter 5 using the simple substitutions $\rho_{12}(t) = \tilde{\rho}_{12}\exp(i\omega t)$ and $\dot{\tilde{\rho}}_{12} = 0$.

Appendix H
Irreducible Spherical Tensor Operators and the Wigner–Eckart Theorem

In our introductions to various sciences, most of us encounter problems that seem to have a set of "natural" coordinates associated with them. By choosing "natural" coordinates one can simplify the mathematical analysis. Good choices correspond to descriptions that mimic one or more symmetry elements of the problem. "Unnatural" coordinates give rise to unwieldy expressions that are unnecessarily complicated. Consequently, an entire subject is devoted to identification and exploitation of the simplest mathematical formalism to use in analyzing any problem, namely group theory. Because group theory is an important subject that transcends the limited objectives of this book, and can be a very powerful companion in the analysis of optical problems we shall not diminish it by attempting to introduce it here. However some exposure to key results that can be explained succinctly is beneficial for application to the advanced topics considered in Chapters 6 and 7 of this book. Consequently this appendix covers the essentials of tensor analysis to explain what is known as the "irreducible" representation of physical problems.

A pure rotation R transforms a state vector ψ into ψ' via a unitary transformation U_R that acts on each component in the same way that spatial coordinates transform:

$$\psi' = U_R \psi \tag{H.1}$$

In an analogous fashion, operators $V_i (i = 1, 2, 3)$ are said to form the Cartesian components of a vector operator V if under every rotation their expectation values transform like components of a vector. Thus a vector operator is required to have expectation values that transform according to the equation

$$\langle \psi | V_i | \psi \rangle' = \sum_{j=1}^{3} R_{ij} \langle \psi | V_j | \psi \rangle, \quad (i, j = 1, 2, 3), \tag{H.2}$$

The entries in the rotation matrix $R_{ij} = \partial x_i' / \partial x_j$ are cosines of the angles between the rotated positive x_i' axis and the positive x_j axis prior to rotation [H.1] describing the transformation of the coordinates.

Substitution of Eq. (H.1) into Eq. (H.2) yields

$$\begin{aligned} \langle \psi | U_R^+ V U_R | \psi \rangle &= \sum_{j=1}^{3} R_{ij} \langle \psi | V_j | \psi \rangle, \\ V_i' = U_R^+ V U_R &= \sum_{j=1}^{3} R_{ij} V_j, \end{aligned} \tag{H.3}$$

since the relation should hold for arbitrary ψ.

We now proceed to generalize Eq. (H.3) to operators of higher rank than vectors. Vectors are tensors of rank one ($k=1$) that have three spatial components related to x, y, and z. Tensors of rank two ($k=2$) can be thought of as the most general product of two vector operators $\bar{V}_{1m}$ and $\bar{V}_{2n}$. They therefore consist of the set of all nine products of the vector components (i.e., $V_{1m}V_{2n}$ for $m, n = x, y, z$). Irreducible spherical tensors are constructed in a basis that reflects the rotational symmetry of a sphere. So a spherical tensor operator of arbitrary rank k will be defined more generally as a set of quantities $\left[T^{(k)}\right]_j$ which obey a transformation law similar to Eq. (H.3), except that we shall assume it is of covariant form, which transforms as the inverse of Eq. (H.3), and write it in a basis that distinguishes clockwise from counterclockwise rotations.

As a first step in determining the conventional expression for the tensor transformation law, consider extending Eq. (H.3) to the case of a second rank tensor with all Cartesian indices written out explicitly. The required transformation has the form

$$(T)'_{ij} = \sum_{k,l} \frac{\partial x_k}{\partial x'_i} \frac{\partial x_l}{\partial x'_j} T_{kl} = \sum_{k,\,l} D_{klij} T_{kl}, \tag{H.4}$$

where the transformation coefficients are $D_{klij} = \dfrac{\partial x_k}{\partial x'_i} \dfrac{\partial x_l}{\partial x'_j}$. Now, the notation applicable to tensors of all ranks uses an index q (where $q = \{-k, -k+1, -k+2, \ldots + k\}$) to replace multiple Cartesian subscripts referring to individual components of the tensor. In addition we note a distinction is introduced between D and R. We shall insist that D be the simplest possible representation of rotations, or in other words that it be "irreducible" in addition to producing the desired transformation for rank k tensors. Irreducible tensors consist of the minimum number of independent elements that span a given space, formed into linear combinations of Cartesian components that present the least mathematical complexity, and they mimic the properties and dynamics of objects and processes in the real world with which we are familiar.

Thus the tensor transformation law is a notationally simplified version of Eq. (H.4) which uses the index q and coefficients $D^{(k)}_{q'q}$ from a unitary rotation based on the angular momentum operator:

$$\left(T^{(k)}_q\right)' = \sum_{q'=-k}^{k} T^{(k)}_{q'} D^{(k)}_{q'q}(R). \tag{H.5}$$

A direct comparison with vector transformation components is possible after inversion of Eq. (H.3):

$$(V_j)' = \sum_{i=1}^{3} V_i R_{ij}. \tag{H.6}$$

Notice the summation in Eq. (H.6) is over the first subscript of R, just as the summation in Eq. (H.5) is over the first subscript of $D^{(k)}_{q'q}$.

Equation (H.5) now defines tensor operator components that transform like irreducible components of vectors. This equation can be used to test whether quantities of interest have the tensor properties required to provide the simplest mathematical description of system states, energies and dynamics. It can also be used to determine explicit components of tensors in terms of vector Cartesian components. For example, we make use of Eq. (H.5) to determine the components of a rank one tensor $T^{(1)}$. However, it proves to be more convenient to work with commutation relations that are equivalent to the transformation rules than to use Eq. (H.5) itself.

To find the commutation relations of interest, we proceed by substituting the irreducible form for infinitesimal rotations into Eq. (H.5). The axis and magnitude of the rotation are specified by axial vector $\bar{\varepsilon}$, and only terms up to first order will be considered. Borrowing expressions for the lowest order terms of the Wigner D coefficients that represent three-dimensional irreducible rotations (see, for example, Eq. (16.54) of Ref. [H.2]), we have

$$D^{(i)}_{m'm}(\bar{\varepsilon}) = \delta_{m'm} - \frac{i\varepsilon_x + \varepsilon_y}{2}\sqrt{(j-m)(j+m+1)}\delta_{m',m+1} - \frac{i\varepsilon_x - \varepsilon_y}{2}\sqrt{(j+m)(j-m+1)}\delta_{m',m+1} - i\varepsilon_z m\delta_{m'm}. \tag{H.7}$$

The transformation rule (H.5) dictates that

$$\left(T_q^{(k)}\right)' = U_R T_q^{(k)} U_R^+ = \sum_{q'=-k}^{k} T_q^{(k)} D_{q'q}^{(k)}(R). \tag{H.8}$$

Substituting the first-order expression for the rotation given in Eq. (H.7), namely

$$U_R = 1 - \frac{i}{\hbar}\bar{\varepsilon}\cdot\bar{J},$$

into Eq. (H.8), the left-hand side becomes

$$\begin{aligned}\left(1-\frac{i}{\hbar}\bar{\varepsilon}\cdot\bar{J}\right)T_q^{(k)}\left(1+\frac{i}{\hbar}\bar{\varepsilon}\cdot\bar{J}\right) &= T_q^{(k)} - \frac{i}{\hbar}\bar{\varepsilon}\cdot\bar{J}T_q^{(k)} + \frac{i}{\hbar}T_q^{(k)}\bar{\varepsilon}\cdot\bar{J} + \frac{1}{\hbar^2}\bar{\varepsilon}\cdot\bar{J}T_q^{(k)}\bar{\varepsilon}\cdot\bar{J} \\ &\approx T_q^{(k)} - \frac{i}{\hbar}\varepsilon_x\left[J_x, T_q^{(k)}\right] - \frac{i}{\hbar}\varepsilon_y\left[J_y, T_q^{(k)}\right] - \frac{i}{\hbar}\varepsilon_z\left[J_z, T_q^{(k)}\right].\end{aligned} \tag{H.9}$$

The right-hand side yields

$$\begin{aligned}\sum_{q'=-k}^{k} T_{q'}^{(k)} D_{q'q}^{(k)}(R) &= \sum_{q'=-k}^{k} T_{q'}^{(k)}\left\{\delta_{q'q} - \frac{i\varepsilon_x + \varepsilon_y}{2}\sqrt{(k-q)(k+q+1)}\delta_{q,'q+1} - \frac{i\varepsilon_x - \varepsilon_y}{2}\sqrt{(k+q)(k-q+1)}\delta_{q',q-1} - i\varepsilon_z q\delta_{q'q}\right\} \\ &= T_q^{(k)} + \frac{i\varepsilon_x}{\hbar}\left\{-\frac{\hbar}{2}\sqrt{(k-q)(k+q+1)}T_{q+1}^{(k)} - \frac{\hbar}{2}\sqrt{(k+q)(k-q+1)}T_{q-1}^{(k)}\right\} \\ &\quad + \frac{i\varepsilon_y}{\hbar}\left\{-\frac{\hbar}{2i}\sqrt{(k-q)(k+q+1)}T_{q+1}^{(k)} + \frac{\hbar}{2i}\sqrt{(k+q)(k-q+1)}T_{q-1}^{(k)}\right\} \\ &\quad + \frac{i}{\hbar}(-\hbar\varepsilon_z q)T_q^{(k)}.\end{aligned} \tag{H.10}$$

Equating left and right sides, and comparing coefficients of $\varepsilon_x, \varepsilon_y, \varepsilon_z$, we obtain

$$\left[J_x, T_q^{(k)}\right] = \frac{\hbar}{2}\left\{\sqrt{(k-q)(k+q+1)}\,T_{q+1}^{(k)} + \sqrt{(k+q)(k-q+1)}\,T_{q-1}^{(k)}\right\}, \tag{H.11}$$

$$\left[J_y, T_q^{(k)}\right] = \frac{\hbar}{2i}\left\{\sqrt{(k-q)(k+q+1)}\,T_{q+1}^{(k)} - \sqrt{(k+q)(k-q+1)}\,T_{q-1}^{(k)}\right\}, \tag{H.12}$$

$$\left[J_z, T_q^{(k)}\right] = \hbar q T_q^{(k)}. \tag{H.13}$$

Now $J_\pm = J_x \pm iJ_y$, so that

$$\left[J_+, T_q^{(k)}\right] = \hbar\sqrt{(k-q)(k+q+1)}\,T_{q+1}^{(k)}, \tag{H.14}$$

$$\left[J_-, T_q^{(k)}\right] = \hbar\sqrt{(k+q)(k-q+1)}\,T_{q-1}^{(k)}. \tag{H.15}$$

Hence Eqs. (H.13)–(H.15) are completely equivalent (for infinitesimal rotations) to the required transformation relations in Eq. (H.8) for irreducible tensor operators. This is sufficient to establish the equivalence between Eq. (H.8) and the commutation relations for all rotations, and make it clear that the rotation properties of any operator are determined by its commutator with the angular momentum. This is fortunate, because in practice we could not have applied Eq. (H.8) for *all* possible rotations R! We now turn to the problem of relating irreducible tensor components to the Cartesian components of rank zero, rank one, rank two, etc. tensors, using Eqs. (H.13)–(H.15).

The procedure for constructing spherical tensors is illustrated for rank one (vectors). Because the tensor components $T^{(k)}$ and the Cartesian vector components V_i both span the space, and obey linear transformation equations, they must be linear combinations of one another:

$$\left.\begin{aligned} T_1^{(1)} &= aV_x + bV_y + cV_z \\ T_0^{(1)} &= dV_x + eV_y + fV_z \\ T_{-1}^{(1)} &= gV_x + mV_y + nV_z \end{aligned}\right\}. \tag{H.16}$$

Substitution of Eq. (H.16) into Eqs. (H.13)–(H.15) yields nine equations in nine unknowns. We assume that the vector V satisfies the usual angular momentum commutation relations. Each equation can then be simplified by using the relations $\left[J_i, V_j\right] = i\hbar\varepsilon_{ijk}V_k$, where ε_{ijk} is the Levi–Civita pseudotensor. Its value is +1 for cyclic permutations of the indices, -1 for anti-cyclic permutations, and zero if two or more indices are equal:

$$\left[J_z, T_{-1}^{(1)}\right] = \left[J_z, (gV_x + mV_y + nV_z)\right] = -\hbar T_{-1}^{(1)},$$

$$i(gV_y - mV_x) = T_{-1}^{(1)}, \tag{H.17}$$

$$\left[J_z, T_0^{(1)}\right] = \left[J_z, (dV_x + eV_y + fV_z)\right] = 0,$$

$$dV_y - eV_x = 0, \tag{H.18}$$

$$\left[J_z, T_1^{(1)}\right] = \left[J_z, \left(aV_x + bV_y + cV_z\right)\right] = \hbar T_1^{(1)},$$

$$i\left(aV_y - bV_x\right) = T_1^{(1)}, \tag{H.19}$$

$$\left[J_+, T_{-1}^{(1)}\right] = \left[J_+, \left(gV_x + mV_y + nV_z\right)\right] = \sqrt{2}\hbar T_0^{(1)},$$

$$-n\left(V_x + iV_y\right) + (g + im)\, V_z = \sqrt{2} T_0^{(1)}, \tag{H.20}$$

$$\left[J_+, T_0^{(1)}\right] = \left[J_+, \left(dV_x + eV_y + fV_z\right)\right] = \sqrt{2}\hbar T_1^{(1)},$$

$$-f\left(V_x + iV_y\right) + (d + ie)\, V_z = \sqrt{2} T_1^{(1)}, \tag{H.21}$$

$$\left[J_+, T_1^{(1)}\right] = \left[J_x + J_y, \left(aV_x + bV_y + cV_z\right)\right] = 0,$$

$$-c\left(V_x + iV_y\right) + (a + ib)\, V_z = 0, \tag{H.22}$$

$$\left[J_-, T_{-1}^{(1)}\right] = \left[J_x - iJ_y, \left(gV_x + mV_y + nV_z\right)\right] = 0,$$

$$n\left(V_x - iV_y\right) + (-g + im)\, V_z = 0, \tag{H.23}$$

$$\left[J_-, T_0^{(1)}\right] = \left[J_-, \left(dV_x + eV_y + fV_z\right)\right] = \sqrt{2}\hbar T_{-1}^{(1)},$$

$$f\left(V_x - iV_y\right) + (-d + ie)\, V_z = \sqrt{2} T_{-1}^{(1)}, \tag{H.24}$$

$$\left[J_-, T_1^{(1)}\right] = \left[J_-, \left(aV_x + bV_y + cV_z\right)\right] = \sqrt{2}\hbar T_0^{(1)},$$

$$c\left(V_x - iV_y\right) + (-a + ib)\, V_z = \sqrt{2} T_0^{(1)}. \tag{H.25}$$

These relations are completely general. But notice from Eq. (H.18) that

$$dV_y - eV_x = 0 \rightarrow d = e = 0, \tag{H.26}$$

since $V_y \neq V_x$. Hence from Eq. (H.16) we find

$$T_0^{(1)} = f\, V_z. \tag{H.27}$$

Also from Eq. (H.21) we can write

$$T_1^{(1)} = -f\frac{\left(V_x + iV_y\right)}{\sqrt{2}}, \tag{H.28}$$

and from Eq. (H.24) we find

$$T_{-1}^{(1)} = f\frac{\left(V_x - iV_y\right)}{\sqrt{2}}. \tag{H.29}$$

The simplest choice is $f = 1$, for which we obtain

$$T_1^{(1)} = -\frac{(V_x + iV_y)}{\sqrt{2}}, \tag{H.30}$$

$$T_0^{(1)} = V_z, \tag{H.31}$$

$$T_{-1}^{(1)} = \frac{(V_x - iV_y)}{\sqrt{2}}. \tag{H.32}$$

Based on Eqs. (H.30)–(H.32), the general form of irreducible rank one spherical tensors is

$$V_1^{(1)} = -\frac{(V_x + iV_y)}{\sqrt{2}}, \tag{H.33}$$

$$V_0^{(1)} = V_z, \tag{H.34}$$

$$V_{-1}^{(1)} = \frac{(V_x - iV_y)}{\sqrt{2}}. \tag{H.35}$$

Inverse relations are provided by solving Eqs. (H.33)–(H.35) for V_x, V_y, and V_z:

$$V_y = +i\left(\frac{V_1^{(1)} + V_{-1}^{(1)}}{\sqrt{2}}\right), \tag{H.36}$$

$$V_z = V_0^{(1)}, \tag{H.37}$$

$$V_x = -\left(\frac{V_1^{(1)} - V_{-1}^{(1)}}{\sqrt{2}}\right). \tag{H.38}$$

From Eqs. (H.36)–(H.38) we can now derive an expression for the scalar product of two vectors $\bar{A}$, $\bar{B}$ in terms of tensor components A_q, B_q adapted to spherical coordinates.

Scalar product:

$$\begin{aligned}\bar{A}\cdot\bar{B} &= A_xB_x + A_yB_y + A_zB_z \\ &= \left(\frac{A_1^{(1)} - A_{-1}^{(1)}}{\sqrt{2}}\right)\left(\frac{B_1^{(1)} - B_{-1}^{(1)}}{\sqrt{2}}\right) - \left(\frac{A_1^{(1)} - A_{-1}^{(1)}}{\sqrt{2}}\right)\left(\frac{B_1^{(1)} - B_{-1}^{(1)}}{\sqrt{2}}\right) + A_0^{(1)}B_0^{(1)} \\ &= -A_1^{(1)}B_{-1}^{(1)} - A_{-1}^{(1)}B_1^{(1)} + A_0^{(1)}B_0^{(1)},\end{aligned} \tag{H.39}$$

$$\bar{A}\cdot\bar{B} = \sum_{q=-1}^{1} (-1)^q A_q B_{-q}. \tag{H.40}$$

To specify individual vectors like $\bar{A}$ or $\bar{B}$ completely in spherical tensor notation, expressions for the basis vectors as well as the amplitudes are needed. Because their magnitude is unity, the spherical basis vectors may be obtained directly from Eqs. (H.33)–(H.35), and are denoted by

$$\hat{\varepsilon}_1^{(1)} = -\frac{1}{\sqrt{2}}\left(\hat{x} + i\hat{y}\right), \tag{H.41}$$

$$\hat{\varepsilon}_0^{(1)} = \hat{z}, \tag{H.42}$$

and

$$\hat{\varepsilon}_{-1}^{(1)} = \frac{1}{\sqrt{2}}\left(\hat{x} - i\hat{y}\right). \tag{H.43}$$

The coefficients of a general vector $\bar{A}$ can then be obtained by replacing all the Cartesian components introduced in the defining expression for $\bar{A}$ with their spherical tensor equivalents given by Eqs. (H.36)–(H.38).

$$\begin{aligned}
\bar{A} &= A_x\hat{x} + A_y\hat{y} + A_z\hat{z} \\
&= \left(\frac{-A_1^{(1)} + A_{-1}^{(1)}}{\sqrt{2}}\right)\hat{x} + i\left(\frac{A_1^{(1)} + A_{-1}^{(1)}}{\sqrt{2}}\right)\hat{y} + A_0^{(1)}\hat{z} \\
&= A_1^{(1)}\left(-\frac{\hat{x} - i\hat{y}}{\sqrt{2}}\right) + A_{-1}^{(1)}\left(\frac{\hat{x} + i\hat{y}}{\sqrt{2}}\right) + A_0^{(1)}\hat{z} \\
&= (-1)^1 A_1^{(1)}\hat{\varepsilon}_{-1} + (-1)^{-1}A_{-1}^{(1)}\hat{\varepsilon}_1 + (-1)^0 A_0^{(1)}\hat{\varepsilon}_0 \\
&= \sum_{q=-1}^{1} (-1)^q A_q \hat{\varepsilon}_{-q}
\end{aligned} \tag{H.44}$$

Irreducible representation of V:

The electric dipole moment operator is

$$\begin{aligned}
\bar{\mu} &= e\bar{r} = e\left[x\hat{x} + y\hat{y} + z\hat{z}\right] \\
&= er\left[\sin\theta\left(\hat{x}\cos\phi + \hat{y}\sin\phi\right) + \hat{z}\cos\phi\right] \\
&= er\left(\frac{1}{2}\sin\theta\right)\left[\left(\hat{x} - i\hat{y}\right)\exp(i\phi) + \left(\hat{x} + i\hat{y}\right)\exp(-i\phi)\right] + \hat{z}er\cos\theta \\
&= er\left\{\frac{\sin\theta}{\sqrt{2}}\left[\hat{\varepsilon}_{-1}\exp(i\phi) - \hat{\varepsilon}_1\exp(-i\phi)\right] + \hat{\varepsilon}_0\cos\theta\right\} \\
&= er\left\{-\hat{\varepsilon}_{-1}Y_1^{(1)}\sqrt{\frac{4\pi}{3}} - \hat{\varepsilon}_1 Y_{-1}^{(1)}\sqrt{\frac{4\pi}{3}} + \hat{\varepsilon}_0 Y_0^{(1)}\sqrt{\frac{4\pi}{3}}\right\} \\
&= er\left\{\sum_q (-1)^q\left(\frac{4\pi}{3}\right)^{1/2} Y_q^{(1)}\hat{\varepsilon}_{-q}\right\}.
\end{aligned} \tag{H.45}$$

Consequently, it has three irreducible tensor components given by

$$\mu_{\pm 1} = -erC^{(1)}_{\pm 1}, \tag{H.46}$$

$$\mu_0 = erC^{(1)}_0, \tag{H.47}$$

where

$$C^{(1)}_q \equiv \left(\frac{4\pi}{3}\right)^{1/2} Y^{(1)}_q \tag{H.48}$$

is the Racah tensor. In the same way, we can write out the electric field in irreducible form as

$$\bar{E} = \frac{1}{2}\left\{E_{+1}\hat{\varepsilon}_{-1} + E_{-1}\hat{\varepsilon}_1\right\}\exp[-i(\omega t - kz)t] + c.c. \tag{H.49}$$

This describes a transverse wave that is real if $E_{+1} = -E^*_{-1}$. Furthermore, the wave is linearly polarized along $\hat{x}$ if $E_{\pm 1} = E^*_{\pm 1}$. Using these results, the interaction Hamiltonian becomes

$$\begin{aligned}
V(t) &= -\bar{\mu}\cdot\bar{E}(t) \\
&= er\sum_q (-1)^q\left(\frac{4\pi}{3}\right)^{1/2} Y^{(1)}_q\hat{\varepsilon}_{-q}\cdot\left(\frac{1}{2}\right)\left\{\left(E_+\hat{\varepsilon}_- + E_-\hat{\varepsilon}_+\right)\exp[-i(\omega t - kz)t] + c.c.\right\} \\
&= er\frac{1}{2}\sqrt{\frac{4\pi}{3}}\left\{\left(E_+Y^{(1)}_{-1} + Y^{(1)}_1 E_-\right)\exp[-i(\omega t - kz)t] + c.c.\right\} \\
&= -\frac{1}{2}\left\{\mu_- E_+ + \mu_+ E_-\right\}\exp[-i(\omega t - kz)t] + c.c. \\
&= (V_+ + V_-)\exp[-i(\omega t - kz)t] + c.c.
\end{aligned} \tag{H.50}$$

We need matrix elements for V starting from a state i and ending in state f. Let the states themselves be represented by

$$|i\rangle = |\alpha' j' m'\rangle,$$

$$|f\rangle = |\alpha j m\rangle,$$

where j, j' are the angular momenta of the two states and m, m' specify their projections on the axis of quantization. α, α' denote other quantum numbers (like the principle quantum number n) required to specify each state completely. Then

$$\begin{aligned}
\langle f|V_-|i\rangle &= -\frac{1}{2}\langle\alpha jm|\mu_+ E_-|\alpha' j' m'\rangle \\
&= \frac{1}{2}eE_-\left\langle\alpha jm\left|rC^{(1)}_+\right|\alpha' j' m'\right\rangle.
\end{aligned} \tag{H.51}$$

Using the Wigner–Eckart (W-E) theorem (see Eq. (H.63)), this reduces to

$$\langle f|\, V_{-}\, |i\rangle = (-1)^{J-m}\frac{1}{2}eE_{-}\left\langle \alpha j \left\| rC^{(1)} \right\| |\alpha' j'\rangle \right. \begin{pmatrix} j & 1 & j' \\ -m & 1 & m' \end{pmatrix}, \tag{H.52}$$

where $\left\langle \alpha j \left\| rC^{(1)} \right\| \alpha' j' \right\rangle$ is the so-called reduced matrix element (independent of m, m'), and $\begin{pmatrix} j & 1 & j' \\ -m & 1 & m' \end{pmatrix}$ is a $3j$ symbol which assures that momentum is conserved. This particular $3j$ symbol is non-zero, and specifies that the interaction can take place only if

$$\Delta j = j - j' = \pm 1, \tag{H.53}$$

$$\Delta m = m - m' = +1. \tag{H.54}$$

Hence Eqs. (H.53) and (H.54) constitute selection rules for electric dipole transitions excited by circularly polarized light. Transitions from $j' = 0$ to $j = 0$ are also excluded.

In writing Eq. (H.52) we used the W-E theorem to clarify as much as possible the dependence of the matrix element on the J and m values of the states involved. This theorem is readily proved starting from the transformation relation (H.8):

$$U_R T_q^{(k)} U_R^{+} = \sum_{q'=-k}^{k} T_{q'}^{(k)} D_{q'q}^{(k)}(R).$$

Evaluating both sides for the states considered, we find

$$\left\langle \alpha' j' m' \left| U_R T_q^{(k)} U_R^{+} \right| \alpha j m \right\rangle = \sum_{q'=-k}^{k} \left\langle \alpha' j' m' \left| T_q^{(k)} \right| \alpha j m \right\rangle D_{q'q}^{(k)}(R). \tag{H.55}$$

On the left side we can make the replacements

$$U_R^{+}\, |\alpha j m\rangle = \sum_{\mu} |\alpha j \mu\rangle D_{m\mu}^{(j)}(R) \tag{H.56}$$

and

$$\langle \alpha' j' m'|\, U_R^{+} = \sum_{\mu'} D_{m'\mu'}^{(j')}(R)\, \langle \alpha' j' m'|\,. \tag{H.57}$$

These relations follow directly from the definition of $D_{m'\mu}^{(j')}$ as the expansion coefficients of a rotated representation of the wavefunction. (See, for example, the defining relation and form of $D_{m'\mu}^{(j')}$ given by Eqs. (16.43) and (16.50) of Ref. [H.2], respectively).

Using Eqs. (H.56) and (H.57) in Eq. (H.55), we find

$$\sum_{\mu\mu'} D^{(j')}_{m'\mu'}(R) \langle \alpha' j' \mu' | T^{(k)}_q |\alpha j \mu\rangle \, D^{(j)}_{m\mu}(R) = \sum_{q'} \langle \alpha' j' m' | \, T^{(k)}_q |\alpha j m\rangle D^{(k)}_{q'q}(R). \tag{H.58}$$

This equation has exactly the same form as an important transformation identity in which Clebsch–Gordan coefficients replace the matrix elements on both sides of Eq. (H.58) (see, for example, Eq. (16.91) in Ref. [H.2]).

Clebsch–Gordan coefficients furnish the unitary transformation from basis states $|j_1 j_2 m_1 m_2\rangle = |j_1 m_1\rangle \, |j_2 m_2\rangle$ to basis states $|j_1 j_2 j m\rangle$. That is,

$$|j_1 j_2 j m\rangle = \sum_{m=m_1+m_2} C^{j}_{m_1 m_2} \, |j_1 j_2 m_1 m_2\rangle \, . \tag{H.59}$$

From Eq. (H.59) we see that the $C^{j}_{m_1 m_2}$ have the values

$$C^{j}_{m_1 m_2} = \langle j_1 j_2 m_1 m_2 | \, j_1 j_2 j m\rangle \, . \tag{H.60}$$

These coefficients have been extensively tabulated, but are often listed instead in terms of $3j$ symbols which have simpler permutation properties. The $3j$ symbols are defined by

$$\langle j_1 j_2 m_1 m_2 | \, j_1 j_2 j m\rangle = (-1)^{-j_1 + j_2 - m} \sqrt{2j+1} \begin{pmatrix} j_1 & j_2 & j \\ m_1 & m_2 & -m \end{pmatrix} . \tag{H.61}$$

The fact that the form of Eq. (H.58) is the same as the formula derived from the Clebsch–Gordan series simply means that the matrix element $\langle \alpha' j' m' | \, T^{(k)}_q |\alpha j m\rangle$ we are interested in is proportional to the Clebsch–Gordan coefficient. That is,

$$\langle \alpha' j' m' | T^{(k)}_q |\alpha j m\rangle \propto \langle j k m q \, | j k j' m' \rangle . \tag{H.62}$$

The constant of proportionality in Eq. (H.62) does not depend on m, m', or q and is called the "reduced" matrix element, denoted $\langle \alpha' j' \| T^{(k)} \| \alpha j\rangle$. With this notation, the W-E theorem can be stated as

$$\langle \alpha' j' m' | \, T^{(k)}_q |\alpha j m\rangle = \langle j k m q \, | j k j' m' \rangle \langle \alpha' j' \| \, T^{(k)} \| \alpha j\rangle \, . \tag{H.63}$$

Equation (H.59) has two important implications. First, since transition probabilities depend on $|\langle \alpha' j' m' | T^{(k)}_q |\alpha j m\rangle|^2$, it is apparent from the right-hand side of Eq. (H.59) that the relative strengths of optical transitions between different magnetic substates of the same $\alpha j \rightarrow \alpha' j'$ transition depend only on squared ratios of Clebsch–Gordan coefficients. Second, since the reduced matrix element $\langle \alpha' j' || T^{(k)} || \alpha j\rangle$ does not depend on m, m', or q, it represents a transition multipole moment that is independent of the choice of origin. This resolves the issue that standard integral definitions of individual moments yield values of the moments that change with a shift of coordinates.

Our main interest in this book is in the tensor describing the interaction of light with matter as Hamiltonian operator $\hat{V} = -\bar{\mu} \cdot \bar{E}$. How can reduced matrix elements like the one in Eq. (H.52) be evaluated in the simplest, most general way? To explore this question, we first factor out and perform the radial integration for the case of an electric dipole operator. All the radially dependent quantities in the integral are well defined so one can write

$$\begin{aligned}\langle \alpha' j' \| T^{(1)} \| \alpha j \rangle &= \langle \alpha' j' \| rC^{(1)} \| \alpha j \rangle \\ &= \int_0^\infty \psi_{\alpha j}^* r \psi_{\alpha' j'}^* r^2 dr \langle \alpha' j' \| C^{(1)} \| \alpha j \rangle_{\theta\phi} \\ &\equiv \langle \alpha' j' \| r \| \alpha j \rangle_r \langle \alpha' j' \| C^{(1)} \| \alpha j \rangle_{\theta\phi} \,. \end{aligned} \tag{H.64}$$

In the final expression, $\langle \alpha' j' \| r \| . \alpha j \rangle ._r$ is a purely radial integral of r between the specified states. Note however that the angular integral over the undetermined quantity $C^{(1)}$ must be worked out by inverting the W-E theorem. This is necessitated by the fact that $\langle \alpha' j' \| C^{(1)} \| \alpha j \rangle_{\theta\phi}$ is not just the angular integral of a known combination of $Y_m^{(1)}$ or other functions. As mentioned previously, it is in fact independent of q, m, and m'.

The radial dependence can be factored out of the general W-E relation (H.63) in the same way. From

$$\langle \alpha' j' m' | T_q^{(k)} | \alpha j m \rangle = \langle jkmq \| jkj'm' \rangle \langle \alpha' j' \| T^{(k)} \| \alpha j \rangle \,,$$

we thus obtain

$$\langle \alpha' j' m' | C_+^{(1)} | \alpha j m \rangle_{\theta\phi} = (-1)^{J'-m'} \langle \alpha' j' \| C^{(1)} \| \alpha j \rangle_{\theta\phi} \begin{pmatrix} j' & 1 & j \\ -m' & 1 & m \end{pmatrix}. \tag{H.65}$$

Since we don't know the exact form of $\langle \alpha' j' \| C^{(1)} \| \alpha j \rangle_{\theta\phi}$, we must proceed by evaluating the angular integral on the left of expressions like Eq. (H.65) in order to find the reduced matrix element. Following Sobelman [H.3], the left-hand side of Eq. (H.65) is

$$\int Y_{lm}^* Y_q^{(1)} Y_{l'm'} \sin\theta d\theta d\phi = (-)^m \sqrt{\frac{(2l+1)(2k+1)(2l'+1)}{4\pi}} \begin{pmatrix} l & k & l' \\ 0 & 0 & 0 \end{pmatrix} \begin{pmatrix} l & k & l' \\ -m & q & m' \end{pmatrix},$$

so that the reduced matrix element obtained by inverting Eq. (H.65) is

$$\langle \alpha' j' m' | \left| C^{(1)} \right| | \alpha j m \rangle = (-)^{j'} \sqrt{(2j'+1)(2j+1)} \begin{pmatrix} j' & 1 & j \\ 0 & 0 & 0 \end{pmatrix}. \tag{H.66}$$

The 3j symbol in the expression (H.66) for the reduced matrix element dictates that $\Delta j = \pm 1$. The selection rule with respect to magnetic quantum number m arises from the $3j$ symbol in Eq. (H.65) and dictates that $\Delta m = m' - m = 1$ in the example of circularly polarized light. For linearly polarized light, the $3j$ symbol in Eq. (H.65) would be

$$\begin{pmatrix} j' & 1 & j \\ -m' & 0 & m \end{pmatrix},$$

changing the magnetic quantum number selection rule to $\Delta m = m' - m = 0$. Additional discussion of the relationship between Clebsch–Gordan coefficients and $3j$ symbols can be found in Ref. [H.4] and formulas for their evaluation are on pp.60–6 of Ref. [H.3]. The choice of phase in Refs. [H.3] and [H.4] is that of Condon and Shortley [H.5], which renders the Clebsch–Gordan coefficients real.

REFERENCES

[H.1] See, for example, G. Arfken, *Mathematical Methods for Physicists*, 2nd ed. New York: Academic Press, 1970, pp.8–11 and 121–36.

[H.2] E. Merzbacher, *Quantum Mechanics*, 2nd ed. New York: John Wiley & Sons, 1970.

[H.3] I.I. Sobelman, *Atomic Spectra and Radiative Transitions*. New York: Springer-Verlag, 1979, p.78.

[H.4] L.I. Schiff, *Quantum Mechanics*, 3rd ed. New York: McGraw-Hill, pp.214–24.

[H.5] E. Condon and G. Shortley, *The Theory of Atomic Spectra*. Cambridge: Cambridge University Press, 1951.

Appendix I
Derivation of Effective Hamiltonians

To treat the dynamics of fully quantized optical interactions consistently in a single reference frame, one can perform a transformation that takes the unperturbed atom into the rotating frame of the light. The atom–field interaction of cavity quantum electrodynamics, for example, calls for transformation of the static atomic Hamiltonian into the rotating frame of the cavity mode. This unitary transformation must also account for the fermionic character of the two-level atom (see Bloch vector model), and the pseudospin rotation matrix must therefore be written in terms of half-angles.

Consider rotating the wavefunction of the atom into the frame of a cavity field at frequency Ω_c. This is given by

$$\psi(t)' = \begin{bmatrix} \psi_1(t)\exp(-i\Omega_c t/2) \\ \psi_2(t)\exp(i\Omega_c t/2) \end{bmatrix}. \tag{I.1}$$

According to Eq. (2.4.12) of Chapter 2, which describes the temporal evolution of ψ for a time-independent Hamiltonian H_0, the components of ψ' in Eq. (I.1) progress according to

$$|\psi_1(t)\rangle = C_1(t)\exp(-i\omega_1 t)\,|1\rangle, \tag{I.2}$$

$$|\psi_2(t)\rangle = C_2(t)\exp(-i\omega_2 t)\,|2\rangle. \tag{I.3}$$

Hence the first component changes to

$$|\psi_1(t)'\rangle = C_1(t)\exp(i\Delta t/2)\,|1\rangle, \tag{I.4}$$

where $\Delta \equiv \omega_2 - \omega_1 - \Omega_c$. Here the zero of energy has been chosen such that $(\omega_1 + \omega_2)/2 = 0$. Similarly, the other component of the wavefunction transforms to

$$|\psi_2(t)'\rangle = C_2(t)\exp(-i\Delta t/2)\,|2\rangle. \tag{I.5}$$

The frequency factor in the argument of the exponential functions in Eqs. (I.4) and (I.5) can now be interpreted as new effective frequencies of the energy levels in the rotating frame. Hence we can write an effective Hamiltonian in this frame that is given by

$$H_{eff} = \frac{\hbar}{2}\begin{bmatrix} -\Delta & 0 \\ 0 & \Delta \end{bmatrix} = -\frac{\hbar}{2}\Delta\sigma_z, \tag{I.6}$$

where σ_z is one of the Pauli spin matrices.

It is left to the reader to verify that the use of Eq. (I.6) in the equation of motion for the density matrix gives the same relation for the slowly varying amplitude $\tilde{\rho}_{12}$ of the coherence as that in Eq. (7.7.18). This establishes the validity of calculations based on effective Hamiltonians such as that in Eq. (I.6). The derivation also provides a general approach to deriving simplified Hamiltonians for fully quantized systems in a single reference frame. Another example of this approach was given in Chapter 7 where the effective Hamiltonian method was used to reduce the more serious level of complexity of coherent population transfer in a two-field interaction (Eq. (7.3.2)).

Appendix J
Irreducible Representation of Magnetic Dipole Interactions

Magnetic dipole moments and magnetic fields are examples of axial vectors. Their irreducible representations and scalar products differ from those of electric dipoles and fields which are polar vectors. To extend the treatment of Appendix H to magnetic interactions, the corresponding representations for axial vectors and their products are derived here.

An axial vector is a pseudovector that transforms as the cross product of two polar vectors. Let us write

$$\bar{V}_{axial} = \bar{V}_1 \otimes \bar{V}_2, \tag{J.1}$$

where $\bar{V}_1$ and $\bar{V}_2$ represent polar vectors. Each polar vector transforms according to the rules discussed in Appendix H, namely

$$(V_1)' = \sum_{j=1}^{3} R_{ij}(V_1)_j, \tag{J.2}$$

$$(V_2)' = \sum_{m=1}^{3} R_{km}(V_2)_m. \tag{J.3}$$

Hence the individual vectors $\bar{V}_1$ and $\bar{V}_2$ have irreducible representations of the form determined by the previous analysis to be

$$V_1^{(1)} = -\frac{V_x + iV_y}{\sqrt{2}}, \tag{J.4}$$

$$V_0^{(1)} = V_z, \tag{J.5}$$

$$V_{-1}^{(1)} = \frac{V_x - iV_y}{\sqrt{2}}, \tag{J.6}$$

and inverse expressions which are

$$V_x = -\frac{V_1^{(1)} - V_{-1}^{(1)}}{\sqrt{2}}, \tag{J.7}$$

$$V_y = i\left(\frac{V_1^{(1)} + V_{-1}^{(1)}}{\sqrt{2}}\right), \tag{J.8}$$

$$V_z = V_0^{(1)}. \tag{J.9}$$

Using Eqs. (J.7)–(J.9), $\bar{V}_{axial}$ can be written in terms of spherical tensors immediately as

$$\begin{aligned}
\bar{V}_{axial} &= \bar{V}_1 \otimes \bar{V}_2 \\
&= \hat{x}\left[V_{1y}V_{2z} - V_{1z}V_{2y}\right] - \hat{y}\left[V_{1x}V_{2z} - V_{1z}V_{2x}\right] + \hat{z}\left[V_{1x}V_{2y} - V_{1y}V_{2x}\right] \\
&= \hat{x}\frac{i}{\sqrt{2}}\left[\left[\left(V_1^{(1)}\right)_1 + \left(V_{-1}^{(1)}\right)_1\right]\left(V_0^{(1)}\right)_2 - \left(V_0^{(1)}\right)_1\left[\left(V_1^{(1)}\right)_2 + \left(V_{-1}^{(1)}\right)_2\right]\right] \\
&\quad + \hat{y}\frac{1}{\sqrt{2}}\left[\left[\left(V_1^{(1)}\right)_1 - \left(V_{-1}^{(1)}\right)_1\right]\left(V_0^{(1)}\right)_2 - \left(V_0^{(1)}\right)_1\left[\left(V_1^{(1)}\right)_2 - \left(V_{-1}^{(1)}\right)_2\right]\right] \\
&\quad - \hat{z}\frac{i}{2}\left[\left[\left(V_1^{(1)}\right)_1 - \left(V_{-1}^{(1)}\right)_1\right]\left[\left(V_1^{(1)}\right)_2 + \left(V_{-1}^{(1)}\right)_2\right]\right. \\
&\quad \left. - \left[\left(V_1^{(1)}\right)_1 + \left(V_{-1}^{(1)}\right)_1\right]\left[\left(V_1^{(1)}\right)_2 - \left(V_{-1}^{(1)}\right)_2\right]\right] \\
&= i\left\{\left(V_1^{(1)}\right)_1\left(V_0^{(1)}\right)_2\hat{\varepsilon}_- - \left(V_{-1}^{(1)}\right)_1\left(V_0^{(1)}\right)_2\hat{\varepsilon}_+ - \left(V_0^{(1)}\right)_1\left(V_1^{(1)}\right)_2\hat{\varepsilon}_- + \left(V_0^{(1)}\right)_1\left(V_{-1}^{(1)}\right)_2\hat{\varepsilon}_+\right\} \\
&\quad - i\left\{\left(V_1^{(1)}\right)_1\left(V_{-1}^{(1)}\right)_2\hat{\varepsilon}_0 - \left(V_{-1}^{(1)}\right)_1\left(V_1^{(1)}\right)_2\hat{\varepsilon}_0\right\}.
\end{aligned} \tag{J.10}$$

If we now simplify the notation by relabeling the vectors $\bar{V}_1 = \bar{A}$ and $\bar{V}_2 = \bar{B}$, Eq. (J.10) can be written

$$\bar{V}_{axial} = i\{A_1B_0\hat{\varepsilon}_- - A_{-1}B_0\hat{\varepsilon}_+ - A_0B_1\hat{\varepsilon}_- + A_0B_{-1}\hat{\varepsilon}_+\} + i\{-A_1B_{-1} + A_{-1}B_1\}\hat{\varepsilon}_0. \tag{J.11}$$

Upon grouping the coefficients for each spherical basis vector, it becomes evident that this expression is already in irreducible form. Introducing the definitions

$$C_\pm \equiv B_0A_\pm, \tag{J.12}$$

$$D_\pm \equiv A_0B_\pm, \tag{J.13}$$

$$F_\pm \equiv A_\mp B_\pm, \tag{J.14}$$

one obtains

$$\begin{aligned}
\bar{V}_{axial} &= \bar{A} \otimes \bar{B} \\
&= i\{C_+\hat{\varepsilon}_- - C_-\hat{\varepsilon}_+ - D_+\hat{\varepsilon}_- + D_-\hat{\varepsilon}_+\} + i\{F_+ - F_-\}\hat{\varepsilon}_0 \\
&= i\{(C_+ - D_+)\hat{\varepsilon}_- - (C_- - D_-)\hat{\varepsilon}_+\} + \{iF_+ + c.c.\}\hat{\varepsilon}_0.
\end{aligned} \tag{J.15}$$

The last result makes use of the replacement $F_\pm = A_\mp B_\pm = (-A_\pm^*)(-B_\mp^*) = (F_\mp)^*$, valid for real vectors $\bar{A}$ and $\bar{B}$. The irreducible form of the vector can therefore be written as

$$\bar{V}_{axial} = iV_+\hat{\varepsilon}_- - iV_-\hat{\varepsilon}_+ + V_0\hat{\varepsilon}_0, \tag{J.16}$$

where $V_\pm \equiv (C_\pm - D_\pm)$ and $V_0 \equiv (iF_+ + c.c.)$ is entirely real. Because $\bar{V}_{axial}$ is defined here in terms of a cross product, there is a sign ambiguity associated with whether its definition should be $\bar{A} \otimes \bar{B}$ or $\bar{B} \otimes \bar{A} = -\bar{A} \otimes \bar{B}$. Consequently the irreducible representation has an overall sign ambiguity that must be determined by other considerations, such as the direction of positive flow of energy associated with an axial field or the convention for positive energy itself.

As an example of the determination of sign for axial vectors, consider the magnetic flux density $\bar{B}$. Given the irreducible form of $\bar{E}$ in Eq. (H.49) for an x-polarized wave traveling in the $+\hat{z}$ direction, the overall sign of the irreducible representation for $\bar{B}$ must be chosen to ensure that the energy flow along $+\hat{z}$ is positive. This requires that the Poynting vector $\bar{S} = \bar{E} \otimes \bar{H}$ points along $+\hat{z}$. For this to be the case, $\bar{B}$ must have the same sign and form as the transverse components of the axial vector in Eq. (J.16):

$$\bar{S} \propto -\left(E_+\hat{\varepsilon}_- + E_-\hat{\varepsilon}_+\right) \otimes \left(iB_+\hat{\varepsilon}_- - iB_-\hat{\varepsilon}_+\right).$$

Simplifying the cross products by using the results

$$\hat{\varepsilon}_\pm \otimes \hat{\varepsilon}_\pm = 0,$$

$$\hat{\varepsilon}_\pm \otimes \hat{\varepsilon}_\mp = -\hat{\varepsilon}_\mp \otimes \hat{\varepsilon}_\pm = \pm i\hat{z},$$

one can easily verify that $\bar{S} \propto +\hat{z}$.

Equation (J.16) displays the tensorial form suitable for representing axial vectors such as magnetic dipole moments and magnetic fields. With this result in hand, the magnetic field interaction Hamiltonian which is the scalar product of two axial vectors can readily be obtained. First, note that

$$\bar{A}_{axial} \cdot \bar{B}_{axial} = \pm\left[-iA_+\hat{\varepsilon}_- + iA_-\hat{\varepsilon}_+ + A_0\hat{\varepsilon}_0\right] \cdot \left[-iB_+\hat{\varepsilon}_- + iB_-\hat{\varepsilon}_+ + B_0\hat{\varepsilon}_0\right]. \tag{J.17}$$

The $\pm$ sign on the right-hand side reflects the sign ambiguity mentioned previously. Since different considerations may determine the overall sign of each axial vector comprising the product, the product itself may in principle be positive or negative. Next, Eq. (J.17) can be simplified by evaluating the scalar products of the basis vectors.

$$\hat{\varepsilon}_\pm \cdot \hat{\varepsilon}_\pm = \hat{\varepsilon}_0 \cdot \hat{\varepsilon}_\pm = 0 \tag{J.18}$$

$$\hat{\varepsilon}_\pm \cdot \hat{\varepsilon}_\mp = -1 \tag{J.19}$$

$$\hat{\varepsilon}_0 \cdot \hat{\varepsilon}_0 = 1 \tag{J.20}$$

Upon substitution of Eqs. (J.18)–(J.20) into Eq. (J.17) one finds

$$\bar{A}_{axial} \cdot \bar{B}_{axial} = \mp\left[(A_+B_- + A_-B_+) + \left(A_0B_0 + c.c. - A_0B_0^* + c.c.\right)\right]. \tag{J.21}$$

In electric dipole interactions of light with isolated atoms, the induced dipole is parallel to and in the same direction as the electric field. However, magnetic dipoles induced by light are anti-parallel to the magnetic flux density $\bar{B}$. Hence explicit forms for the optical magnetic field and its induced dipole are

$$\bar{B} = \frac{1}{2}\left[(iB_+\hat{\varepsilon}_- - iB_-\hat{\varepsilon}_+)e^{i(\omega t-kz)} + c.c.\right], \tag{J.22}$$

$$\bar{\mu}^{(m)} = \frac{1}{2}\left[(-i\mu_+^{(m)}\hat{\varepsilon}_- + i\mu_-^{(m)}\hat{\varepsilon}_+)e^{i(\omega t-kz)} + c.c.\right]. \tag{J.23}$$

Index

T